Contribution of Clusters Physics to Materials Science and Technology

NATO ASI Series

Advanced Science Institutes Series

A Series presenting the results of activities sponsored by the NATO Science Committee, which aims at the dissemination of advanced scientific and technological knowledge, with a view to strengthening links between scientific communities.

The Series is published by an international board of publishers in conjunction with the NATO Scientific Affairs Division

A	**Life Sciences**	Plenum Publishing Corporation
B	**Physics**	London and New York
C	**Mathematical and Physical Sciences**	D. Reidel Publishing Company Dordrecht and Boston
D	**Behavioural and Social Sciences**	Martinus Nijhoff Publishers Dordrecht/Boston/Lancaster
E	**Applied Sciences**	
F	**Computer and Systems Sciences**	Springer-Verlag Berlin/Heidelberg/New York
G	**Ecological Sciences**	

Series E: Applied Sciences – No. 104

Contribution of Clusters Physics to Materials Science and Technology

From Isolated Clusters to Aggregated Materials

Edited by

J. Davenas

Département de Physique des Matériaux
Université Claude Bernard Lyon I
43 Bd du 11 Novembre 1918
69622 Villeurbanne Cedex
France

P.M. Rabette

Laboratoire de Chimie des Solides
Université Pierre et Marie Curie
4 Place Jussieu
75230 Paris Cedex 05
France

1986 **Springer-Science+Business Media, B.V.**

Proceedings of the NATO Advanced Study Institute on Impact of Clusters Physics in Materials Science and Technology, Cap d'Agde, France, June 13-25, 1982

Library of Congress Cataloging in Publication Data

ISBN 978-94-010-8444-4 ISBN 978-94-009-4374-2 (eBook)
DOI 10.1007/978-94-009-4374-2

Originally published by Martinus Nijhoff Publishers, Dordrecht in 1986
Softcover reprint of the hardcover 1st edition 1986

FOREWORD

During the last decade there has been an increasing interest in clusters and small particles because of the peculiar properties induced by their large area to volume ratio. For that reason small particles are often considered as an intermediate state of matter at the border between atomic (or molecular) chemistry, and physics of the condensed matter. The importance of the surface effect can explain the anomalous properties, for example the existence of the five fold symmetry observed in different circumstances (beams of rare gas clusters, gold particles deposited on a substrate). However the question of the critical size at which the transition to bulk properties occurs cannot be simply answered, since the reply depends on the peculiar property which is studied. The importance of the size effect was emphasized in the last International Meetings. However the situation remains confused in most cases since the exact role of the cluster environment cannot be clearly elucidated and is a main difficulty, except in cluster beam experiments. In fact ideally free clusters constitute a laboratory exception. In most applications small particles must be supported on a surface or embedded in a matrix, in order to be stabilized, which obviously shows the role of the environment. This idea led to controversies among catalysis people, but the role of preservation of the support is evidenced by the instability of organometallic clusters, which are considered as a model of free clusters. The discussion of the metal-support, or metal-additive, effect was one of the purposes of the Summer School. Another one was to consider systems of high concentrated clusters, in which case U. Kreibig showed that the single particle properties could be completely concealed by the interaction effects. The extreme case of this situation is the percolation problem where a collection of metallic particles dispersed in an insulating matrix may acquire the properties of a bulk metal for a critical concentration.

A number of meetings have been devoted in the last years to small particles, some others to the properties of the granular matter. The aim of this two weeks Summer School held in the

Cap d'Agde in summer 1982, was to sum up the background now available on small particles in order to use this knowledge for a better understanding of the properties of the granular matter, which is a very common state of matter met in such important fields of application as surface science and catalysis, photographic process, composite materials, solar energy conversion ....

The NATO Advanced Study Institutes use in constituting a natural focal point for specialists coming from different areas of activity, gathered around a central idea. From that point of view this ASI was highly enriched by the complementary participation of chemists who have greatly improved the tools used for the characterization of clusters in complex environments and physicists for their contribution to a better description of clusters properties. The living and eating arrangements available at the Hotel Palméria in le Cap d'Agde maximized these interactions and information exchanges, which led to new ideas of experiments. It was in particular the starting point of new collaborations on cluster beam experiments, which were catalysed by the enthusiasm of our friend Ludger Wöste.

This book provides an interdisciplinary overview of the state of the art for physicists and chemists interested in fundamental studies on small aggregates and also for engineers looking for a better comprehension of granular matter properties. Lectures have been gathered around three main themes.

o Metal clusters and aggregates :

"Clusters in the gas phase, Metal cluster beams, Molecular clusters and catalysis, Quantum chemistry for metal clusters, Electronic structure of metal clusters".

In this part, anomalies due to the small size of the particles are emphasized. The cluster is studied through a peculiar property (appearance potentials, molecular structure, catalytic activity ...) which is strongly connected to the mean of production. Theoretical efforts are important in this domain since the electronic structure is strongly dependent upon the size of smaller clusters.

o Small particles :

"Characterization of supported metal particles, Heterogeneous nucleation and growth processes, Formation-action and properties of clusters in the photographic process, Formation of clusters in bulk materials, Optical properties of small particles in insulating matrices".

Small particles constitute the most common case characterized by a bulk like electronic structure and a competition between interfacial and volume energies. For that reason effects due to cluster-support interactions become important and several complementary techniques are required for the characterization of the state of the cluster.

o Granular materials :

"Percolation theory, Electronic and transport properties of granular materials, Optical properties and solar selectivity of metal insulator composites, Adhesion and sintering of small particles".

In this extreme situation, the single cluster properties are completely concealed by the interaction effects, which may induce a sort of phase transition known as percolation effect. Theoretical descriptions of these processes are macroscopical and need a number of assumptions on the nature of the inhomogeneities. The microscopical characterization of the clusters is then of great help in these problems.

Finally this ASI has shown the importance of the perturbation of the cluster properties by the environment in most cases of practical interest, which requires an adequate characterization for each specific case, but also a need for new experiments on perfectly isolated clusters where great strides are to be expected in the future with the development of spectroscopy on cluster beams.

It is with the aid of many persons that this Institute was able to be organized and I want to acknowledge in particular J. BULLOCK, M. GACHET, P. MELINON and X.L. XU for their technical assistance, and also to C. DUPUY, M. CHE, F. CYROT LACKMANN, C. NACCACHE and J. VEDRINE for their advice during the preparation of this ASI.

The participants and myself are indebted to the Scientific Affairs Division of NATO, to N.S.F. and C.N.R.S. for their financial support.

J. DAVENAS

CONTENTS

PROPERTIES OF CLUSTERS IN THE GAS PHASE

L. Wöste

Institut de Physique Expérimentale,
Ecole Polytechnique Fédérale de Lausanne,
PHB-Ecublens, CH-1015 Lausanne

1. INTRODUCTION

In chemistry there is a large number of interesting particles, whose existence is known from mass spectrometric detection. The properties of these particles, however, are almost completely unknown, because they are very unstable and they only occur in very low abundances in mixtures. Particles of this kind are - for example - fragment ions, van der Waals complexes and clusters.

Among these particles, clusters certainly play a most important role : The properties of the metallic bulk, for example, are today quite well understood with solid state physics. Atom and molecular physicists on the other hand can quite well explain the behaviour of atoms and dimers. Nobody knows, however, how larger and larger compounds are being formed from these atoms and small molecules, which then finally show bulk behaviour. The comprehension of these phenomena may lead to the same description for the covalent and metallic bound. It is of highest interest for fundamental research therefore, to study this "missing link" between solid state physics and quantum chemistry.

From the scope of this Nato-ASI we see, that the interest in clusters is not just purely in fundamental research, but that there is also an important "Impact of Cluster Physics in Material Science and Technology". The main field of application of clusters is certainly in the field of catalysis, where a classical example is given by silver photography. As light hits the sensitive photoplate, photochemical processes generate the latent image. During the development process the latent image is then converted to a real picture by reduction of the photographic material to silver metal. So far, it is not completely understood, how this reduction process takes place at different speeds - according to the contents of the latent image. There are models, which attribute this phenomenon to catalytical properties of silver clusters, stored in the latent image. It would certainly be very interesting to examine this in detail (1).

Catalysis, however, is not the only important application for clusters. Let us just think of meteorology and the phenomena of cloud formation, rain and hail. Maybe cluster physics one day will contribute to successful cloud injections in areas where rain is badly needed, or prevent damaging hail. There are encouraging experiments which indicate that clusters might also be used as thermonuclear fuel for fusion plants or as molecular systems in isotope separation processes (2,3).

2. CLUSTER PRODUCTION

Being aware of the important aspects of cluster physics in the field of fundamental and applied research, it might be astonishing to some that the discipline is relatively young. One reason for this is certainly the difficulty in producing clusters. Furthermore they are very unstable, and can only be produced in size mixtures. In other words : it is impossible to buy a bottle of Na_7. There are several methods to synthesize clusters. The best way to produce them, however, depends on the intended application.

2.1 Chemical Methods

Several approaches to chemically synthesize metal clusters have been carried out so far. Some of them look very promising :

- A. Hermann filled molecular sieve with sodium and obtained zeolites of the formula $Na_{86}\,[(Al\,O_2)_{86}\,(SiO_2)_{106}\cdot 264\,H_2O]$. He prepared these molecules by distilling the sodium into the molecular sieve and observed a coloration. When heating the material up inside a mass spectrometer he observed explosions and simultaneously the appearance of cluster peaks (4).
- R. Tschudin reacted Ferrocen with metastable Argon and Hydrogen and obtained iron - vapor plus hydrocarbons due to (5)

$$Fe\,(C_5\,H_5)_2 \rightarrow Fe + 2\,C_5\,H_5\,\cdot$$

$$2\,C_5\,H_5\cdot + H_2 \rightarrow 2\,C_5\,H_6$$

- In similar reactions Ruthenium and Uranium vapors were obtained in experiments performed by Trent and Zare (6).
- Carbonyls like $Fe(CO)_5$ vaporize at very low temperatures and decomposite in electric discharges. They may be considered as ideal candidates to produce metal vapors like Ni, Fe, Cr, Mo and W.

2.2 Cluster formation in gas cells

It is a well known phenomenon, that under high pressure stagnation conditions, polymerisations take place. The effect is used to industrially produce synthetic polymers. Accordingly clusters as well are present in gas cells. In a sodium gas cell, for example, at 500 °K a Na_2 concentration of 0.5 % is present. Wu could detect Li_3 and Li_4 (7) from a Knudsen cell. Gingerich et al. found lead and tin up to Pb_5 and Sn_7 (8). Adding an inert gas may considerably favour the cluster formation. Kimoto and Nishida, for example, found Lithium aggregates up to Li_{15} (9). Remarkably, however, the size distribution was not monotonicly decreasing with growing cluster size as is usually the case when no inert gas is present. So, for example, they did not observe any Li_3 and Li_5.

A heat pipe is a special type of gas cell. It consists of a metallic tube with a cylindrical mesh inserted inside. The tube is heated in the center and cooled at both ends. It is sealed with two windows to allow spectroscopic observation. The tube is filled with the metal of interest and a carrier gas. Thus the metal is vaporized in the center, carried away towards the cold region, where is condenses, and returns as a liquid along the mesh to the center. The advantage of the heat pipe is that relatively high partial vapor pressures can be obtained without coating the window. Scheingraber and Vidal (10) used it for the spectroscopic investigation of Li_2. The presence of the carrier gas contributes to the particle formation : It is possible to observe little metal droplets at the boundary of the cold zone even by bear eye ! This inspired U. Even to propose a cluster source as shown in figure 1 (11).

In the beginning of the cold zone a skimmer is inserted at variable distance. Thus condensating particles can escape at a desired status of condensation. This technique will possibly allow the production of clusters of well defined size distribution.

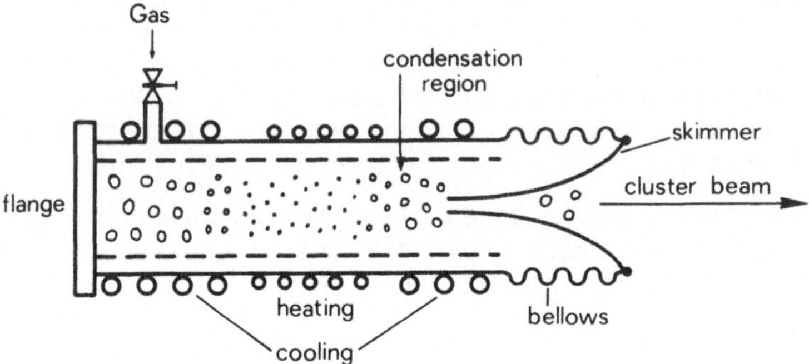

Fig. 1 : Cluster extraction out of a heat pipe.

2.3 Supersonic Molecular Beam Sources

Gas phase clusters are most commonly produced in adiabatic gas expansions; a phenomenon, which can quite frequently be observed in the sky, when jet aircrafts produce condensation trails. Under laboratory conditions these gas expansions are commonly produced as molecular beam expansions into the vacuum.

The formation of clusters in supersonic molecular beams was first observed in 1961 by Bentley and Henkes on $(CO_2)_n$ - aggregates, when they measured velocity distributions and found discontinuities in the speed distribution (12). Soon after that, Leckenby and Robbins found that this cluster formation was possible with practically all gases, which they reported as : "Condensation embryos in an expanding gas beam" (13). They also detected sodium clusters up to Na_8 and performed rough experiments to measure ionization potentials.

For many scientists in the early days, the effect of cluster formation was more disturbing than helpful and they tried to prevent it. Many investigations of elastic or reactive scattering experiments, for example, were very much disturbed by this effect. Thus a part of the cluster beam research was directed at studying the onset of condensation, so one could know the range of nozzle source conditions to safely produce nozzle beams without having unwanted clusters. From the very beginning, however, the research efforts were not limited to the avoidance of clusters. On the contrary, to produce cluster beams and to study and utilize their unique properties soon developed into an interesting project based on its own merits. A great part of this cluster activity was concentrated on the possible use of deuterium - tritium cluster beams as fuel for a thermonuclear plasma (Becker 1960) (14). Pioneering work in this regard has been performed by the group in Karlsruhe, where the formation of hydrogen clusters and other gases has extensively been studied (15).

The basic parameters for cluster formation that were studied in Karlsruhe were the nozzle geometries, the supply pressure and the temperature. The results showed that the lower the temperature and the higher the pressure, the greater is the cluster formation. The comparison of a diverging conical and a sonic nozzle for Argon also showed that the conical nozzle is more efficient.

The cluster formation takes place in the collision area behind the nozzle during the process of adiabatic expansion. The following reactions must be taken into account :

$$A_N + A_1 \rightarrow A_{N+1} \qquad \text{(growth)}$$

$$A_N^* + A_1 \rightarrow A_N + A_1^* \qquad \text{(energie exchange)}$$

$$A_N + A_1 \rightarrow A_{N-1} + 2\,A_1 \qquad \text{(sputtering)}$$

A deeper insight into the thermodynamic properties during the cluster formation is given by G. Stein (16).

The experimental observation of the transition from the covalent to the metal bond and catalysis applications are the main reasons for a special interest in metal clusters. Metal cluster sources however, have to overcome several difficulties :

- Oven constructions that allow metal vapor pressures of several atmospheres, as applied in gaseous sources, are very difficult to build, since they require very high temperatures, and temperature and chemically resistant materials.
- The same high temperature has to be applied to the nozzle, or else it would clog. Cluster formation, however, considerably decreases with an increasing nozzle temperature.
- The nature of the molecular bond, is considerably weaker for metal clusters than for gaseous clusters, as we will see later. Consequently the formation is more delicate.

A typical metal cluster source for alkali aggregates, mercury and other metals of a low boiling temperature is shown in figure 2.

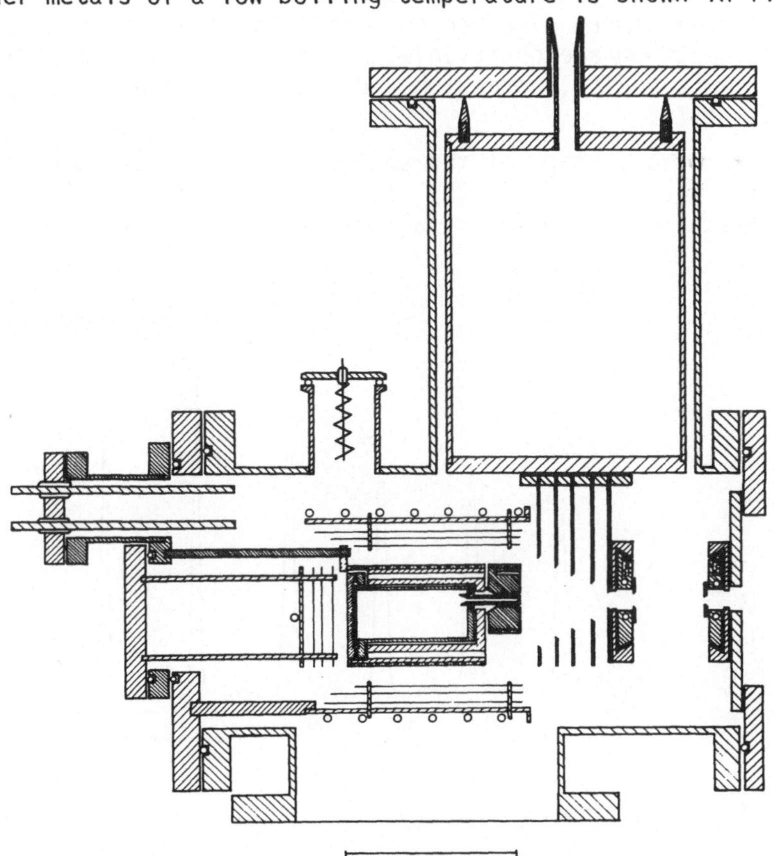

Fig. 2 : The Berne oven source.

An oven cartridge, with a nozzle welded on, is placed inside a heating jacket, which independently heats oven and nozzle. The nozzle is placed excentrically in the upper part of the cartridge. The heating system consist of a thermocoax heating wire, also welded to the oven. The whole oven system is surrounded by a 3-fold tantalum heat reflector and a water cooled heating jacket. The oven design makes no use of a skimmer. Instead a set of 5 large collimators cuts off a beam. The collimators are cryogenically cooled by a LN_2-cold trap. Behind these, are two collimators of adjustable width. They can be manipulated from outside of the vacuum. In addition to the efficient cryogenic pumping arrangement, a 1000 ℓ/sec. diffusion pump is used. This is mainly for pumping of the high rate of outgasing particles, when the oven is hot. The nozzle shape has 0.2 mm channel diameter, 0.4 mm channel length and an opening cone of 60^0. At the entry of the nozzle a metallic edge prevents creeping of liquid metal into the nozzle volume. The nozzle always has to be kept at least 50 ^oC hotter than the body to prevent clogging. The cartridge itself is sealed by a tantalum plate and backed up by a copper disk, which is pressed against a rounded sealing edge.

A typical mass spectrum of a mercury cluster beam from the Berne-type oven source ("source Bernoise") is shown in figure 3. Aggregates up to Hg_8 are observable. Furthermore, the system was used to produce sodium, potassium and cadmium aggregates up to Na_{16}, K_{12} and Cd_4 (17).

A major disadvantage of the source is the limited temperature range, which does not permit the operation beyond 1000 ^oC. Major improvements have been made by Sattler, Andres and Stein (18-20).

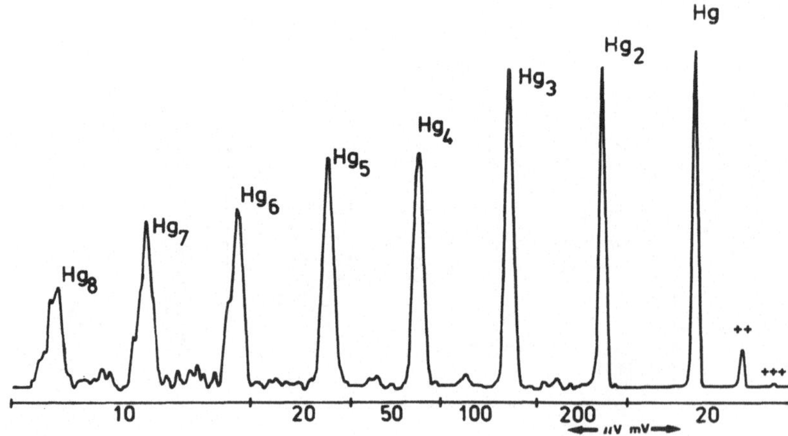

Fig. 3 : Mass spectrum of mercury aggregates.

Their sources make use of an inert carrier gas and a collision chamber in front of the nozzle, in order to cool down the heat of formation of the particles produced. Metal aggregates containing over 1000 atoms of silver, indium, lead or tin were generated this way. A detailed explanation is given by G. Stein (16).

A major advantage of the molecular beam techniques is the collision free environment, and the possibility of Doppler-free observation : Spectroscopic experiments, for example, which are carried out in a gas cell, have to take the Doppler broadening for normal thermal motion of $\sim 10^4$ cm/sec into account. This computes to linewidths of $\sim 10^{-2}$ A, which - by collisions - still can be broadened up to 10^{-1} A. This is a typical value, for example, for spectral lamps. Collisions inside the molecular beam, however, can be neglected and the line broadening due to the Doppler effect, when observing perpendicular to the beam, is well below the natural linewidth (17). Spectral resolutions $< 10^{-4}$ A may be obtained therefore. This is especially important for resolving rotational structures or hyperfine splittings.

2.4 Laser Vaporization

Although it is possible now, to reliably produce metal aggregates by means of molecular beam expansion, materials with high boiling temperatures are still difficult to be vaporized. This is the case, for example, for most of the very interesting transition metals. Alternative methods, like laser vaporization, have been tried out therefore : When a laser pulse, of several Joules energy, is focussed on nearly any material, such as graphite, tungsten, etc..., a small hole is drilled. The diameter can range from 10 to 100 μ, with a depth greater than a millimeter. This corresponds to an amount of $\sim 10^{-6}$ g ejected per pulse, and a temperature of about 4000 °C. If we consider a duration of 10^{-4} second for the event, we have a mass flux of 10^{-2} g/sec, which is comparable to a high performance cluster source. Accordingly, we performed laser vaporization on iron with a Nd-YAG laser, and observed the ejected material by means of mass spectrometry (21). Similar experiments have been performed as well by Berkowitz on graphite (22) and by Smally on aluminium (23). It should be noted, that the laser pulse duration has to be set sufficiently long, to prevent plasma formation. The mass spectra obtained show metal aggregates, but also many impurities. This is due to the fact that the preparation of an absolutely clean surface is rather difficult. Further work, therefore, is still in progress (21).

2.5 Field Ion Emission

In many experiments the production of charged metal clusters is of particular interest. A very interesting approach in this regard has recently been reported by Dixon et al. (24) : They inserted a sharp pin of tungsten (point radius ∿ 1 - 10 μm) into a reservoir containing the molten metal of interest. A film of molten metal flowed up to the tip. A sufficient high voltage was applied to this metal - film covered tip, which resulted in field ion emission of metal clusters. The mass spectrum of this field ion emission showed for example aggregates of tin up to Sn_7^+ . An interesting observation is that the relative abundance of large clusters with respect to small ones decreases, as the total emitted current grows. This phenomenon could perhaps be usuable to produce clusters of a desired size.

2.6 SIMS and FAB

SIMS is a synonym for "Secondary Ion Mass Spectrometry". The technique employs a high energy ion beam (typically 5 - 15 keV) to bombard a metallic target. This sputters off negatively and positively charged clusters. The secondary ions are then analyzed with a mass spectrometer. Hortig and Müller found silver clusters containing up to 30 atoms with a strong difference between the intensities of clusters of respectively odd and even numbers of atoms (15). The same effect was found by Leleyter and Joyes, and they gave a theoretical description of this process (26). Los and co-workers discovered that neutral particles were ejected as well and they studied alkali-dimers by means of a Stern Gerlach deflection (27).

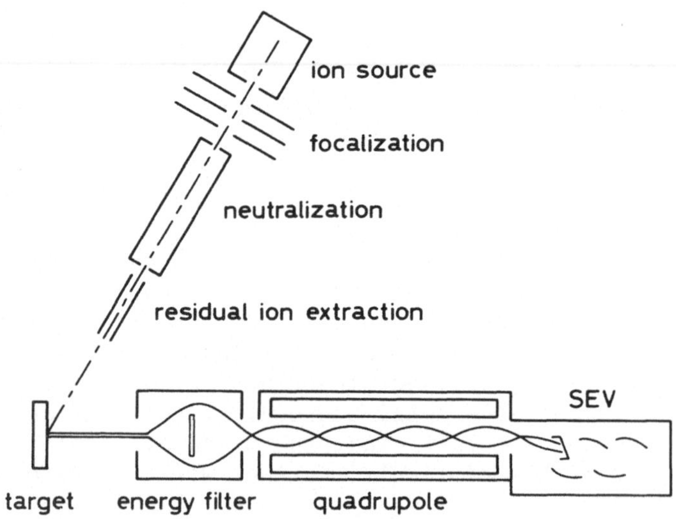

Fig. 4 : Secondary Ion Mass Spectrometer.

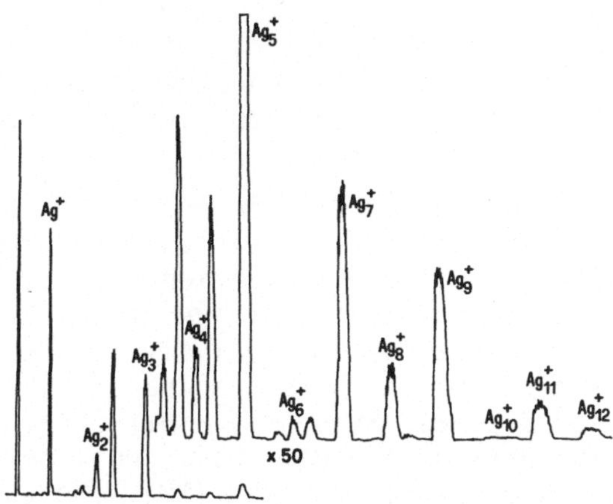

Fig. 5 : SIMS of silver.

Devienne and Roustan modified a SIMS device by neutralizing the primary ion beam (FAB = "Fast-Atom-Bombardment"). This is done by resonant gas exchange in a gas chamber which contains the same gas as the ion beam. The conversion efficiency is about 10 %. Devienne et al. obtained Ag, Au, U, Ta, Fe - clusters containing up to 41 metal atoms (29).

FAB-results are very similar to SIMS measurements. A special advantage of the technique is, however, the possibility of using amorphous and nonconductive target materials, which - under ion beam bombardment - would get highly charged and create undefined focussing conditions.

- A typical mass spectrum for silver aggregates, obtained in our laboratory, is shown in figure 4. Particles up Ag_{12}^+ and characteristic intensity alternations are observable (28).

3. MASS SPECTROMETRIC DETECTION

Mass spectrometry is most commonly applied to determine size distributions of cluster sources and to filter out certain particle sizes. The interpretation of signal intensities, however, is rather difficult, because particle dependent ionization cross sections, size dependent ion discrimination and fragmentation processes exhibit a large source of artefacts. Ionization, mass separation and ion detection should therefore well be selected. Whenever possible however, further experiments to determine absolute cluster intensities should be performed.

3.1 Ionization

As outlined before, a careful analysis of the ionization process is necessary to obtain a correlation between the cluster ion spectrum and the neutral size distribution. Various processes can be applied :

Surface ionization is about the most efficient ionization process. It is based on the Langmuir effect : Particles, that strike a hot filament can get ionized, if the ionization potential of the particles is lower than the work function of the filament. Unfortunately, this restricts the use almost exclusively to alkali aggregates and Rhenium or Tungsten filaments. Mass spectra of the emitted ions, however, only show atomic peaks. Evidently, therefore, the clusters fragmentize completely. Parrish and Herm have confirmed, that every atom of a cluster is independently ionized (30). The surface ionization detector, therefore, is an ideal instrument to determine the absolute mass flux of a beam.

Electron impact ionization is the most commonly used type of ionization : An electron beam, created on a hot filament, is accelerated to electron energies of typically 30 - 100 eV, which allow electron currents of some milliamperes. The electron beam crosses the neutral beam, which partially gets ionized by electron impact. Typical ionization efficiencies are about 10^{-4}. Figure 6 shows a special, highly performing e^- - impact ionizer proposed by Weiss (31). It reaches ionization efficiencies up to 10^{-2}. The mass spectrum in figure 2 was obtained with this type of ionizer.

It was impossible, however, to obtain an extended mass spectrum of alkali aggregates under similar conditions. The features are explained in figure 7, where we measured the ion signals for Na^+, Na_2^+ and Na_3^+ as a function of the electron energy. The electron current and the beam conditions were kept constant. At an electron energy of 5 eV the signal for Na^+ appears.-This corresponds to a rough ionization potential measurement-. The signal reaches a maximum value at about 25 eV. For Na_2^+ and Na_3^+, however, the signal starts decreasing at about 12 eV, and at 30 eV the Na_3^+-signal vanishes completely.

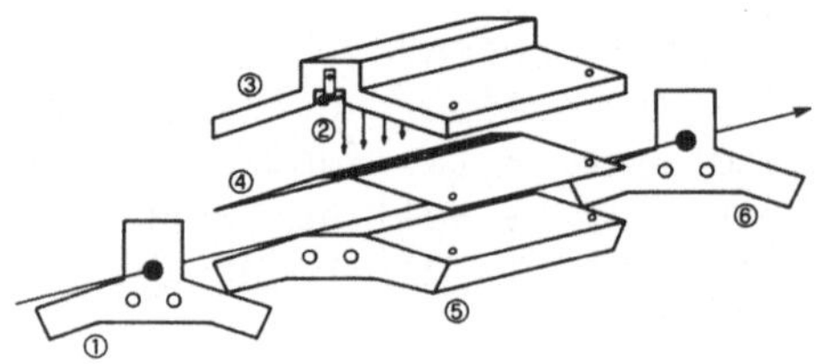

Fig. 6 : Weiss type ionizer. 1) entrance grid, 2) filament, 3) cathode, 4) grid, 5) anode, 6) exit grid.

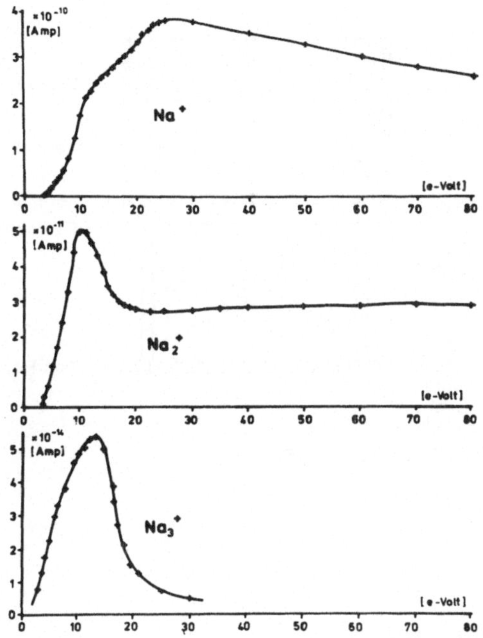

Fig. 7 : e^- - ionization of Na-aggregates as a function of the electron energy.

Evidently strong fragmentation processes occur, induced from excess electron energy. Consequently, the ionizer should be set to very low electron energies right above the ionization threshold. At these energies, however, the electron current is very low due to space charge problems. Consequently the sensitivity of the instrument decreases. Aggregates larger than Na_3, therefore, were never observed in our laboratory by means of electron impact ionization.

Photoionization proved to be best method to ionize alkali aggregates. Figure 8 shows a photoionization source, which we applied to inject the ions into a quadrupole mass spectrometer. We used Hg- or Xe- arc lamps and lasers as light sources.

A typical photoionization mass spectrum is shown in figure 9. Typically no background gas peaks appear. This is due to their higher ionization potential which is not reached with the UV-light of the arc lamp.

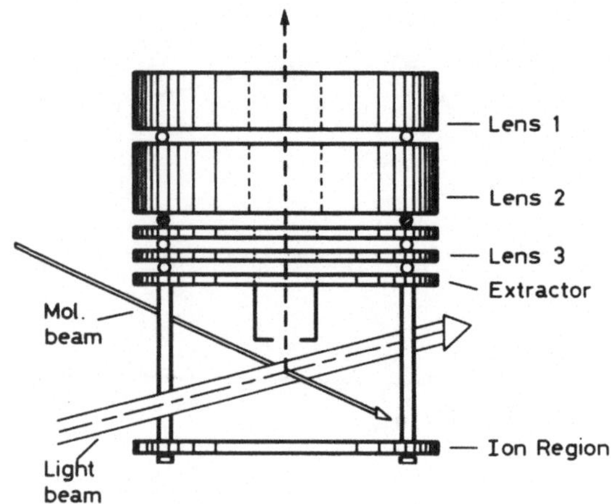

Fig. 8 : Photoionization source.

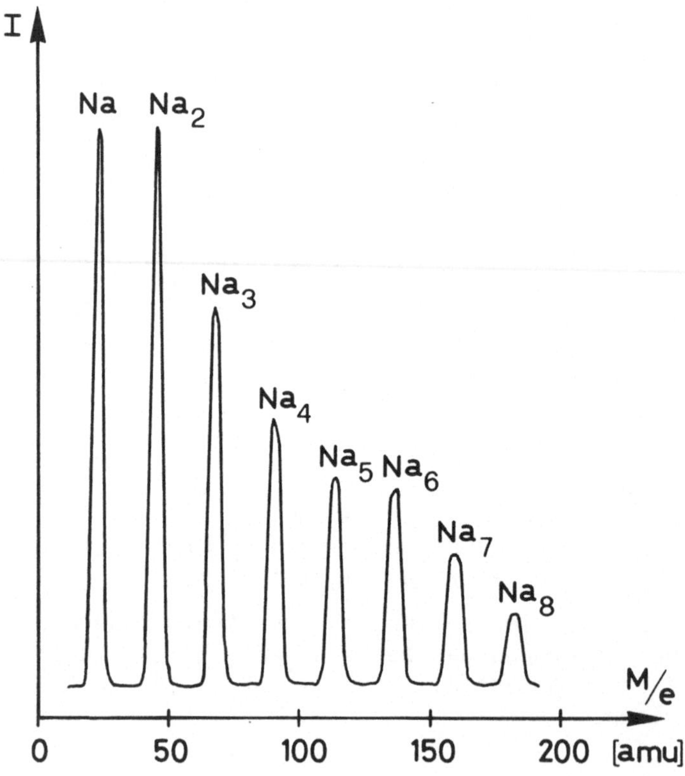

Fig. 9 : Photoionization mass spectrum of a sodium cluster beam.

The photoelectrons, which are also generated, may very well cause secondary e^- - impact ionization processes, if the acceleration potential between the first two electrodes is too high. The photo-ions, on the other hand, may well cause Penning ionization processes. Consequently, the ion density should not become too high.

3.2 Mass separation

The palette of mass spectrometers ranges from simple rest gas analyzers to highly sophisticated analytical instruments, and various separation schemes are applied : magnetic deflection, dynamic mass separation or Time-of-Flight analysis (TOF). The ideal system for cluster applications does not exist, it has to be tailored to the experiment. The most important parameters are : mode of operation, mass range, sensitivity and resolution.

Magnetic mass separators are commercially available for analytical purposes in a large variety, reaching resolutions up to 100'000. The mass range of these instruments, however, is very limited, since it is a well known fact that it is difficult to generate unfragmentized ions of an M/e exceeding 800. R. Rechsteiner therefore constructed a special magnetic instrument that was capable to reach masses up to M/e = 30'000 at still reasonable transmission values (32). The resolution of the instrument was kept relatively low at about 600 which, however, is fully sufficient for cluster applications.

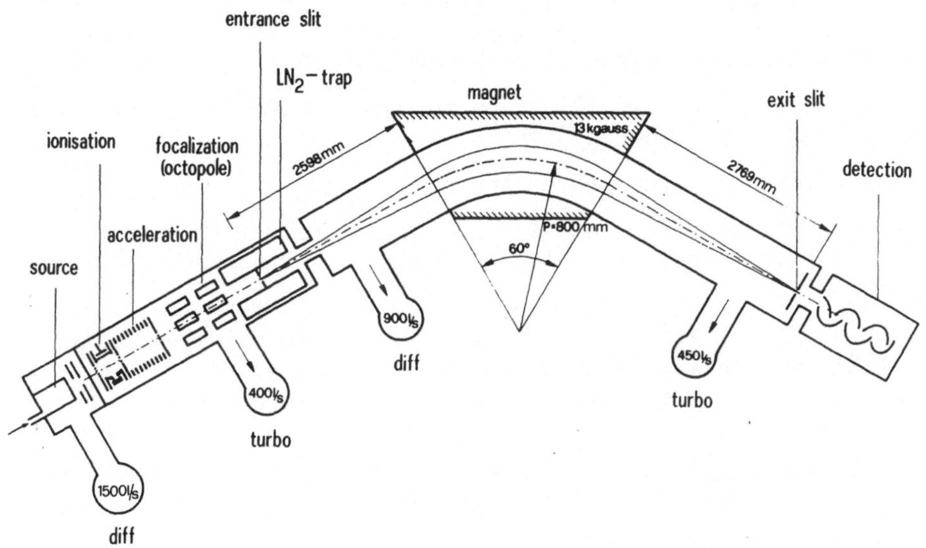

Fig. 10 : Magnetic mass spectrometer for cluster analysis (32).

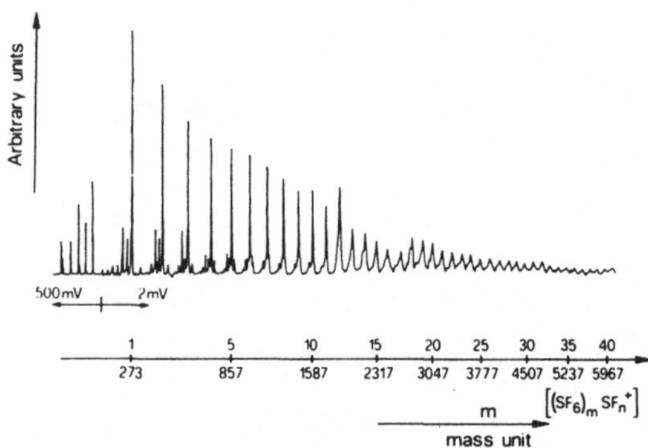

Fig. 11 : Magnetic mass spectrum of a $(SF_6)_n$ - cluster beam.

The design of the instrument is shown in figure 9: Clusters that are generated in the source are ionized, accelerated and focalized on the entrance slit of the instrument. From there they enter after a free path of 2,60 m the magnetic sector field of 60° on a radius of 80 cm. The magnet reaches a field up to 13 KGauss. Correspondingly it weighs 2,6 tons. After passing another free path of 2,77 m the ions are focalized on the exit slit, where they are detected. The focalization of the instrument is performed in the horizontal and vertical plane with the help of a static focalization octopole. The performance of the mass spectrometer is indicated with the SF_6-cluster mass spectrum in figure 11. Particle sizes up to $(SF_6)_{42}$ are observable, which corresponds to M/e = 6132 (33).

In spite of the excelllent performance, however, magnetic instruments have two disadvantages :

- The hardwave is large and expensive.
- The transmission is relatively poor (typically $< 10^{-3}$) due to the fact that most ions are cut off by the narrow entrance and exit slits.

Quadrupole mass filters, therefore, do have several advantages: They are relatively small, inexpensive and reach excellent transmission values of > 10 %. The schematic setup of a quadrupole instrument is shown in figure 12.

The ions enter in a system of 4 cylindrical rods, which are simulatenously connected to a HF and DC electric source. This way a potential of hyperbolical shape is formed across the rods, which is described by :

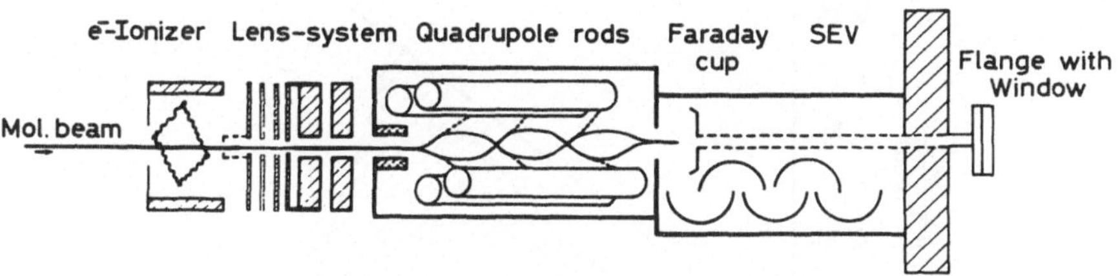

Fig. 12 : Schematic setup of a quadrupole mass spectrometer.

$$\phi = (U + V \cos \omega t)\, \frac{x^2 - y^2}{r_o^2}$$

U = DC-voltage
V = RF-voltage
ω = frequency

Consequently, the ions follow the equations of motion :

$$m\ddot{x} + 2e\,(U + V\cos\omega t)\,\frac{x}{r_o^2} = 0$$

$$m\ddot{y} - 2e\,(U + V\cos\omega t)\,\frac{y}{r_o^2} = 0$$

$$m\ddot{z} = \qquad = 0$$

With the following transformation

$$\omega t = 2\,\xi, \quad a = \frac{8\,eU}{m\,r_o^2\,\omega^2}\,, \quad q = \frac{4\,eV}{m\,r_o^2\,\omega^2}$$

a Mathieu differential equation is obtained :

$$\frac{d^2 U}{d\,\xi^2} + (a - 2q\cos\xi)\;U = 0$$

The equation has converging and diverging solutions, depending on the parameters a and q. A part of the solution is shown in figure 13 : The dashed aerea shows the converging part : If the conditions are set accordingly, the ions will cross the filter by performing a modest oscillation. Out of this range the solution is diverging; accordingly the ion amplitude will constantly increase until the rods are hit. We see from the graph that the solution is practically always stable, if no DC-voltage is connected. This means we have a broadband ion trap. If, however, we position our quadrupole

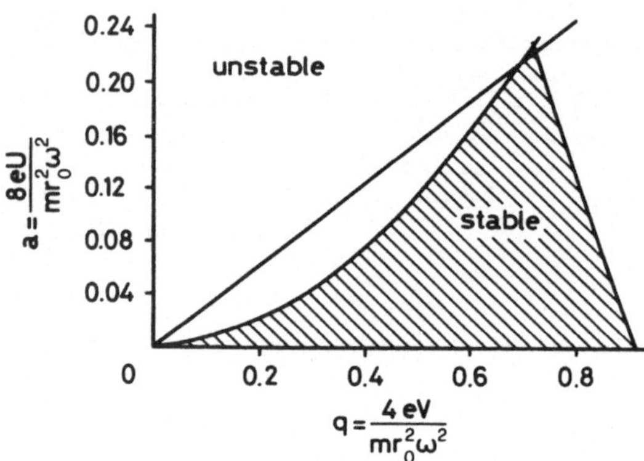

Fig. 13 : Stability diagram of a quadrupole filter.

parameters along the axis A, only a very finite mass is transmitted. The quadrupole,therefore,allows to choose the resolution. Values up to 2000 can be reached. The highest mass is determined by electrical restrictions from the RF-source. Masses up to M/e = 2000 can be reached. A typical mass spectrum from a quadrupole mass filter is shown in figure 9. Disadvantages of the quadrupole filter are the limited mass range and a non linear mass discrimination, which complicates the interpretation of cluster intensities.

Time-of-Flight analysis (TOF) is indicated, when the mass range of a quadrupole filter is too limited, while the transmission of a magnetic instrument is too low, or the mode of the experiment is pulsed (pulsed laser, pulsed nozzle, etc.). A typical setup of a TOF instrument, developed by J.D. Ganière (34), is shown in figure 14.

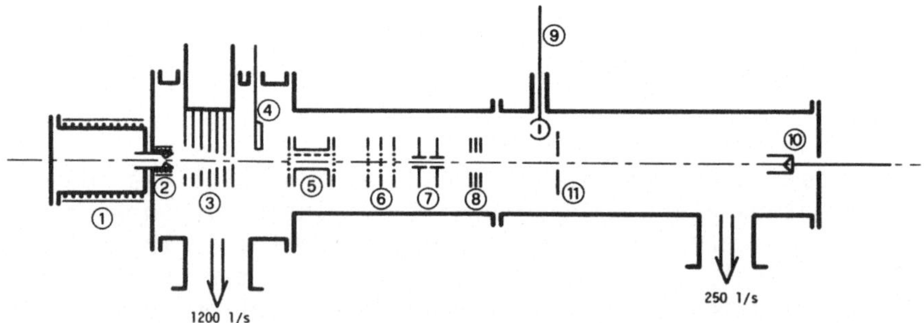

Fig. 14 : Time-of-Flight mass spectrometer for cluster beam analysis: 1-oven, 2-nozzle, 3-collimators, 4-beam flag, 5-ionizer, 6-8 ion optics, 9-LT-detector, 10- FC-cup, 11- beam gate.

The particle beam is axially or perpendicularly injected into the ionization region. The ionization is pulsed, for example, by means of a pulsed laser. The ionized particles are then smoothly accelerated into the acceleration region where they are accelerated up to 3 keV into the drift tube. The typical length of a drift tube may be in the order of 1 m, then the ions strike the detector. Since heavy ions are accelerated to lower velocities then light ions, the particles spread up in the drift tube. The whole spectrum, therefore, appears in a time-resolved sequence.

The ionization region is usually spread over a finite volume. Consequently a drift length dispersion and an acceleration dispersion occurs. At proper chosen potentials, however, both dispersions can be compensated to a first order spatial focus (35).

Fig. 15 shows the TOF mass spectrum of a sodium cluster beam, ionized with the 4th harmonic of a Nd-YAG laser. Particles up to Na_{21} can be observed. The mass range of a TOF is principally unlimited, the resolution, however, decreases with growing mass. It can be considered as a great advantage that each pulse provides a complete mass spectrum. In the mass spectrum additional broad peaks occur. This phenomenon is due to metastable particles, which cannot be observed with quadrupole mass filters.

3.3 Ion detection

Due to the lack of large ionic species (M/e >> 800) in analytical mass spectrometry, the response of electron multiplication systems is relatively unknown in the higher mass range.

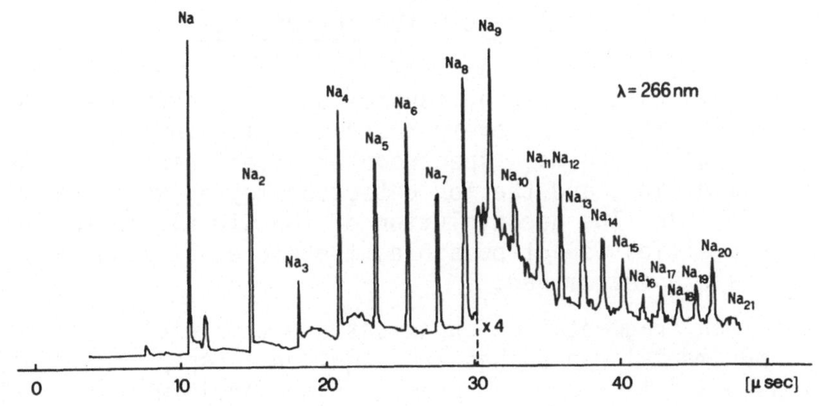

Fig. 15 : TOF mass spectrum of Na-aggregates photoionized at λ = 266 nm.

Faraday cup detectors have a linear response to the ion current. Therefore, they are recommended to be used, whenever the ion signal is sufficiently strong ($I > 10^{-14}$ Amp.). They also allow to calibrate gain curves of electron multiplication systems at different cluster sizes.

Channeltron e^- - multiplication tubes are well suited for analog measurements and ion counting. Under suitable conditions their background noise is below 1 cps, and gain factors > 10^8 are reached. They are well suited to be used in magnetic instruments and quadrupole systems. At high ion currents ($I > 10^{-12}$ Amp.), however, space charge problems occur due to the finite diameter of the tube (ϕ < 1 mm). Also, the entrance surface of a channeltron is relatively small ($\sim$ 1 cm^2) and requires a well focussed ion beam.

Venitian blind multipliers are considerably larger. Therefore, they are well suited to be used in TOF mass spectrometers, where widely spread ion packages at temporarily high currents may occur. They are, however, very sensitive to light, and must carefully be shielded, when photoionization experiments are performed.

Daly-detectors (36) are recommended, when laser spectroscopic experiments require a detection system, which is insensitive to light, but combines the advantages of an e^- - multiplication system : The incident ion beam is directed to a metal target on high potential ($U \approx$ 30 keV), which causes secondary electron emission. The 30 keV-electrons easily envade a scintillator plate at ground potential, which is coated with a thin metal layer. Scintillation events are then recorded with an optically coupled photomultiplier tube. The system is insensitive to light and air, background counts are negligible and gain factors > 10^8 are reached.

3.4 Measurement of absolute cluster intensities

Clusters of different particle size have different ionization potentials (37) and different ionization cross sections. Since they occur at different speeds, their residence time in the ionizer is different as well. The mass spectrometer itself has a size dependent mass discrimination, and the ion detection may as well change with the particle size. The deconvolution of ion signals to absolute cluster intensities is not possible, therefore, as long as these parameters are undetermined.

Even when fragmentation processes are completely avoided, different experimental approaches are required : Gspann and Vollmar (38) crossed a potassium oven beam at right angle with a N_2-cluster beam and measured the relative beam deflection with high angular resolution. From the results they could derive the mean cluster size N.

Gordon, Lee and Herschbach (39) separated in a Stern-Gerlach apparatus the paramagnetic monomers of an alkali beam from diamagnetic dimers. The beam intensity of deflected and undeflected particles was determined with a surface ionization detector. The results showed that their beam contained a 30 % - dimer mole fraction. An inbuilt velocity selector allowed furthermore independently to determine the beam velocity for monomers and dimers, which they found to be quite similar. The signal of a surface ionization detector represents the total mass flux of a beam. This allows one to calibrate the detection efficiency of a mass spectrometer on the monomer peak under low stagnation pressures, while no clusters are formed yet. When later at higher stagnation pressures clusters are formed, the total mass flux is still measured with the surface ionization detector, while the mass spectrometer, which is still set on the monomer peak, discriminates all cluster ions. The difference between both signals, therefore, represents the cluster mole fraction of the beam. Measurements, which were accordingly performed with a sodium cluster beam at various oven pressures are plotted in figure 16. The results shows, that a cluster mole fraction of > 50 % can be obtained.

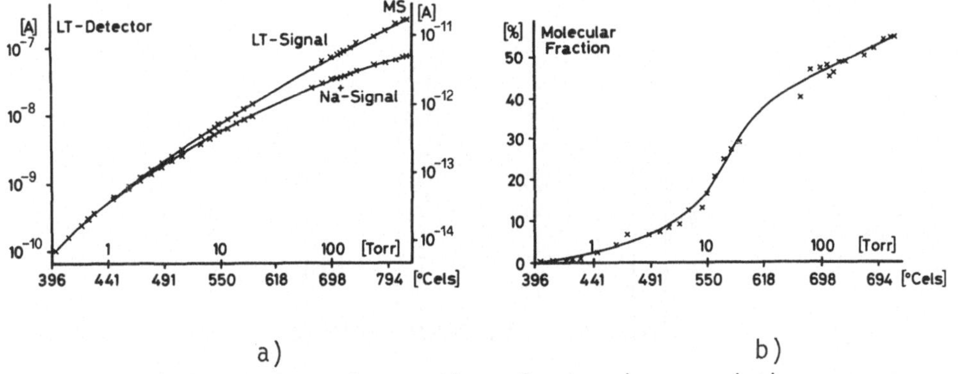

a) b)

Fig. 16 : a) Intensity of a sodium cluster beam and the monomer signal at various oven pressures. b) Cluster mole fraction of the beam at various oven pressures.

4. EXPERIMENTS ON CLUSTERS IN THE GAS PHASE

4.1 Electron scattering experiments

The structure of clusters embedded in a rare gas matrix can well be determined by electron scattering experiments (40). Early experiments on gas phase clusters were performed by Farges (41). He crossed a supersonic noble gas beam with a 50 keV electron beam, and recorded the diffraction patterns. From the results he concluded that the clusters of his noble gas beam were mainly built up in icosaedric structures. Similar experiments on metal aggregates will be discussed in detail by G. Stein (16).

4.2 Coulomb explosion of doubly charged clusters

It is a well known phenomenon in mass spectrometry that doubly charged particles occur, when sufficient ionization energy is available. A very interesting effect in this regard, was recently found by Sattler et al., while observing cluster beams : The mass peaks of doubly charged clusters only appeared starting from a certain cluster size. Similar findings were confirmed by R. Rechsteiner in our laboratory on $(CO_2)_n$ - clusters (42), as figure 17 shows : The mass peak of doubly ionized clusters only appears at M/e = 22,5 which corresponds to $(CO_2)_{45}^{++}$.

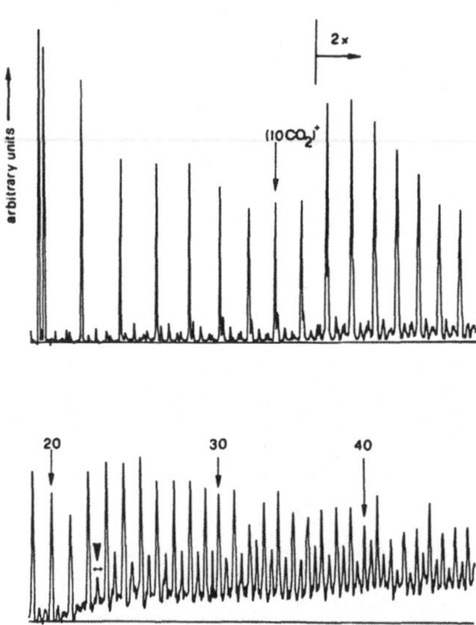

Fig. 17 : Mass spectrum of a CO_2 - cluster beam and the occurance of doubly charged particles beginning at M/e = 22.5

The phenomenon is independent from variations of the electron energy and does not represent a measurement of an ionization potential. The effect, therefore, must be explained as followed : The charges on a double ionized cluster tend to localize on opposite extreme sides. Whenever the cluster size becomes too small, the charge locations get very close, and the cluster explodes due to Coulomb repulsion. The critical size, where the effect starts, is therefore a direct information about the bond strength. A theoretical description of the pheonomenon has recently been published by Gay and Berne (43).

4.3 Ion-molecule reactions of gas phase clusters

Ion molecule reactions of clusters are of significant interest in the chemistry of the troposphere. Interesting experiments - in this regard - were performed by Castleman et al. (44). They measured reaction constants for a variety of cluster-ion reactions as a function of the temperature. Some of the reactions studied were :

$$Na^+ (SO_2)_n + SO_2 \rightleftharpoons Na^+ (SO_2)_{n+1}$$

$$Ag^+ (H_2O)_n + H_2O \rightleftharpoons Ag^+ (H_2O)_{n+1}$$

$$Ag^+ (NH_3)_n + NH_3 \rightleftharpoons Ag^+ (NH_3)_{n+1}$$

The experimental procedure is the following : Ions like Na^+ are formed on a hot wire. Cluster formation occurs, as these ions drift in a low electric field through a gas reactant mixture, which is well pressure-and temperature-controlled. Then they exit through a small orifice (typ. 50 μ) into a QMS chamber, where they are mass-analyzed. Equilibrium constants $K_{n,n+1}$ are obtained from

$$K_{n,n+1} = I_R/P$$

where I_R is the concentration ratio of product and reactant ions, and P is the reactant pressure in atmospheres. For each reaction $K_{n,n+1}$ was measured for a range of temperatures T. The results show remarkable size dependencies. They are extremely interesting with respect to atmospheric implications (45).

4.4 Free hydrated electrons

Among chemical reactions with clusters, reactions with water clusters are certainly most exciting, because they give an insight into solvatisation problems, answering the question : How many water molecules are necessary to bring another particle into solution. Haberland tried to answer this question for electrons. The experimental arrangement is as followed : Water vapor of 100 to 1000 Torr and 400 K is irradiated with electrons from a β^- - radioactive ^{63}Ni foil. The 66 keV electrons are all quickly thermalized, and some are getting solvated. Then the water particles get expanded, and they are measured by means of TOF-or QMS-analysis. In an early experiment, the group observed the first negatively charged water clusters at $(H_2O)_{11}$ (46). This value, however, could experimentally not yet be reconfirmed again (47).

4.5 Magnetic measurements

Early experiments on separating diamagnetic Cs - dimers from paramagnetic monomers have been performed by Gordon, Lee and Herschbach (39). Similar experiments on larger alkali aggregates were later performed by W. Knight and coworkers (48). They constructed a Rabi-type machine consisting of two inhomogeneous magnetic fields (A-field and B-field) with a homogeneous field (C-field) in between. The C-field contains a microwave cavity in order to induce electron spin flips on particles, which were magnetically filtered in the A-field. Transitions can then be analyzed with the B-field. A mass spectrometric detection system allowed to take deflection profiles of alkali aggregates containing up to seven atoms (49). ESR-measurements, however, have not yet produced conclusive results, which may be due to random spin flips, or undetermined rotational conditions (50). The precise knowledge about particle temperatures, therefore, is required.

4.6 Appearance potential measurements

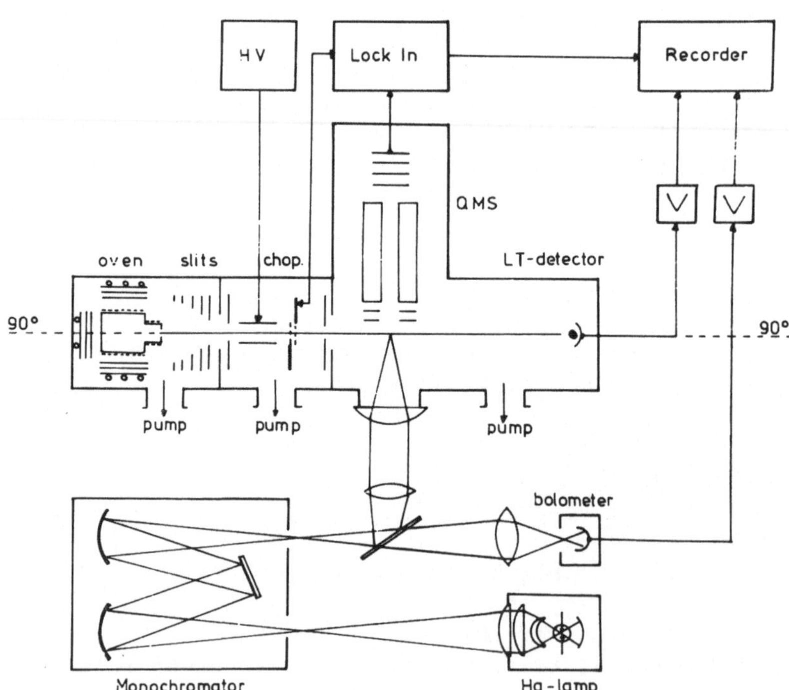

Fig. 18 : Experimental set-up for measuring ionization potentials.

The ionization energy of atomic sodium is 5.14 eV, while the work function of the metallic bulk is only 2.3 eV. Appearance potential measurements of the clusters can provide an insight for well understanding this gap. An experimental arrangement for measuring the ionization potentials of alkali aggregates is shown in figure 18: The aggregates are produced in a supersonic jet expansion and collimated to a beam. An electric field extracts all ionic species from the beam, which accidentally may occur from surface ionization on the hot nozzle. The cluster beam is modulated by means of a tuning fork chopper. The particles are irradiated with the light of a 1 kW mercury arc lamp, which is passed through an illumination monochromator. The spectral density of the lamp is measured with a bolometer. Photoions are mass selectively recorded with a quadrupole mass spectrometer. Measurements were performed by setting the mass spectrometer on a particular cluster size, while scanning the transmission wavelength of the illumination monochromator. Results for the aggregates Na_3 - Na_5 are shown in figure 19.

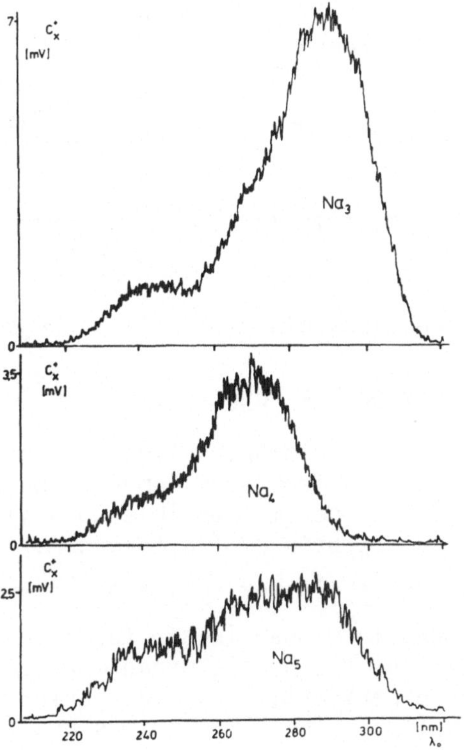

Fig. 19 : Uncompensated appearance potential measurements for Na_3, Na_4 and Na_5.

The measurements do not yet take the spectral variations of the mercury lamp spectrum into account. The typical decrease of the ion signal to zero beyond the ionization wavelength, however, is well observable. The complete results for the ionization thresholds of Na - Na_{14} and K - K_{12} are shown in figure 20. The results in figure 20 indicate very well, how the ionization potential of a growing cluster size changes from the atomic value towards the work function of the bulk. The behaviour can roughly be described with a model, where an elementary charge is extracted from a sphere of growing size (51). The model, however, does not explain the extraordinary low value for trimers, and alternations, which are observable. These effects are certainly due to quantum phenomena. Their understanding requires high resolution experiments with informations about the rotational and vibrational structure of the particles.

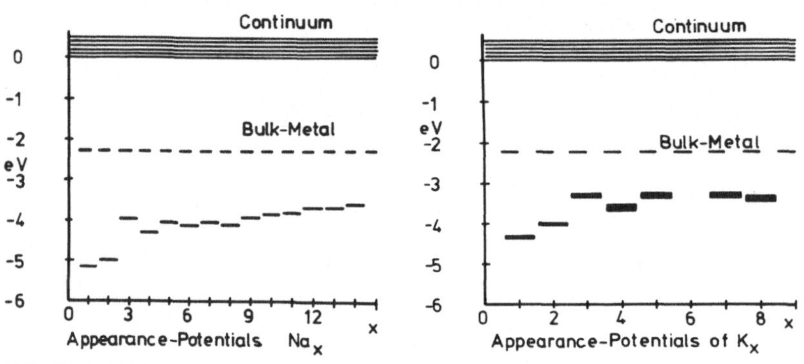

Fig. 20 : Ionization threshold of alkali aggregates.

5. LASER SPECTROSCOPY OF FREE CLUSTERS

The molecular beam techniques allow to study particles in a collision - and Doppler - free environment; and with the advent of the tunable laser it has become possible, to well resolve the vibrational and rotational structure of a molecule.

5.1 Infrared probing of cluster formation

A deeper insight into the cluster formation process can be obtained by irradiating the nozzle expansion zone with a laser. Figure 21 shows an experiment, as it was performed in our laboratory on SF_6 - aggregates (52). The SF_6-clusters are formed in a temperature controlled nozzle expansion, skimmed and collimated to a molecular beam, and injected into a mass spectrometer. A typical mass spectrum of the aggregates has been shown previously in figure 10. When the expansion zone is now irradiated with resonant infrared

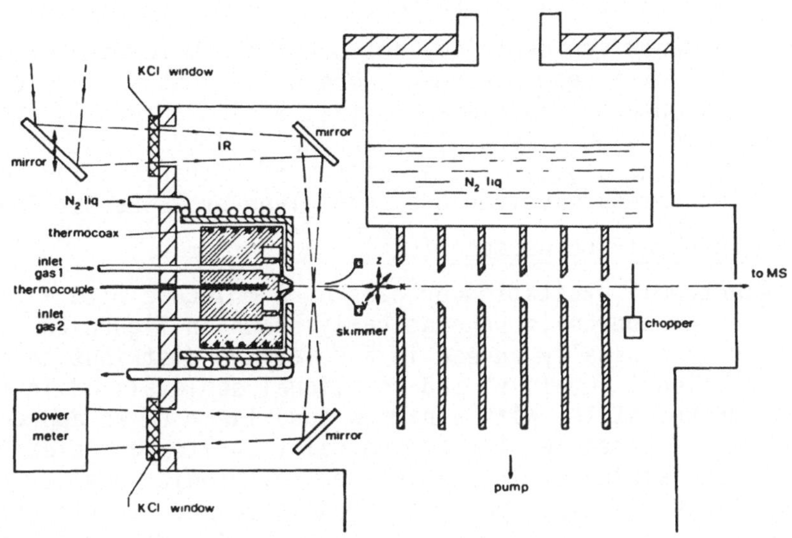

Fig. 21 : Laser probing of SF_6 - cluster formation.

light of a CO_2 - laser, the particles get vibrationally heated. Eventually they get so hot, that they cannot form clusters anymore. This is well shown in figure 22, where the attenuation of the dimer

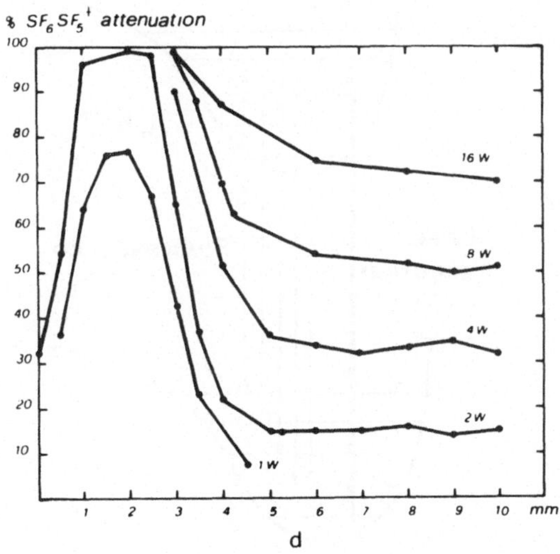

Fig. 22 : Attenuation of the dimer signal as a function of the laser power and the distance from the nozzle.

signal at the mass spectrometer is plotted at different laser powers. The result shows, that laser powers of 16 Watts are already sufficient to completely prevent cluster formation.

The phenomenon can certainly not occur any more further downstream the molecular beam, where no more collisions and no more cluster formation occur. But even in that region a certain attenuation of the dimer signal is still observable. This is due to photodissociation of the clusters. The attempt of a quantitative description of all phenomena involved has been given by R. Rechsteiner (32).

5.2 Laser-Induced-Fluorescence (LIF)

The electronic excitation process of a molecule with a narrow band laser light source is schematically shown in figure 23. The particles are usually spread in a Boltzmann distribution over various vibrational levels v" and rotational sublevels J" in the electronic ground state. With a narrow band laser of an appropriate wavelength it is possible to electronically excite particles from a specific ground state level to a defined rovibronic level of the excited state. Certainly, the transition is subject to vibrational transition probabilities f (v', v") (Franck-Condon factors) and the rotational selection rule $\Delta J = 0, \pm 1$ (P-,Q-,R- branches). After a characteristic lifetime τ of the excited state the molecule relaxes back to the electronic ground state under emission of a fluorescence photon. This can directly be detected with a photomultiplier. Thus by scanning the laser wavelength, an excitation spectrum is obtained.

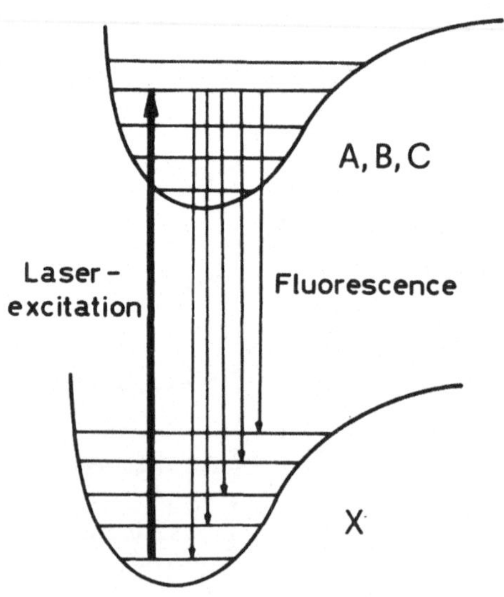

Fig. 23 : Principle of Laser-Induced-Fluorescence.

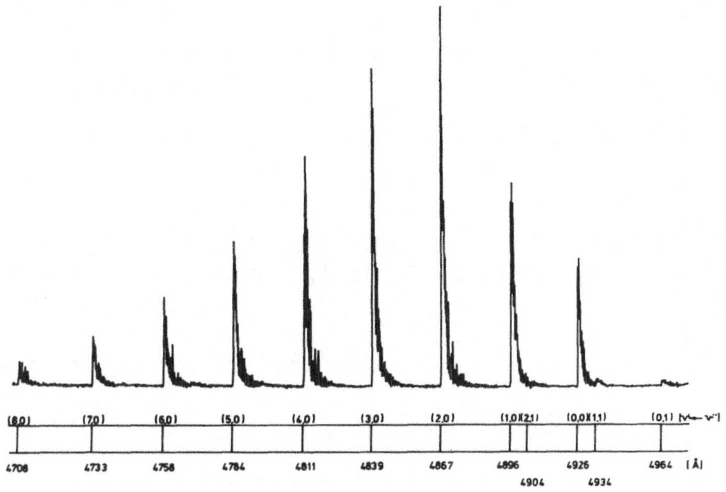

Fig. 24 : LIF-spectrum of the system Na_2 B←X recorded under molecular beam conditions.

Alternatively the laser may be kept at a fixed wavelength, while the fluorescence light is analyzed with a monochromator in order to obtain a fluorescence spectrum.

A typical LIF-excitation spectrum of the Na_2 - molecule is shown in figure 24. The system has electronically been excited with a tunable Coumarin 4 - laser in the range from 4700 to 4950 A, the photons were directly detected with an RCA 7565 PM-tube. The spectrum clearly shows the vibrational transitions of the system Na_2 B←X. The peak intensities correspond very well with the Franck-Condon factors (53).

5.3 Photon recoil spectroscopy

The LIF technique is very powerful for probing the internal state distribution of molecular beam particles (54). The method, however, cannot provide size specific informations, when particle mixtures like cluster beams are investigated. This requires the use of a mass spectrometric detection system, which - for example - is possible in a photon recoil experiment : When a free particle absorbs a photon, it obtains a linear momentum

$$p_0 = \frac{h}{\lambda}$$

This momentum transfer can be recompensated, when a fluorescence photon is reemitted into the same direction, which is usually the case for stimulated emission processes. Spontaneous emissions, however, do have an angular distribution. The total momentum transfer of a laser excitation, and a spontaneous emission process is therefore :

$$0 < p < 2\,p_o.$$

This will statistically average to

$$\Sigma\, p_o = \Sigma\, p$$

A molecular beam, which is irradiated perpendicularly with resonant laser light will therefore be deflected sidewards at a total deflection velocity v_d :

$$v_d = \frac{n\,p}{m} = \frac{n\,h}{m\,\lambda} \qquad n = \text{number of excitation processes}$$

This results to a deflection angle α with :

$$\tan\alpha = \frac{v_d}{v_m} \qquad v_m = \text{molecular beam velocity}$$

The total number of excitation processes can therefore be determined by :

$$n = \frac{m\,v_m\,\tan\alpha}{h}$$

An experimental setup for measuring this effect is shown in figure 25. A mass spectrometer serves as detector for the molecular beam. It can be moved perpendicularly to the molecular beam profile.

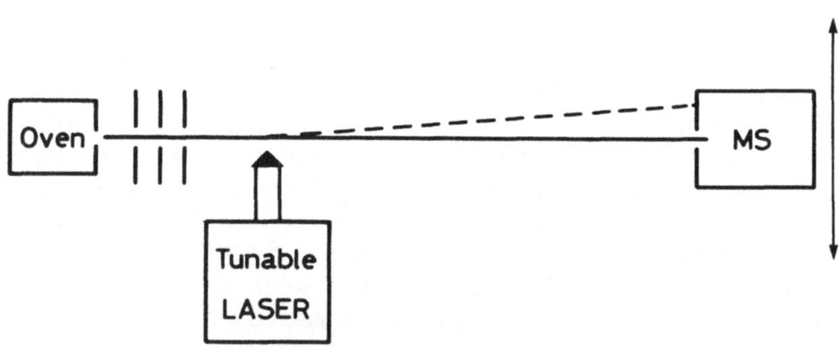

Fig. 25 : Experiment for performing mass selective photon recoil spectroscopy.

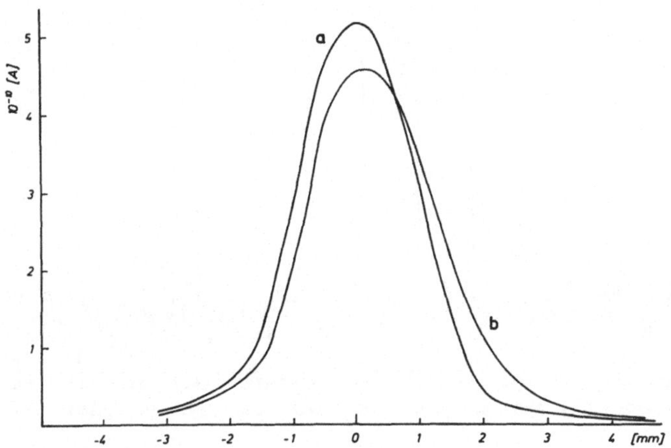

Fig. 26 : Profiles of a sodium atomic beam. Curve a shows the undeflected profile, when the laser is off resonance, curve b shows the deflected profile with the laser switched on the D_2 - line.

Curve a in figure 26 shows the undeflected profile of a sodium atomic beam; curve b presents the deflected profile, when the beam is irradiated at right angles with resonance light at λ = 5889 A.

For the result, which is shown in figure 26, about 60 % of all sodium atoms were in resonance with the laser, and each of these atoms was excited about 60 times (55). Experiments on molecules are more difficult to perform :

- The ground state distribution of molecules is considerably larger than the hyperfine splitting of atomic systems. Only a very small percentage of molecules can get in resonance with the laser therefore.
- After an electronic excitation process a molecule fluoresces back into different vibrational ground state levels (see figure 23). Therefore a molecule can only be excited once in a collision free environment (optical pumping).

With a well collimated molecular beam and a phase sensitive detection on the slope of the molecular beam profile, the deflection process can still be observed for molecules as well (56). Figure 27 shows a spectroscopic example, as it was obtained for Na_2 - molecules.

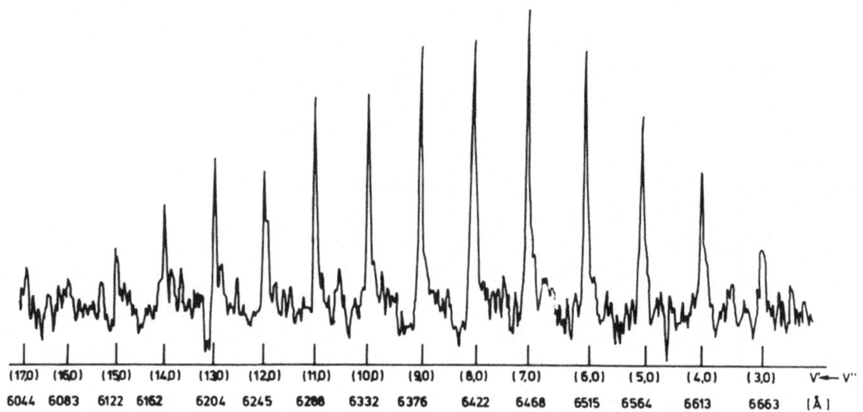

Fig. 27 : Vibrational sequence of the Na_2 A←X system measured by means of molecular beam deflection.

The result in figure 27 shows well the vibrational sequence of the system Na_2 A←X. It was obtained with a 150 mWatt Rh B - laser system. The signal to noise ratio of the measurement is about 10. It could not be improved by applying more laser power, because the system would saturate. This is documented in the power dependency measurement in figure 28.

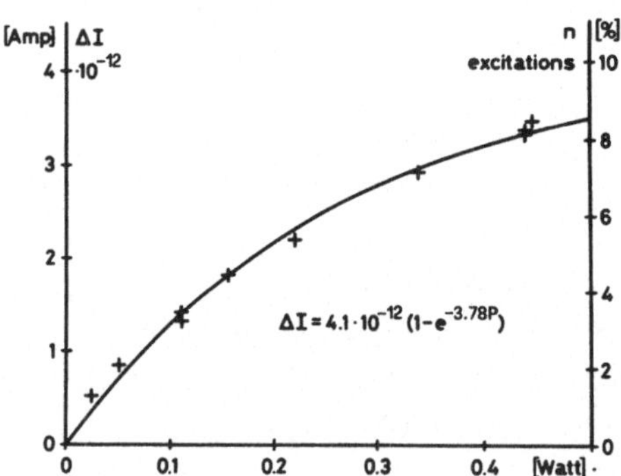

Fig. 28 : Power dependency of the molecular beam deflection measured on the transition Na_2 A←X (9'←0").

The experimental result of the power dependency curve can numerically be fitted to the formula

$$\Delta I = 4{,}1 \cdot 10^{-12} \, (1 - e^{-3{,}78\,p})$$

ΔI = deflection effect on the slope of the beam profile
p = laser power (Watt)

Due to the formula given before, this result gives a direct measure for the number of excitations per molecule. At 150 m Watt laser power, for example, 5 % of the Na_2 - molecules did get excited.

The experiment mentioned here did not make use of a mass selective detection system. This would certainly improve the signal to noise ratio. From these results one may conclude, however, that ultimately cluster sizes up to Na_7 can be investigated with this method (53).

5.4 Two-Photon-Ionization

The LIF - method does not allow a mass selective detection, the deflection method, on the other hand, does not provide very excellent signal to noise levels. For this reason the method of Two-Photon-Ionization (TPI) was developped.

5.4.1 The principle of the experiment

The principle of the TPI-process is indicated in figure 29 : The molecules are electronically excited with a tunable laser $h\nu_1$ from the ground state S_0 to an excited state S_1. From there they may fluoresce back to the ground state (LIF). If, however, a sufficiently strong second light source $h\nu_2$ of an appropriate wavelength is irradiated simultaneously, the excited particles can get ionized. The energies of $h\nu_1$ and $h\nu_2$ are such that none of them alone, but only their sum reaches the ionization potential :

$$h\nu_1 + h\nu_2 \geq E_{ion}$$

The ionization energies of the particles investigated are well known from appearance potential measurements (37).

An experimental setup for measuring TPI-processes is shown in figure 30 : The metal clusters are formed in an adiabatic expansion and well collimated to a molecular beam. Then the cluster beam is injected into a second-differentially pumped vacuum chamber, where it is monitored with an ion gauge or a surface ionization detector. Both lasers $h\nu_1$ and $h\nu_2$ are irradiated on the same spot of the molecular beam, which is placed right below the entrance of a mass spectrometer. Preferably the laser with the longer wavelength is modulated for permitting phase sensitive detection. This way, two photon processes, which may occur from one laser alone, are well discriminated.

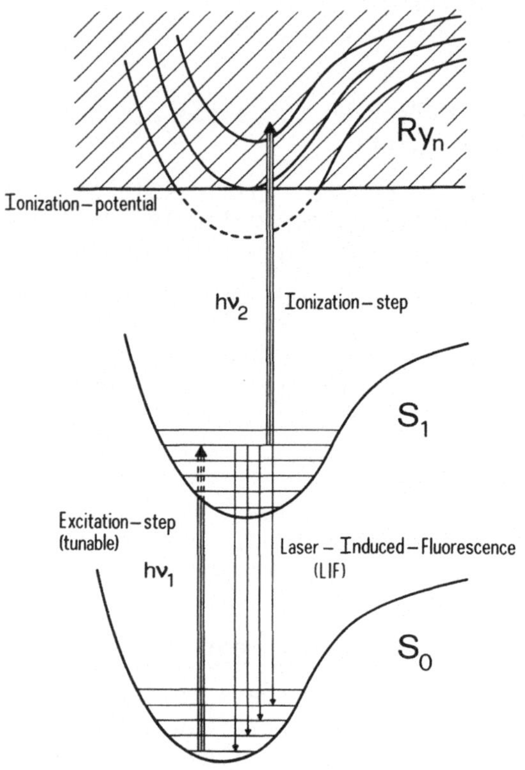

Fig. 29 : The Two-Photon-Ionization process.

A typical result, as it was obtained for potassium dimers, is shown in figure 31. The vibrational sequence shows very well individual electronic transitions from the vibronic ground state levels v" = 0, 1, 2 to vibronic levels in the excited state between v' = 0 and v' = 11.

5.4.2 Isotope shift measurements

The mass selective detection of the experiment easily allows to perform isotope selective measurements. A typical example for the aggregates $^{39,41}K_2$ and $^{39,39}K_2$ is shown in figure 32. In both cases a beam of natural potassium was used, in the first case, however, the mass spectrometer was set on M/e = 80, while in the second case the spectrum was recorded at M/e = 78.

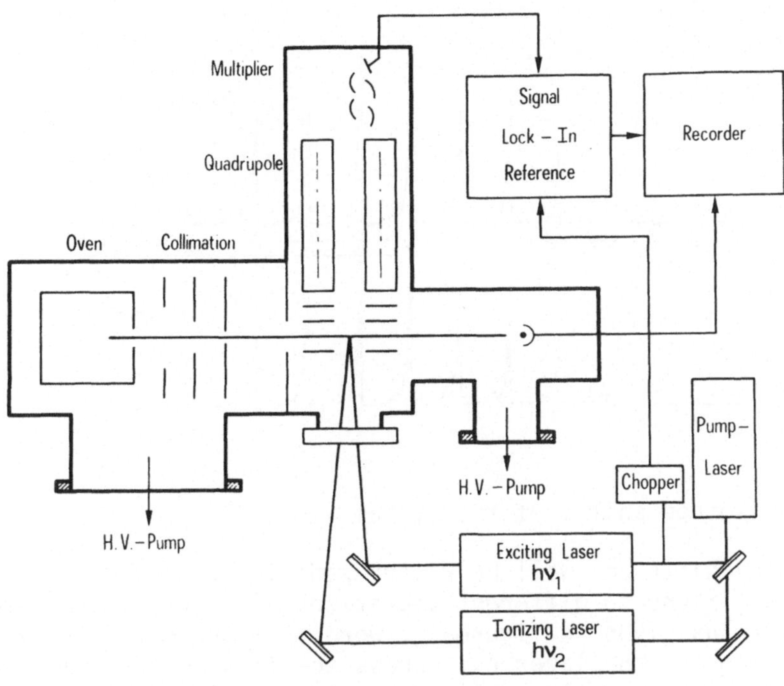

Fig. 30 : Experimental setup for TPI-measurements.

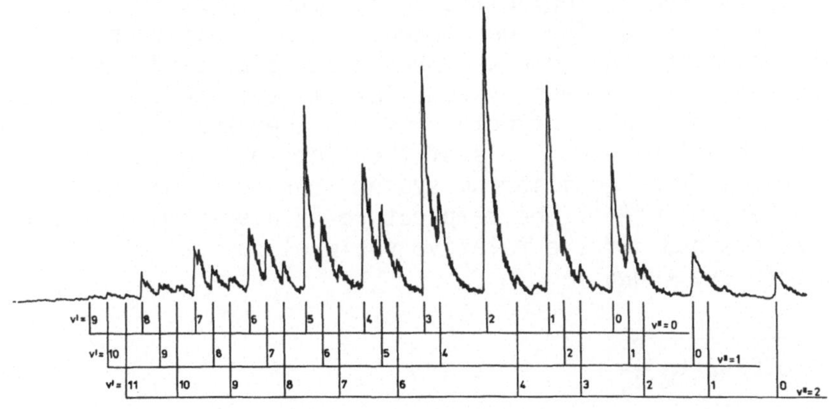

Fig. 31 : TPI-measurement of the system K_2 A←X.

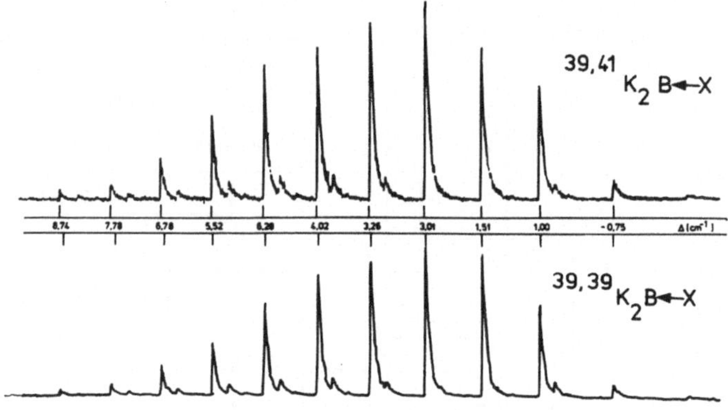

Fig. 32 : Isotope shifts of the system K_2 B←X.

The isotope shifts can well be recognized at the bandhead of individual transitions to different vibrational levels v'. The precise values of Δ, as given in figure 32, were interferrometrically determined by scanning the laser $h\nu_1$ across the fringes of a Fabry-Perot etalon. The measured values agree very well with calculations obtained from a mathematical framework given by Herzberg (57,53).

5.4.3 Beam temperatures

The measurements shown in figures 31 and 32 were all performed on the system K_2 B←X. Very different relative peak intensities, however, are observed. The phenomenon is due to different beam temperatures, since the results were obtained at different stagnation conditions of the oven system. This is systematically shown in figure 33: The number of molecules, as they are distributed over different vibrational sublevels of the electronic ground state is - for an undisturbed equilibrium system - proportional to the Boltzmann-Factor $e^{-E/kT}$. The temperature of a system, therefore, can well be determined, if the relative sublevel populations of the ground state are known :

$$T = \frac{\Delta E}{0.6952 \ \ln \frac{N_b}{N_a}}$$

N_b, N_a = population of two sublevels a and b.

ΔE = energy difference between a and b (cm^{-1}) .

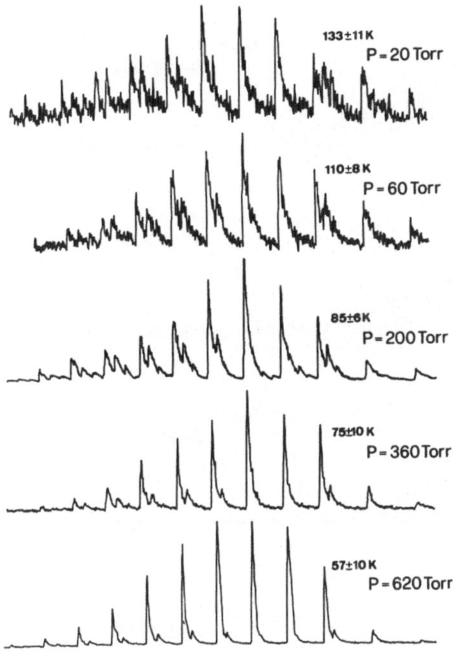

Fig. 33 : Vibrational sequences of the system K_2 B ← X at various oven pressures.

The relative vibrational sublevel populations N_n can be obtained from the intensities I_{mn} of individual transitions and their relating FC-factors q_{mn} :

$$N_n = N_{mn} = \frac{I_{mn}}{q_{mn}}$$

The systematic interpretation of the relative peak-intensities averaged over seven different transitions resulted to the vibrational temperatures given in figure 33. The strong cooling of the adiabatic expansion process can well be seen as a function of the oven pressure.

At a higher resolution each individual vibrational band shows the rotational structure of the system. Here again strong cooling due to the adiabatic expansion process occurs. A series of measurements is shown in figure 34. The observed rotational temperatures, however, are well below the corresponding vibrational temperatures. This is due to the fact, that the supersonic beam cannot be considered as an equilibrium system.

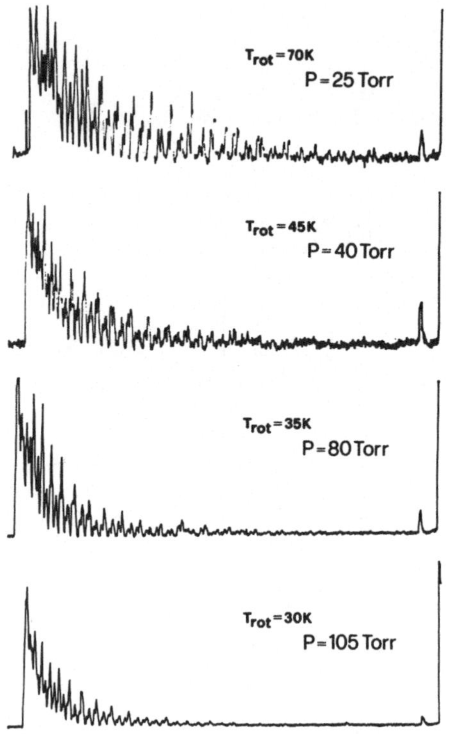

Fig. 34 : Rotational sequences of the system K_2 B←X (9'←0") at various oven pressures.

5.4.4 Spectroscopy of Rydberg states

A deeper insight into the ionization process can be obtained, when the exciting laser $h\nu_1$ is kept at a fixed transition, while the ionizing laser $h\nu_2$ is scanned across the ionization potential. Measurements of this kind are shown in figure 35. The X-axis in figure 35 indicates the total photon energy $h\nu_1 + h\nu_2$ (eV). Both measurements, therefore, give very precise values for an ionization potential. The spectral behaviour of the curves after reaching the ionization continuum, however, is very different : Curve a) shows a high density of individual peaks with no continuum below. Curve b), on the other hand, has a continuum. The individual peaks, however, which are superposed on, do not show any correspondance to the peaks in curve a). Evidently two different processes occur (58) :

The continuum is generated by direct photoionization due to :

$$M_2^* + h\nu_2 \longrightarrow M_2^+ + e^-$$

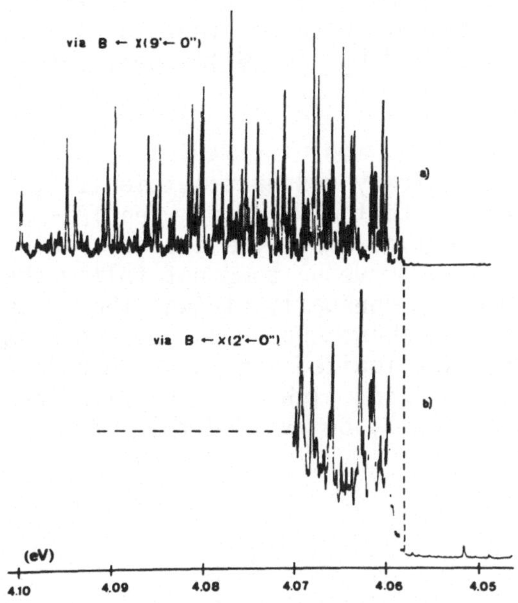

Fig. 35 : Spectral behaviour of the TPI-process when tuning the ionizing laser across the ionization potential, while the exciting laser is set on a fixed transition of the system K_2 B←X. a) via 9'←0", b) via 2'←0".

No resonance is required in this case, because the electron can take any amount of kinetic energy. The individual peaks are generated by resonant excitation of autoionizing Rydberg states due to :

$$M_2^* + h\nu \longrightarrow M_2^{**} \qquad \text{(excitation)}$$

$$M_2^{**} \longrightarrow M_2^+ + e^- \qquad \text{(autoionization)}$$

The Rydberg states are highly excited electronic states of the molecule, which asymptoticly approach the ionization potential. Since the molecule contains in addition to the electronic energy T_e vibrational energy G and rotational energy F , the total energy of a neutral particle in a highly excited Rydberg state can therefore be well above the ionization threshold. Herzberg and Jungen succeeded first to observe and identify such Rydberg states on H_2 (59). Due to their framework Martin et al. were able to interpret the Rydberg series of Na_2 (60).

Since the Rydberg state approaches with growing quantum number n asymtotically the potential of the ionic molecule, the extrapolation $n \to \infty$ of a well identified Rydberg series allows therefore precise conclusions on molecular constants of the ion. This was recently done for Na_2^+ and K_2^+ (60,61), experiments on Na_3^+ are in progress (62).

5.4.5 Perturbations

The existance of highly resonant autoionizing Rydberg states explains perturbations that may occur, when TPI-spectra are recorded. An example is shown in figure 36. The intensity distribution of the vibrational sequence in curve a) does not follow the FC-factors, the transition $6' \leftarrow 0''$ is unusually strong. The system was ionized with an Ar^+ - laser. The phenomenon does not occur any more in curve b), where the system was ionized with a broad band arc lamp. Strong coincidences of the ionizing light source $h\nu_2$ with an autoionizing Rydberg state can obviously be avoided this way.

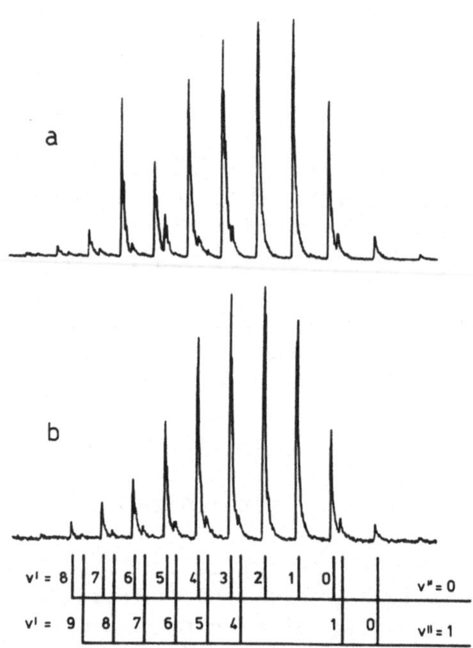

Fig. 36 : Vibrational sequence of the system K_2 $B \leftarrow X$. a) ionizing with an Ar^+ - laser, b) ionizing with an arc lamp ($h\nu_2 < E_{ion}$).

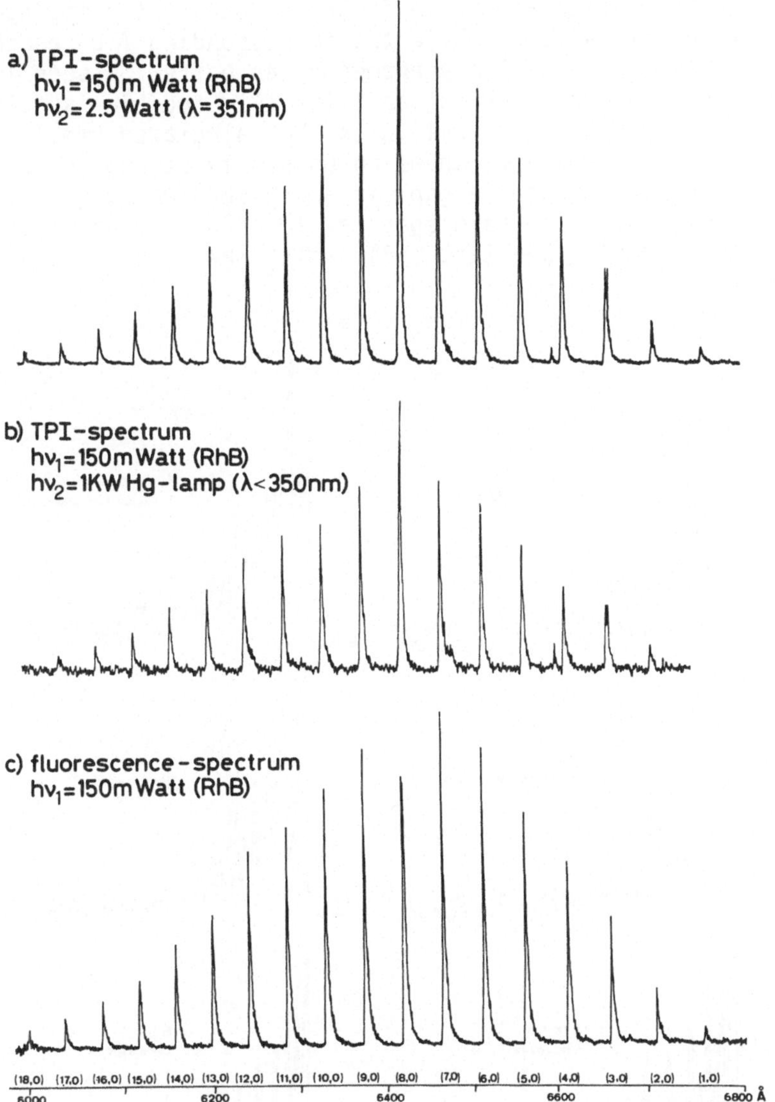

Fig. 37 : TPI - and fluorescence spectra of the system Na_2 A←X.

Figure 37 shows an example, where the use of a broad band arc lamp as ionization light source did not resolve the observed perturbations : The transition 8' ← 0" in figure 37 occurs too strong even when inoizing with a lamp, while the corresponding fluorescence signal appears too weak. The phenomenon can better be seen in figure 38 in rotational resolution : The TPI - and LIF - spectra of the 9' ← 0" transition agree well with the calculated result. The 8' ← 0" transition, however, shows an unusually strong peak, exactly, where the LIF-sequence has a gap. In a previous work, Kusch and Hessel have found at the same spectral location an intersystem crossing to a metastable $^3\Pi$ - state (63), which makes the molecule stay

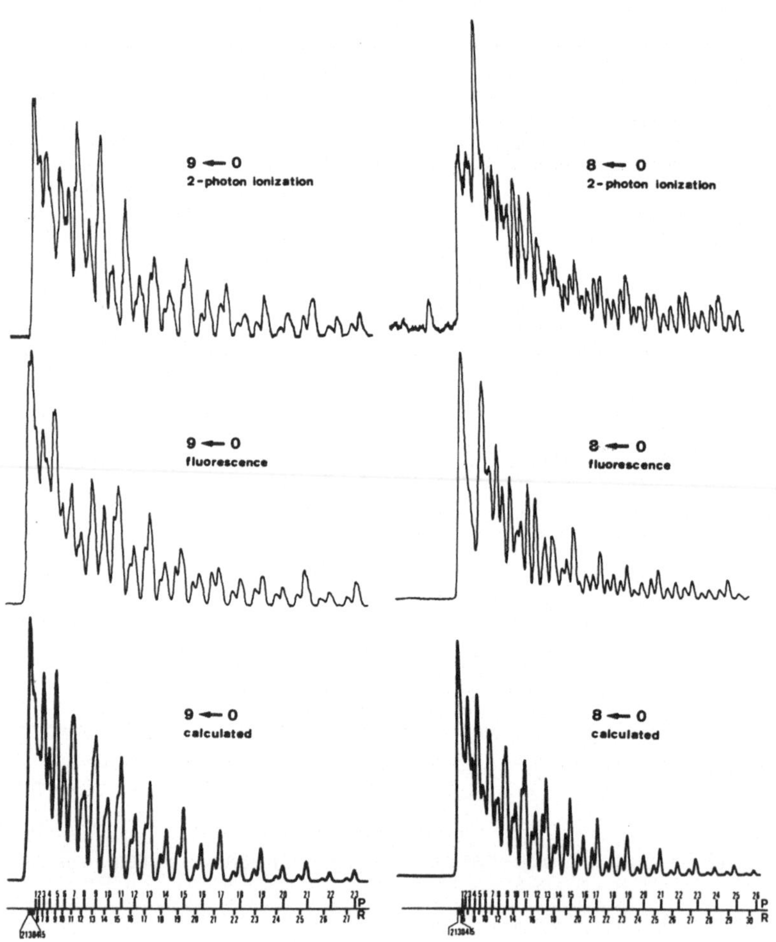

Fig. 38 : Rotational sequences and lifetime perturbations of the Na_2 A ← X band. a) taken by TPI, b) taken by LIF, c) unperturbed simulation.

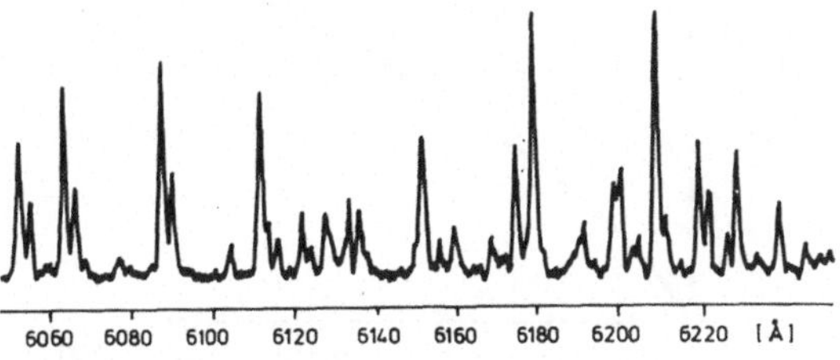

Fig. 39 : 3-photon transitions of the K_2 - molecule when simultaneously irradiating the system with a Rh B - and a Rh G - laser.

longer in the excited state. Consequently the ionization probability increases, while the fluorescence decreases (64).

Further perturbations of TPI-spectra may occur by 3- and multi-photon processes. The probability of such processes increases very much at higher laser powers. A typical example of a 3-photon spectrum is shown in figure 39.
In order to avoid such undesired multi-photon processes one should always verify that no ion signal occurs, when only one of the two lasers $h\nu_1$ and $h\nu_2$ is irradiated. A final proof for a clean TPI-process, however, can only be given by power dependency measurements.

5.4.6 Power dependencies

For an efficient and clean TPI-process it is desirable, to excite an ionize nearly all resonant particles. One would like, therefore, to approach saturation for the irradiated transitions without exceeding the laser power too much, in order to prevent undesired multi-photon effects. These conditions can be determined in power dependency measurements,

The power dependency measurements in figure 40 were performed with cw-laser systems. The measurment for $h\nu_1$ in figure 40 a) shows an early saturation. It corresponds well with the result that was obtained in the photodeflection experiment, where 5 % of the Na_2 - molecules were excited at 150 m Watt laser power. The power dependency of the ionization process (figure 40 b) is strictly linear and does not show any saturation.

Differential measurments with the surface ionization detector gave evidence that the ionization efficiency of the total TPI-process was 10^{-4} when Na_2 - molecules were excited with 150 m Watt laser power, and the system was ionized with 10 Watts. Correspondingly one molecule out of twenty was excited, and one excited

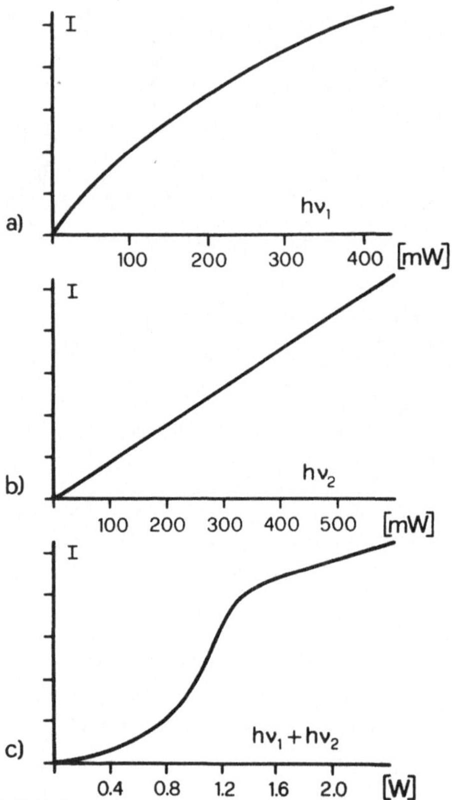

Fig. 40 : Power dependencies of the TPI-process : a) when changing the power of the exciting laser, b) when changing the power of the ionizing laser, c) when exciting and ionizing just with one laser ($h\nu_1 = h\nu_2 > 1/2\, E_{ion}$).

molecule out of 500 was ionized. An efficient ionization process would therefore require about 5 kWatt laser power.

Continous laser systems do not reach the required power level. Pulsed laser systems - like Nd-YAG pumped dye lasers - easily reach Megawatt power levels; their repetition rate, however, is so slow (typ 10 Hz) that no efficient duty cycle of the experiment can be reached. For this reason we developped two copper-vapor-laser (CVL) pumped dye lasers, which typically provide about 10 kWatt peak powers at 6 kHz repetition rate (65). This way we were able to improve our detection efficiency by a factor of 250, and we got close to the capability of "single molecule detection". Since the method additionally exhibits an extreme selectivity, it seems to be well suited for the analysis of trace amounts (66).

5.4.7 Spectroscopy of larger aggregates and fragmentations

The high sensitivity of the method allowed selectivity to record electronic excitation spectra of the Na_3 - molecule (67). Examples are given in figure 41 a-d. The spectra presented in figure 41 are not yet fully understood. This is certainly due to the high density of states superimposed in the electronic ground state. It is necessary, therefore, to further cool down the particles emerging from the jet expansion, in order to clean up the spectra.

Since the ionization potential of Na_3 is below the values of all neighbour aggregates, it should well be possible - at proper conditions of $h\nu_1$ and $h\nu_2$ - only to generate a mass peak of Na_3^+.

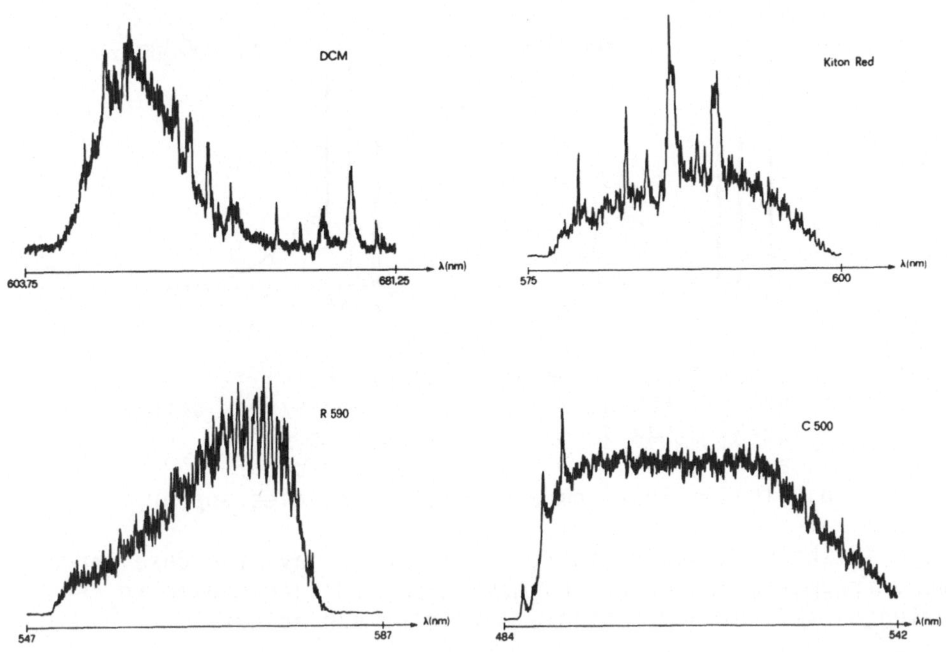

Fig. 41 : TPI-spectra of Na_3 at different spectral locations.

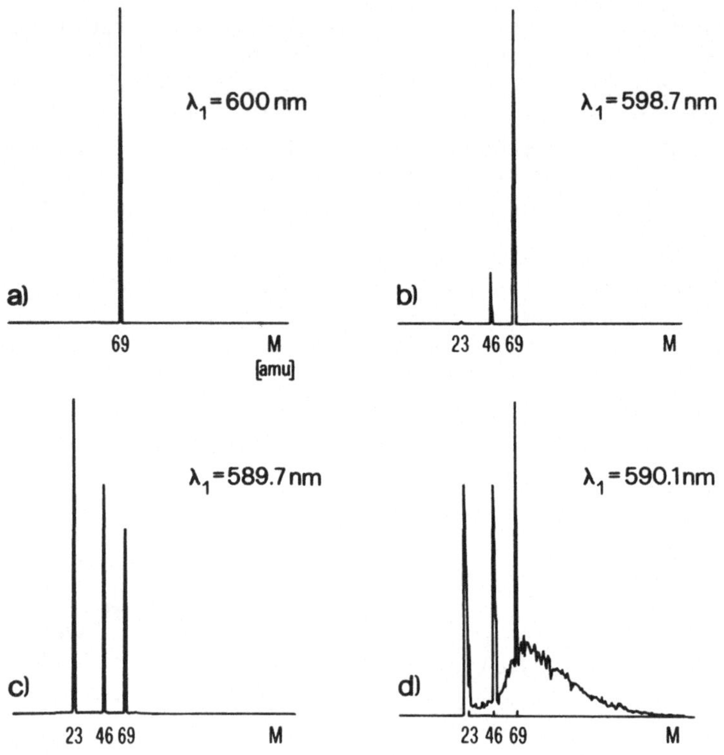

Fig. 42 : Mass spectrum of the Na_3 - TPI-process :
a) at λ_1 = 600 nm a proper Na_3^+ - mass peak occurs
b) at λ_1 = 548,7 nm Na_2^+ - fragments occur
c) at λ_1 = 589,7 nm Na^+ - fragments occur
d) at λ = 590,1 nm metastable particles appear

Figure 42 shows, however, that this is not always the case, and strong fragmentation processes may occur (68). According to the wavelength set, even metastable particles can appear.

The most plausible processes for the occurence of these fragments is

$$Na_3 \begin{cases} \rightarrow Na_2^+ + Na + e^- \\ \rightarrow Na^+ + Na_2 + e^- \\ \rightarrow Na^+ + Na_2^- \\ \rightarrow Na^- + Na_2^+ \end{cases}$$

The threshold energies E_{th} for these processes compute to :

$$E_{th}(1) = D(Na_3) + E_{ion}(Na_2)$$
$$= 0.41^{(10)} + 4.93 = 5.34 \text{ eV}$$

$$E_{th}(2) = D(Na_3) + E_{ion}(Na)$$
$$= 0.41 + 5.14 = 5.55 \text{ eV}$$

$$E_{th}(3) = D(Na_3) + E_{ion}(Na) - Ae(Na_2)$$
$$= 0.41 + 5.14 - 0.45^{(11)} = 5.1 \text{ eV}$$

$$E_{th}(4) = D(Na_3) + E_{ion}(Na_2) - Ae(Na)$$
$$= 0.41 + 4.93 - 0.55^{(12)} = 4.79 \text{ eV}$$

where D : dissociation energy, A = electron affinity.

None of these processe,however, is energetically below the 2-photon-energy,which is $E \approx 4{,}2$ eV,when exciting and ionizing with $\lambda = 590$ nm. No explanation can be given so far. Negative ion detection and an energy analysis of arriving metastable mass peaks will be necessary. Possibly the Na_3 particles in the beam are not as much cooled down, as this has been found for Na_2; possibly they even occur in metastable states or as very unstable isomers. Possibly also, accelerated photoelectrons in the ionization area are responsible for the observed fragmentation, or uncontrolled multiphoton processes occur. Further experiments are in progress.

The tendency of the aggregates to break apart while being excited or ionized still seems to increase with growing particle size. We are reluctant, therefore, to attribute those spectra, which we recorded on the mass peak 92, exclusibely to the Na_4 - molecule. An interpretation requires a systematic deconvolution of all fragments at all wavelengths, as shown in the 2-dimensional optical mass spectrum in figure 43.

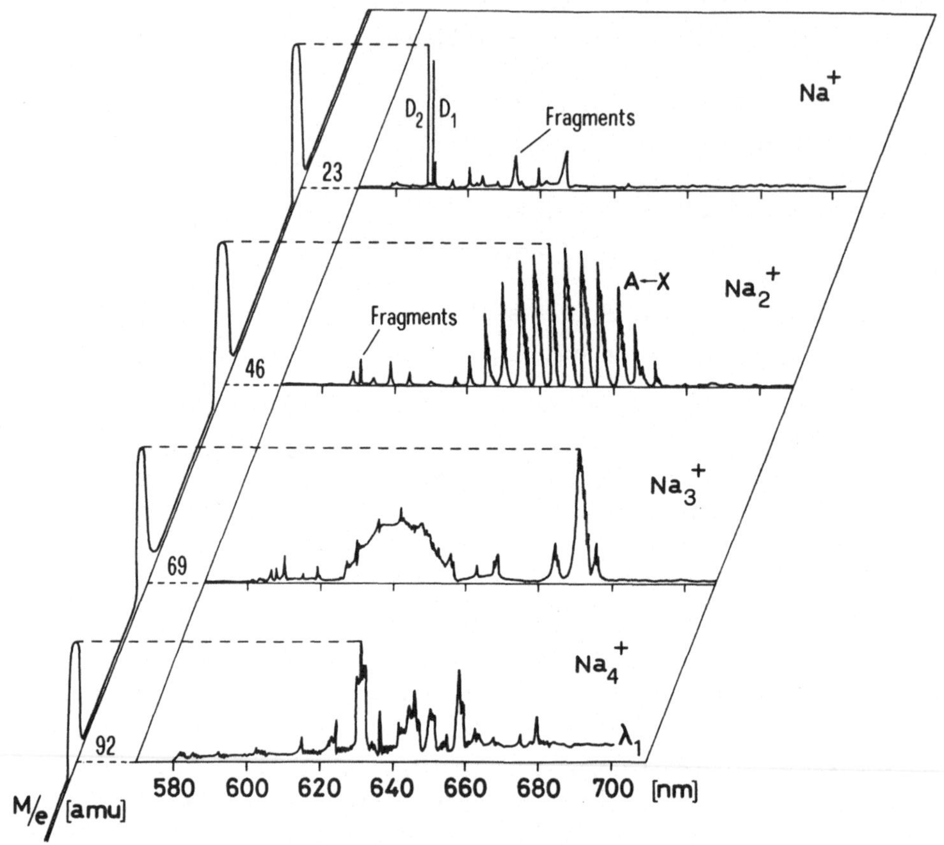

Fig. 43 : 2-dimensional optical mass spectrum of the alkali aggregates. The direction of the X-axis indicates the wavelength of the exciting laser, the Y-axis corresponds to the particle size and the Z-axis relates to the ion intensity.

6. FUTURE DEVELOPMENTS

Great progress was made in recent years in the development of efficient metal cluster sources, and the optical properties of the particles can well be characterized by means of laser spectroscopy. Fragmentation problems, however, still make it difficult to obtain size specific informations, even when mass spectrometric detection systems are used. For this reason we propose an apparatus, which will allow to perform experiments with clusters just of a single size. The schematic setup of the apparatus is shown in figure 44.

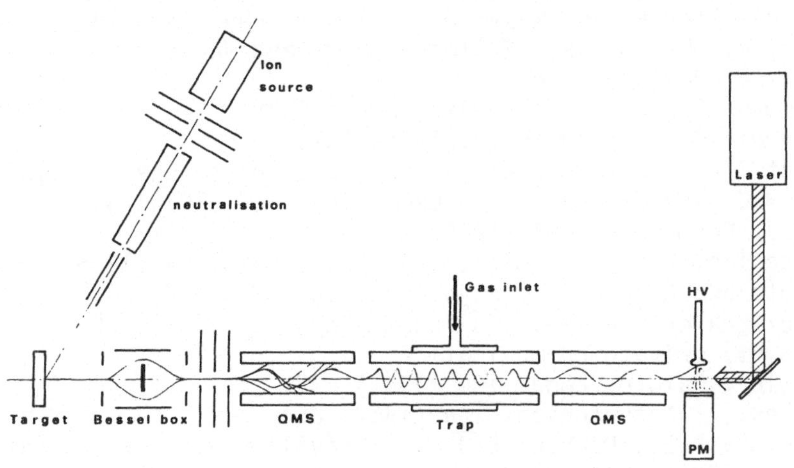

Fig. 44 : Experimental setup of the cluster-ion trap.

Metal cluster ions are formed by means of "fast-atom-bombardment" (FAB). The emitted particles are then energy filtered, mass analyzed and injected into a HF-ion trap. There the particles are stored for several milliseconds until they drift out into a second mass analyzer. The geometry of the experiment is such that the particles in the trap can axially be irradiated with a laser, in order to induce photofragmentation processes. All fragments, however, do only originate from a single particle size and can well be analyzed therefore.

The stored particles can also be used for chemical reactions with a gas, that may be injected into the trap. Alternatively size dependent catalysis experiments may be performed, when different reactive components are injected.

The experimental setup is restricted to charged particles only. Similar experiments on neutral clusters are very difficult. Hopefully, however, new experimental ways will be found, to continue building the bridge from the atom to the bulk.

REFERENCES

1. Moisar E., Photogr. Science and Engin. 25 (1981) 45.
2. Klingelhöfer R. and Moser H., J. Appl. Phys. 43 (1972) 4575.
3. Monot R., Melinon P., Zellweger J.M. and Van den Bergh H., private communication.
4. Herrmann A., doctoral thesis, Berne 1978.
5. Tschudin R., Diplomarbeit, Berne 1976.
6. Trent D.E., Paris B. and Krause H.H., Inorg. Chem. 3 (1964) 1057. Lee H.U. and Zare R.N., J. Chem. Phys. 64 (1976) 431.
7. Wu, J. Chem. Phys. 65 (1976) 3181.
8. Gingerich, Cocke and Miller, J. Chem. Phys. 64 (1976) 4027.
9. Kimoto and Nishida, J. Phys. Soc. Jpn 42 (1977) 2071.
10. Scheingraber and Vidal, J. Chem. Phys. 66 (1977) 3694.
11. Even U., private communication.
12. Bentley, Nature, Lond. 190 (1961) 432. Henkes, Z. Naturforsch. 15a (1961) 851.
13. Leckenby R.E., Robbins E.J. and Trevalion P.A., Proc. Roy. Soc. Lond. A280 (1964) 409.
14. Becker E.W., Bier K. and Henkes W., Z. Phys. 146 (1956) 333.
15. Hagena O., in Wegener, Molecular Beams and Low Density Gasdynamics, Marcel Dekker Inc., New York 1974.
16. Stein G.D., Proceedings of the Nato ASI, Cap d'Agde 1982.
17. Wöste L., doctoral thesis, Berne 1978.
18. Sattler K., Mühlbach J. and Recknagel E., Phys. Rev. Lett. 45 (1980) 821. Sattler K., Mühlbach J., Echt O., Pfau P. and Recknagel E., Phys. Rev. Lett. 47 (1981) 160.
19. Bowles R.S., Kolstad J.J., Carlo J.M. and Andres R.P., Surf. Science 106 (1981) 117.
20. De Boer B.G. and Stein G.D., Surf. Science 106 (1981) 84.
21. Delacrétaz G., Monot R. and Wöste L., work in progress.
22. Berkowitz J. and Chupka W.A., Chem. Phys. 40 (1964) 2735.
23. Dietz T.G., Duncan M.A., Powers D.E. and Smalley R.E., J. Chem. Phys. 77 (1982) 4417.
24. Dixon A., Colliex C., Ohana R., Sudraud P. and Van de Walle J., Phys. Rev. Lett. 46 (1981) 865.
25. Hortig and Müller, Zeitschr. Physik 221 (1969) 119.
26. Joyes P. and Leleyter M., J. Phys. B 6 (1975) 150.
27. Baede A.P.M., Jungmann W.F. and Los J., Physica 54 (1971) 459.
28. Fayet P. and Wöste L., work in progress.
29. Devienne F.M. and Roustan J.C., Org. Mass Spectr. 17 (1982) 173.
30. Parrish D.D. and Herm R.R., J. Chem. Phys. 51 (1969) 5467.
31. Weiss R., Rev. Sci. Instr. 32 (1961) 397.
32. Rechsteiner R., doctoral thesis, Lausanne 1982.
33. Rechsteiner R., Monot R., Wöste L., Zellweger J.M. and Van den Bergh H., Helv. Phys. Acta 54 (1981) 282.

34. Ganière J.D., doctoral thesis, Lausanne 1982.
35. Wiley W.C. and Mc Laren C.H., Rev. Sci. Instr. 26 (1955) 1150.
36. Daly N.R., Rev. Sci. Instr. 31 (1960) 264.
37. Herrmann A., Leutwyler S., Schumacher E. and Wöste L., Helv. Chim. Acta 52 (1977) 418.
38. Gspann J. and Vollmar H., 8th Int. Symp. on Rarefied Gas Dynamics, Stanford 1972.
39. Gordon R.J., Lee Y.T. and Herschbach D.R., J. Chem. Phys. 54 (1971) 2393.
40. Solliard C., doctoral thesis, Lausanne 1983.
41. Raoult B. and Farges J., Rev. Sci. Instr. 44 (1973) 430.
42. Rechsteiner R., doctoral thesis, Lausanne 1982.
43. Gay J.G. and Berne B.J., Phys. Rev. Lett. 49 (1982) 194.
44. Lee N., Keesee R.G. and Castleman Jr. A.W., J. Chem. Phys. 72 (1980) 1089.
45. Fehsenfeld F.C., Howard C.J. and Schmeltekopf A.L., J. Chem. Phys. 63 (1975) 2835.
46. Armbruster M., Haberland H. and Schindler H.G., Phys. Rev. Lett. 47 (1981) 323.
47. Haberland H., private communication.
48. Knight W., Monot R., Dietz E.R. and George A.R., Phys. Rev. Lett. 40 (1978) 1324.
49. De Heer W.A., George A.R., Gerber W.H. and Knight W.D., submitted for publication.
50. Gerber W.H., private communication.
51. Herrmann A., Schumacher E. and Wöste L., J. Chem. Phys 68 (1978) 2327.
52. Rechsteiner R., Monot R., Wöste L., Zellweger J.M. and Van den Bergh H., Helv. Phys. Acta 54 (1981) 282.
53. Demtröder W. and Stock M., J. Mol. Spectroscopy 55 (1975) 476.
54. Sinha C., Schulz A. and Zare R.N., J. Chem. Phys. 58 (1973) 549.
55. Schieder R., Walther H. and Wöste L., Opt. Comm. 5 (1972) 337.
56. Herrmann A., Leutwyler S., Wöste L. and Schumacher E., Chem. Phys. Lett. 62 (1979) 444.
57. Herzberg G., Spectra of Diatomic Molecules, van Nostrand Reinhold Company, New York.
58. Leutwyler S., Herrmann A., Schumacher E. and Wöste L., Chem. Phys. 48 (1980) 253.
59. Herzberg G. and Jungen Ch., J. Mol. Spectroscopy 41 (1972) 425.
60. Martin S., Chevaleyre J., Valignat S., Perrot J.P. and Broyer M., Chem. Phys. Lett. 87 (1982) 235.
61. Broyer M., Chevaleyre J., Delacrétaz G., Martin S. and Wöste L., submitted for publication.
62. Broyer M., Chevaleyre J., Delcrétaz G. and Wöste L., work in progress.
63. Kusch P. and Hessel M.M., J. Chem. Phys. 63 (1975) 4087.
64. Herrmann A., Leutwyler S., Schumacher E. and Wöste L., Chem. Phys. Lett. 52 (1977) 418.

Broyer M., Chevaleyre J. and Wöste L., to be published.
66. Wöste L., Laser und Optoelectronik 1 (1983) 9.
67. Herrmann A., Hoffmann M., Leutwyler S., Schumacher E. and Wöste L., Chem. Phys. Lett. 62 (1979) 216.
68. Delacrétaz G., Ganière J.D., Monot R. and Wöste L., Appl. Phys. B 29 (1982) 55.

METAL CLUSTER BEAMS AND ELECTRON DIFFRACTION : DEVIATIONS FROM THE BULK STATES OF MATTER

Gilbert D. STEIN*

Northwestern University, Gasdynamics Laboratory,
Evanston, Illinois 60201, USA

ABSTRACT

The salient features of cluster production are outlined and several specific cluster beam sources presented. The mean cluster size is seen to be dependent on oven temperature, related to the evaporation rate; the carrier gas pressure, related to the metal cooling and diffusion rate away from the evaporating surface; the carrier gas temperature, related to the metal vapor cooling rate; and on the carrier gas molecular weight, related to the metal-carrier gas collision cross section and therefore to the energy and mass transport processes.

Results of high energy (40 to 100 keV) electron diffraction from these beams reveal that as cluster size is reduced, changes such as a decrease in unit cell parameter, a transition from one crystalline structure to another, appearance of a progressively larger fraction of non-crystalline phase, and gradual transition to minimum energy configurations such as icosahedra, are seen. Under some conditions the diffraction patterns are used to estimate cluster size and temperature. Production of amorphous clusters from pure metal vapor expansions has been observed and is a surprise to this investigator, in view of the extremely low predicted nucleation rates.

---

* On sabbatical leave at the Institut de Physique Expérimentale, Ecole Polytechnique Fédérale de Lausanne, Switzerland

1. INTRODUCTION

The investigation into the kinetics of nucleation and the physical properties of small atomic and molecular clusters has been the primary activity in our laboratory. This area of research is attracting a rapidly growing cadre of new investigators, partially because it has numerous important implications to questions of a fundamental physical nature, as well as abundant possibilities for very significant practical applications. The goal of this work is to probe the properties of matter in the interesting transition size regime between that of bulk and the monomeric gas phase. Homogeneous clusters are produced in molecular beams at final densities so low that they are in the "splendid isolation" which leaves them free of interactions with solvants, matrix, surfaces, or each other.

The experimental price that must be paid for these low sample densities is decreased signal and increased signal-to-noise ratios. Thus, particular configurations must be carefully chosen in order to ensure a reasonable chance of success. The prospect of producing samples of narrow or even delta function size distributions are goals important enough to lure some researchers into the limbo of high vacuum.

The genesis of our continued interest in small cluster properties lies in prior studies of the dynamics of gas phase nucleation. In this theory there arises a "critical" size cluster, in a supersaturated environment, that is just large enough to survive as a site for continued growth of the new, condensed phase. For virtually all phase changes, this size is very small, with $g \leqslant 500$, where g is the number of molecules (atoms, if monatomic species) in the cluster. The energy of formation of a critical size cluster, ΔG^*, is central to any nucleation rate theory and appears in an exponential term. Because the critical size cluster is so small, ΔG^* is very sensitively dependent on the surface energy contribution. Enormous variation in rate occurs for relatively small changes in surface tension σ. Of course, the clusters are so small that many investigators question the validity of macroscopic properties such as σ.

Thus arose our interest in the properties of small aggregates: When do bulk properties change as cluster size is reduced, and what are the specific changes ? Theoretical estimates for the size at which bulk properties appear range from g = 8 for metal electronic structure to g = 2,000 to 5,000 for latent heat and surface tension (1). Experiments on the decline of melting point temperatures show deviation from the bulk melting point at g = 200 to 500 (2) and crystallographic structure deviation from diffraction experiments show progressive changes from g = 200 to 500 for noble gases (3) and g = 2,000 to 5,000 for metal clusters (4).

Molecular beam techniques were investigated as a potential nucleation research device at the outset of my cluster studies, but dropped in favor of continuum, compressible flow Laval nozzles. However, beams are an excellent choice for the study of cluster properties, per se, and is a natural extension of our use of supersonic, continuum flow. As for diagnostic techniques for probing cluster properties, laser intracavity scattering had been used with the Laval nozzle (5), but lasers were not intense or tunable enough at that time for the low densities encountered in molecular beams. (The picture today is, of course, greatly changed with many impressive successes). Also, mass spectrometry for clusters had already been employed by several research groups (6,7) and while there were some carefully conceived experiments that yielded interesting results, the omnipresence of cluster fracture upon ionization was wrecking havic on attempts at quantitative interpretations. Thus, mass spectrometry, as a primary investigation technique for neutral clusters, did not appear to be an attractive approach. The use of mass spectrometry to study ion clusters avoids the problems of ionization and fracture and has enjoyed great success (8,9). (There has been a tremendous resurgance in this technique for neutral clusters, with some very clever applications of time-of-flight instruments having mass ranges in excess of 20,000 AMU (10,11)).

The use of diffraction techniques to study cluster structure in beams came to our attention with some impressive initial results published by P. Audit and M. Rouault in France (12). Electron diffraction, while having some disadvantages over x-ray diffraction, won out due to its much higher atomic scattering cross section ($O(10^4)$) and its more intense beams ($O(10^4)$ also) (13). Free jet expansions of pure molecular flows into vacuum have been employed since the 1930's as a means of studying molecular structure using high energy electron diffraction ($E = 30$ to 60 keV) (14). Moreover, the phenomena of diffraction, in existence since the work of Bragg, has been developed into a standard analytical method. Voila cluster beam - electron diffraction !

2. THE CLUSTER NUCLEATION PROCESS

The study of the dynamics of gas phase nucleation has been helpful in understanding and designing experiments to investigate cluster structure. The theory of nucleation, begun in the 1920's in Germany (15,16) underwent a more or less steady evolution through the 30's and early 40's (17-19) with relatively little activity after that until the 1960's (20,21). A re-examination of the theory lead to heated controversy and a great resurgance, both theoretical and experimental (21,22). An exposition of this subject is available in the literature (19,23) and will not be recounted here. Only a few important features are outlined with regard to the specific application of producing cluster beam sources.

The so-called classical rate theory, and most versions since this formulation, yield expressions of an Arrhenius type :

$$J = K \exp(-\Delta G^*/kT) , \tag{1}$$

where J is the number of critical (denoted by $*$) size clusters formed per unit volume and time, $cm^{-3}\,sec^{-1}$, K is a large number of the order of the number of binary collisions $cm^{-3}\,sec^{-1}$, ΔG^* is the Gibbs free energy of formation for a cluster of size g^* in its local thermodynamic state (i.e. that of a supersaturated vapor), k is Boltzmann's constant and T the temperature.

The energy of formation, reflecting its large surface energy contribution, for a macroscopic, motionless, spherical, liquid droplet, is given by :

$$\frac{\Delta G^*}{kT} = \frac{g^*}{2}\,\ln S , \tag{2}$$

or, $$\frac{\Delta G^*}{kT} = \frac{16\pi}{3}\left(\frac{\sigma}{kT}\right)^3\left(\frac{v_c}{\ln S}\right)^2 , \tag{3}$$

where S is the saturation ratio, $S \equiv (p_v/p_{v\infty})_T$, p_v being the vapor pressure and $p_{v\infty}$ the equilibrium vapor pressure of the same temperature (see Fig. 1) and v_c is the volume per molecule in the condensed phase.

Two of the most common processes for supersaturating a vapor are shown in the phase diagram of Fig. 1. The upper curve A corresponds to a perfect mixing process between condensable vapor and carrier gas, or to a radiative or slowly conductive cooling of a quiescent gas. (p_v is the total pressure for a pure vapor or the partial pressure for a vapor - carrier gas mixture). If the process is one of diffusion through a temperature gradient, or a combination of diffusion and mixing (the case for the metal sources described below), then the process will fall below the horizontal line, i.e. the dotted line, curve B.

The second,common method of gas phase nucleation is the adiabatic expansion shown as the lower curve C. If this process is, furthermore, isentropic (i.e. reversible adiabatic) then it is easily shown that its slope is $\gamma/(\gamma-1)$ where $\gamma = c_p/c_v$, the ratio of the constant pressure to constant volume specific heats. The value of γ for a mixture depends on the mole fractions in the usual way, and the higher is γ $(1 < \gamma < 5/3)$ the lower the slope and therefore a given supersaturation is attained earlier in the expansion (i.e. lower Mach number $M = u/a$, u being the velocity and a the local sound speed).

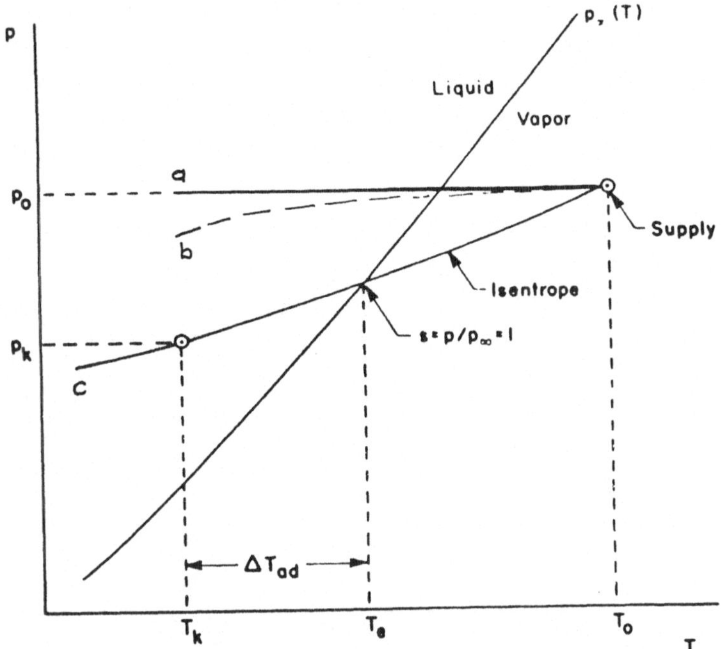

Fig. 1 : A pressure - temperature phase diagram shows three possible processes for vapor nucleation : a - Perfect mixing of a hot vapor with a cold carrier gas, i.e. one possibility for the metal cluster source, b - A combination of mixing and diffusion of a hot vapor with a cold carrier gas, i.e. another possibility for the metal sources, c - Isentropic expansion of a vapor or a vapor in a carrier gas, i.e. the process for most molecular beam sources. Here the onset of nucleation is indicated as point k.

Since the cluster energy of formation $\Delta G^* \sim \sigma^3$ in Eq. (3) which appears as an exponential in the rate expression, Eq. (1), the value of J is very sensitive to variations in σ. For example in a moist air Laval expansion, J varies by a factor of 10^{40} (!) for a factor of 2 change in σ from 75 erg cm^{-2} (bulk value for water) to 150. For a change from 75 to 95, typical of the difference of bulk water and ice, J changes by a factor of 10^{12} (Fig.2). Thus, one readily sees that cluster properties are of paramount importance in phase change nucleation theories.

To emphasize this fact, as applied to materials used in the beam experiments in our laboratory, consider σ and the interatomic (molecular) potential well depth ε for a van der Waals potential (Argon), a hydrogen bonded material (water) and a metallic species (Lead). It is interesting to note the similarity of the range in σ

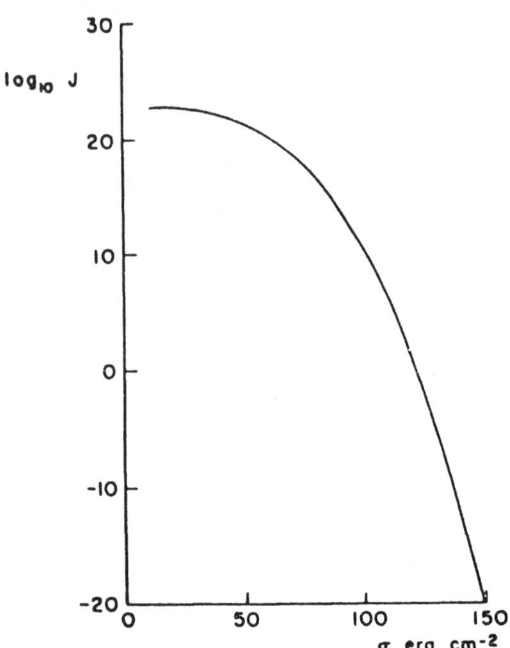

Fig. 2 : The effect of surface tension σ on the nucleation rate J for ice at T = 215 °K and a saturation ratio of S = 120, typical of a moist air expansion (J in units of cm^{-3} sec^{-1}, after Wegener (1)).

and in ε (and there is surely a connection, see Table I). Their order of magnitude variation is the same. Also, consider now the impact on the magnitude of the nucleation rate. Take the case of Argon and Lead (both have face centered cubic bulk crystalline structure) at the same value of T and S (variations in v_C being relatively small):

$$\frac{J_{Pb}}{J_{Ar}} = \frac{\exp(\Delta G^*_{Ar}/kT)}{\exp(10^6\,\Delta G^*_{Ar}/kT)}\,, \tag{4}$$

an increditably small number, for $\Delta G^*/kT$ typically ≅ 10, it is equal to $1/\exp(10^7)$! Well this denominator of course, is an incomprehensible number, since the largest natural number in the universe is between 10^{42} to 10^{46} (e.g. the number of atomic nuclei laid end to end across the "diameter" of the universe). Clearly we are in a regime which is quantitatively meaningless.

species	σ (erg/cm^2)	σ/σ_{Ar}	$\varepsilon/\varepsilon_{Ar}$
Argon	10	1	1
Water	70	7	20
Lead	1,000	100	400

Table I : Surface tension σ and potential well depth ε .

However, the rudimentary qualitative feature is valid, to wit : metals will nucleate much less readily than gases which have a much lower value of σ . Thus one would not expect cluster forming techniques which work well for gases to be applicable to metals, for example comparing the methods needed to obtain g = 1,000 for Argon or Lead. The gas beam technique may form some Lead of size g = 2 to 4 or so, but will not nucleate large clusters, and it is the large ones needed for the present electron diffraction experiments.

To summarise : in order to cluster metals efficiently the characteristic experimental or flow time must be many orders of magnitude greater that those of gas beams which are typically 10^{-6} to 10^{-4} sec, the supersaturation must be orders of magnitude higher than gas beams which are typically 10^2 to 10^4, and the condensable collision frequency kept as high as possible (i.e. high vapor density).

There are often extremes in cluster source conditions, such as short times, low densities, turbulent mixing and transport, which make it unlikely that quantitative application of the nucleation theory is valid. In these cases, we often make use of only the qualitative features of the theory or use a very simple kinetic theory calculation as a relative guide.

2.1 Vapor Collisions in the Supersaturated State

In order to nucleate clusters from the vapor state it is necessary to have :

1.) Supersaturation
2.) Surface
3.) Collisions

Strictly speaking clusters do form in unsaturated environments, but they are unstable to collision. In the supersaturated case the cluster are in relatively higher concentrations for all sizes and become stable beyond the critical size and may then grow via additional vapor collisions. In either case stable clusters of any size may arise as the gas is expanded to densities low enough so that collisions cease (i.e. free molecular flow or collision less regime).

Once a gas is supersaturated it is possible to initiate condensation onto foreign or heterogeneous surfaces. Indeed some materials may condense onto specific foreign sites at saturations less than one (unsaturated). If no foreign surface is present then the vapor must form its own surface, i.e. homogeneous clustering or nucleation. All of the above-mentioned changes require collisions.

With this in mind a qualitative indication of the tendency for a vapor to cluster can be obtained by calculating the total number of binary collisions, N_{coll}, a vapor atom (molecule) undergoes in the particular experiment of interest. Thus in a molecular beam flow, if one rides on an individual molecule and counts the number of collisions from the time the vapor becomes saturated, t_1 to the time it becomes collisionless, t_2 :

$$N_{coll} = \int_{t_1}^{t_2} Z\,dt\ , \qquad (5)$$

where the collision frequency $Z = \sqrt{2}\,\sigma_c n \bar{v}$, σ_c being the molecular collision cross section, n the molecular number density and $\bar{v}$ the mean thermal gas velocity. The density for a perfect gas is $n = p/kT$. The mean velocity $\bar{v} = (8\,kT/\pi m)^{\frac{1}{2}}$ where m is the mass of one molecule. Substituting into Eq. (5) :

$$N_{coll} = \int_{t_1}^{t_2} \sqrt{2}\,\sigma_c \left(\frac{p}{kT}\right)\left(\frac{8\,kT}{\pi m}\right)^{\frac{1}{2}} dt\ ,$$

$$= C \int_{t_1}^{t_2} P\,T^{-\frac{1}{2}}\,dt\ , \qquad (6)$$

for fixed σ_c and m, $C = 4\sigma_c/(\pi km)^{\frac{1}{2}}$. Thus N_{coll} increases and therefore also the amount of clustering, as p and $(t_2 - t_1)$ increases and T decreases. Converting from time, t, to position, x, as coordinate, i.e. from moving to laboratory frame, or from Lagrangian to Eulerian coordinates, by the relation $dt = dx/u$, u being the flow velocity; incorporating a characteristic dimension, D usually a flow diameter; and considering the flow as isentropic, one obtains:

$$N_{coll} = \frac{4\sigma_c}{(\pi\gamma)^{\frac{1}{2}}} \left(\frac{p_o}{kT_o}\right) D \int_{(x_1/0)}^{(x_2/0)} M^{-1} \left(1 + \frac{\gamma-1}{2} M^2\right)^{-1/(\gamma-1)} d(x/D)$$

$$= K \int f(\gamma,M)\, d(x/D) \,. \qquad (7)$$

For an explicit relationship of Mach number to dimensionless flow coordinate

$$M = M(\gamma, x/D) \,, \qquad (8)$$

the integral in Eq. (7) can be evaluated. For free jet beam sources this is a known universal function, see Fig. 3 (24). Also the upper integration limit (point of reaching free molecular flow) can be obtained analytically for noble gases by using the expression for terminal Mach number.

Any isentropic flow that can be characterized with Eq. (8) can then be used to determine N_{coll}. If the flow is not isentropic then

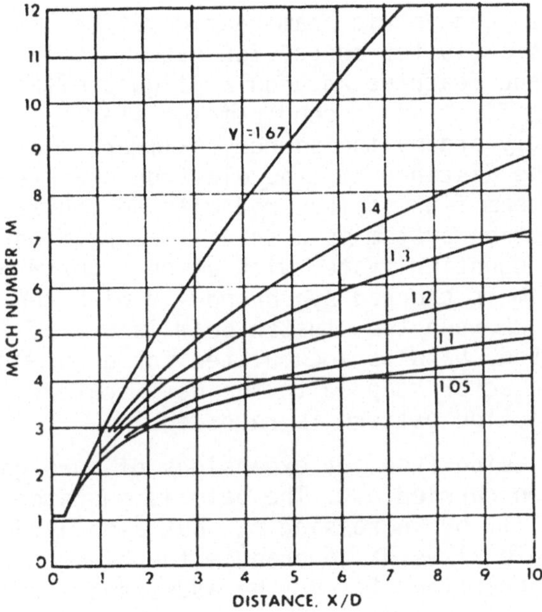

Fig. 3 : Free jet Mach number on the centerline axis as a function of the nondimensional flow coordinate x/D, as predicted by a method of characteristics solution (after Anderson (24)).

Eq. (6) would have to be used, which may entail a great deal more effort to evaluate.

In summary, the total number of binary collisions per molecule, that can lead to clustering, scales with the starting or stagnation density n_0 (or P_0/T_0), the collision cross section σ_c, and the characteristic experimental dimension, often a diameter D . The distance from saturation x_1 , to the point of interest or the point where the flow no longer has collisions x_2 , the specific heat ratio γ and the Mach number histroy between x_1 and x_2 all affect N_{coll}. As stressed above, this is only a qualitative estimate for cluster nucleation. For instance, collisions at high supersaturation will lead to more clustering that the same number of collisions at low S , and there is no distinction in this regard in the calculation of N_{coll}.

3. CLUSTER BEAMS

Molecular beam designs have been specialized for the production of cluster beams in which the mean size can be varied and the cluster density maximized while minimizing background density. The primary diagnostic technique is high energy electron diffraction.

3.1 Gas Beam - Adiabatic Expansion

A multistage, concentric beam configuration has been used in our laboratory as a cluster source for gases. It is shown schematically in Fig. 4 and features an annular liquid nitrogen trap which can pump both the first and second pumping stages. The pumping is folded concentricly around the source, skimmer, and columnator so as to minimize the distance to the electron beam (about 7 cm). Total molecular beam flux is measured with an ionization gage. The diffraction detection system is a scintillator, fiber optic, photomultiplier pulse counting system with stepper motor position control. The molecular beam is chopped synchronously with the counting system which counts up when the cluster beam is on and down when the beam is interrupted. Data is accumulated for a preset number of cycles and the detector stepped to the next scattering angle, until a complete diffraction pattern is recorded.

Experiments using free jet expansions of pure gases and gas mixtures have been carried out. The beam is operated at increasing density n_0 (usually by increasing p_0, but also by decreasing T_0) and at various hole sizes D to produce the cluster beams. The details of beam formation will not be addressed here, in detail, at it appears in this volume in Dr. Wöste's manuscript. One example, of beam intensity characteristics, is given here to show the onset of nucleation in the beam and to indicate that the nozzle to skimmer distance is very important for beam optimization (25). A free jet Argon beam in Fig. 5 has the typical features for starting

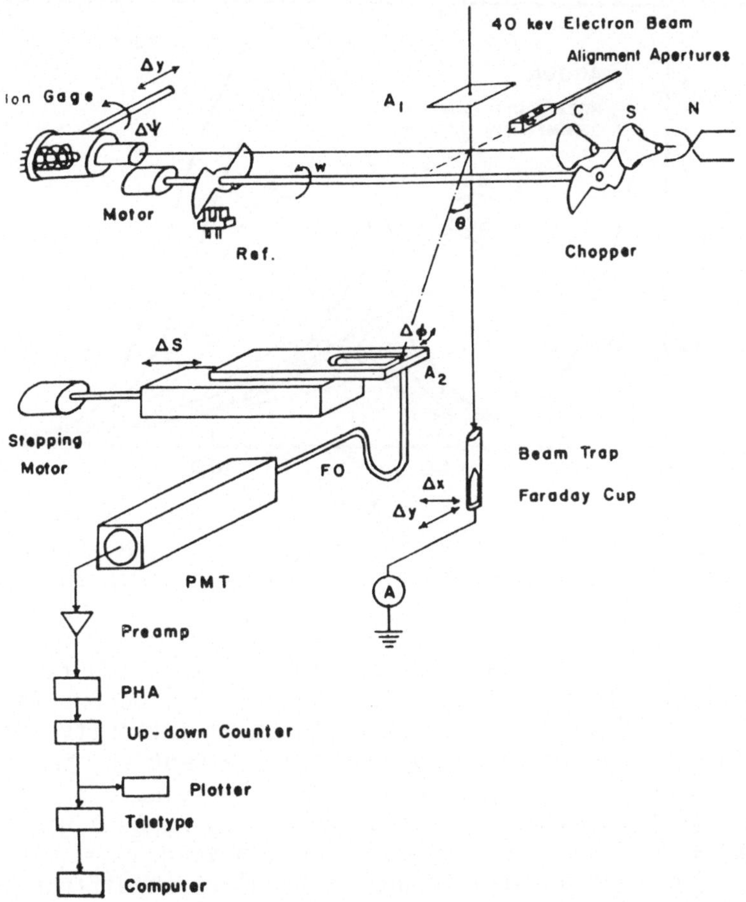

Fig. 4 : The single-channel pulse counting detection system has a synchronous motor driving the molecular beam chopper at an angular speed w. The aperture A_2 is driven to specific angular positions, with respect to the electron beam direction, with a stepping motor programmed to dwell at a given r (related to angle θ) for a specified number of chopper cycles n then index a distance Δr repeatedly from a minimum to a maximum angle r_{min} to r_{max}. The electrons passing through A_2 strike a Pitot B scintillator producing photons which pass through the vacuum system via a fiber optic, FC, to a photomultiplier. The Up-Down Counter counts up signal + background during the open half cycle and down for the background during the closed half cycle, thus accumulating signal. The counter controls the stepping motor and provides variation to r_{min}, r_{max}, Δr and n .

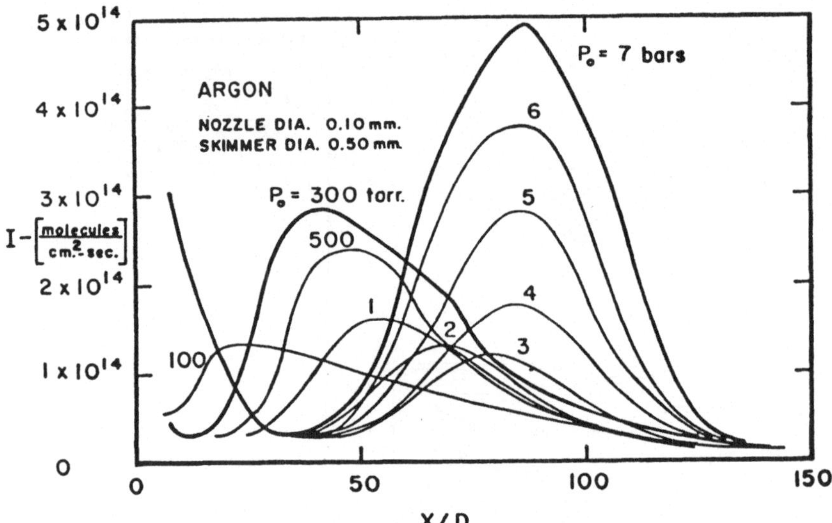

Fig. 5 : Molecular beam intensity I as a function of the distance from free jet source to skimmer, x/D, is shown as a function of pressure.

pressure up to about 1 Bar. The optimum for an unclusted beam for this particular beam configuration and pumping, occurs for p_0 = 300 Torr. Beyond this pressure the beam intensity deteriorates and would continue to do so monotonically except for the onset of massive clustering as p_0 exceeds 2 Bar. The clusters grow to sizes large enough that skimmer interaction and background gas scattering do not attenuate the cluster beam. Thus, beyond p_0 = 2 Bar, there is sufficient beam density to undertake electron diffraction.

An important development in the use of cluster molecular beams is the use of small Laval nozzles. They constitute a "controlled expansion" in contrast to the free jet whose expansion is determined once a particular size, D, and gas, γ, are chosen. It has been known for many years that Laval nozzles are excellent devices to promote and study the nucleation process (26-30). A nozzle was proposed in the original Kantrowitz and Grey high pressure beam apparatus, but did not perform well due to some design problems (31,32). Most beam researchers, subsequently chose the free jet for its simplicity. The Laval nozzle was revived for cluster beam work in Karlsruhe (33-35) and has been used to great advantage in our laboratory (36-38).

To illustrate the great advantage of the nozzle, Argon cluster beams are shown in Fig. 6 for a free jet and a Laval, diverging nozzle having the same minimum or throat diameters and, therefore, the same mass flow rates (39). The beam intensity is measured with

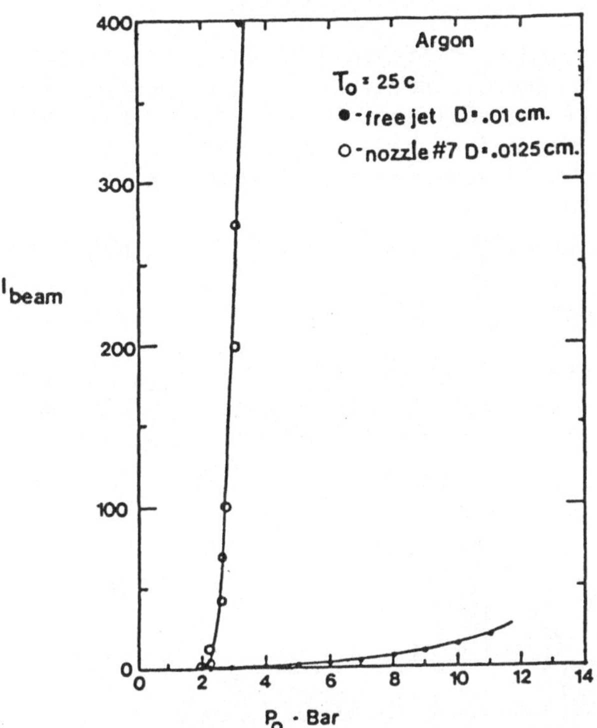

Fig. 6 : Argon beam intensity, I_{beam}, measured with the ionization gage in Fig. 4 for free jet and nozzle having nearly the same throat size. The nozzle is orders of magnitude more efficient as a cluster source.

the ion gage (see Fig. 4) and goes off scale for the nozzle near p_o = 2 Bar. A calculation of N_{coll} reveals that it is greater for the nozzle by about three orders of magnitude ($O(10^3)$). This is about the order of magnitude of the increase in beam intensity for the nozzle compared to the free jet. Note that when the beam intensity reaches a value of 100 on this scale, the beam cluster density is high enough to obtain electron diffraction patterns. This free jet source is not intense enough at 12 Bar to conduct diffraction experiments. Similar results are seen in seeded beams such as SF_6 in Ar (36).

Diffraction data for such things as Ar, Kr, Xe, CO_2, H_2O and SF_6 have been obtained and reported in the literature (40-42,3). However, since the subject of this summer Institute is cluster physics applied to materials science, the gas phase cluster structure will be omitted in favor of some of our metal results.

3.2 Metal Cluster Beams

Due to the enormous difference in the nucleation tendencies of metals compared to gases, the cluster source design must be quite different. (Recall the goal here is not metal dimers or trimers). Indeed adiabatic expansions have been tried and failed, even for situations that should give far more nucleation that typical molecular beam geometries (43).

The net result is a source design shown in Fig. 7, which shall be referred to here as the Northwestern source (4). Metal is vaporized from a filament, wire basket, or crucibles of Aluminium oxide or boron nitride. A cold carrier gas mixed with the metal for a relatively long time before being pumped through nozzles N_1 and N_2. The flow is still subsonic ($n < 1$) between N_1 and N_2 so that there is no shock wave system to complicate the sample extraction through

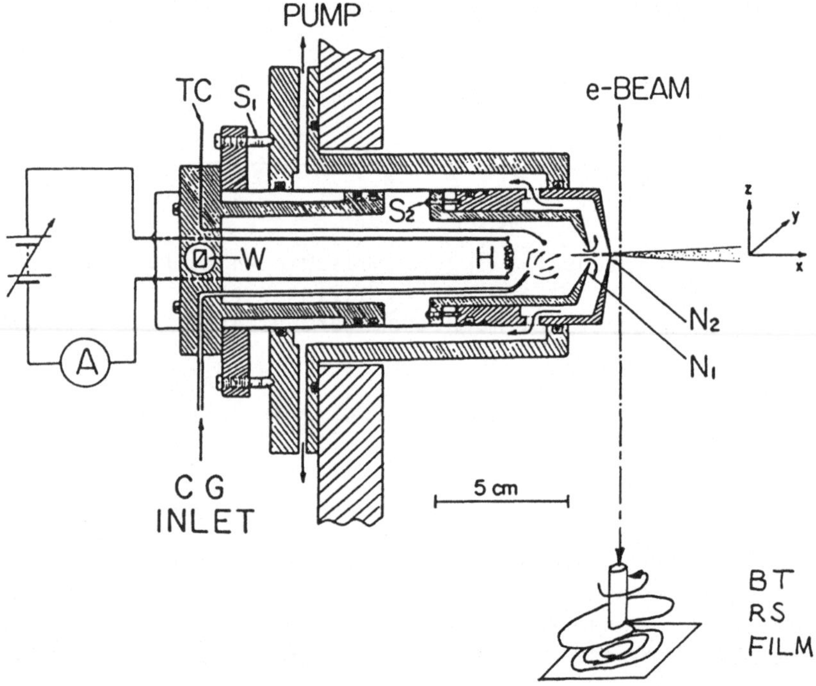

Fig. 7 : Schematic diagram of a metal cluster generator for electron diffraction : TC - thermocouple, W - prism mirror and window, H - heater for metal vaporization, S_1 and S_2 - sets of screws for configuration adjustments, Ar - argon carrier gas, N_1 and N_2 - flow orifices or nozzles, BT - beam trap, and RS - rotating sector.

N_2. Controlling the gas inlet flow rate and first stage pumping, it is possible to vary the gas velocity, u, and density n_o (see Fig. 7). The metal evaporation rate and thus its density n_{om}, the distance from oven to N_1, the aperture sizes for N_1 and N_2, the carrier gas species, i.e. collision cross section, and temperature are all potential variables to be exploited as a means of variation in N_{coll} and thus cluster size and temperature. The metal nucleation can be viewed as a combination of diffusion, energy transport and either luminar or turbulent mixing. Application of the nucleation theory is valid for a know process such as the diffusion cloud chamber (44,45) but is risky at best for the unknown flow conditions here. If there were no macroscopic mixing of evaporating metal with carrier gas, or if there is a region near the evaporating surface which is controlled by diffusion processes, then the following properties would affect the metal supersaturation and nucleation process :

1. Carrier gas density, n_o (mainly p_o but also T_o variation) which controls species diffusion.
2. Metal vapor density, n_{om} which controls its diffusion, Ficks law.
3. Carrier gas temperature, T_o, which controls thermal diffusion (energy transport) and species (Soret) diffusion.
4. Metal temperature, T_{om}, which controls evaporation rate and Soret diffusion.
5. Metal-gas collision cross section, which controls the metal vapor mean free path affecting both mass and energy transport.

Typical operating characteristics are given in Table II. High supersaturation and long flow time from filament to N_1 constitutes the nucleation and growth regime, the time from N_1 to N_2 and from N_2 to the electron beam is very short. Thus the source is described as a pre-expansion device with N_2 sampling the flow through N_1. Estimates of the cluster concentration and size have been made as a function of p_o T_{om} and shows the trend seen in continuum flow Laval nozzle nucleation (5) : Smaller size and higher concentration at lower condensable mole fraction (4). A more complete size dependence for several metals and carrier gases is shown in Fig. 8.

As the carrier pressure p_o is lowered, one expects that the nucleation region moves further from the evaporating surface (Effect 1, above) causing lower vapor density n_{om} , and consequently higher supersaturation and nucleation rate lead to higher concentrations of smaller clusters. As T_{om} is increased the vapor density is everywhere increased, lowering the supersaturation and thus the

TABLE II Oven operating characteristics.

Dimensions		
Heater—argon supply jet	0.6 cm	
Heater—thermocouple	0.93 cm	Argon supply jet diameter d_s 0.096 cm
Heater—nozzle N_1	1.3—2.6 cm	Nozzle N_1 diameter, dN_1 0.075 cm
N_1-N_2	0.3 cm	Nozzle N_2 diameter, dN_2 0.05 cm
N_2—electron beam	0.075 ± 0.025 cm	

Typical oven properties
(Range of properties in parenthesis)

Location	Pressure (Torr)	Temperature (°K)	Argon mass flow rate (g/sec)	Mean free path (cm)	Knudsen No.—Kn	Velocity (cm/sec)	Reynolds No.—Re	Mach No.—M	Flow time (sec)
Exit of argon supply jet	1.0 (0.5—1.2) -	300	0.033	5.0×10^{-3}	0.050	15 000	20	0	0.2 (0.015—0.4)
Entrance to nozzle N_1	1.0 (0.5—1.2)	(340—650)	0.033	6.6×10^{-3}	0.088	~ 30 000	18	~ 0.01	10×10^{-4}
Entrance to nozzle N_2	0.6 (0.3—0.7)	315 (268—512)	0.0265	1.1×10^{-2}	0.214	27 000[a]	9.5	0.9	2×10^{-4}
At electron beam	1×10^{-5} (1×10^{-5} to 1.4×10^{-5})	117 (100—190)	0.0265	500[b]	$\gg 1$	46 800	. . .	2.5[c]	

[a] Terminal Mach No. in the free jet (Ref. 6) and occurs at $x/D_{n2} \cong 1$ or 0.05 cm from nozzle N_2.
[b] Values are for the background gas.
[c] Upper limit; it could be as much as an order of magnitude lower.

Table II : Oven operating characteristics.

nucleation rate (Effect 4), leading to lower concentrations of larger final size. For fixed p_o T_{om} but decreased carrier gas collision cross section (going from SF_6 to At to He, Effect 5), the vapor diffuses more readily from the evaporating surface, the saturation increases, and with it, the nucleation rate, leading to higher concentrations of smaller clusters. Figure 9 shows the effect of different carrier gases as well as several metals. Carrier gas temperature has not been varied (Effect 3) here, but has been successfully used by other researchers (10).

Another cluster source, which has been very successfully used is the one from Konstanz, shown in Fig. 10. Recall that the goal of this research is to make the very smallest clusters in the range g = 1 to 50, while the Northwestern source had as its goal g = 500 to 5,000. Thus they are exploring the very small size range, for use with time-of-flight spectrometry, and therefore use very cold temperatures T_0 (Effect 3) and low collision cross section (Effect 5). In fact their source only produces clusters within the detection range if helium is used at liquid nitrogen conditions. The helium chamber is pumped slowly so that the flow is very low velocity and conditions for a diffusion cloud chamber operating regime are approached. This group has recently produced numerous interesting results with this source (10).

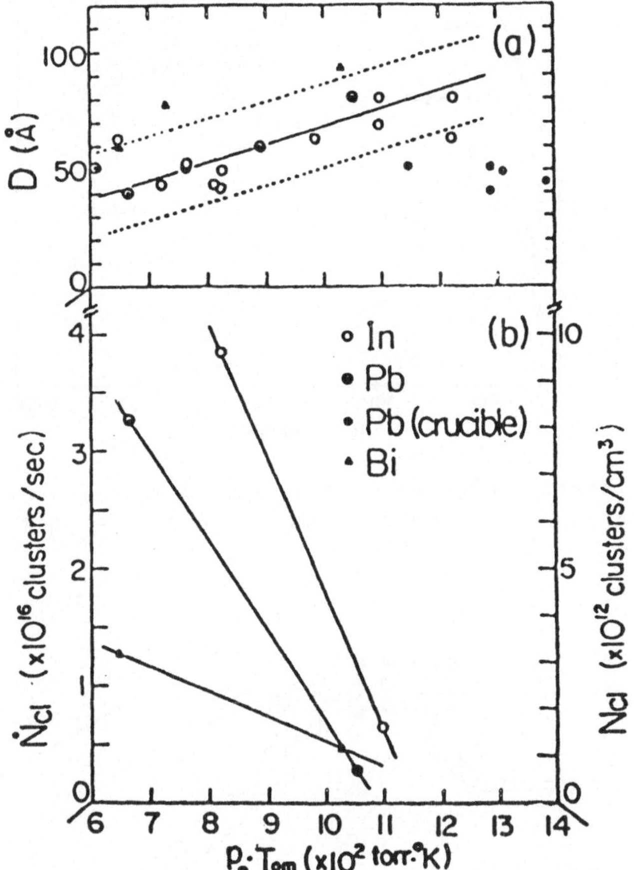

Fig. 8 : (a) Average cluster size D and (b) the cluster production rate $\dot{N}_{cl}$ and the cluster concentration at the crossed-beam intersection N_{cl} are plotted as a function of $p_o T_{om}$, the product of argon stagnation pressure and the temperature at the surface of the evaporating metal. The general trend is that as $p_o T_{om}$ is increased clusters of larger diameter and lower concentration are formed.

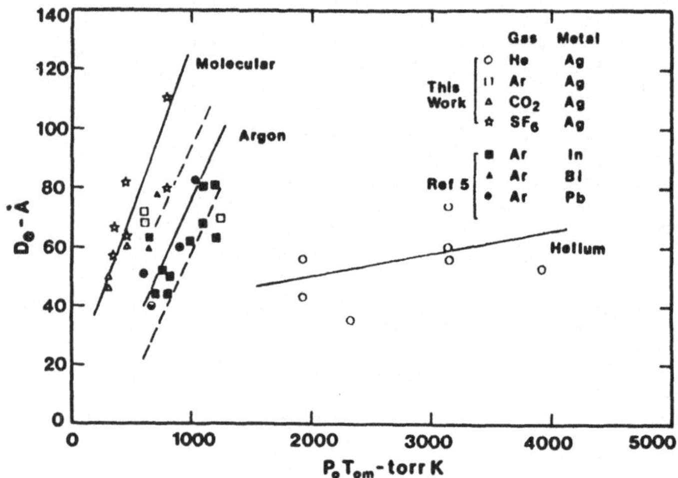

Fig. 9 : Experimental cluster size D_e, as a function of oven pressure and temperature for several metals and carrier gas.

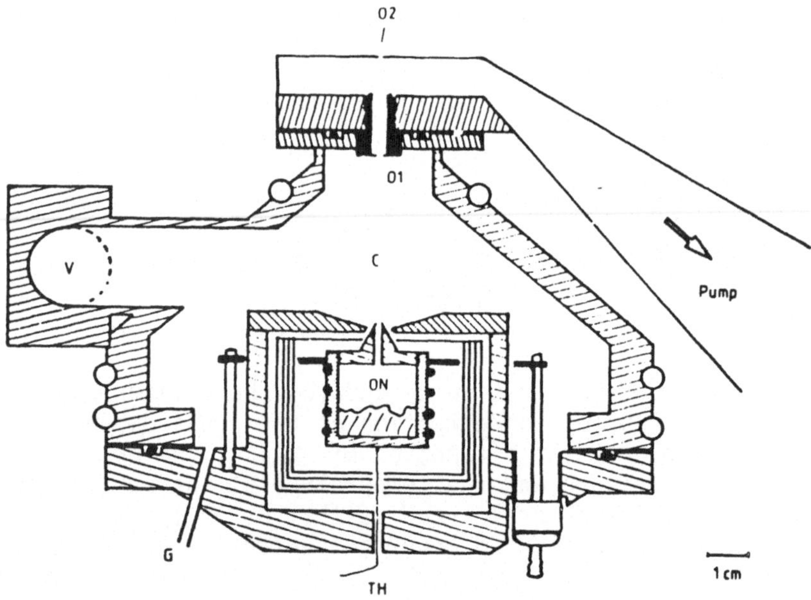

Fig. 10 : This cluster source consists of a metal oven, ON heated resistively having a thermocouple, TH, insulated from a liquid nitrogen cooled housing, shown cross hatched, with the Helium gas inlet, G, condensation zone, C, flow apertures, 01 and 02 and valve for vacuum connection. The cluster beam emerging from 02 enters the ionizer of a time-of-flight mass spectrometer (10).

4. CLUSTER ELECTRON DIFFRACTION

The theory of electron diffraction, with many similarities both physically and mathematically to x-ray diffraction, has been used for 50 years as a method of determining molecular structure. The electron de Broglie wavelength λ is :

$$\lambda = h/p = h/(2\, m_e\, E)^{\frac{1}{2}} \;, \tag{9}$$

where h is Planck's constant, p is the electron momentum , m_e is the electron mass and E its energy. For $E = 40$ keV, $\lambda \cong 0.06$ Å = 0.006 nm. The clusters in the beam have random orientation so that the diffraction pattern is symmetric about the electron beam axis with differential scattering cross section $d\sigma/d\Omega$,

$$d\sigma/d\Omega = \sum_{i \neq j} f_i^{\,B} f_j^{\,B} \frac{\sin(s\, r_{ij})}{s\, r_{ij}} + \sum_i (f_i^{\,B})^2 + \frac{4}{a_n} \sum S_i , \tag{10}$$

with the Bohr radius $a_n = \hbar^2/m_e e^2$, the electron scattering factor $f_i^{\,B} = (8\pi^2 m_e e^2/h^2)(Z_i - F_i)/s^2$, the scattering parameter $s = (4\pi/\lambda)\sin(\theta/2)$, θ being the electron scattering angle, e the electronic charge, Z the atomic number, S the inelastic scattering factor (46), F the x-ray scattering factor (46), and r_{ij} the distance from atom i to atom j in the cluster. For uncorrelated cluster positions in the beam, the sum of the differential cross sections from all of the clusters in the electron beam, is related to the experimental parameters,

$$\sum_k (d\sigma/d\Omega)_k = R^2\, I_\theta/I_0 , \tag{11}$$

where R is the distance from the crossed beams to the detector and I_0 and I_θ are respectively, the electron beam and scattered beam intensities.

Examination of the diffraction can be used to determine the following:

1.) Crystalline, amorphous or liquid state.
2.) Structure and unit cell dimensions if crystalline.
3.) Cluster size D from peak broadening.
4.) Cluster temperature - Debye-Waller factor.
5.) Cluster structure using model building.
6.) Radial distribution function.

Cluster size is approximated with the relation (47),

$$D = R\,\lambda\,(B^2 - B_0^2)^{\frac{1}{2}} , \tag{12}$$

where $L\lambda$ is the camera constant determined from a known thin film sample and B is the peak full width at half maximum (FWHM). B_0 is FWHM for very large clusters, which is theoretically zero but finite due to non idealities in the electron beam apparatus, referred to as instrument broadening.

The Debye-Waller effect is the progressively greater reduction in peak height with scattering angle relative to the theoretical peak heights at 0 K, i.e. no thermal motion of the atoms about their equilibrium positions in the lattice.

As an example consider the diffraction from microcrystals of perfect body centered cubic-BCC unit cells of dimension a. Only the coherent interatomic term is plotted (for x-rays in this case) in abcissa. The patterns, though perfectly undistorted, yield very different results. Note how the peaks get higher and more narrow (FWHM) in accordance with the Scherrer Eq. (12). Morozumi and Ritter (49) pointed out that about 60 unit cells (for example, particles 20 A diameter) will show diffraction rings characteristic of macroscopic crystals out to about t = 25.

There has been a natural evolution in the level of sophistication in the detection techniques used in our laboratory. Systems 1-4 below have been used previously, 5, is in the design stage. They include :

1. Photographic film
2. Rotating sector and film
   (to extend the dynamic range of the film due to the rapidly decreasing scattered electron signal with increasing angle, see Fig. 7 and Ref. (14)).
3. Single channel, synchronous detection
   (scintillator, fiber optic, photomultiplier, counting system, see Fig. 4 and Ref. (48)).
4. Multichannel detection, two dimensional
   (fluorescent screen, optics, Optical Multichannel Analyzer - OMA).
5. Multichannel detection, two dimensional and synchronous
   (cooled OMA with chopped signal - approaching unit electron counting efficiency).

Systems 1 to 4 have been used previous to this writing. System 5 is still in the design stage.

A diffraction pattern using system 3 is shown in Fig. 12. The clustered species is SF_6 in an Ar carrier gas and Laval nozzle 7. The peaks are intense, narrow and well resolved indicating rather large cluster size. As size is reduced the signal-to-noise also decreases, but no pattern at all is obtained without using the synchronous detection method.

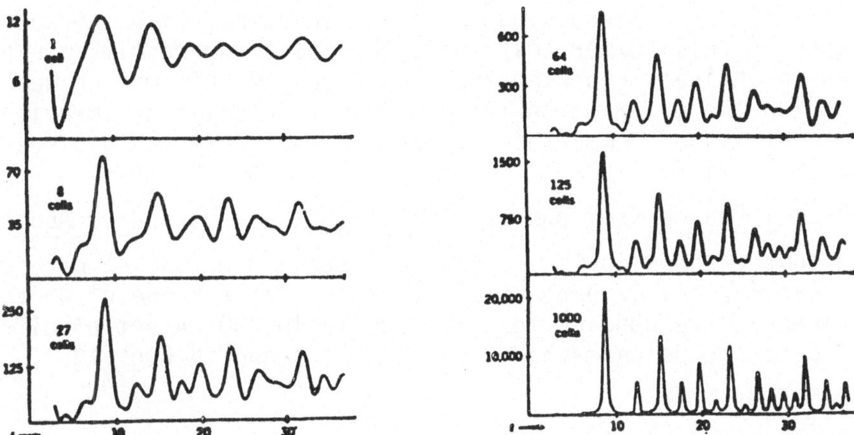

Fig. 11 : Theoretical x-ray diffraction patterns are calculated for perfect, undistorted microcrystals of BCC unit cells with dimension a, and $t = a\,s = (a\,4\pi/\lambda)\,\sin(\theta/2)$. (After Morozumi and Ritter (49)).

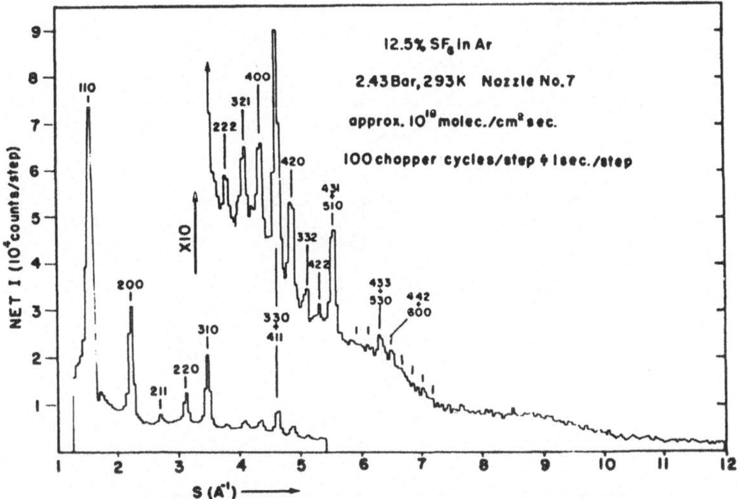

Fig. 12 : Electron diffraction pattern for an intense SF_6 cluster beam produced in Laval nozzle No. 7 from a 12.5 mole per cent mixture of SF_6 in Ar, Nozzle No. 7 has $D_o = 0.0127$ cm, $D_e = 0.218$ cm and L = 2.56 cm. For details see Ref. (36).

5. METAL CLUSTER RESULTS

Restricting the discussion to metals in accord with the subject matter of this summer Institute, the following results are presented as an indication of the type and scope of information we have obtained with the diffraction method. In the interests of brevity the treatment is not complete, but most of this research has been reported in detail.

5.1 Indium : Change from one Crystallographic Structure to Another

The source shown in Fig. 7 is used with a wire basket oven and argon as carrier gas to produce clusters in a size range of 40 to 80 Å diameter corresponding to g = 1,000 to 10,000. A set of diffraction patterns using method 2 above, are shown in Fig. 13.

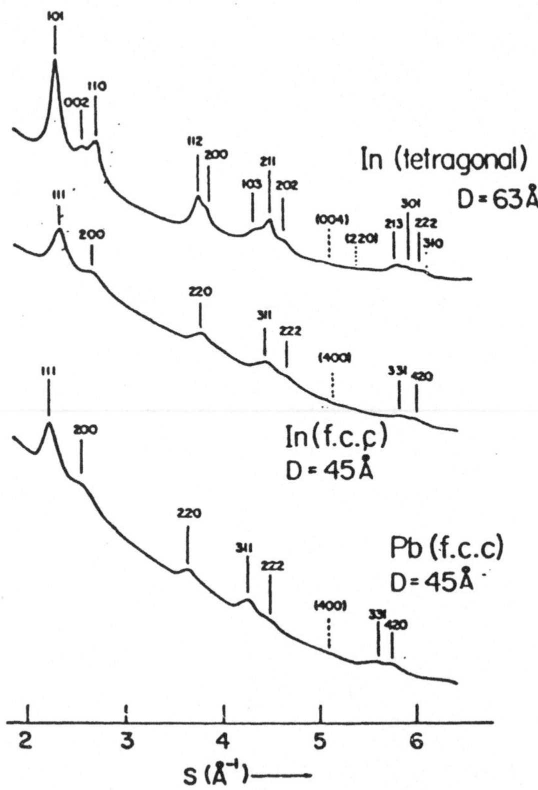

Fig. 13 : Microdensitometer traces of electron diffraction photographs for indium and lead microclusters show tetragonal In of an average cluster size of ∿ 63 Å in diameter and is compared with cubic (fcc) In of D ≃ 45 Å. A typical fcc pattern of lead is shown with an average size of ∿ 45 Å. The abscissa S is the scattering parameter $S = (4\pi/\lambda)\sin^{\theta}$ with θ the Bragg angle.

Notice the two peaks 002 and 110 for the upper diffraction pattern obtained for the largest cluster size. As size decreases the structure deforms from face centered tetragonal to face center cubic-fcc, both of which are crystallographic structures (4). It is interesting to note that this cannot be explained simply as an effect of increased pressure inside the cluster due to a classical surface tension effect (i.e. $\Delta p = 2\sigma/r$). The bulk structure, under pressure, remains fc tetragonal with the unit cell height c increasing relative to the square base dimension a . The variation in unit cell dimensions with size, Fig. 14, shows a decrease in height c and an increase in base a with c = a, i.e. fcc structure. The transition occurs gradually over the range g = 5,000 to 1,000.

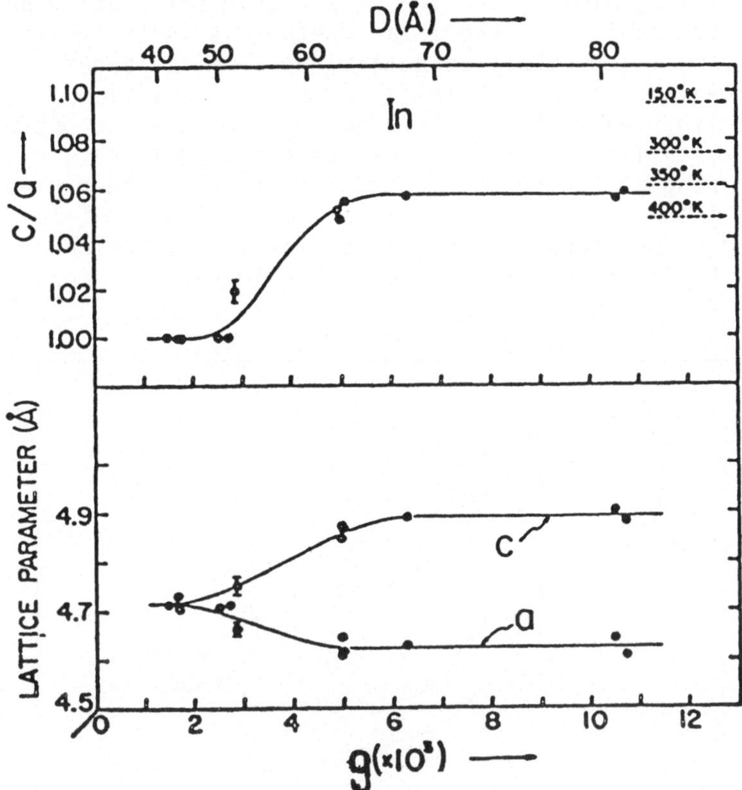

Fig. 14 : Lattice parameters and their ratio for the indium microclusters are plotted as a function of the number of atoms per cluster (n) and the average cluster diameter D . Broken lines indicate the corresponding values in bulk crystalline materials at various temperatures. Typical experimental error bars are shown by vertical lines. Deviation from bulk structure begins as the diameter approaches 60 Å.

5.2 Lead and Silver : Change from Crystallographic to Non-Crystallographic Structure

Typical diffraction patterns, for Lead with the same source and diffraction method as for Indium, are seen in Fig. 15 (50). There is only a very small reduction, if any, of the unit cell parameter as size is decreased. Moreover, a cursory examination indicates nothing of particular interest, and no apparent structure change. However, upon closer examination of the peak intensities, using the Blackman two-beam correction for multiple scattering (51), a slight deviation from the theoretical peaks heights is seen for the largest size in Fig. 16. As size is reduced this deviation increases and appears to be systematic. A theoretical model consisting of bulk fcc, plus a fraction of a liquid diffraction pattern makes a reasonable fit to the data. Whether is is truly liquid or amorphous cannot accurately be determined from these data. What is certain, however, is that there is a gradual transition from a bulk crystallographic (fcc in this case) structure, to one which is not a crystalline arrangement. It is presumed to be a rearrangement occuring at the surface althought it cannot be proved explicitly with these results. A transition to the liquid state, i.e. the melting temperature of metal particles as a function of size has been a topic of research interest for 70 years or more with all

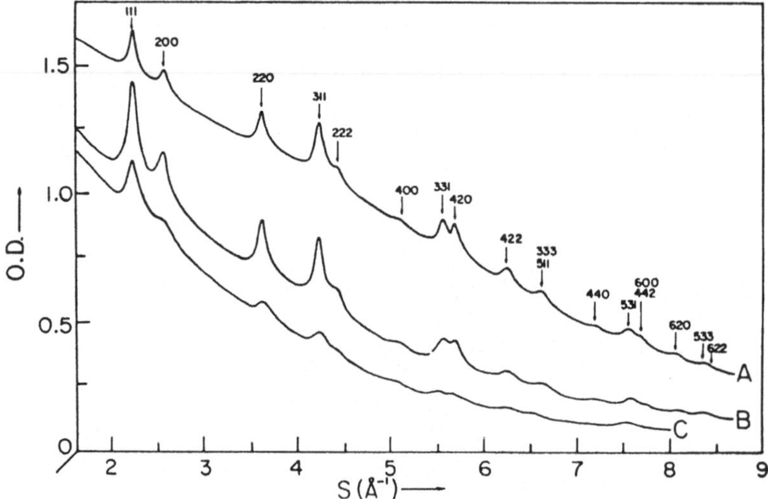

Fig. 15 : Photographic density (OD) of the diffraction patterns of Pb clusters for plates A, B and C, plotted as a function of the scattering variable s; Miller indices are given in A.

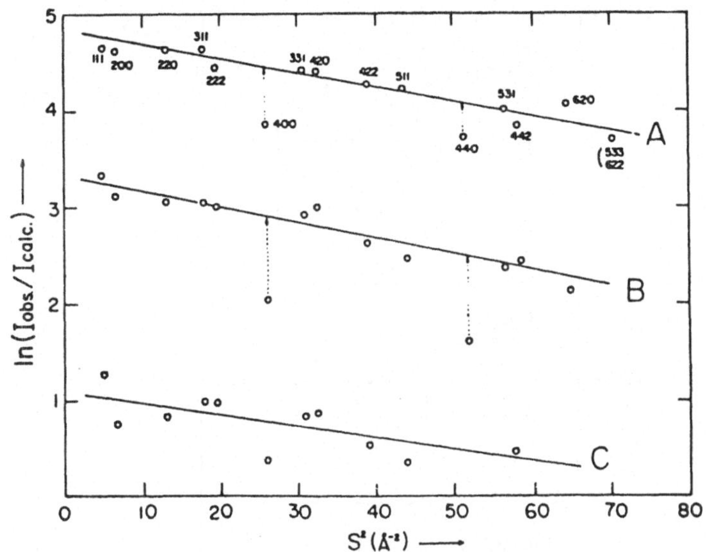

Fig. 16 : Logarithmic plots of the ratio of the observed Bragg-peak intensity I_{obs} (plate A) to the calculated intensity I_{calc}, plotted as a function of s^2 (Debye-Waller factor analysis). The best fit diameters are 80 Å, 60 Å and 50 Å for A, B and C respectively.

measurements and predictions giving a decrease in melting temperature with size (2,52). It has even been predicted that some metals will be liquid at all temperatures below a size of 10 Å.

Investigation of silver clusters has been carried out using several different carrier gases and method 1. for pattern detection (i.e. film). In addition,a 12 beam multiple scattering analysis and computer program was developed to assess the severity of this potential complication in the analysis (53). The results are shown as the difference between the theoretical and measured peak heights as a function of cluster size (Fig. 17). As with lead (both, bulk structures are fcc) there is a deviation from fcc which increases with decreasing size. It is also systematic and similar to the Lead results. The conclusions are therefore similar with, perhaps, the additional conclusion that, at most, the Blackman two beam correction is required for these very small aggregates.

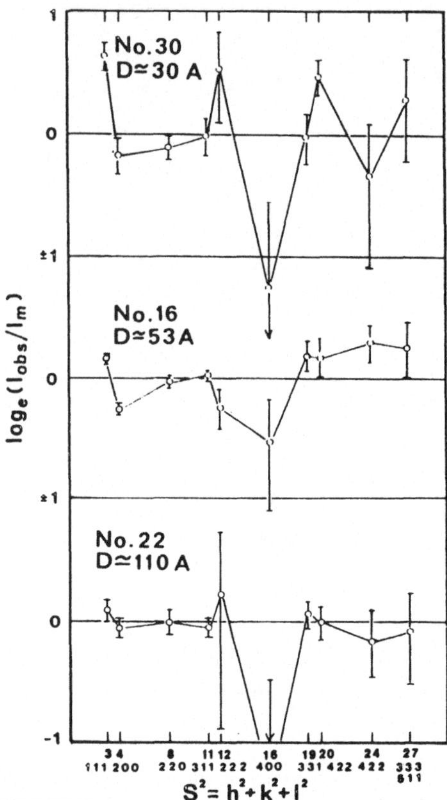

Fig. 17 : A detailed comparison, peak by peak, of the observed and calculated intensities for Ag plotted versus S^2 with Miller indices indicated.

5.3 Cluster Temperature

An example of cluster temperatures T_{cl} for the metals bismuth, Lead, and indium, are shown in Fig. 18. They were produced in the pre-expansion oven of Fig. 7 with argon as the carrier gas. Because the argon is expanding from a rather low pressure, typically 0.5 to 2 Torr, the amount of adiabatic cooling is not great. Thus even though the pressure ratio is high (microscope pressure $< 2 \times 10^{-5}$ Torr), the free jet expansion becomes collisionless quickly due to the low stagnation chamber density n_o (low p_o, high T_o). Since the carrier gas is monatomic its terminal temperature, T_T, has been measured for free jet expansions and found to be a function of the stagnation chamber Knudsen number $Kn_o = \lambda_o / D_{N_2}$ (24,54) :

$$M_T = c\,\sigma_c\,Kn_o^{\,-\gamma/(\gamma-1)} \tag{13}$$

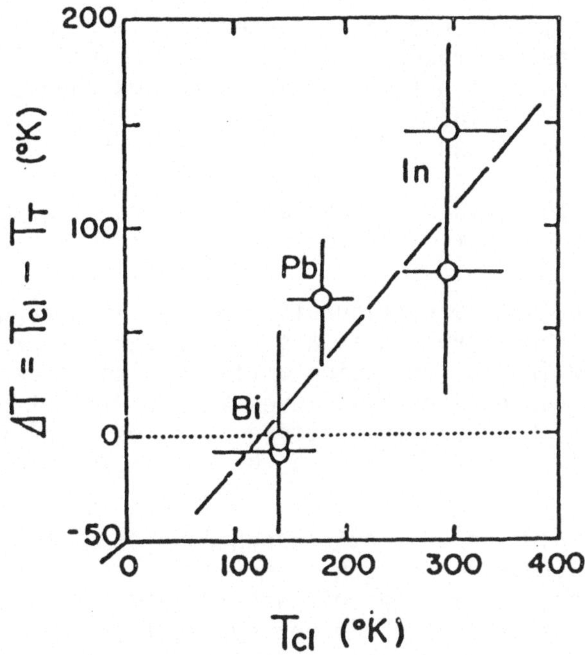

Fig. 18 : Typical cluster temperatures at the crossed-beam location, as determined from Debye-Waller factors and from unit-cell parameters are plotted as a function of the difference in temperature between the clusters and the argon carrier gas.

where λ_0 is the argon mean free path in the stagnation chamber ($c = c(\gamma)$, $\gamma = 5/3$ giving $c\,\sigma_c = 1.17$. The terminal temperature obtained from M_T is :

$$T_T = T_0 \left(1 + \frac{\gamma-1}{2} M_T^2\right)^{-1}. \qquad (14)$$

The metal cluster temperature is estimated using the Debye-Waller technique, and for those diffraction patterns with cluster size approaching bulk, published values of the temperature variation of the unit cell dimension have been used (4). The error bars indicate that these measurements are not precise. Nevertheless, it is interesting to speculate why some metals have little or no temperature difference from that of the final value of the expanding gas (Bi), while others are as much as 50 to 150 ^{0}C warmer. Since all the metals have about the same starting vapor pressures, their starting temperatures from the evaporating oven are hotter in the order Bi, Pb, In. Thus, perhaps it is related to the logical argument with respect to the starting temperature T_{om} or T_0, i.e. the hotter they start the hotter they remain.

5.4 Amorphous Structure

An investigation of antimony cluster diffraction was initiated due to some interesting research on its use for producing thin films. Experiments were carried out in our now standard format using the pre-expansion source (Fig. 7) with argon at 1 Torr as the carrier gas. The results are shown as the upper, diffraction pattern in Fig. 19, with the lower pattern obtained from a polycrystalline thin film having the bulk rhombohedral structure. The gas cluster pattern is a weak one but displays the qualitative crystalline features.

A second, new source configuration was tried by Prof. I. Yamada on leave in our laboratory. It is a single stage pre-expansion configuration, shown in Fig. 20, using antimony with no carrier gas. For reasons outlines earlier this investigator did not expect significant clustering. However, the pattern obtained is plotted as the middle curve in Fig. 19. It displays the undulations characteristic of a liquid.

It is well known in adiabatic expansions of pure vapors that clusters temperatures can be much warmer than those formed in a low mole fraction, carrier gas mixture. In addition the pure vapor in these pre-expansion sources do not supersaturate as highly or as rapidly as the inert carrier gas case. These are the most probable reasons for the appearance of the non-crystalline cluster structure from the pure vapor oven, Source B.

When the liquid-like pattern, Exp. 2 in Fig. 20,is Fourier transformed to obtain its radical distribution function G(r), the result is not that of a liquid. The most telling characteristic is the knee in the curve at 4 A^{-1} (see Fig. 21). This is typical of an amorphous solid. There must, therefore, have been sufficient cooling via radiation and some amount of adiabatic expansion to permit the condensing clusters to sublime or to condense as a liquid and then solidify.

The use of the diffraction peak broadening, after Scherrer (47) and the Debye-Waller (3,51) factor to obtain cluster size and temperature at the electron beam location, which are used with crystalline particles, are not usable for amorphous structures. Thus we are not able to determine how much below the bulk melting temperature the cluster temperatures are. Previous results in our laboratory, however, have shown that cluster beams of Pb and of Ag, nucleated in carrier gases, have traces of an amorphous type structure in a predominantly crystalline (f.c.c.) arrangement, even though the cluster temperature is in the neighborhood of 150 K, well below the melting point (50,53).

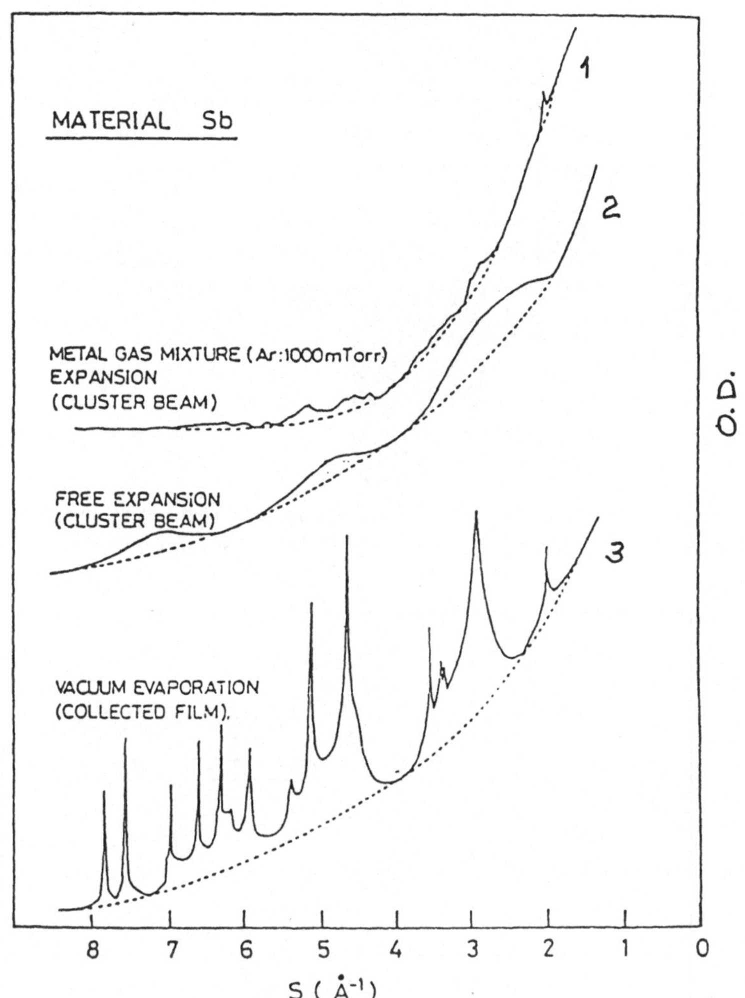

Fig. 19 : Diffraction patterns of Sb are plotted for an argon carrier gas in the source of Fig. 7, as the upper curve, Exp. 1, and for the pure vapor Sb source of Fig. 20, as the middle curve, Exp. 2. For reference, a polycrystalline thin film, grown with the conventional vacuum evaporation (i.e. unclustered) technique, shows the bulk rhombohedral structure.

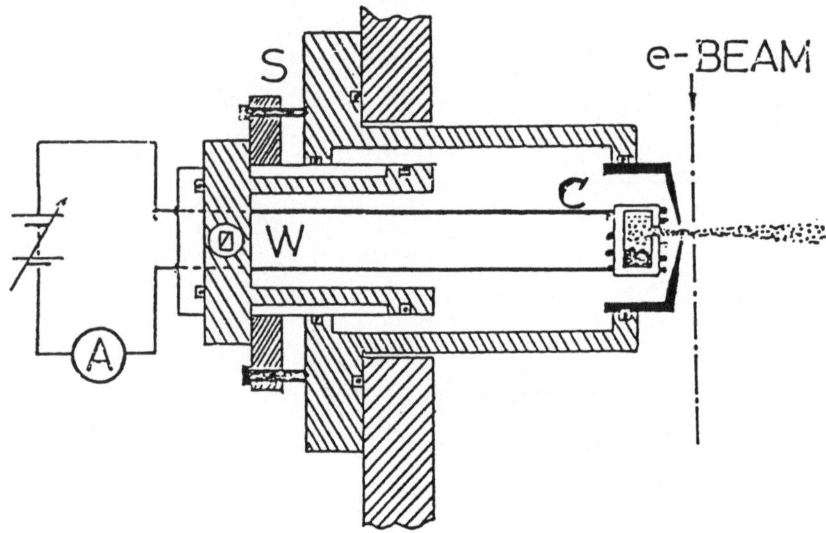

Fig. 20 : The pure vapor source uses no carrier gas and one stage of pumping with A - ammeter, C - crucible, S - adjustment screws, and W - window.

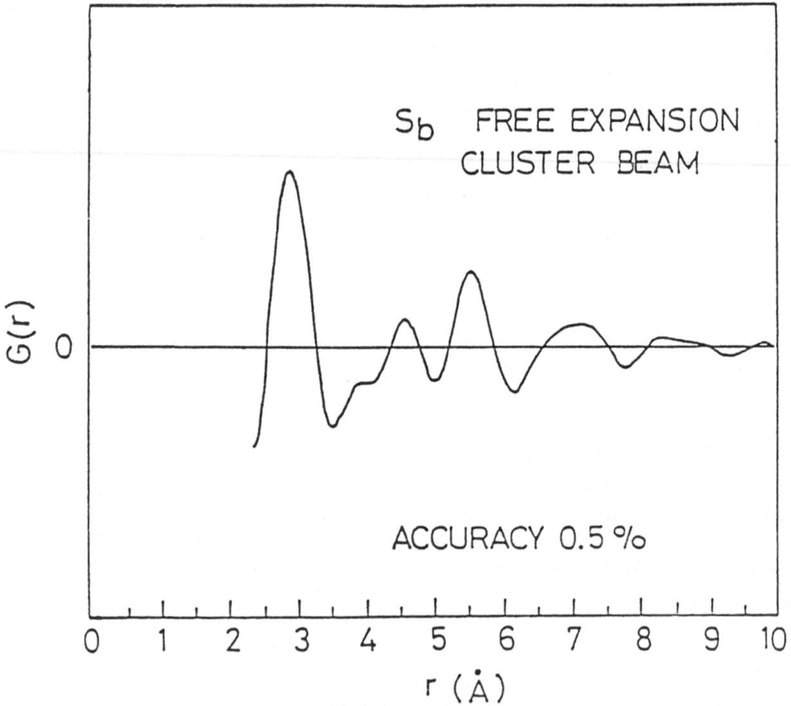

Fig. 21 : The radial distribution function for the liquid-like pattern of Exp. 2 is characteristic of an amorphous solid.

Nevertheless it is significant that this pure Sb, free jet expansion, having a bulk surface tension as high as 350 ergs/cm^2, can nucleate clusters large enough and in sufficient concentration to produce diffraction patterns.

The author is greatly indebted to his co-workers mostly from the past two to four years : Drs. J.A. Armstrong, B.G. DeBoer, A. Yokozeki, S.S. Kim, Messrs. J.H. Binn, O. Abraham, and D.C. Shi on leave of absence from the National Academy of Sciences, Peking, China and Prof. I. Yamada on Sabbatical from Kyoto University, Japan. He wishes also to thank the funding agencies that provided the support essential for this research: the Energy Energetics section of the National Science Foundation, and the Power Branch and Chemistry Division of the Office of Naval Research.

REFERENCES

1. Wegener, P.P. "Nonequilibrium Flows", Gasdynamics of Expansion Flows with Condensation and Homogeneous Nucleation of Water Vapor, P.P. Wegener, Ed., Vol. 1, Part I. Marcel Dekker, New York (1969) 163-243.
2. Buffat, Ph.-A. and J.P. Borel. Size Effects on the Melting Temperature of Gold Particles, Phys. Rev. A 13 (1976) 2287-2298.
3. Kim, S.S. and G.D. Stein, Creation and Structure Study of Vacuum Isolated Clusters of Argon, Krypton, and Xenon, J. Coll. Inter. Sci. (1982) 180-203.
4. Yokozeki, A. and G.D. Stein, A Metal Cluster Generator for Gas-Physe Electron Diffraction and its Application to Bismuth, Lead, and Indium: Variation in Microcrystal Structure with Size. J. Appl. Phys. 49 (1978) 2224-2232.
5. Stein G.D. and P.P. Wegener, J. Chem. Phys. 46 (1967) 3658.
6. Leckenby R.E., E.J. Robbins and P.A. Trevalion, Proc. Roy. Soc. (London) A 280 (1964) 409.
7. Milne, T.A. and F.T. Greene, J. Chem. Phys. 47 (1967) 4095.
8. Searcy J.Q. and J.B. Fenn, J. Chem. Phys. 47 (1974) 5282.
9. Castleman Jr., A.W., B.D. Kay, V. Hermann, P.M. Holland and T.D. Mark, Studies of the Formation and Structure of Homomolecular and Heteromolecular Clusters, Surface Sci. 106 (1981) 179-187.
10. Sattler, K. Diagnostics of Clusters in Molecular Beams, to appear in Rarefied Gas Dynamics, Proceedings of 13th International Symposium, Novosibirsk (1982).

11. Delacrétaz, G., J.-D. Ganière, P. Melinon, R. Monot, R. Rechsteiner, L. Wöste, H. van den Bergh and J.M. Zellweger. Laser Probing of Cluster Formation and Dissociation in Molecular Beams, to appear in Rarefied Gas Dynamics, Proceedings of the 13th International Symposium, Novosibirsk (1982).
12. Audit, P. and M. Rouault, Compt. Rend. 265 (1967) 1100.
13. Pirenne, M.H., the Diffraction of X-Ray and Electrons by Free Molecules, Cambridge Univ. Press, London (1946).
14a. Schäfer, L. Electron Diffraction as a Tool of Structural Chemistry, Applied Spectroscopy 30 (1976) 123-144 and references therein.
14b. Cowley, J.M., "Diffraction Physics", North-Holland, Amsterdam (1975).
15. Volmer, M. and A. Weber, Z. Phys. Chem. (Leipzig) 119 (1926) 277.
16. Farkas, L., Z. Phys. Chem. (Leipzig) A 125 (1927) 236.
17. Becker, R. and W. Döring, Ann. Phys. (Leipzig) 24 (1935) 719.
18. Zeldovich, J.B., Acta Physicochim. (URSS) 18 (1943) 1.
19. Frenkel, J. "Kinetic Theory of Liquids", Dover, New York (1946).
20. McDonald, J.E., Homogeneous Nucleation of Vapor Condensation I Thermodynamics Aspects, Amer. J. Phys. 30 (1962) 870-877; also 31 (1963) 31-41.
21. Lothe J. and G.M. Pound, J. Chem. Phys. 36 (1962) 2080.
22. Lothe J. and G.M. Pound, J. Chem. Phys. 48 (1968) 1849.
23. Abraham,F.F., "Homogeneous Nucleation Theory" Academic Press, New York (1974).
24. Anderson, J.B., Molecular Beams for Nozzle Sources, "Molecular Beams and Low Density Gasdynamics", P.P. Wegener, Ed., Marcel Dekker, Inc. New York (1974) 1-92.
25. Armstrong, J.A. and G.D. Stein. Nucleation Experiments in Molecular Beams, "Rarefied Gas Dynamics", K. Karamcheti (Ed.) Academic Press, New York (1974) 279.
26. Stodola, A. "Steam and Gas Turbines" McGraw-Hill, New York (1927).
27. Oswatitsch, K., ZAMM 22 (1942) 1.
28. Wegener, P.P. and L.M. Mack. Condensation in Supersonic and Hypersonic Wind Tunnels, Adv. Appl. Mech. Vol. V, Academic Press, New York (1958) 307.
29. Wegener, P.P. and A.A. Pouring, Phys. Fluids 7 (1964) 352.
30. Wegener, P.P. and G.D. Stein, "Twelfth Symposium (International) on Combustion, The Combustion Institute, Pittsburgh, (1969) 1183.
31. Kantrowitz, A. and J. Grey, Rev. Sci. Instr. 22 (1951) 328.
32. Kistiakowsky, G.B. and W.P. Slichter, Rev. Sci. Instr. 22 (1951) 333.
33. Hagena, O.F. and W. Obert, J. Chem. Phys. 56 (1972) 1793.
34. Hagena, O.F., in "Molecular Beams and Low Density Gas Dynamics" P.P. Wegener, Ed., Marcel Dekker, New York (1974).
35. Obert, W., in "Rarefied Gas Dynamics", R. Campargue, Ed., p. 1181. Commissariat à l'Energie Atomique, Paris (1979).

36. DeBoer, B.G., S.S. Kim and G.D. Stein, Molecular Beam Studies of Sulfur Hexafluoride Clustering in an Argon Carrier Gas from both Free Jet and Laval Nozzle Sources, Rarefied Gas Dynamics, edited by R. Camparague, p. 1151-1160. Commissariat à l'Energie Atomique, Paris (1979).
37. Abraham, O., S.S. Kim and G.D. Stein, Homogeneous Nucleation of Sulfur Hexafluoride Clusters in Laval Nozzle Molecular Beams, J. Chem. Phys. 75 (1981) 402-411.
38. Abraham, O., J.H. Binn, B.G. DeBoer and G.D. Stein, Gasdynamics of very Small Laval Nozzles, Phys. Fluids 24 (1981) 1017-1031.
39. Kim, S.S., D.C. Shi and G.D. Stein, Noble Gas Condensation in Controlled-Expansion Beam Sources, in Rarefied Gas Dynamics, edited by S.S. Fisher, Amer. Inst. Aeronaut. and Astronaut., New York (1981) 1211-1224.
40. Stein, G.D. and J.A. Armstrong, J. Chem. Phys. 58 (1973) 1999.
41. Farges, J., M.F. de Feraudy, B. Raoult and G. Torchet, J. Phys. (Paris) 38 (1977) C2-47.
42. Torchet, G., "Structure et Propriétés Physiques des Agrégats Formés par Détente en Jet Libre des Gaz N_2, O_2, CO_2, H_2O". Thesis, Université de Paris-Sud, Centre d'Orsay (1978).
43. Hill, P.G., H. Witting and E.P. Demetri, J. Heat Transfer 85, (1963) 303; G.E. Merritt and R.C. Weatherston, AIAA J. 6 (1967) 721.
44. Franck, J.P. and H.G. Hertz, Z. Phys. 143 (1956) 559.
45. Katz, J.L. and B.J. Ostermier, J. Chem. Phys. 47 (1967) 478.
46. "International Tables for X-Ray Crystallography", (C.H. Mac Gillavry and G.D. Rieck, Eds.) Vol. III Kynoch Press, Birmingham, England, 1962; (J.A. Ibers and W.C. Hamilton, Eds.), Vol. IV, Kynoch Press, Birmingham, England.
47. Scherrer, P., The Space Lattice of Aluminium, Z. Physik 19 (1918) 23.
48. Kim, S.S. and G.D. Stein, Scattered Electron Counting System for Nozzle Beam Studies of Small Clusters, Rev. Sci. Instr. 53 (1982) 838-844.
49. Morozumi, C. and H.L. Ritter, Acta Crystallogr. 6 (1953) 588.
50. Yokozeki, A., Lead Microclusters in the Vapor Phase as Studied by Molecular Beam Electron Diffraction: Vestige of Amorphous Structure, J. Chem. Phys. 68 (1978) 3766-3773.
51. Blackman, M., on the Instensities of Electron Diffraction Rings, Proc. Roy. Soc. London 173, (1939) 68.
52. Pawlow, P. Z. Physik. Chem. 65 (1909) 545.
53. DeBoer, B.G. and G.D. Stein, Surface Sci. 106 (1981) 84.
54. Anderson, J.B. and J.B. Fenn, Phys. Fluids 8 (1965) 780.

GENERATION OF BEAMS OF REFRACTORY METAL CLUSTERS*

S. Wexler, S. J. Riley, E. K. Parks, C-R. Mao and L. G. Pobo

Chemistry Division,
Argonne National Laboratory
Argonne, Illinois, 60439, USA

Interest in the physical and chemical properties of small metal clusters has recently stimulated the development of sources for the generation of molecular beams of metal clusters, since the collision-free environment of a beam has the advantage of permitting in-flight study of isolated species free of interference from surroundings. For example, spectroscopic studies utilizing tunable lasers may be performed in the molecular beam environment. The objectives of our research program are the elucidation of the physical and chemical properties of clusters of refractory metal atoms, in particular those of the catalytically active transition metals. For these purposes we have built and tested two sources suitable for generation of cluster beams of refractory metals, one for continuous beams and the other for pulsed beams.

The continuous source combines a very high temperature vaporization oven (operated up to 2000°C) with a cryogenically cooled (to -189°C) quench cell. A scaled drawing appears in Figure 1. The metal sample is usually contained in a graphite or tungsten basket suspended in a vertically mounted, resistively heated graphite tube surrounded by heat shields and a water-cooled box. Inert gas is flowed down through the graphite tube, the volatized metal vapor being entrained in the gas and carried out through an exit aperture. Measured oven operating temperatures correspond to metal vapor pressures of 0.1 to 1.0 torr. 2 to 8 mmol of metal can be placed in the source, so that steady beams can be maintained for many hours. The metal-carrier gas mixture that exits the source enters an 8.25 cm long liquid N_2-cooled copper quench cell, at the beginning of which is a manifold of 8 annular holes which directs cold quench gas downstream into

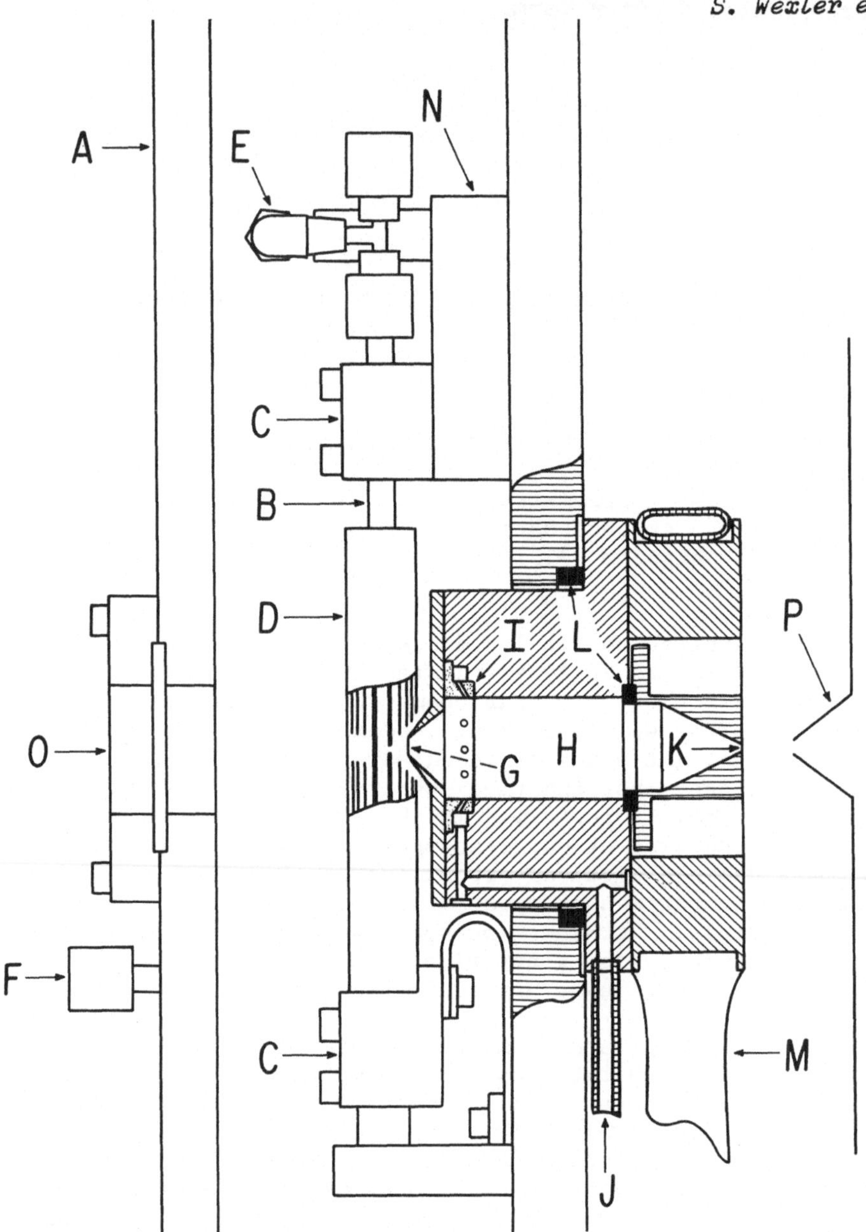

Figure 1. Scale drawing of the continuous cluster beam source. A, copper box (water cooling not shown); B, graphite tube; C, high current clamps; D, heat shield assembly; E, nozzle gas inlet; F, box gas inlet; G, quench cell entrance aperture; H, quench cell; I, quench gas manifold; J, quench gas inlet; K, quench cell exit aperture (heater not shown); L, Teflon gaskets; M, Liquid-N_2 inlet; N, Kel-F insulator; O, view port; P, skimmer separating source chamber from differential pumping chamber.

the cell. The cluster beam exits the quench cell and then passes through four successive differentially pumped stages to reach a time-of-flight mass spectrometer, whose ionization region is 45.5 cm from the exit of the quench cell. In an earlier configuration, an electron impact ionization quadrupole mass spectrometer was used in the detector chamber. In a typical experiment, Ar passes through the graphite tube at a 50 sccm flow rate and N_2 is added to the quench cell at 250 sccm, which produces a quench cell pressure of 15 to 20 torr, and successive chamber pressures of $1.2x10^{-3}$, $1x10^{-4}$, $4x10^{-7}$ and $1x10^{-7}$ torr.

With the above arrangement continuous cluster beams of such relatively refractory metals as Al, Cr, Ni, Cu and Ag have been generated. The observed distributions of cluster ions produced by ionization of the neutral clusters by ∿10 eV electrons and quadrupole mass analysis are shown in Fig. 2 for Ag, Al, and Ni. These distributions can be considered only qualitative, because of the unknown relative ionization cross sections and the possible fragmentation of cluster ions. Cluster beams of Cu and Cr have also been generated with this source, but analyzed by multiphoton laser ionization and time-of-flight mass analysis. The Cu_n^+ distribution extends up to Cu_{12}^+ with the intensities of Cu^+, Cu_2^+ and Cu_3^+ being ∿50 times greater than those of the higher multimers. However, with Cr only several percent of Cr_2^+ in addition to the predominant Cr^+ has been observed. A measurement of the chromium flux gave an equivalent Cr atom density of $10^8/cm^3$ in the detector region.

In addition to the continuous cluster source, a laser evaporation source similar to that of Smalley and collaborators (1) has been developed for generating pulsed refractory metal cluster beams. In this source a 0.63 cm rod of the metal of interest is the target of a focussed beam of 249 nm photons from a pulsed (20 Hz) KrF excimer laser (∿100 mj/pulse). The rod extends through a channel (at 90°) carrying a high pressure of He carrier gas, and the evaporation photons pass through a quartz window before striking the target rod. The He and the pulse of evaporated metal exit the channel through a 1.5 mm aperture. Although the source will shortly be fitted with a pulsed valve, continuous gas flow at a pressure of ∿0.1 atm was maintained for the results described here. The diagnostic laser is triggered from the evaporating laser but delayed by the appropriate flight time.

With this source replacing the thermal quench continuous source in the same apparatus, a pulsed beam of Cu clusters has been generated, as indicated by multiphoton ionization-produced Cu_n^+ species with n up to 10. The intensities of Cu^+, Cu_2^+ and Cu_3^+ are each at least two orders of magnitude higher than those

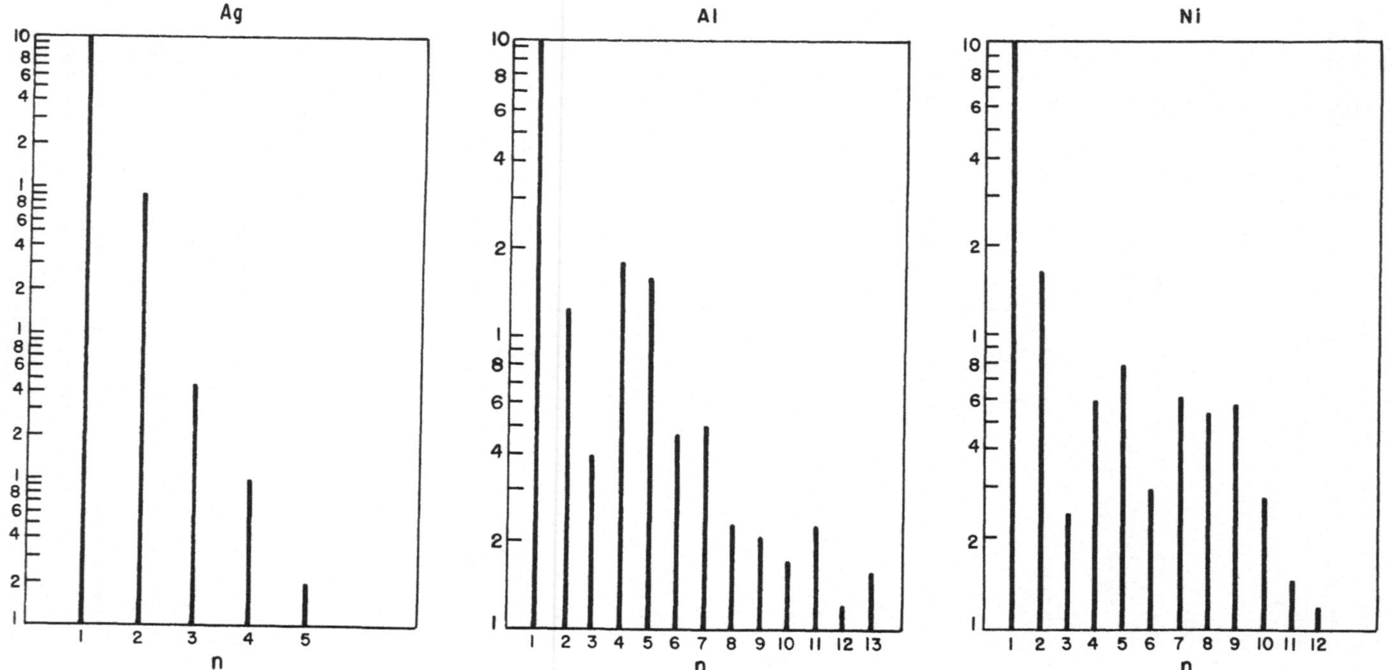

Figure 2. Cluster distributions in Ag, Al, and Ni beams as measured by low energy electron impact ionization and quadrupole mass spectrometry. The ordinates are relative intensities and n is the number of monomer units in each cluster of the distribution. Nozzle temperatures were 1295, 1465, and 1780°C for Ag, Al, and Ni, respectively. Corresponding nozzle flows (Ar) were 25, 35, and 50 sccm, and quench flows (N_2) were 300, 250, and 250 sccm.

of the higher multimers, while the intensity of the Cu_2^+ ion itself is at least an order of magnitude higher than that measured from the continuous thermal quench source. In addition to Cu cluster ions, small yields of copper oxide ions, e.g. CuO^+, Cu_2O^+, Cu_3O^+, have been observed from both sources as well as CuO_2^+, $Cu_2O_2^+$, $Cu_3O_2^+$, $Cu_4O_2^+$ and $Cu_5O_2^+$ from the pulsed source. These oxides are believed to be due to small concentrations of oxygen in the copper target.

Acknowledgment

*Work performed under the auspices of the Office of Basic Energy Sciences, Division of Chemical Sciences, U. S. Department of Energy, under Contract W-31-109-Eng-38.

References

1. Dietz, T. G., Duncan, M.A., Powers, D.E. and Smalley, R.E. Journal of Chemical Physics 74, 6511 (1981).

AN INTRODUCTION TO THE FIELD OF CATALYSIS BY MOLECULAR CLUSTERS

J.M. BASSET

Institut de Recherches sur la Catalyse
2, Avenue Albert Einstein - 69626 Villeurbanne Cédex (France)

SUMMARY

The purpose of this lecture is not to make an extensive survey of the field of molecular clusters in coordination chemistry, field which is expanding very rapidly in various directions, but to present a short introduction of the field of catalysis by clusters for the scientific community of surface science. This community is not always aware of the progresses in molecular chemistry and more specifically in molecular clusters which are at the boarder line between solid state and molecular state. The examples which will be given will be taken from recent reviews in the field of molecular clusters (1-5) in the field of catalysis by molecular clusters (6-7) as well as in the field of surface organometallic chemistry (8-14). For a deeper approach of the considered fields it is suggested to refer to those specialized review articles.

1 AN INTRODUCTION TO MOLECULAR CLUSTERS

1.1 Definition Molecular metal clusters represent a very large family of compounds in which some elements form a framework structure that has a polyhedral form or can be considered a fragment of a polyhedron. In these polyhedra or polyhedral fragments, the framework elements are generally bonded to each other in a multicenter form that is the molecular orbitals that contribute largely to the framework bonding, typically involve orbitals from more than two framework atoms.

The various classes of clusters include (5) :

a. Polyhedral boranes (plus eventually elements of groups III, IV, V and transition metals.)

b. Lithium clusters which is a much smaller class. Be and Mg tend to form chain like structures.

c. Naked clusters class which contains no other element or ligands (they essentially include post transition metals or metalloidal elements like Ge, Sn, Pb, P, As, Sb, Se, Te, Hg.

d. Transition metal clusters (plus Cu, Ag, Au) (vide infra).

e. Molecular metal oxides and metal sulfide clusters in which both the metal and the O or S atoms comprise the framework structure (early transition only).

f. A diverse class of clusters may be generated by expansion of gaseous molecules or atoms through a high speed nozzle. It is possible to generate in the gas phase Mx neutral molecules with M a metallic element but Mx will typically have a range of integer x values.

g. Hypothetical Mx clusters represent a much investigated area by quantum mechanical calculations.

In this lecture we shall consider only Transition Metal clusters.

Transition metal clusters have the general composition where :

- x is an integer : the established values are 3, 4, 5, 6, 7, 8, 9, 10, 12, 13, 15, 17, 18, 19, 26, 30, 38.

- y is an integer never less than x and typically is 2 to 3 times larger than x.

- L is a ligand that may be a molecule like CO, NO, R_3P, etc... or may be a radical like H, halide, alkyl, etc...

- The formal charge Z of the cluster may be zero or may be positive or negative (anionic clusters are more common than cationic clusters in the carbonyl subclass).

- Depending upon cluster size there may be one or more atoms enclosed or encaged by the cluster polyhedron. If the enclosed atoms are non metallic, e.g. H, C, N, P, As, these clusters are refered to as interstitial complexes. With large clusters a transition metal atom can be enclosed or many transition metal atoms can be enclosed. Those clusters represent a fragment of close packed metallic arrays.

The following figures will illustrate examples of molecular clusters of increasing nuclearity and (or) complexity which correspond to the above mentionned definitions :

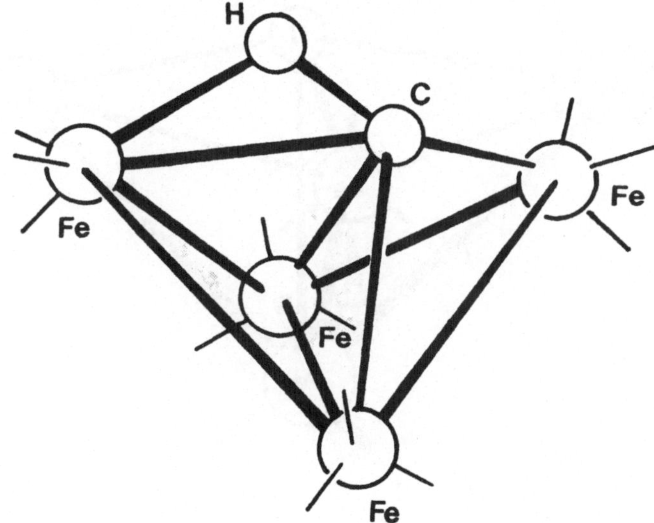

Figure 1 :Xray structure of $HFe_4(\eta^2\text{- CH})\ (CO)_{12}$ according to ref (7)

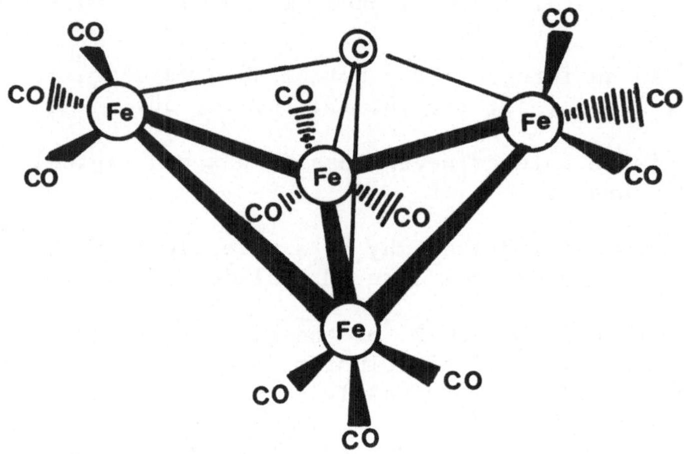

Figure 2 : X ray structure of $Fe_4C(CO)_{12}$ according to 7.

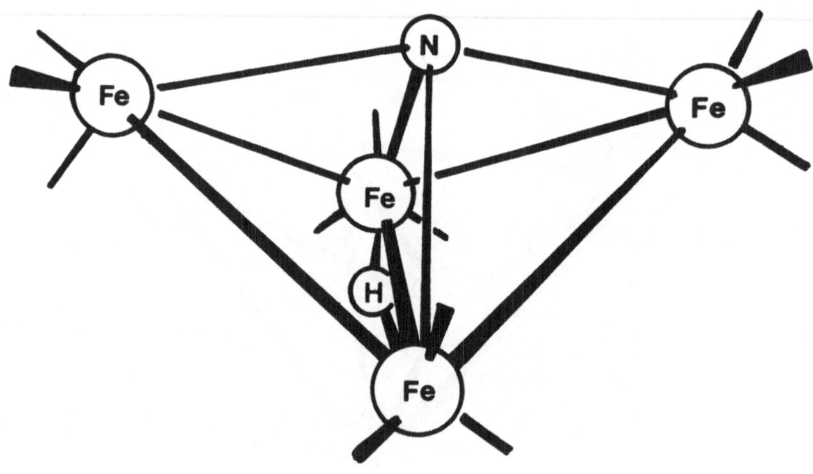

Figure 3 : X ray structure of $HFe_4N(CO)_{12}$

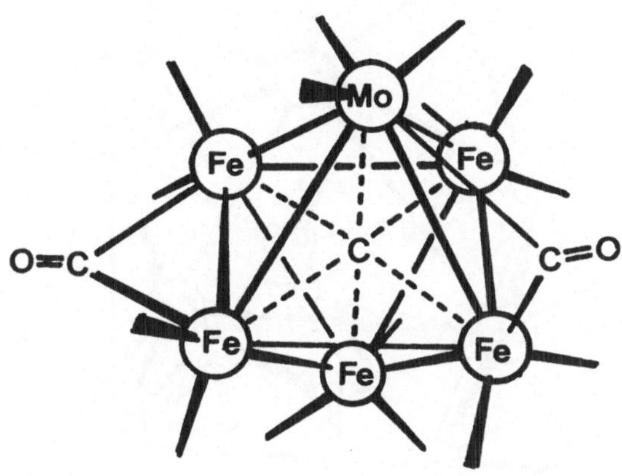

Figure 4 : Molecular structure of $[MoFe_5(CO)_{17}]^{2-}$ (according to ref. 7).

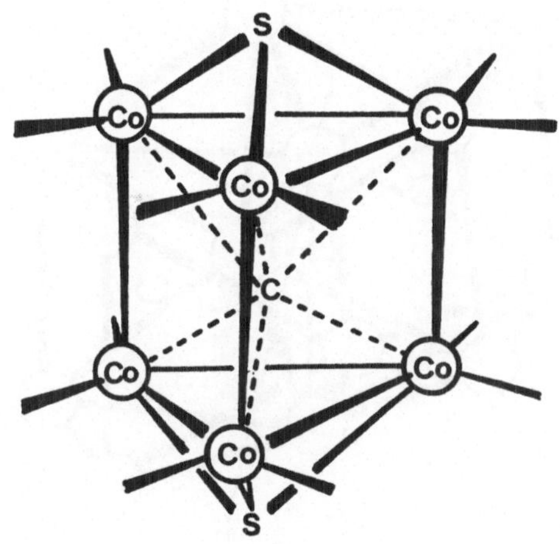

Figure 5 : Molecular structure of $Co_6C(CO)_{12}S_2$ (according to ref. 7).

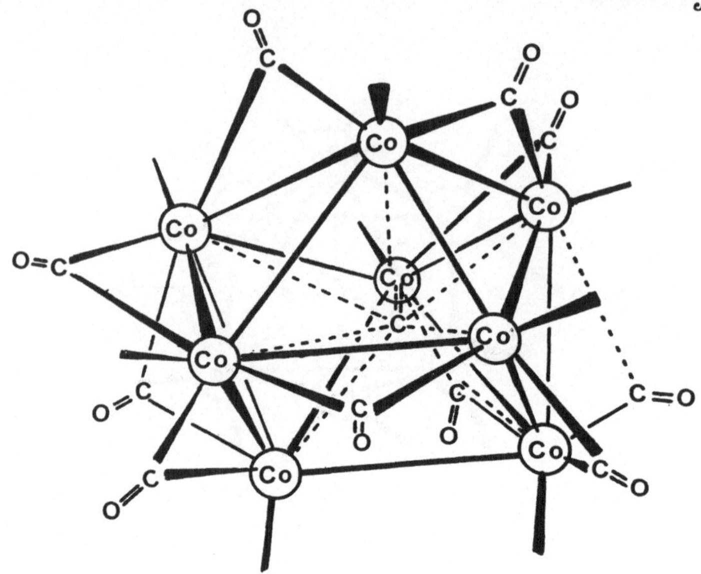

Figure 6 : Molecular structure of $[Co_8C(CO)_{18}]^{2-}$ according to ref. 7.

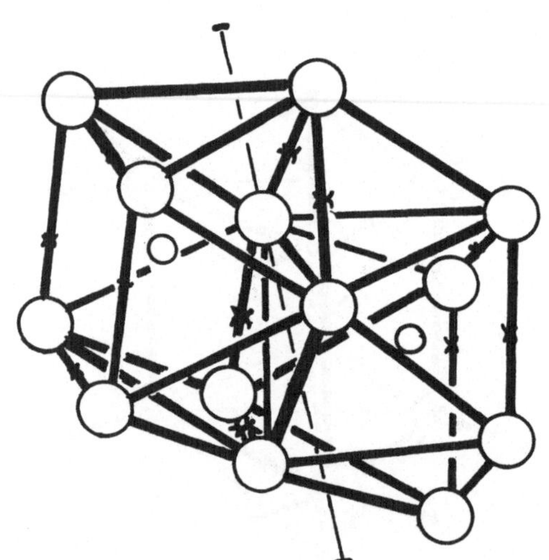

Figure 7 : Molecular structure of $[Co_{13}(CO)_{12}(\mu\text{-}CO)_{12}C_2H]^{n-}$ according to ref. (1).

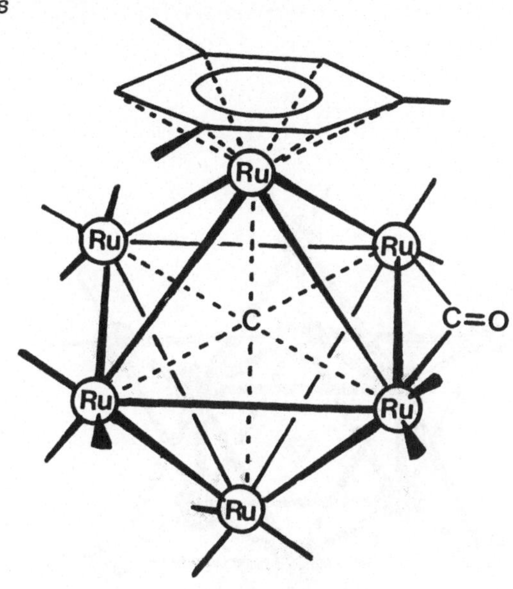

Figure 8 : Schematic representation of $Ru_6\ C(CO)_{14}$ (mesitylene) (according to ref. (7)).

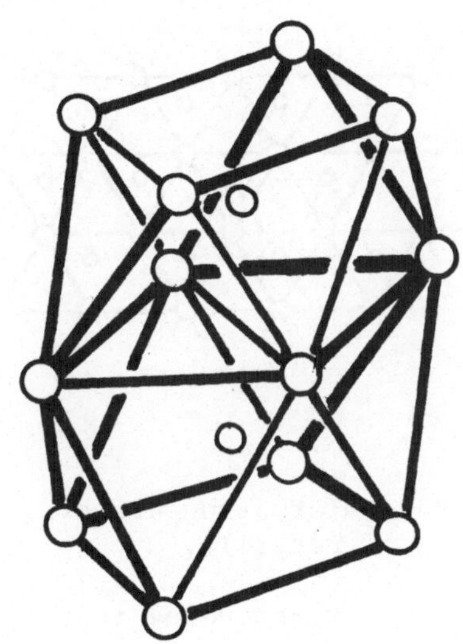

Figure 9 : Schematic representation of $[Rh_{12}(CO)_{16}(\mu\text{-}CO)_8C_2]^{2-}$ (according to ref. (1)).

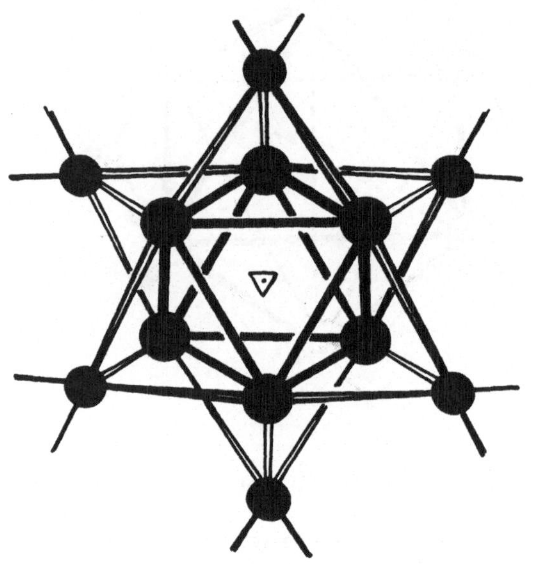

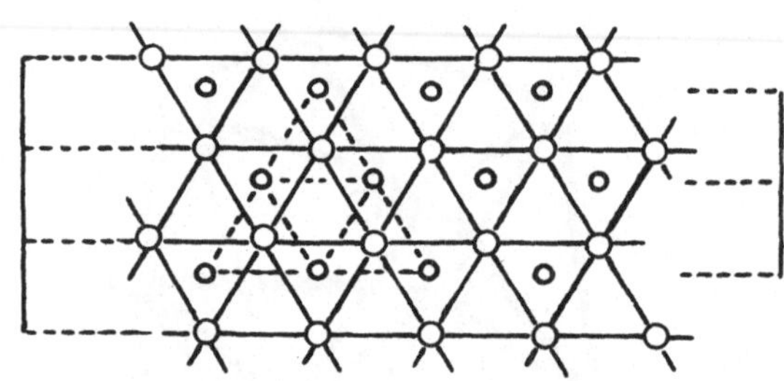

Figure 10 : Above : Molecular structure of $[Fe_6Pd_6(CO)_{24}H]^{3-}$ (according to ref.(1)).

Below : The corresponding representation of two layers in M'_3M'' alloy such as Cu_3Au (ccp) and Ni_3Sn (hcp) (see ref.(1)).

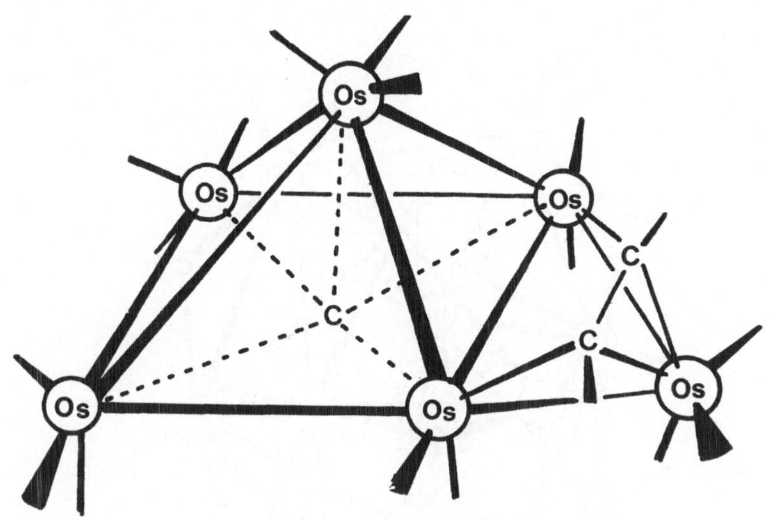

Figure 11 : Molecular structure of $Os_6C(CO)_{16}(CH_3C_2CH_3)$ according to ref. (7).

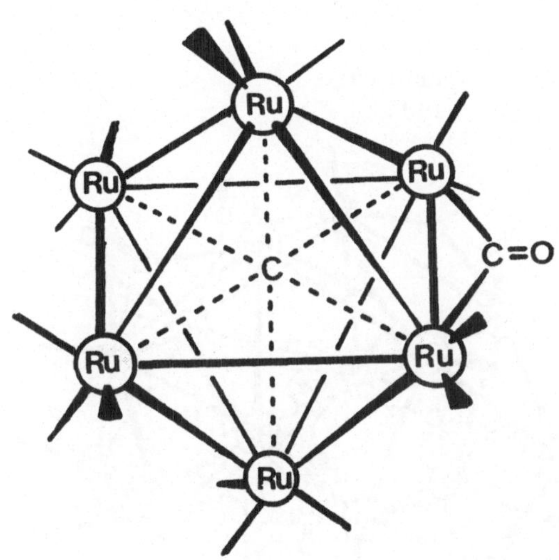

Figure 12 : Molecular structure of $Ru_6C(CO)_{17}$ (according to ref. (7)).

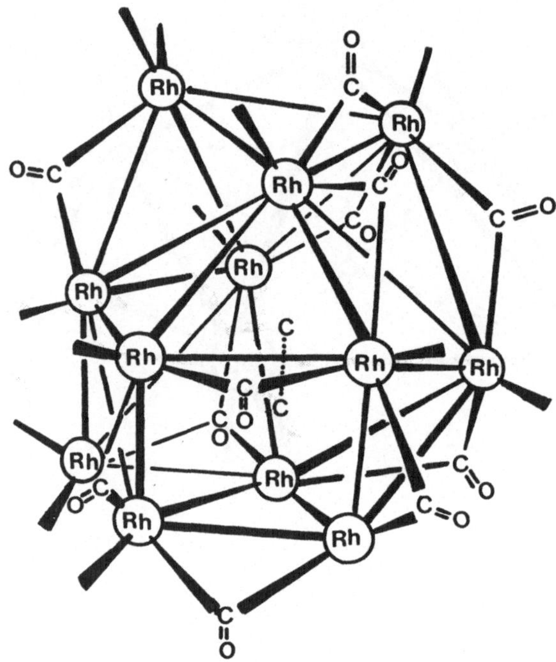

Figure 13 : Molecular structure of $Rh_{12}C_2(CO)_{25}$ (see ref. (1)).

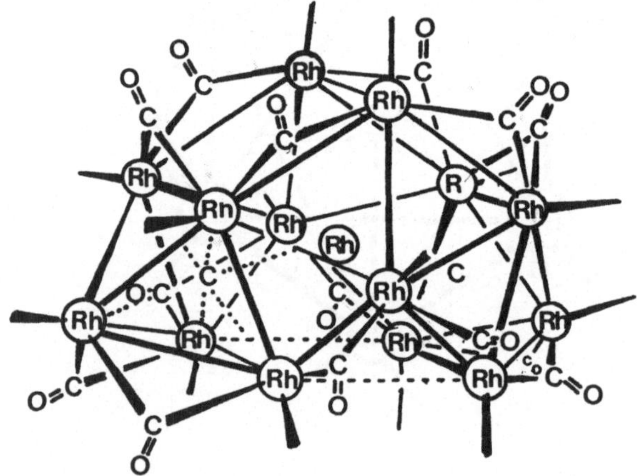

Figure 14 : Molecular structure of $[Rh_{15}C_2(CO)_{28}]^-$ (see ref. (1)).

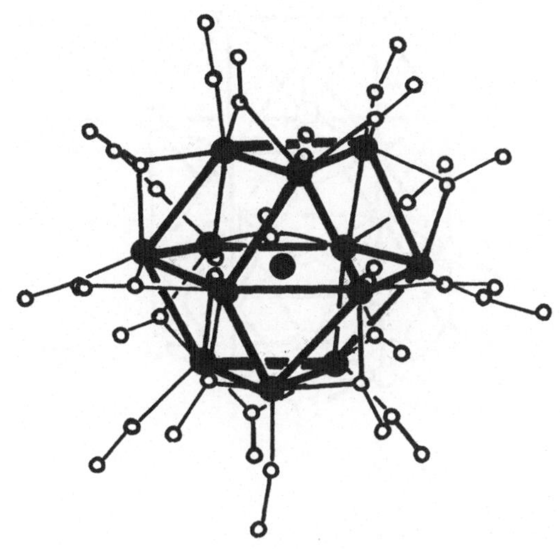

Figure 15 : X ray structure of $\left[H_{5-n}Rh_{13}(CO)_{24}\right]^{n-}$ (n=2,3,4) (anticubo-octahedral metal skeleton (ref. (1))).

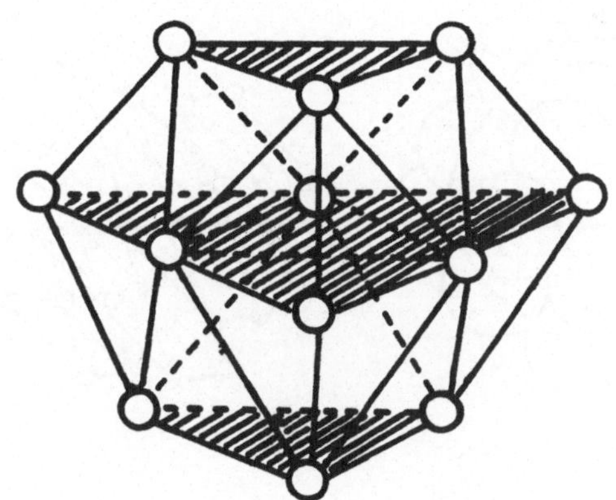

Figure 16 : X ray structure of $\left[Ni_{12}(CO)_{2}H_{4-n}\right]^{h-}$ (n=2,3,4) (ref. (1)).

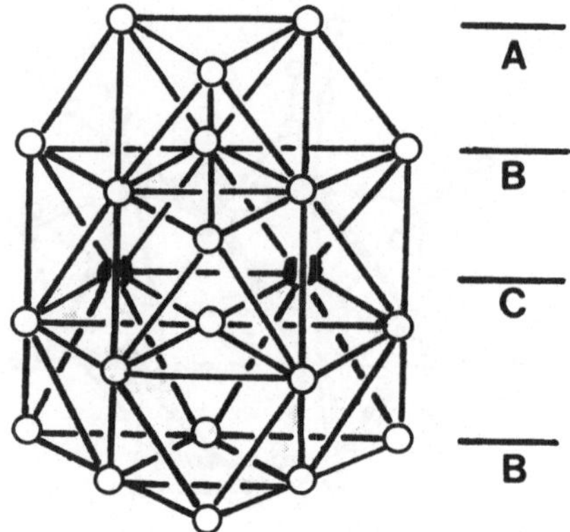

Figure 17 : X ray structure of $[Rh_{22}(CO)_{12}(\mu_2\text{-}CO)_{18}(\mu_3\text{-}CO)_7]^{4-}$ (according to ref. (1)).

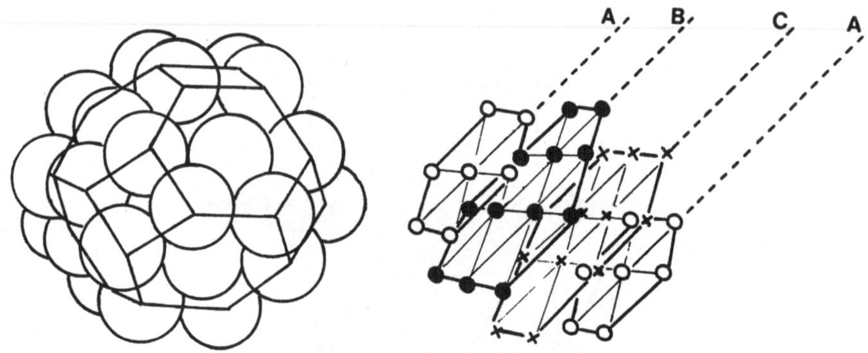

Figure 18 : The truncated octahedron of platinum atoms found in the $[Pt_{38}(CO)_{44}]^{2-}$ and the sequence of cpp layers (ref. (1)).

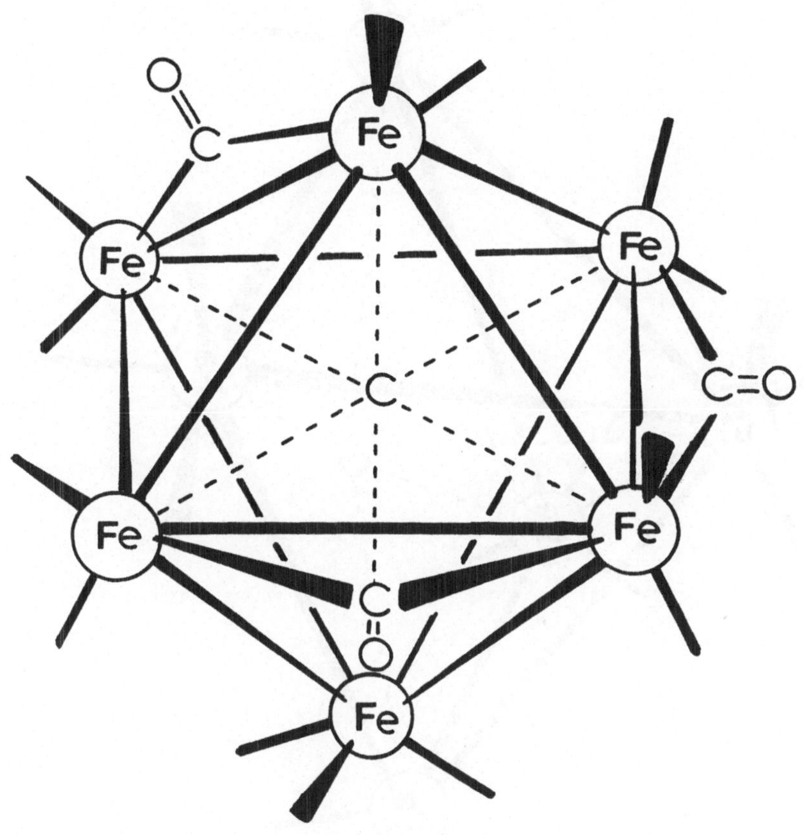

Figure 19 : Schematic representation of $[Fe_6\ C\ (CO)_{16}]^{2-}$ (according to ref. (7)).

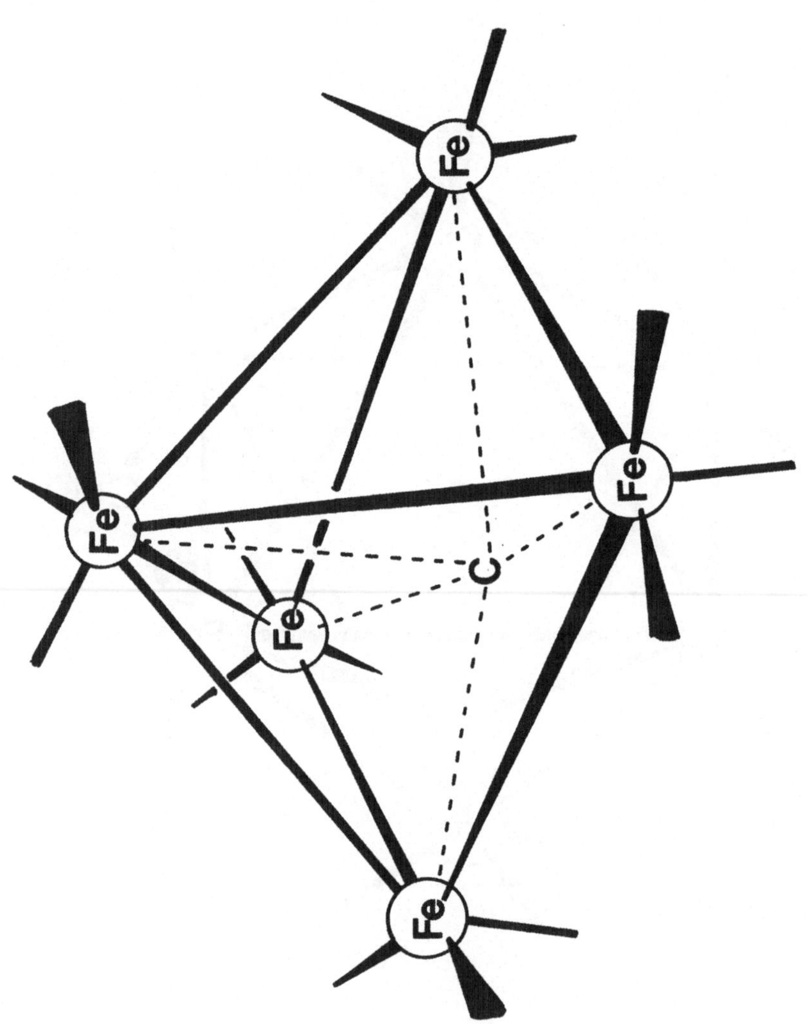

Figure 20 : Schematic representation of $Fe_5C(CO)_{15}$ according to ref.(7.)

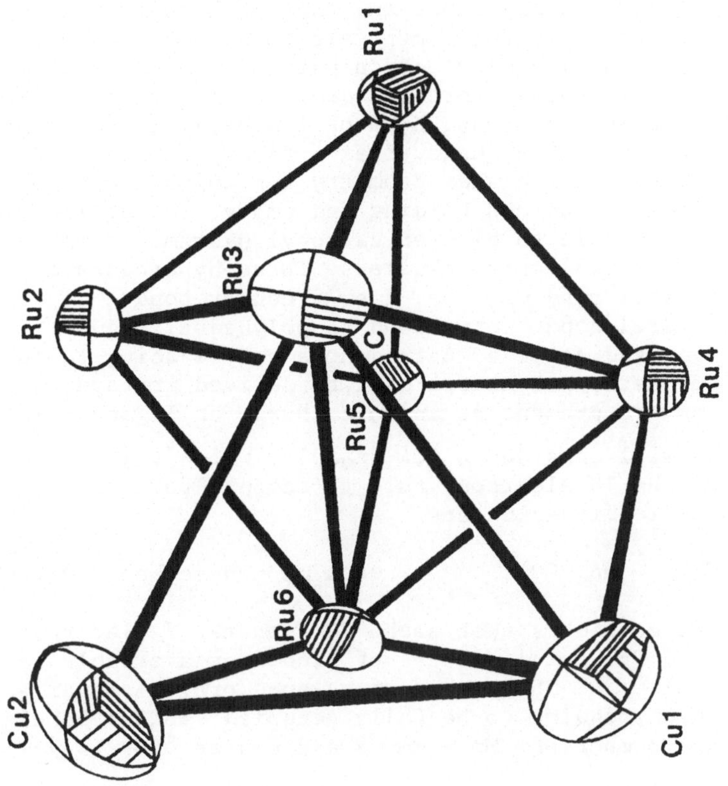

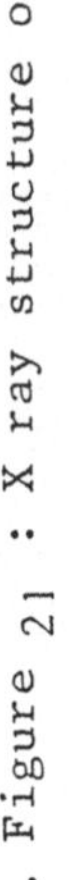
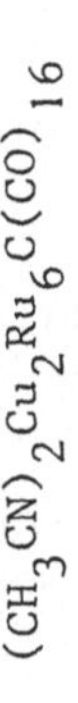
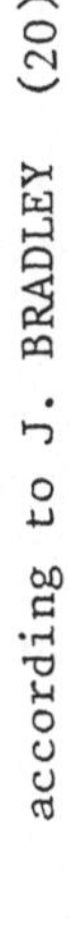

A. Figure 21 : X ray structure of $(CH_3CN)_2Cu_2Ru_6C(CO)_{16}$ according to J. BRADLEY (20).

1.2 Bonding in molecular clusters

1.2.1 The metal-metal bond in clusters   Two main approaches have been used to rationalize metal-metal bonding in transition metal carbonyl clusters : The 18 electrons rule and the skeletal electron pair theory (S.E.P.) (For a complete survey of this field it is necessary to consider the review of K. Wade (3) or B.F.G. Johnson (2)). In this lecture we will consider only the 18 electrons rule which accounts for the geometry of the metallic frame of simple clusters of small nuclearity. In this rule one assumes that the skeletal atoms are held together by a network of two electron pair/ two center (2e/2 center) bonds and that each individual cluster atom utilizes its nine atomic orbitals to accomodate both metal valence electrons and ligand electron pairs and also to form two electrons metal-metal bonds. Therefore each transition metal can accomodate a total of 18 electrons. Most of low oxidation diamagnetic complexes obey this rule. However exceptions occur especially with d^8 metal ions with square planar geometry for which the high lying p_z orbital is found to be non bonding and empty. The application of the 18 electrons rule to cluster carbonyl systems is in general restricted to small nuclearity clusters. Three hypotheses are made :

- The metal-metal bond will be 2 e /2 center bond
- The metal-metal bond correspond to polyhedral edges
- Ligands are used as 2 e pair donors only leading to the view that the same metal polyhedron will be derived irrespective of wheter electrons are present as anionic charge or ligand pairs.

We will consider a few examples which illustrate the application of the 18 electron rule to account for the geometry of simple low nuclearity clusters.

a. In $Mn_2(CO)_{10}$, each manganese has 7 valence electrons and one considers that each CO donates 2 electrons to the cluster frame. The total number of electrons available is (2x7) + (10x2) = 34. The number of valence orbitals available is 9x2 = 18, which require to be fully occupied 18x2 = 36 electrons. Therefore the number of Mn - Mn 2 e/2 center bond is 1

b. $Ru_3(CO)_{12}$: Each ruthenium has eight valence electrons.

1. Total number of electrons	=	48
2. Valence orbitals available	=	27
3. Required number of electrons to fill 27 A.O	=	54
Difference item 3 - item 1	=	6

Therefore the number of Ru - Ru 2e/2 center bond = 6/2 = 3 which corresponds to a triangular geometry.

c. $Rh_4(CO)_{12}$: Each cobalt has nine valence electrons

1. Total number of electrons :(9x4) + (2x12) = 60
2. Valence atomic orbitals available : 9x4 = 36
3. Required number of electrons to fill 36 A.O. = 72

   Difference item 3 - item 1 = 12

Therefore the number of Co-Co 2e/2 center bonds =12/2=6 which corresponds to a tetrahedral arrangement.

d. $Os_5(CO)_{16}$: Each Osmium has eight valence electrons

1. Total number of electrons : (8x5) + (16x2) = 72
2. Valence orbitals available : 9x5 = 45
3. Required number of electrons to fill 45 A.O. = 90

   Difference item 3 - item 1 = 18

It follows that the number of Os-Os $2e^-/2$ center bonds is equal to 9 corresponding to a trigonal bipyramidal (9 edges) unit.

The 18 electrons rule may also be applied to electron rich or electron poor clusters which are of great interest in catalytic processes.

$H_2Os_3(CO)_{12}$

1. Total number of electrons : (3x8) + (12x2) + (2x1)= 50
2. Valence orbitals available : 9x3 = 27
3. Required number of electrons to fill 27 A.O. = 54

   Difference item 3 - item 1 = 4

Therefore the number of Os-Os $2e^-/2$ center bonds = 2 corresponding to a linear or open triangle geometry.

$H_2Os_3(CO)_{10}$

1. Total number of electrons : (3x8) + (10x2) + (2x1)= 46
2. Valence orbitals available : = 27
3. Required number of electrons to fill 27 A.O. = 54

   Difference item 3 - item 1 = 8

As a consequence the number of Os-Os $2e^-/2$ center bonds is 4.

The cluster may be represented as a triangle with two Os-Os single bond and one Os = Os double bond (Figure 22).

4. Triply bridging (M_3) between three metal atoms the CO axis being perpendicular or nearly perpendicular to the M_3 plane.

O
C
M M
M

5. Triply bridging (M_3) between three metal atoms while coordinating di-hapto to a fourth.

M·····O
C
M M
M

- When a CO molecule is coordinated to a single metal atom, the metal-carbon distance is shorter than a metal-carbon distance of a metal-alkyl bond. Besides the C-O distance is longer than in free CO. These observations can be rationalized in terms of resonance between the canonical forms $M = C = O$ and $\overset{-}{M} - \overset{+}{C} = \overset{+}{O}$ indicating on M-C bond order between 1 and 2 and a C-O bond order beetwen 2 and 3. The more commonly bonding description is represented on Figure. The lone pair electrons on carbon can be donated to a suitable vacant metal orbital (e.g. the d_{z^2} A.O.) while the ligand metal electron drift is compensated by a back donation from metal to ligand (e.g. from filled metal d_{xz} and d_{yz} A.O. into the ligand π^* orbitals). (Figure 23).

A mono-hapto doubly bridging carbonyl still functions as a 2 electrons ligand : the lone pair orbital on carbon can overlap with a suitable in phase combination of filled metal orbitals, while an out of phase combination of filled metal orbitals can interact with the CO orbital (Figure 23). The net result is to reduce the CO bond order to about 2.

A doubly bridging di-hapto carbonyl ligand is an unusual type of bonding. Nevertheless it can be observed in the dinuclear manganese complex $Mn_2(CO)_5(Ph_2PCH_2PPh_2)_2$. In this complex a carbonyl ligand is terminally bound to one manganese atom while coordinating η^2 to the other. As a result this carbonyl ligand behaves as a poor electron ligand. (Figure 23).

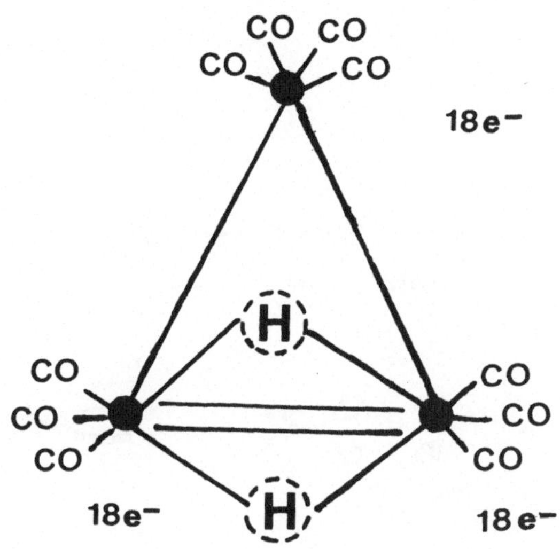

Figure 22 : $H_2Os_3(CO)_{12}$ structure deduced from the application of the 18 electrons rule.

1.2.2 The metal-ligand bond We shall consider here only the CO ligand which is the most commonly observed ligand in low valent cluster compounds. The main ways in which a carbonyl ligand can coordinate to one or more of the metal atoms of a cluster are as fellows (3) :

1. Terminal attachement to one metal atom by a linear or nearly linear M - CO unit. (Usually the angle MCO ranges between 165 and 180°) : M - C = O

2. Doubly bridging (M_2) between two metal atoms coordinating exclusively through the carbon atom (η'), the CO axis beeing perpendicular or nearly perpendicular to the M - M axis.

O
C
M — M

3. Doubly bridging in such a way as to imply a di-hapto coordination to one of the metal atoms with some ... O bonding.

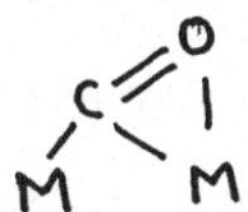

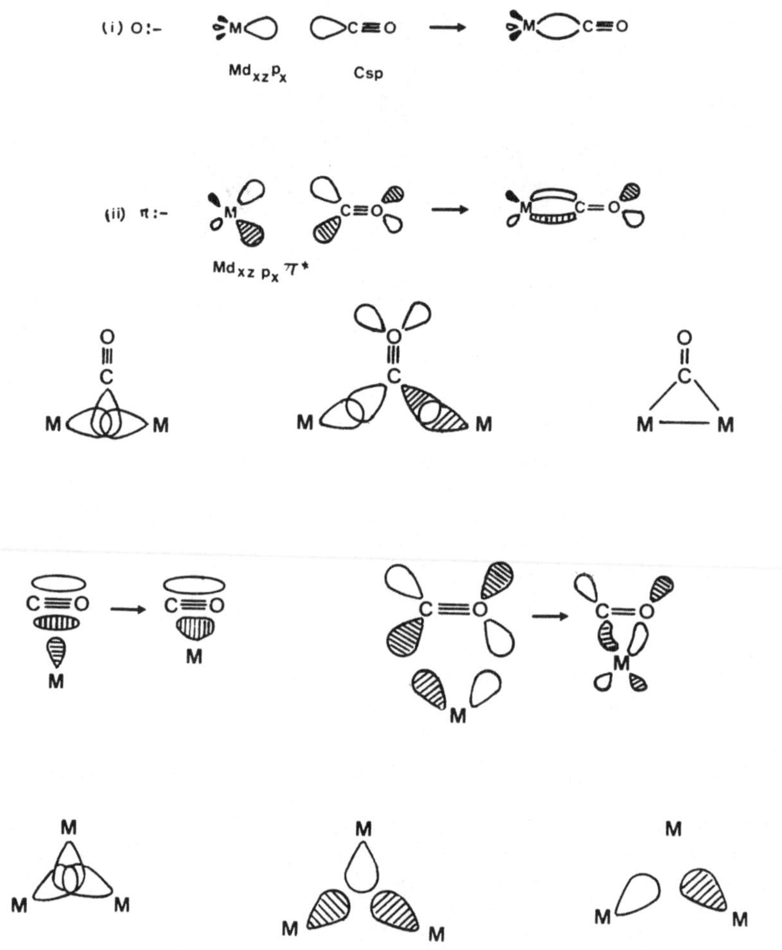

Figure 23 : Molecular orbitals involved in CO bonding.

A. Linear CO ; B. μ_2 bridging CO ; C. η^2-μ_2 bridging CO ; D. $\eta^1\mu_3$ bridging (according to W. WADE (ref. 3)).

In a triply bridging carbonyl ligand the lone pair orbital on carbon can combine with a suitable in phase combination of metal orbitals. Similarly to the other cases a Π back-donation from suitable E symmetry combinations of metal orbitals into the Π^* orbitals of CO. The metal ligand bonding can thus be regarded as involving three pairs of electrons, enough to allow the alternation description of three metal-carbon single bonds holding the carbon atom to the metal triangle, with the CO bond order formally reduced to one (Figure 23).

1.3. Dynamic behaviour of molecular clusters In a molecular cluster the migration of metal atoms or ligands over or inside the cluster frame is an area of very intensive investigation. Not only this kind of studies try to elucidate the mechanistic pathways of these fluxional processes in cluster chemistry but also it gives new concepts in surface science about the mobility of metal atoms as well as chemisorbed molecules on or inside a metallic lattice Three different classes of fluxional processes have been observed in molecular clusters (2, 7) :

- Ligand migration over the cluster surface.
- arrangement of the metal cluster polyhedron.
- Ligand migration within the metallic cluster unit.

1.3.1 Ligand migration over the cluster surface These kind of studies have been restricted to carbon monoxide ligands, hydride ligands and some organic ligands as isocyanides. With carbon monoxide the ligand migration may occur between :

- two metal centres
- three metal centres
- four metal centres

In Figure 24 is represented the pairwise exchange mechanism in which ligand bridges are opened and closed in a pairwise fashion. In bridged dinuclear complexes this process involves the formation of a non bridged intermediate. This intermediate may be of sufficient life time to allow rotation about the metal-metal bond prior to bridge reformation to bring about cis-trans isomerisation.

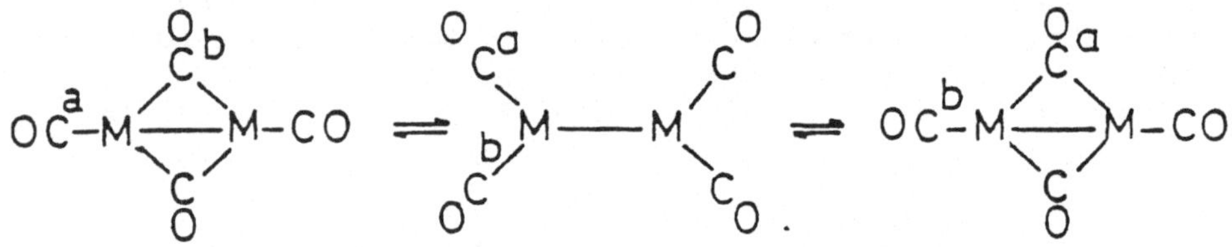

Figure 24 : The pairwise exchange mechanism according to B.F.G. Johnson (ref. 3).

The simplest mechanism for CO migration in a tetranuclear carbonyl is represented on Figure 25. This process involves a $C_{3v} \rightleftharpoons$ Td interconversion of the $M_4(CO)_{12}$ clusters. There is a simple bridge-break, bridge make mechanism of the type employed in binuclear complexes.

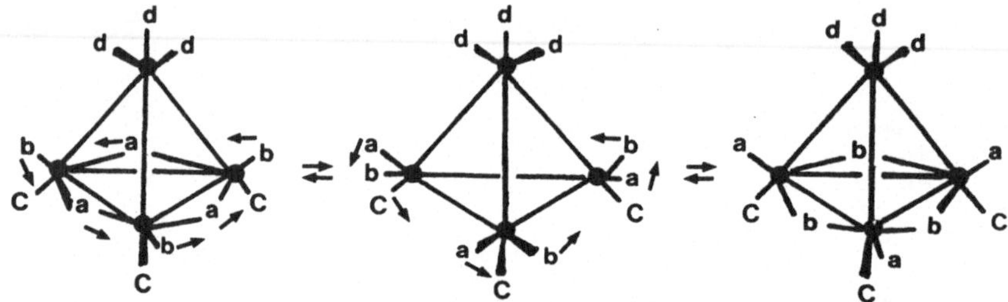

Figure 25 : The interconversion of the C_{3v} and T_d forms of the tetrametal dodecacarbonyls according to E. Muetterties (ref. 7).

The cluster $Ir_4(CO)_{11}$ PPh_2Me exhibits three distinct fluxional processes. The process of lowest activation energy is a cyclic permutation of the three bridging and three terminal carbonyl ligands about the basal plane. At higher temperatures all carbonyls but one undergo exchange. Finally at higher temperatures all eleven carbonyls equilibrate :

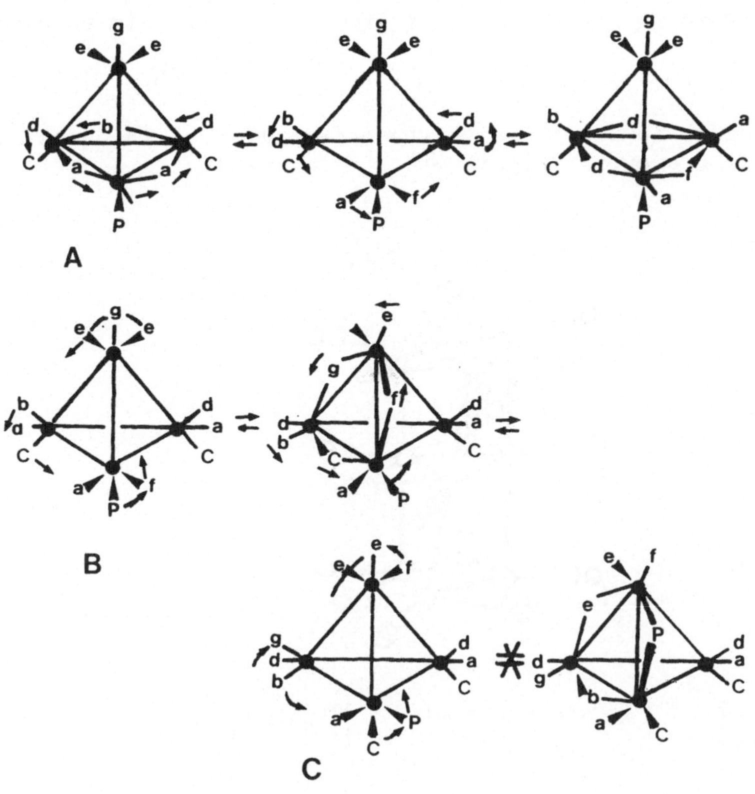

Figure 26 : Fluxional behaviour of $Ir_4(CO)_{11}PPh_2Me$.

1.3.2 Structural rearrangement within the metal core. Two types of such structural rearrangement occur which may be relevant to surface science :

- rearrangement with retention of the ligand arrangement.
- rearrangement with simultaneous ligand mobility. In the cluster $[Pt_9(CO)_{18}]^{2-}$ (structure shown below), it has been shown by ^{195}Pt NMR spectroscopy that the outer Pt_3 triangles rotate

about the pseudo three fold axis with respect to the inner one. The larger dianions $[Pt_9(CO)_{18}]^{2-}$ and $[Pt_{12}(CO)_{24}]^{2-}$ have been shown to exchange $Pt_3(CO)_6$ units in solution. With the mixed metal cluster $H_2FeRuOs_2(CO)_{13}$ the CO and H migration occur in concert with the deformation of the quasi tetrahedral metal framework (Figure 28).

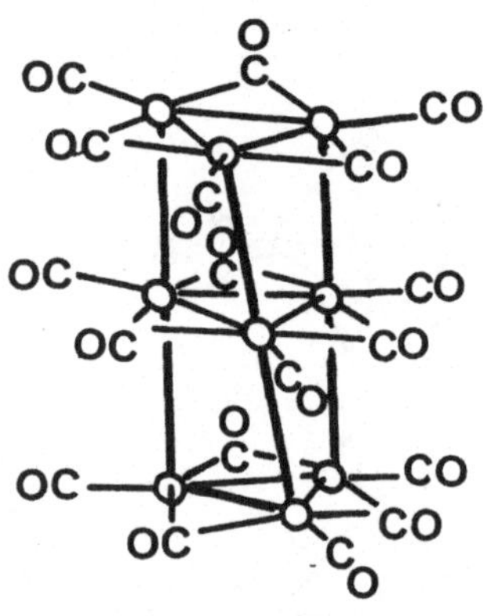

Figure 27 : X ray structure of $[Pt_9(CO)_{18}]^{2-}$

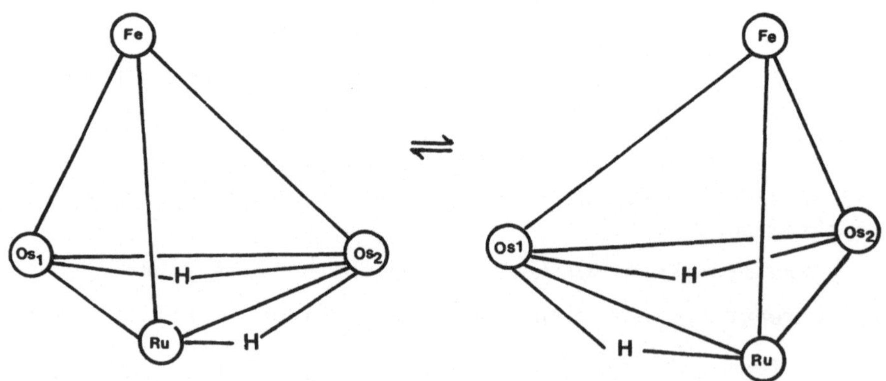

Figure 28: Skeletal rearrangement in $H_2FeRuOs_2(CO)_{13}$.

1.3.3 Ligand migration within the metal cluster unit Hydride ligands are able to occupy interstitial sites inside the metallic clusters. The carbonyl hydride anions

$\left[H_2Rh_{13}(CO)_{24}\right]^{3-}$ and $\left[H_3Rh_{13}(CO)_{24}\right]^{2-}$ have an anti-cubo octahedral metal skeleton with the thirteenth rhodium atom occupy the interstitial site ; this structure represents a fragment of hexagonal close packing. In these clusters the hydride ligands migrate about the whole cluster and couple to all thirteen rhodium nuclei. In this migration process the hydride occupy both octahedral and tetrahedral interstitial sites.

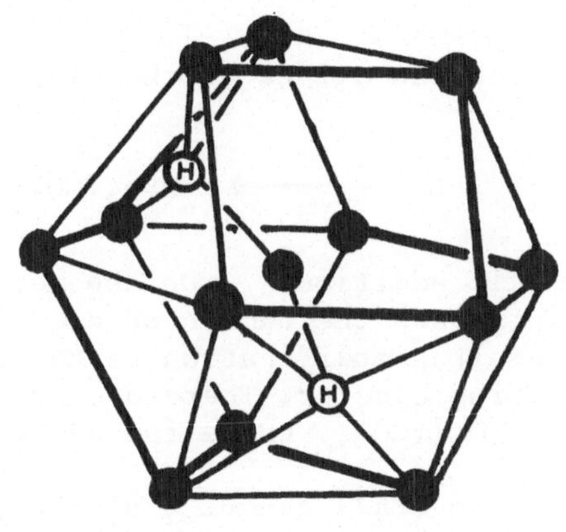

Figure 29: X ray structure of $\left[H_2\ Rh_{13}(CO)_{24}\right]^{3-}$: fluxional behaviour of the hydride ligands.

1.4. <u>Reactivity of molecular clusters</u> A recent review Deeming (4) has considered in details and in a very comprehensive way, the large field of the reactivity of molecular clusters. One can classify reactions of molecular clusters in two areas depending on the fact that the reaction is followed or not by a major modification of the metal skeletton. One can also consider the reactivity in terms of electrophilic or nucleophilic attack, oxidative addition, etc...

1.4.1 Electrophilic attack. The electrophilic attack may occur on the metal center, which is the usual case since the metals have usually the highest electron density and can therefore be protonated. In most cases the resulting hydrido ligand will occupy a bridging position :

$$\left[Re_3(CO)_{12}\right]^{3-} \underset{-H^+}{\overset{+H^+}{\rightleftharpoons}} \left[HRe_3(CO)_{12}\right]^{2-} \underset{-H^-}{\overset{+H^+}{\rightleftharpoons}} \left[H_2Re_3(CO)_{12}\right]^{-}$$

$$\rightleftharpoons H_3Re_3(CO)_{12}$$

In some cases the electrophilic attack can occur at the oxygen atom of a bridging carbonyl. Thus $AlBr_3$ may react wíth a linear carbonyl ligand of $Ru_3(CO)_{12}$ which becomes bridged :

$$Ru_3(CO)_{12} + AlBr_3 \longrightarrow Ru_3(CO)_{11}\ CO \longmapsto AlBr_3$$

Similarly a proton can attack at the oxygen atom of a bridging carbonyl :

$$\left[HFe_3(CO)_{11}\right]^{-} + H^{+} \longrightarrow HFe_3(CO)_{10}(C\ O\cdots H)$$

1.4.2 Nucleophilic addition On the metallic frame
This type of attack increases the number of electrons associated with the cluster. A a result a modification of the geometry or of the nuclearity of the starting cluster. There are however electron deficient clusters as $H_2Os_3(CO)_{10}$ where the addition of an electron pair does not modify the overall geometry of the cluster. This cluster will therefore be able to be a catalyst (vide infra) for olefin isomerisation and (or) hydrogenation. On Figure we have given two examples of nucleophilic addition with and without modification of the overall geometry of the cluster.

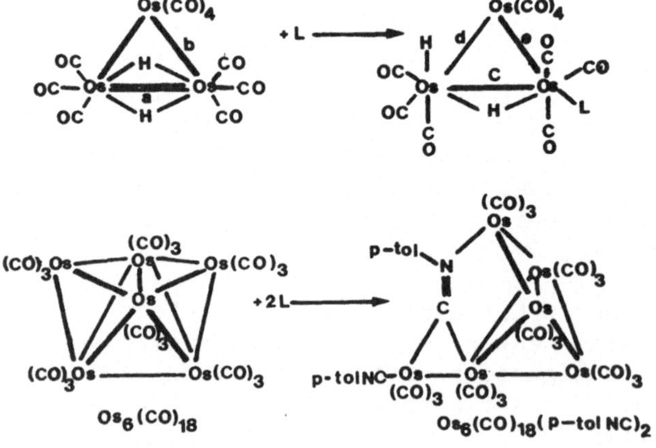

Figure 30 : Nucleophilic addition to $H_2Os_3(CO)_{10}$ (up) and $Os_6(CO)_{18}$ (down).

1.4.3 Nucleophilic attack at the ligands. It usually occurs at the carbon atom of a linear carbonyl ligand. For example the nucleophilic attack of an OH^- group on a coordinated CO gives rise, via β-H elimination, to the formation of anionic hydrido clusters :

$$M_x(CO)_y + OH^- \longrightarrow M_x(CO)_{y-1}(CO_2H) \longrightarrow M_x(CO)_{y-1}H^- + CO_2$$

This kind of attack is frequently encountered in water gas shift reactions catalyzed by molecular clusters.

Another possibility of nucleophilic attack may occur with carbido cluster where the carbonyl can be coordinated to the "surface" carbon atom. Typical examples are given on the Figure where the nucleophile can be an alcool, an amine or an hydride.

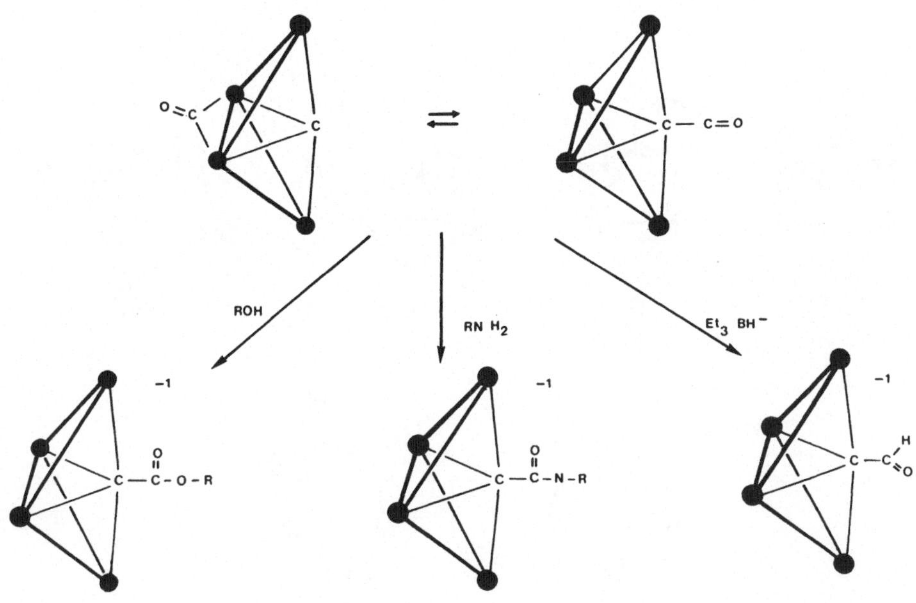

Figure 31 : examples of nucleophilic attack on CO coordinated to an exposed "surface" carbon (according to J.Bradley (20)).

1.4.4 Oxidative addition With mononuclear complexes one can define the oxidative addition of an X-Y molecule to a transition metal M_T by the simple equation

$$L_n M^n + X - Y \rightleftharpoons L_n N^{n+2} (X) (Y).$$

When applied to molecular clusters oxidative addition may concern more than one metal atom. In fact this kind of activation may concern cleavage of H - H bond, C - H bonds of olefinic or acetylenic compounds and various types of ligands. One of the unique aspects of oxidative addition in clusters is the ability of clusters to cleave bonds of a ligand in a very close vicinity to the original point of attachment of this ligand (Figure 32). Thus, the first oxidative addition of ethylene to an Os_3 cluster is likely to be as in (A) (Figure 32) and that of pyridine as in B. Other mechanism of vinylic C - H activation are represented on Figure . Obviously surfaces probably activate olefins bonds in the same way.

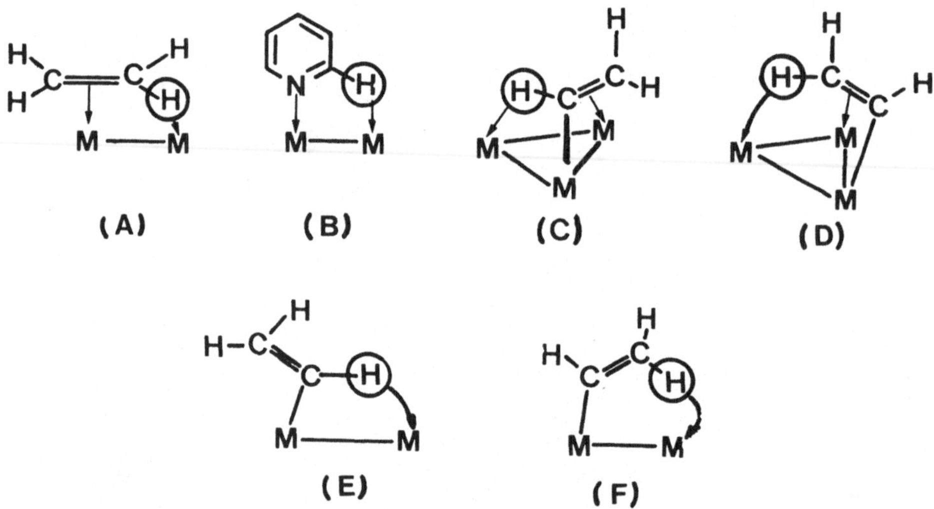

Figure 32 : Oxidative additions with the cluster $Os_3(CO)_{12}$ after Deeming (3).

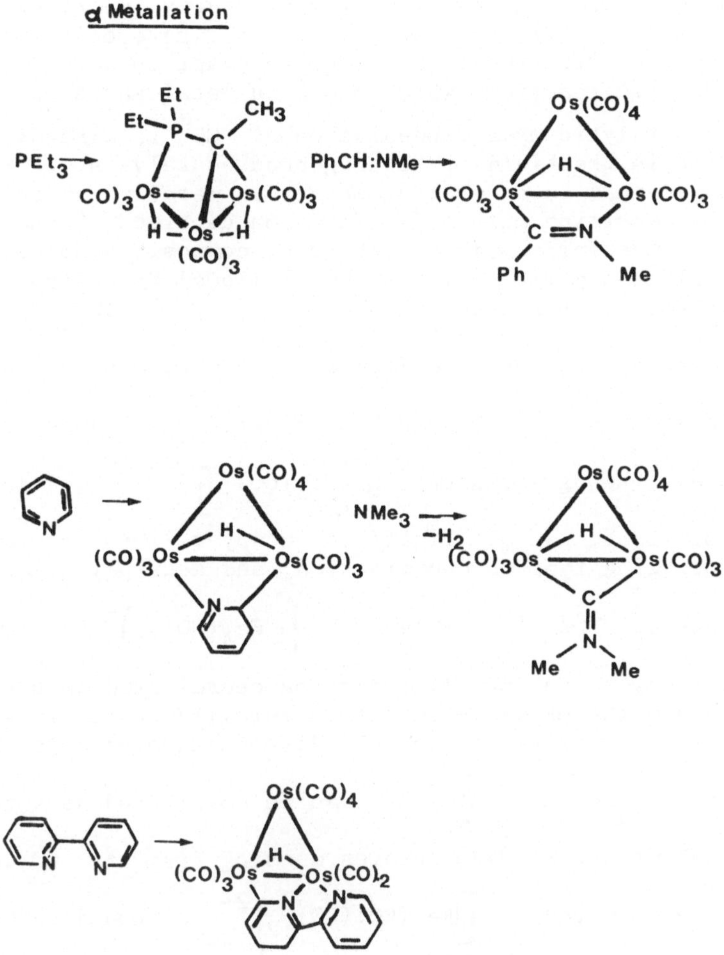

Figure 33 : Activation of C - H or N - H bond by $Os_3(CO)_{12}$ in a close vicinity of the original point of attachment (after Deeming (3)).

1.5. Molecular clusters as structural models of intermediates or chemisorbed species in surface science One of the most important aspect of cluster chemistry is the possible relationship existing between the structure of chemisorbed species and that of ligands coordinated to the metallic cluster frames. There are many examples where the type of bonding between a ligand and a cluster frame is also postulated to occur on a metallic surface. A general review in this field has been written by R. Ugo (8), which describes in a very comprehensive way this very important aspect of surface science. We would like to present here only a few examples of clusters structures which are relevant to surface science and to catalytic reactions which occur on metal surfaces.

The catalytic hydrocondensation of CO is already a very old reaction in the field of heterogeneous catalysis. It is generally assumed that such reaction occurs on metallic surfaces of group VIII transition metals (sometime on carbides of the same metals). The mechanism which is the most commonly admitted, at the moment, implies a dissociation of CO, followed by a stepwise reduction of surface carbon into $\geq C-H$, $>CH_2$, and $-CH_3$ surface species. The Figure 34 illustrates the various intermediates in which CO can be activated and reduced on metallic surfaces. The A cluster $[Fe_4(CO)_{13}]^{2-}$, illustrates the possibility for a CO ligand to occupy a linear coordination, a doubly bridging type of coordination, and a triply bridging type of coordination. The B cluster $[H\,Fe_4(CO)_{13}]^-$ illustrates the possibility for a CO ligand to be coordinated both through the carbon and the oxygen atom onto a metallic frame. This is a kind of precursor state to the CO dissociation process. The C cluster $[Fe_4(CO)_{12}C]^{2-}$ can be considered as a possible model for a surface carbido species arising from the CO dissociation. The D cluster $[HFe_4(CH)(CO)_{12}]^{2-}$ could represent a partially hydrogenated form of surface carbon. The bridging CH_2 ligand in E $(Fe_2(CO)_8(CH_2))$ is probably one of the final state of activation and reduction of CO and many mechanistic paths require either $>CH_2$ coupling or $>CH_2$ reduction to methane :

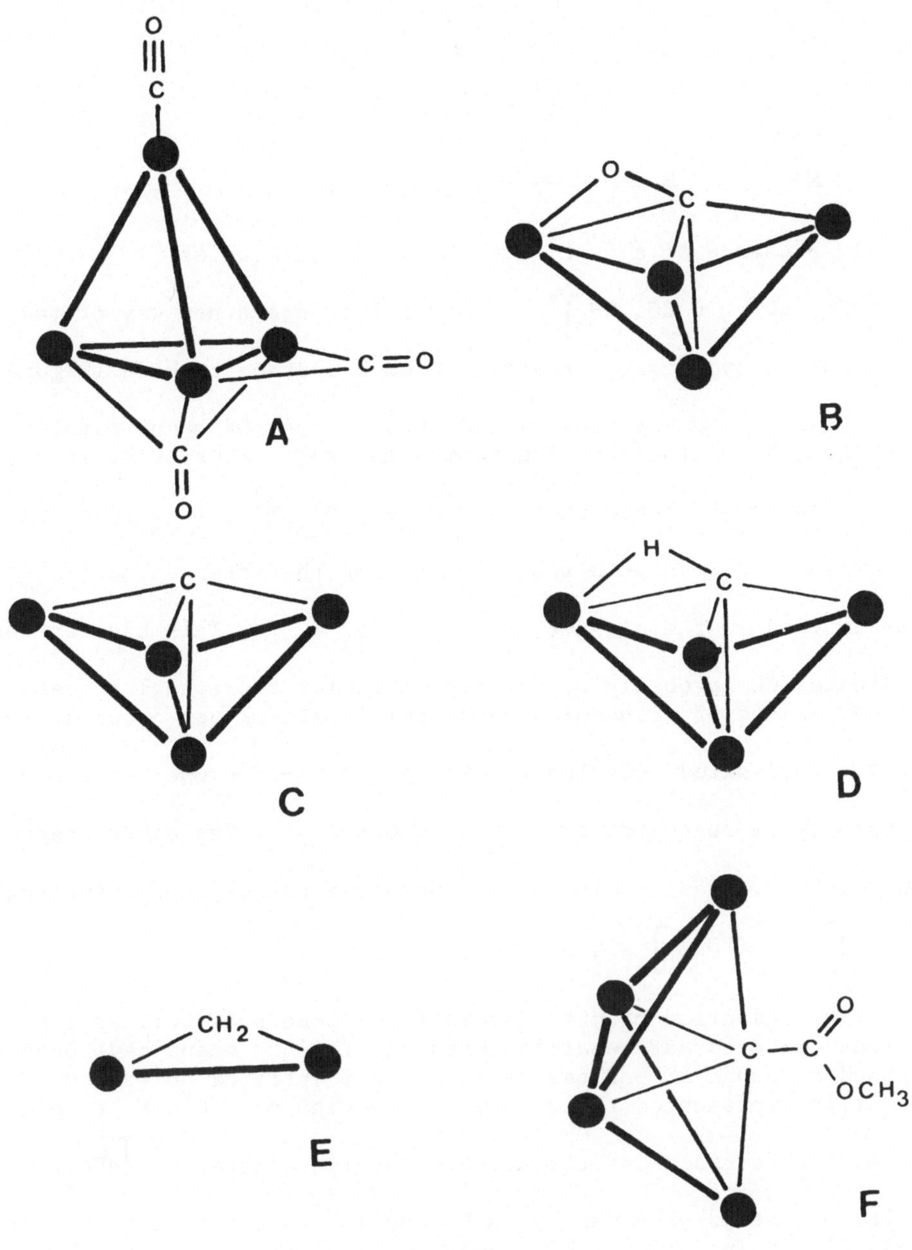

Figure 34 : Iron clusters representing some intermediates postulated in Fischer-Tropsch synthesis.

$$2\ M{-}CH_2{-}M \longrightarrow C_2H_4$$

$$M{-}CH_2{-}M \xrightarrow{H} CH_4$$

Finally presence of a C - CO bond in the cluster

$[HFe_4(CO)_{12}\ C\ CO_2\ CH_3]^-$ may illustrate a new way of making an oxygenated hydrocarbon starting from a surface carbide (Figure 34).

On Figure 35 is represented another type of mechanism for C - C bond formation starting from a molecular cluster having already an internal carbon atom $[Fe_6C\ (CO)_{16}]^{2-}$. By oxidation of such cluster with tropylium cation the internal carbon becomes exposed in the cluster $Fe_4(CO)_{13}\ C$. This cluster can coordinate CO probably by an intramolecular fluxionnal process. In the presence of methanol a nucleophilic attack may occur at this carbide-coordinated CO. The resulting $\geq C - \overset{O}{\overset{\|}{C}} - OMe$ group may then by hydrogenated to $CH_3\ COOCH_3$. The other steps proposed by J. Bradley indicate a possible pathway for returning to $[Fe_6\ C\ (CO)_{16}]^{2-}$

The reduction of nitro-aromatic compounds as well as that of nitriles can occur at metallic surfaces. Various steps have been proposed although no one has been really identified so far. On Figure 35 is represented the stepwise reduction of $C \equiv N$ triple bond at the "surface" of the anionic hydrido cluster $[HFe_3(CO)_{11}]^-$ The various steps observed by H.D. Kaesz are expected to be good models of the intermediatesin these surface reactions. (figure 36)

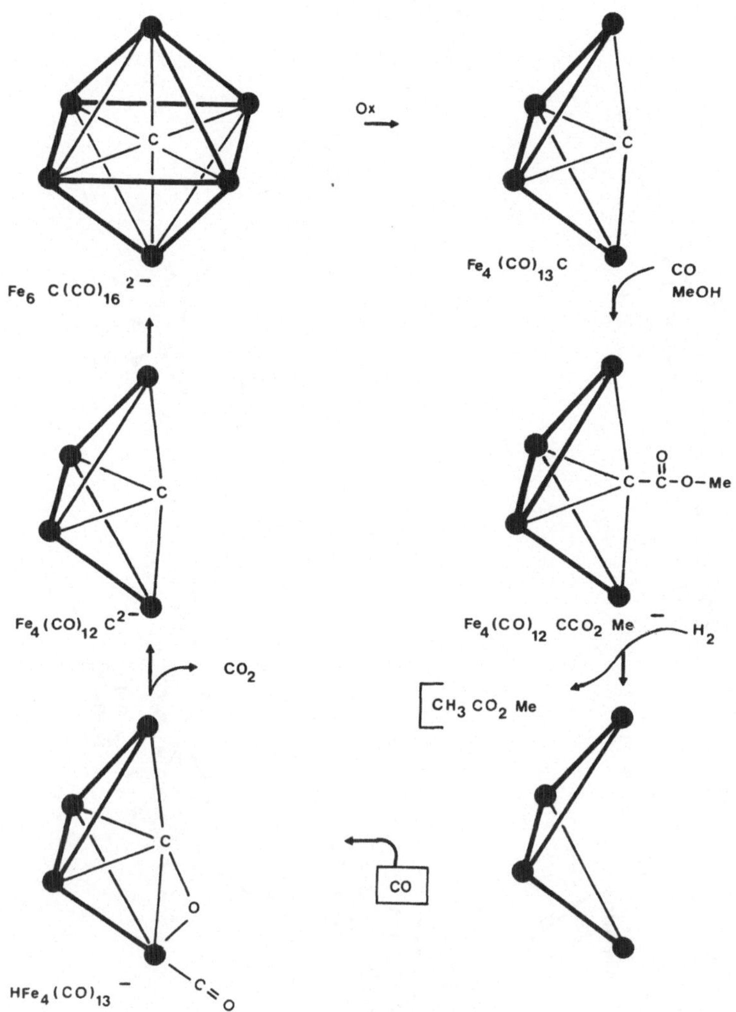

Figure 35 : Various possible steps for making CH_3COOCH_3 from $CO + H_2$ with the cluster $[Fe_6C(CO)_{16}]^{2-}$.

Some steps necessary to return to the starting cluster are hypothetical (according to J. BRADLEY (20).

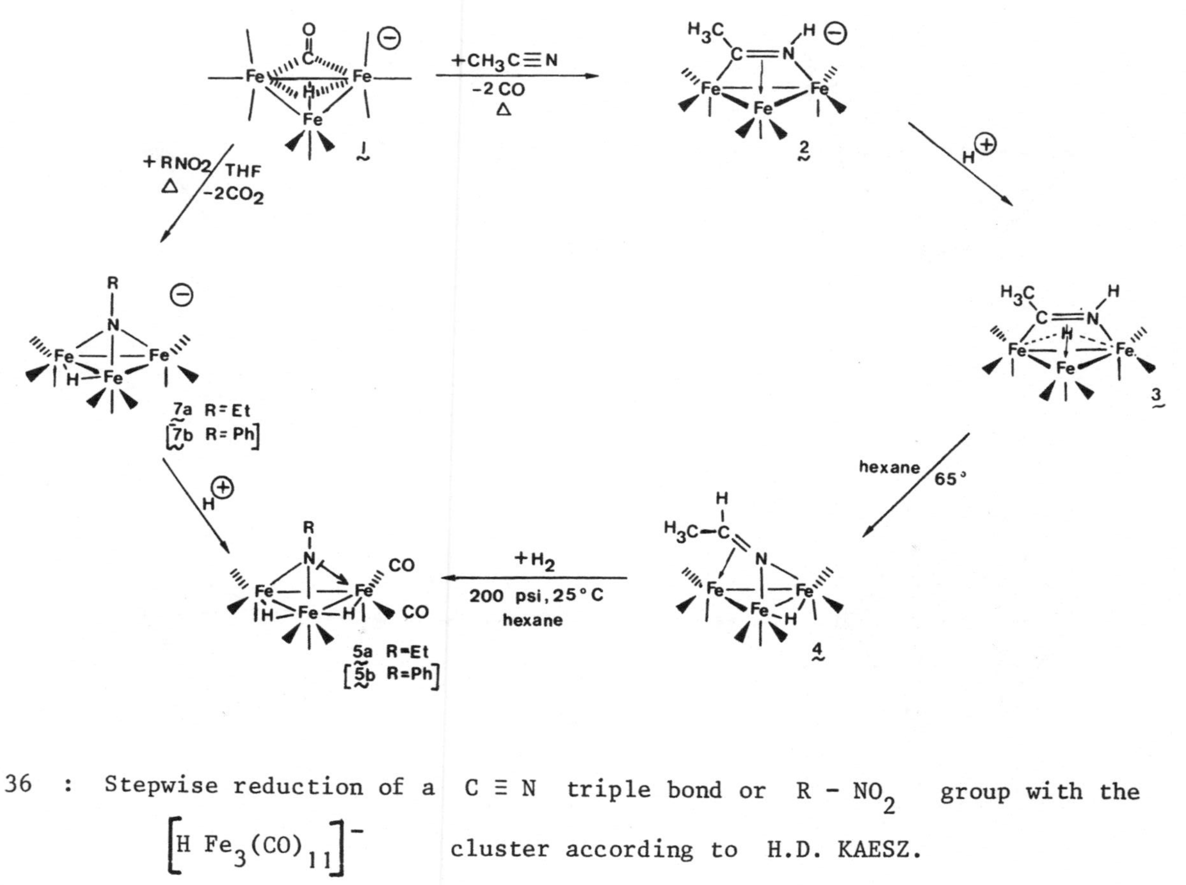

Figure 36 : Stepwise reduction of a C ≡ N triple bond or R - NO_2 group with the $[H\ Fe_3(CO)_{11}]^-$ cluster according to H.D. KAESZ.

2. CATALYSIS BY MOLECULAR CLUSTERS

2.1 The relationship between molecular clusters and small metal particles The most interesting aspect in the study of molecular clusters concerns the frontier situation that they occupy between the molecular state and the metallic state. This frontier situation is also expected in catalysis where the clusters are at the boarder line between molecular catalysis and solid state catalysis. Heterogeneous catalysis on metals is not always well understood. Surfaces are not well defined at a microscopic levels and they contain corner,faces,edges etc... so that the selectivity of a given reaction may depend on the respective amount of such geometric parameters. In the last twenty years homogeneous catalysis has been developped considerably. Usually the catalytic reaction occurs in the coordination sphere of a single transition metal atom surrounded by a variety of well defined ligands which may orientate the reaction in the desired direction depending on the electronic or steric effects of those ligands (Figure 36).

Due to this frontier situation, the molecular cluster presents a considerable interest in catalysis and this for many reasons :

- Its metallic frame presents a geometry very close to that encountered is small metallic particles encaged in the cavities of some zeolithes. A priori one might expect that it will be possible to carry out in a molecular cluster frame the catalytic reaction which occur on a surface but with a better control of the activity and of the selectivity due to the presence of known ligands. The objective is a rather ambitious one since it is well known that the rigidity of the metallic frame of many clusters is weak and the geometry of the metallic frame will depend on the number of electrons brought by the ligands - and on the nature of such ligands -.

- The presence of many metals in a mixed metal cluster offers the possibility of a cooperative effect in reactivity : on may speculate about the possibility of two different metals being responsible for two different types of activations necessary for the overall catalytic reaction.

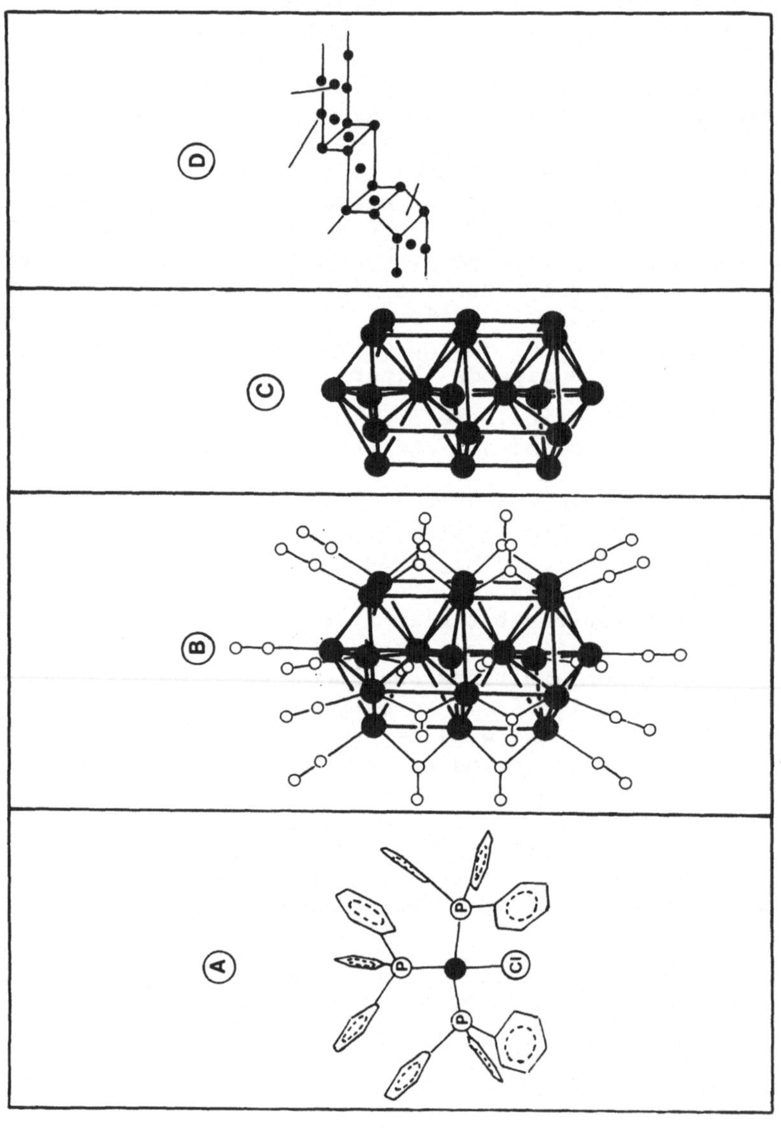

Figure 37 : The molecular clusters at the boarder line between molecular state and metallic state and at the boarder line between homogeneous catalysis and heterogeneous catalysis.

2.2 <u>Homogeneous cluster catalyzed reactions</u> The number of publications related to catalytic reactions using molecular clusters is increasing considerably since the last five years. In 1977, the first review (6) dealing with such aspect of catalysis contained only 50 references. At the moment one can estimate such number to be close to 300. It is not the purpose of this paper to deal with such numerous examples.

Two aspects deserve to be briefly discussed :

- Is there any reaction which can be catalyzed by molecular clusters and which cannot be catalyzed by mononuclear complexes or metal particles ?

- Is there any example of catalytic reaction which occur on a molecular cluster framework ?

Regarding the first question, it is important to mention the works of Union Carbide related to the synthesis of ethylene glycol from syn-gas with Rh or Ru complexes under drastic conditions of pressure (above 500 atm.). With rhodium complexes it has been established that an anionic cluster $[Rh_5(CO)_{15}]^-$ was present in the reaction medium under catalytic conditions.

The same kind of observations was made recently by D. DOMBECK (31) of Union Carbide for the same reaction using ruthenium clusters. In the case of ruthenium there is, under catalytic conditions, an equilibrium between $[H\,Ru_3(CO)_{11}]^-$ and $Ru(CO)_3I_3^-$. An almost complete catalytic cycle has been established by D. DOMBECK. It appears that the hydrido anionic cluster $[H\,Ru_3(CO)_{11}]^-$ makes a nucleophilic attack at CO coordinated to the mononuclear carbonyl Ru^{II} complex to give a formyl species. The reaction, here, would obey a very complex mechanism involving both mononuclear and polynuclear species. This phenomenon seems to be a general rule in many reactions involving CO.

Regarding the second question the number of examples showing unambiguously cluster catalyzed reaction is still rare and is typically of mechanistic character. We would like to give only two examples dealing with the isomerisation of olefins with $H_2Os_3(CO)_{10}$ (Figure 38) and with the water gas shift reaction catalyzed by

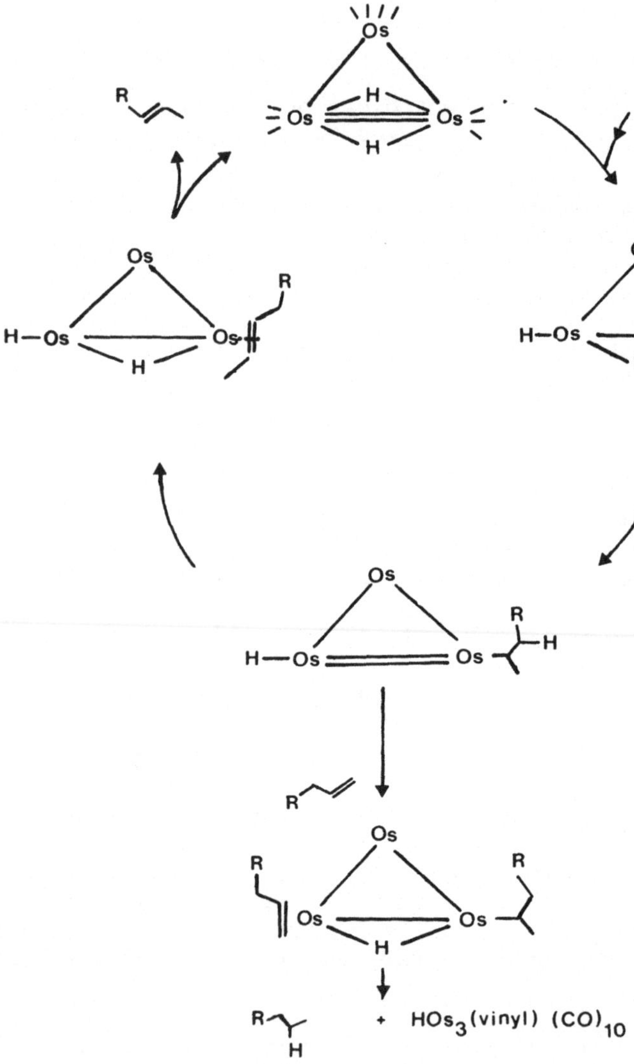

Figure 38 : Isomerization of olefin with $H_2Os_3(CO)_{10}$ according to DEEMING (26).

$Ru_3(CO)_{12}$ (Figure 39). The first reaction, studied in details by DEEMING (23), occurs in the triangle of $H_2Os_3(CO)_{10}$ due to the presence of an electronically unsaturated cluster allowing coordination of the olefin to an osmium atom without loss of a CO ligand. The second reaction of water gas shift, occurs in alcaline solution with $Ru_3(CO)_{12}$ as starting cluster.

Three mechanisms have been postulated by P. FORD which involve in each case anionic hydrides arising from nucleophilic attack of H_2O or OH^- at CO coordinated to the molecular cluster frame.

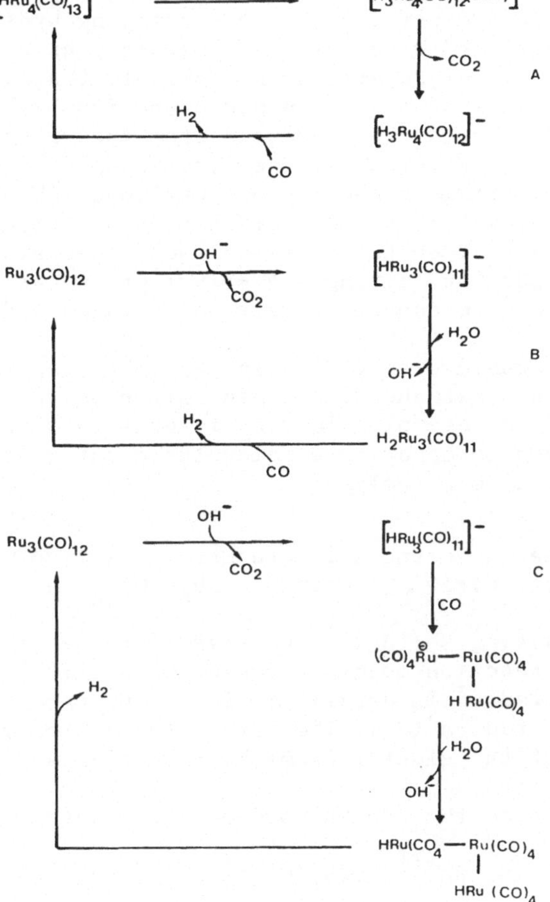

Figure 39 : three possible mechanisms for the water gas shift reaction catalyzed by $Ru_3(CO)_{12}$ according to P. Ford (25)

2.3 Catalysis by supported molecular clusters In the field by supported clusters many cases may occur. In some cases the grafted molecular cluster may remain intact during the complete catalytic cycle. In other cases the molecular cluster may be involved in some steps of the catalytic cycle. Finally, and this is probably the most important aspect of cluster catalysis, the cluster may be decomposed into a small metal particle which is the active species in the catalytic reaction.

2.3.1 The molecular cluster frame remains intact

The reaction of $Os_3(CO)_{12}$ with the silanol groups of silica gives the grafted cluster $(H)\ Os_3(CO)_{10}\ (OSi\leftarrow)$. The grafting occurs by oxidative addition of the silanol group to the Os-Os bond. This grafted cluster contains a bridging hydride and a bridging $3e^-$ oxygen ligand which can be considered as good candidates for giving catalytic properties. Effectively the cluster is a catalyst for the reactions of olefin hydrogenation (27). The mechanism of such reactions has been deduced from spectroscopic as well as kinetic studies (Figure 40). The first step of such mechanism seems to be the opening of one oxygen Os bond which occurs at ca 80°C. This opening favors the coordination of ethylene which is a reversible process. Then the ethylene would reversibly insert into the metal-hydride bond giving a σ-alkyl group. As a result the Os_1 atom would become coordinatively unsaturated and oxidative addition of hydrogen would occur. Finally the last step would be the reductive elimination of ethane with regeneration of the starting cluster. There are other examples in the litterature where catalytic cycles have been shown to occur in a molecular cluster frame supported on an inorganic oxide or polymer.

2.3.2 The supported molecular frame is involved in some steps of the catalytic cycle. When $Rh_6(CO)_{16}$ is supported on alumina, the resulting solid is a catalyst for the water gas shift reaction. The reaction occurs between 25° and 100°C and the mechanism has been studied in details by labelling experiments as well as by infrared studies (28). The first step of the mechanism is the destruction of the cluster frame by oxidative addition of

$>$Al-OH groups to the Rh-Rh bond of the cluster, forming $Rh^I(CO)_2(OAl<)$ and $Rh^{III}(H)(H)(OAl<)$ surface species. The next step is the expulsion of H_2 by molecular CO with for-

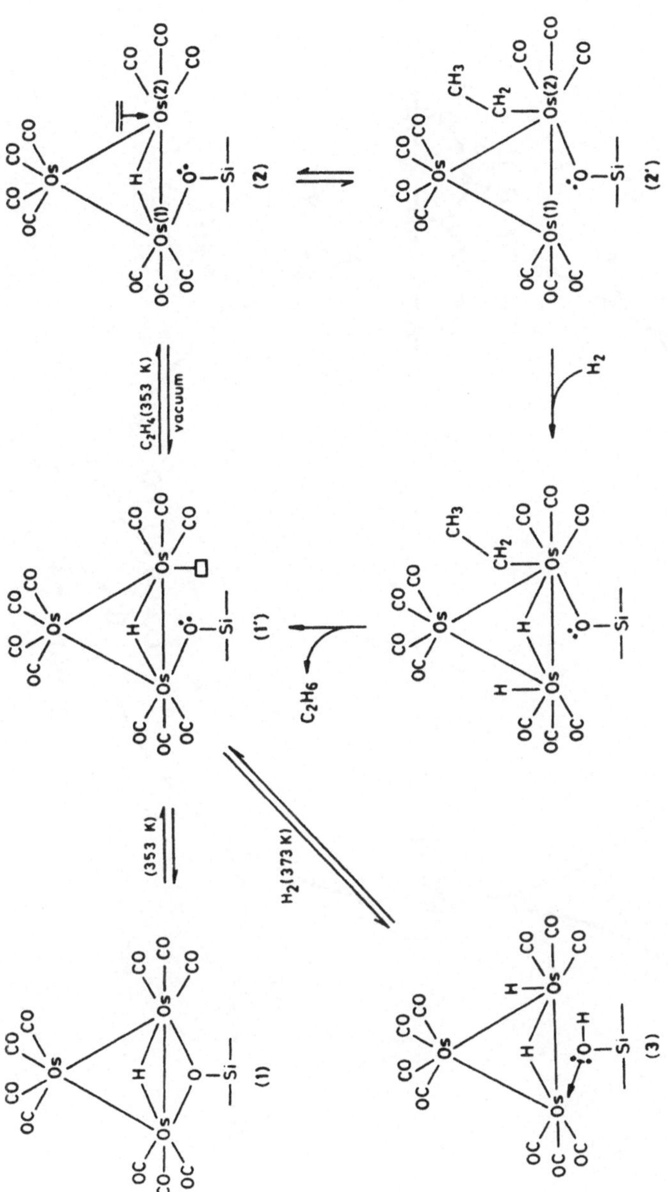

Figure 40 : mechanism for olefin hydrogenation with (H) Os_3 $(CO)_{10}$ (OSi ←) (27)

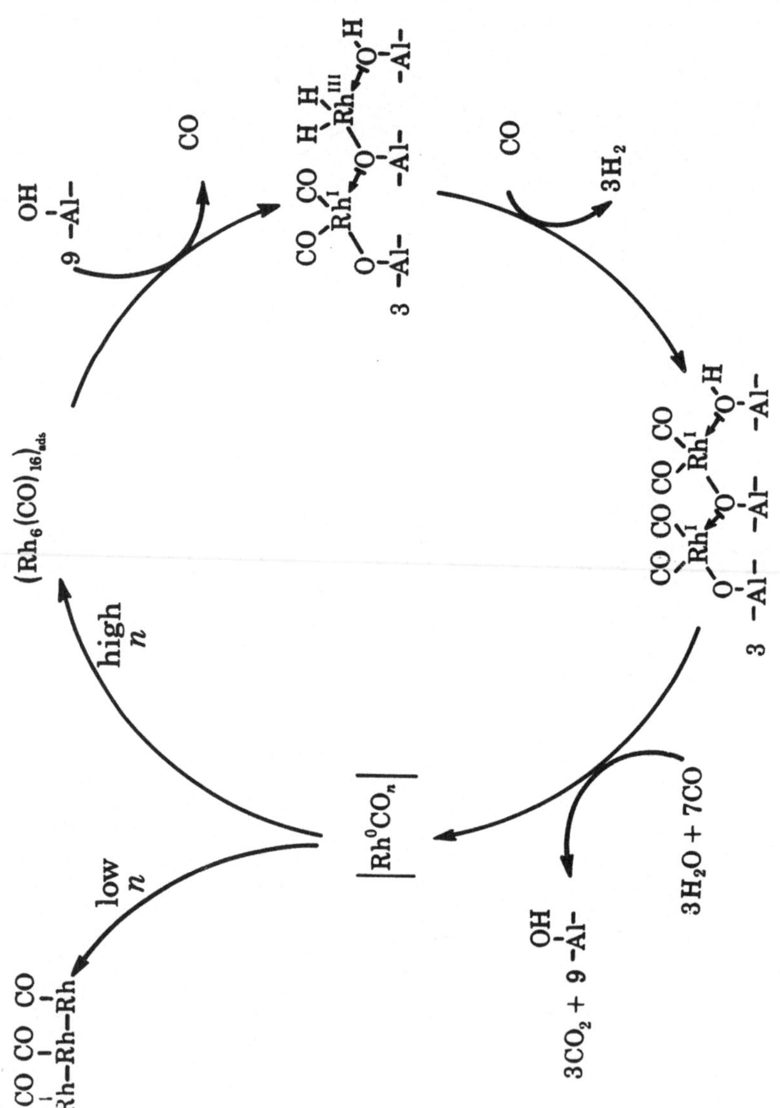

Figure 41 : Mechanism of water gas shift with $Rh_6(CO)_{16}$ supported on alumina (28)

mation of $Rh^{I}(CO)_2(OAl<)$ surface mononuclear complex. The final step is the regeneration of the cluster $Rh_6(CO)_{16}$ under $CO + H_2O$ with formation of CO_2 and H^+. It appears therefore that the mechanism of the water gas shift reaction occurs by stepwise destruction and regeneration of the cluster in the two key steps of the mechanism (Figure41). The involvement of molecular clusters is some catalytic steps seem to be quite general in many reactions which involve molecular CO (23). The reason for this is rather simple. The cluster state is a rather stable zero-valent state when CO is present. This does not necessarily means that the cluster is involved in a complete catalytic cycle but rather in some steps of the catalytic cycle.

2.3.3 The molecular cluster is decomposed into very small particles of metal. Preparation of heterogeneous catalysis can use molecular clusters as starting material. As far as the development of heterogeneous catalysts derived from cluster compounds is concerned (23) such catalysts will be valuable only if they exhibit activities or selectivities that differ from those afforded by material prepared conventionally that is halide impregnation or ion exchange followed by hydrogen reduction. In the case of ruthenium, cluster derived catalysts are shown to display greatly enhanced activity for the complete hydrogenation of straight chain aliphatic hydrocarbons to methane and provide a temperature advantage of 150°C relative to conventionally prepared ruthenium catalysts where only moderate hydrocarbon conversions are noted. The increased activity superficially correlates with the smaller metal crystallite sizes (15 - 20 Å) reproducibly obtained with metal cluster compounds as catalysts precursors (23).

In the case of Fe, the thermal decomposition of $[HFe_3(CO)_{11}]^-$ adsorbed on alumina, leads to the formation of very small particles of Fe (10 Å) which cannot be obtained by any other route. When these catalysts are used in Fischer-Tropsch synthesis, those supported particles are selective for low molecular weight olefins and the selectivity appears to be much larger than that obtained with conventionally prepared catalysts (29).

2.4 <u>Supported clusters and heterogeneous catalysis : Surface organometallic chemistry</u>. One can define surface organometallic chemistry (S.O.M.C.) as the study of the reactivity of organo-metallic complexes with the surface of divided oxides and by extension of zeolithes. It is thus possible to make new catalysts which have no equivalent in homogeneous catalysis and in heterogeneous catalysis. These new surface complexes may be mononuclear, dinuclear, polynuclear (or metal particles). The strategy followed in this area is much closer to that followed in homogeneous catalysis than to that followed in heterogeneous catalysis. Besides, the study of the stoechiometric reaction of organometallic complexes with surfaces introduces the concepts of coordination chemistry to heterogeneous catalysis. The fields of S.O.M.C. has started with the study of the reactivity of mononuclear complexes of the type $Zr(CH_2 - C_6H_5)_4$ with functionnal groups of silica and alumina (30). It has resulted from this approach a new generation of highly active polymerisation catalysts. This field has broadened recently thanks to the use of molecular clusters which give rise to a very interesting reactivity of their metal-metal bond as well as metal ligand bonds with functionnal groups of surfaces.

We would like to illustrate S.O.M.C. by the following examples related to the behaviour of $Os_3(CO)_{12}$ adsorbed on silica and on alumina (Figures 42 and 43).

The first reaction which occurs during the thermal decomposition of $Os_3(CO)_{12}$ on hydroxylated silica is the oxidative addition of a silanol group in the metal-metal bond of $Os_3(CO)_{12}$ with departure of two moles of CO. This grafted cluster has been characterised by infrared spectroscopy, Raman spectroscopy, ^{13}C NMR and EXAFS. A model compound of the type $(H)Os_3(CO)_{10}(OSiPh_3)$ has also been synthetised (32) ; it exhibits spectroscopic data similar to that of the grafted cluster. Due to the presence of a bridging oxygen and bridging hydride, this grafted cluster exhibits various catalytic properties in reactions such a olefin hydrogenation or hydroformylation.

Thermal decomposition of the grafted cluster $(H)Os_3(CO)_{10}(OSi\leftarrow)$ results in a total destruction of the molecular cluster frame. This destruction probably results from a multiple oxidative addition of silanol groups on 3 metal atoms

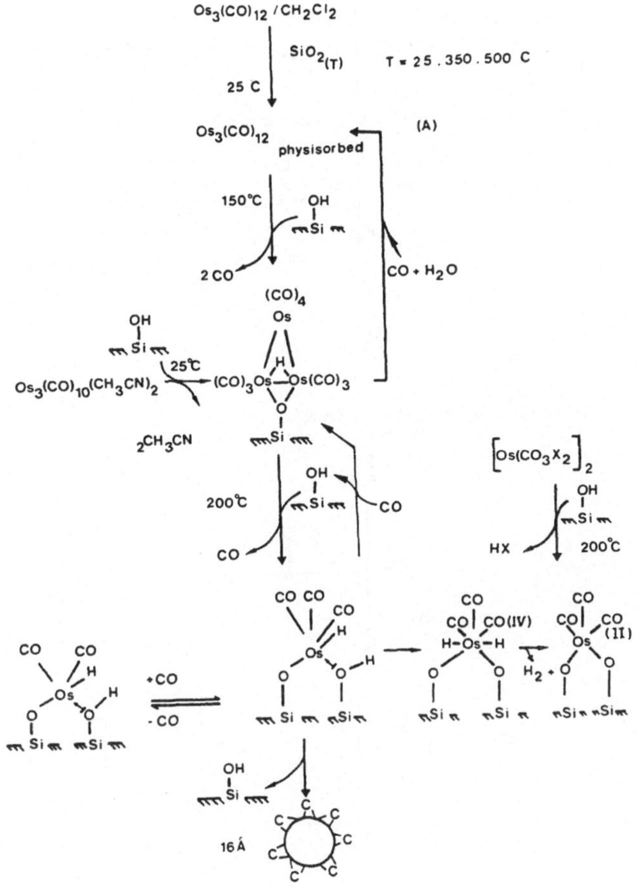

Figure 42 : surface organometallic chemistry of $Os_3(CO)_{12}$ on silica (32)

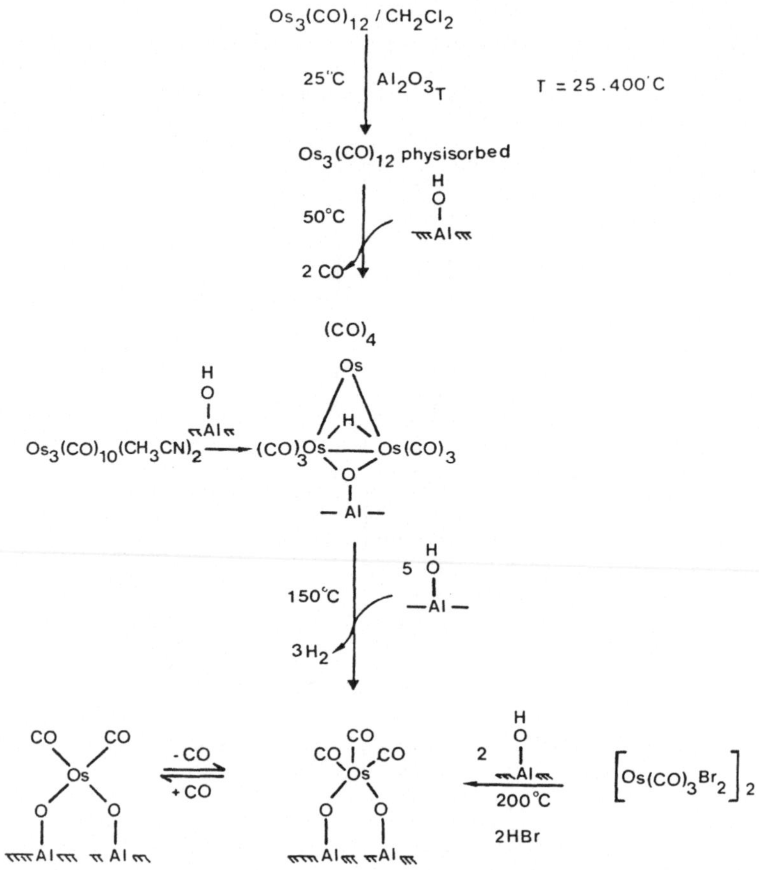

$Os_3(CO)_{12} / CH_2Cl_2$

25°C Al_2O_{3T} T = 25 . 400°C

$Os_3(CO)_{12}$ physisorbed

50°C H O Al

2 CO

$(CO)_4$ Os

H O Al

$Os_3(CO)_{10}(CH_3CN)_2$ → $(CO)_3Os$ H $Os(CO)_3$ O Al

150°C 5 H O Al

$3H_2$

CO CO Os O O Al Al

- CO + CO

CO CO CO Os O O Al Al

2 H O Al 200°C 2HBr

$[Os(CO)_3Br_2]_2$

Figure 43 : surface organometallic chemistry of Os_3 $(CO)_{12}$ on alumina (32) with formation of $(H)\,Os(CO)_3\,(OSi\lessdot)$ which can be reversibly decarbonylated to $H\,Os(CO)_2\,(OSi\lessdot)$. These mononuclear osmium hydrido species are also catalysts for hydroformylation.

It is only by treatment under vacuum above 250°C that small metal particles of Osmium are formed. Those particles have a narrow size distribution around 16 Å. The mechanism of such formation is the reductive elimination of a $\gtrdot Si-OH$ from $H\,Os(CO)_3\,(O\,Si\lessdot)$:

$$(H)\,Os(CO)_3\,(O\,Si\lessdot) \longrightarrow \overset{\circ}{O}s(CO)_3 + \gtrdot Si-OH$$

Above 250°C the equilibrium could be shifted towards the left whereas below 250°C this equilibrium could be shifted to the right and lead to an aggregation phenomenon. Those metallic particles of Os are active in the reaction of methanisation of carbon monoxide.

The reaction of $Os_3(CO)_{12}$ with the alumina surface is different from that observed with the silica surface. The first step is the oxidative addition of an $>Al-OH$ group into the Os - Os bond which gives the species $(H)\,Os_3(CO)_{10}\,(O\,Al<)$. However it is much less stable than on silica ; it decomposes above 150°C to give mononuclear Os^{II} species with liberation of 3 moles of H_2. In contrast to what is observed on silica, these surface species are linked to silica by two covalent bonds $Os\,(OAl<)_2$. Those $Osmium^{II}$ carbonyl complexes are not easily reduced by H_2 to metallic Osmium. In contrast to $(H)\,Os(CO)_3\,(OSi\lessdot)$, $Os(CO)_3\,(OAl<)$ species is not a catalyst for olefin hydroformylation. In order to achieve a reduction to Os metal, it is necessary to reduce Os^{II} with H_2 at 400°C for 20 hours. The metallic particles which are thus obtainned are much smaller than on silica since their size is situated around 8 Å.

Surface organometallic chemistry has already been at the origin of new concepts in surface science on oxides that can be summarized as follows :

- reaction of a metal-alkyl bond with an OH group :

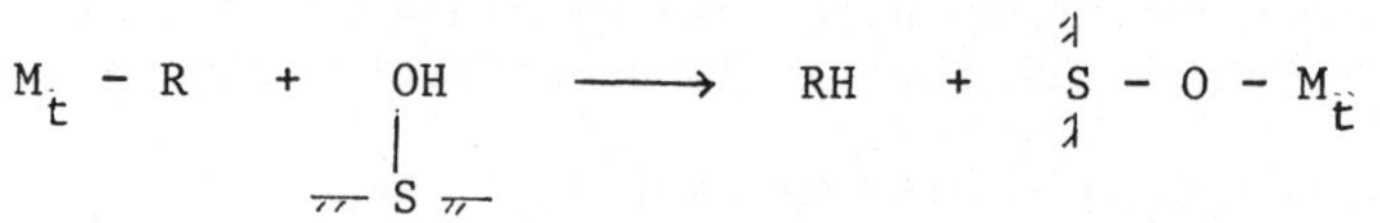

$$M_t - R \;+\; \underset{\text{S}}{\overset{\text{OH}}{|}} \;\longrightarrow\; RH \;+\; S - O - M_t$$

- reaction of a Π allyl group with an OH group

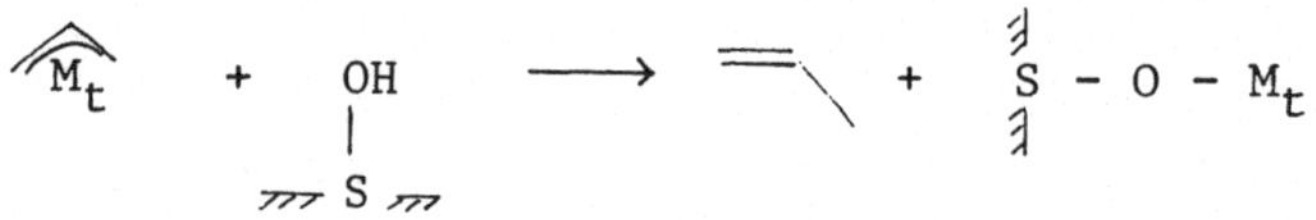

$$(\pi\text{-allyl})M_t \;+\; \underset{\text{S}}{\overset{\text{OH}}{|}} \;\longrightarrow\; (\text{propene}) \;+\; S - O - M_t$$

- oxidative addition of an OH group into a single metal species

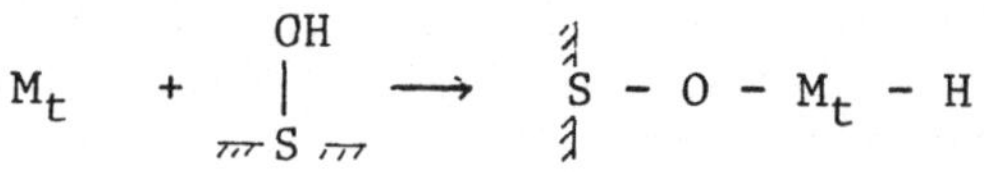

$$M_t \;+\; \underset{\text{S}}{\overset{\text{OH}}{|}} \;\longrightarrow\; S - O - M_t - H$$

- oxidative addition of an OH group into a metal-metal bond

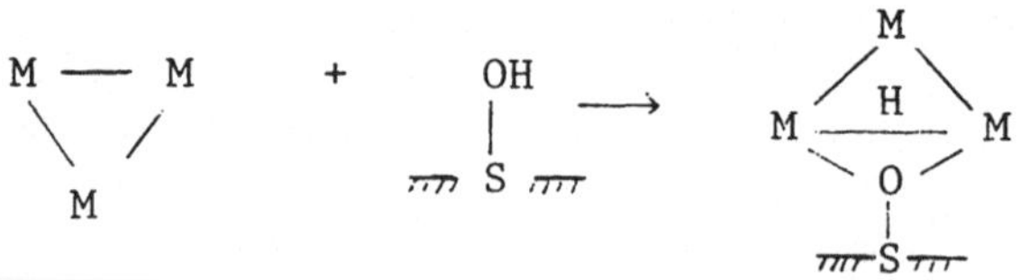

$$M_3 \;+\; \underset{\text{S}}{\overset{\text{OH}}{|}} \;\longrightarrow\; M_3(\mu\text{-H})(\mu\text{-O-S})$$

- multiple oxidative additon with reductive elimination of H_2 :

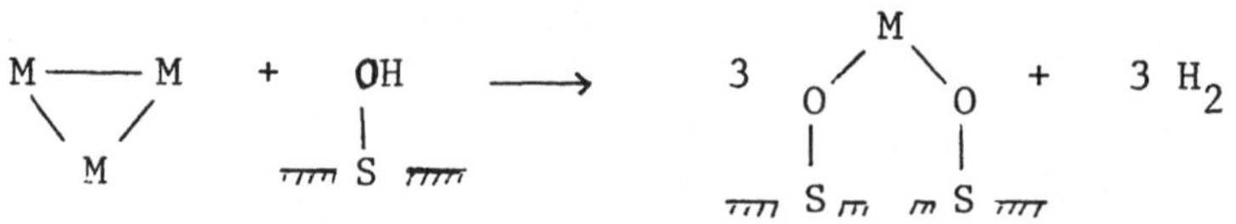

$$M_3 \;+\; \underset{\text{S}}{\overset{\text{OH}}{|}} \;\longrightarrow\; 3\; (S-O)_2M \;+\; 3\;H_2$$

- nucleophilic addition at coordinated CO followed by β - H elimination :

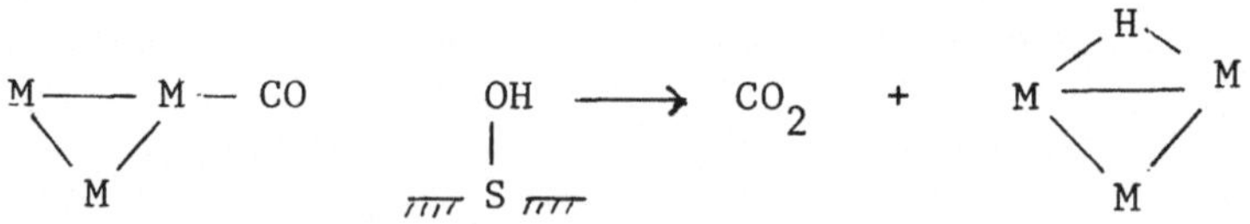

$$M_3-CO \quad \underset{\text{S}}{\overset{\text{OH}}{|}} \;\longrightarrow\; CO_2 \;+\; M_3(\mu\text{-H})$$

- nucleophilic attack by molecular water with construction of a metallic frame :

$$3\,H_2O + 7\,CO + 6\,[(CO)_2Rh(O\text{-}S)(\leftarrow O\text{-}S)] \rightarrow 3\,CO_2 + 6\,(HO\text{-}S) + Rh_6(CO)_{16}$$

- complexation of a carbonyl ligand into a Lewis center of the surface :

$$(CO)_5W-C=O \mapsto Al \quad ; \quad (H)\,Fe_3(CO)_{10}\,CO \longmapsto \overset{+}{Al}$$

- nucleophilic attack of an O^{2-} ligand into a coordinated CO

$$Fe_3(CO)_{12} + O^{2-}\text{–}Mg \longrightarrow (OC)(CO)_3Fe\text{–}C(O\cdots O)\cdots Mg$$

- oxidative addition into a coordinatively unsaturated metal atom :

$$Rh(O\text{–}Al)(\leftarrow O\text{–}Al) + H_2 \longrightarrow H_2Rh(O\text{–}Al)(\leftarrow O\text{–}Al)$$

References

1. P. Chini. J. Organomet. Chem., 200 (1980), 37-61.
2. B.F.G. Johnson. Transition Metal Clusters (B.F.G. Johnson Ed. John Wiley and sons, Pub. 1980).
3. K. Wade. Transition Metal Clusters (B.F.G. Johnson Ed. John Wiley and sons, Pub. 1980) 193-263.
4. A.J. Deeming. Transition Metal Clusters (B.F.G. Johnson Ed. John Wiley and sons, Pub. 1980) 391-469.
5. E.L. Muetterties. La Recherche 117 (1980) 1364-1372 (John Wiley and sons, 1981) 203-238.

6. A.K. Smith and J.M. Basset. J. Mol. Cat., 2 (1977) 229.

7. E. Band and E. Muetterties. Chem. Rev., 78 (1978) 639-658.

8. R. Ugo. Cat. Rev., 11 (1975) 225.

9. G. Martino. Models and precursors for metallic Catalysts (J. Bourdon Ed. 1980) 399.

10. J.M. Basset and R. Ugo. Asp. Homog. Cat. (R. Ugo Ed. 1978) 3, 136.

11. E.L. Muetterties, T.N. Rhodin, E. Band, C.F. Brucker and W.R. Pretzer. Chem. Rev. 79 (1979) 91.

12. D.C. Bailey and S.H. Langer. Chem. Rev., 81 (1981) 109-148.

13. T.L. Brown, 10 (1981) 159-180.

14. J. Evans. Chem. Soc. Rev., 10 (1981) 159-180.

15. H. Vahrenkamp. Phil. Trans. Roy. Soc. Lond. A. 308 (1982) 17-26.

16. R.E. Colborn, A.F. Dyke, S.A.R. Knox, K.A. Macpherson, K.A. Mead, A.G. Orpen, J. Roue and P. Woodward. Phil. Trans. Roy. Soc. Lond. A. 308 (1982) 67-73.

17. A. Fusi, R. Ugo, R. Psaro, P. Braunstein and J. Dehand. Phil. Trans. Roy. Soc. Lond. A. 308 (1982) 125-130.

18. G. Longoni, A. Ceriotti, R.D. Pergola, M. Manassero, M. Perego, G. Piro and M. Sansoni. Phil. Trans. R. Soc. Lond. A. 308 (1982) 47-57.

19. B.F.G. Johnson and J. Lewis. Phil. Trans. R. Soc. Lond. A. 308 (1982) 5-15.

20. J. Bradley. Phil. Trans. R. Soc. Lond. A. 308 (1982) 103-113.

21. B. Heaton. Phil. Trans. R. Soc. Lond. A. 308 (1982) 95-102.

22. D.M.P. Mingos. Phil. Trans. R. Soc. Lond. A. 308 (1982) 75-83.

23. R. Whyman. Phil. Trans. R. Soc. Lond. A. 308 (1982) 131-140.

24. B.F.G. Johnson and J. Lewis. Phil. Trans. R. Soc. Lond. A. 308 (1982) 5-15.

25. P. Ford. Acc. Chem. Res. 14, 31 (1981).

26. Deeming. J. Organomet. Chem., 114 (1976) 313.

27. B. Besson, A. Choplin, L. D'Ornelas and J.M. Basset. J.C.S. Chem. Comm., (1982) 842.

28. J.M. Basset, B. Besson, A. Choplin and A. Theolier. Phil. Trans. R. Soc. Lond. A. 308 (1982) 115-124.

29. D. Commereuc, Y. Chauvin, F. Hugues and J.M. Basset. J.C.S. Chem. Comm., (1980) 154.

30. Ballard. Adv. Cat/ Rel. Subj. 23, (1973), 263

31. Dombeck, (private communication).

32. R.Psaro, R.Ugo, G.M. Zanderighi, B.Besson, A.K. Smith, and J.M. Basset.
J. Organomet. Chem. 213, (1981), 215.

QUANTUM CHEMISTRY FOR METAL CLUSTERS (CAN THE CORRELATION PROBLEM BE CHEATED?)

Dennis R. Salahub

Département de chimie
Université de Montréal
Montréal, Canada H3C 3V1

CONTENTS

1. GENERALITIES-CLUSTER SCIENCE

One has only to read the table of contents of this volume or of previous reports from the cluster meetings at Aix-en-Provence (1) Lausanne (2) or Lyon (3) to appreciate the cross-disciplinary nature of cluster science. I have tried to represent this diversity of interest in clusters in Figure 1 which is perforce an incomplete and also a biased view of the field. In fact the references are to the use of one of the theoretical methods, the SCF-Xα-SW molecular orbital method, which we will discuss later on. The main point of Figure 1 is that a number of subdisciplines of chemistry, physics and materials science are linked

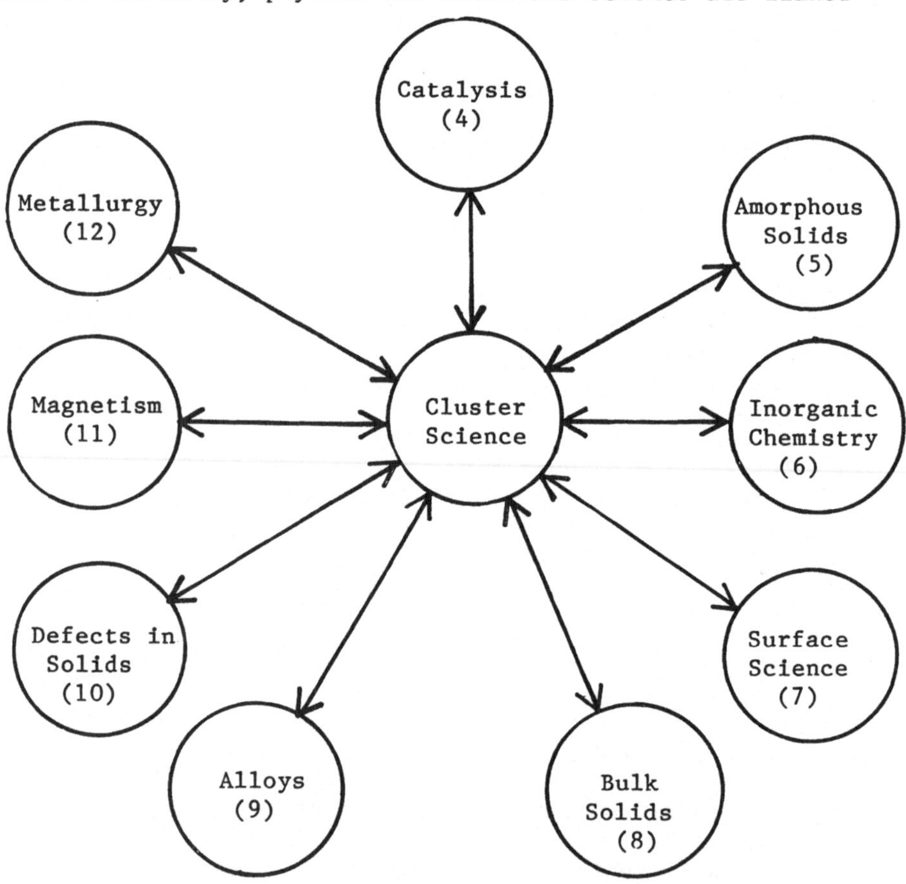

Figure 1. Some fields of application of cluster science. References (in parentheses) are to applications of the SCF-Xα-SW molecular orbital method (13-15).

through their concern with clusters, either as distinct entities or as models for more extended systems. With an organizational diagram such as this, the possibilities for cross-fertilization (represented by the arrows) are very abundant indeed. Unfortunately, with such a diversity of primary interests, problems of language and (mis)understanding are also prevalent. A major purpose of Institutes such as this is to try to remove some of these problems.

In the present case, I was asked, in these lectures, to give an overview and comparison of the quantum chemical methods which are being used to calculate the electronic structure of metal clusters. I will try to do this with the non-expert in mind and have chosen electronic correlation as a sub-theme. "Correlation" is one of those terms which means different things to different people and I will be satisfied if these lectures convey to the "non-quantum chemists" a feeling for what correlation is (as defined by most quantum chemists), why it is important for metal clusters and how the various computational methods take it into account (if they do and if we know how they do it). The formal aspects are treated in Section 2 and the reader who is familiar with the various quantum chemical techniques or who would simply prefer to see some results first may skip to Sections 3 and 4.

A second recurring theme in these notes will be the need for more experimental measurements of the properties of isolated clusters. Electronically, metal clusters are sufficiently complex that the results of different theoretical approaches are all to often in disagreement. In many instances, experimental values for one or a few properties would be sufficient to settle a controversy and allow further progress to be made. Experimental advances in this direction, particularly beam experiments (1-3) have been substantial and so I will try to offer a number of (hopefully not too unrealistic) challenges as we go along.

2. QUANTUM CHEMICAL METHODS 1)

The statement of the problem is simple. Given a system of nuclei (coordinates $\vec{R}_A$, charges Z_A, masses M_A) and electrons

---

1) Much of the material in this section can be found in a number of quantum chemistry texts. Those which I rely on heavily in my courses are: I.N. Levine, Quantum Chemistry (Boston, Allyn and Bacon, 1974); J.P. Lowe, Quantum Chemistry (New York, Academic, 1978) and R. McWeeny, Coulson's Valence (Oxford, Oxford University Press, 1979).

(coordinates $\vec{r}_i$), solve the time independent Schrödinger equation

$$\hat{H}(\vec{R}_A, Z_A, M_A, \vec{r}_i)\ \Psi\ (\vec{R}_A, \vec{r}_i) = E\ \Psi\ (\vec{R}_A, \vec{r}_i) \qquad (2.1)$$

for the many-electron, many-nuclei wave functions and energy levels. Atomic units will be assumed throughout ($m_e=1$, $e=1$, $\hbar=1$) and only spin-free, non-relativistic Hamiltonians will be considered.This much was known in 1926 and since then quantum chemists have been working on methods of approximation to overcome the mathematical horrors that become evident when equation (2.1) is written out in detail for a molecule.

Before looking at the quantum chemical methods in more detail, let us mention briefly three general aspects of the solution of equation (2.1) which are common to the vast majority of methods. First, equation (2.1) involves the (coupled) motion of both the electrons and the nuclei. However, since the electrons are much lighter than the nuclei and consequently have higher speeds, it is usually a good approximation to assume that an electronic Schrödinger equation can be solved for each instantaneous position of the nuclei (the Born-Oppenheimer approximation).

$$\hat{H}_{el}\ \Psi_{el}(\vec{r}_i;\ \vec{R}_A) = E_{TOT}(\vec{R}_A)\ \Psi_{el}(\vec{r}_i;\ \vec{R}_A) \qquad (2.2)$$

where the electronic Hamiltonian is

$$\hat{H}_{el}\ (\vec{R}_A) = \sum_i -\tfrac{1}{2}\nabla_i^2 + \sum_i \sum_A -\frac{Z_A}{r_{iA}} + \sum_{i>j} \frac{1}{r_{ij}} + \sum_{A>B} \frac{Z_A Z_B}{R_{AB}} \qquad (2.3)$$

The electronic wave function $\Psi_{el}\ (\vec{r}_i;\ \vec{R}_A)$ depends explicitly on the electronic coordinates and parametrically on the nuclear coordinates. The nuclear-nuclear repulsion has been included in $\hat{H}_{el}$ so that the energy in equation (2.2) is a total energy. $E_{TOT}(\vec{R}_A)$ defines the energy of the molecule as a function of nuclear coordinates and this is just the potential needed to construct the Hamiltonian for nuclear motion (the vibrational Hamiltonian)

$$\begin{aligned} \hat{H}_{vib} &= \sum_A -\frac{1}{2M_A}\nabla_A^2 + V_{NUC}\ (\vec{R}_A) \\ &= \sum_A -\frac{1}{2M_A}\nabla_A^2 + E_{TOT}\ (\vec{R}_A) \qquad (2.4) \end{aligned}$$

$$\hat{H}_{vib}\,\Psi_{vib}\,(\vec{R}_A) = E_{vib}\,\Psi_{vib}\,(\vec{R}_A) \tag{2.5}$$

All of the calculations we will discuss below are within the Born-Oppenheimer approximation which should be valid as long as there are no "slow" electrons or "fast" nuclei. Significant departures from BO behaviour (i.e. strong vibronic coupling) are expected for "floppy" molecules with nearly flat portions of their potential energy surfaces and high amplitude vibrations. One might wonder whether such situations exist for clusters, given that "soft phonons" and strong electron-phonon coupling can be important for certain solids. However, since no work has been done on this question I will simply leave it as a dangling thought.

A second general feature of electronic structure calculations arises from the fact that electrons are fermions and that hence an N-electron wave function must be antisymmetric to the interchange of space and spin coordinates of any two electrons.

$$\hat{P}_{12}\;\Psi(1,2...N) = \Psi(2,1...N) = -\Psi(1,2...N) \tag{2.6}$$

$$1 \equiv \{x_1,\, y_1,\, z_1,\, \sigma_1\} \tag{2.7}$$

The antisymmetry requirement may be satisfied by constructing the wave function as a determinant or a linear combination of determinants built from a set of one-electron functions (spin-orbitals). This can be written in several forms (taking a 2-electron example)

$$\begin{aligned}\Phi(1,2) &= ||\phi_1(1)\alpha(1)\phi_2(2)\beta(2)|| = \frac{1}{\sqrt{2}}\begin{vmatrix}\phi_1(1)\bar{\phi}_2(1)\\ \phi_1(2)\bar{\phi}_2(2)\end{vmatrix}\\ &= ||\phi_1(1)\bar{\phi}_2(2)||\\ &= A\{\phi_1(1)\bar{\phi}_2(2)\} = \frac{1}{\sqrt{2}}\{\phi_1(1)\bar{\phi}_2(2) - \bar{\phi}_2(1)\phi_1(2)\}\end{aligned} \tag{2.8}$$

Since a determinant vanishes if two rows or two columns are identical, one of the usual statements of the Pauli principle is implicit in equation (2.8), namely that, at the most two electrons, one of each spin, can occupy a given orbital. In practical calculations the Pauli principle is satisfied either by explicitly using determinants or else in an ad hoc manner through the usual restrictions on orbital occupation.

A third guiding principle in quantum mechanics is the variational theorem which states that if $\Phi(\{r_i,\sigma_i\})$ is any normalized, well-behaved function and Ψ exact is the exact ground state function, then

$$\langle\Phi|\hat{H}|\Phi\rangle \equiv \int \ldots \int \Phi^*(1,2\ldots N)\ \hat{H}\ \Phi(1,2\ldots N)\ d\tau_1 \ldots d\tau_N$$
$$\geqslant \langle \Psi^{exact} |\hat{H}| \Psi^{exact} \rangle \equiv E_{ex} \qquad (2.9)$$

The variational integral, $\langle\phi|\hat{H}|\Phi\rangle$ provides an upper bound to the exact ground state energy. This allows one to (energy-)optimize approximate wave functions by varying their form in order to minimize the energy. Of course, if approximations are made to the Hamiltonian then the variational theorem applies only within a "model space" defined by the approximate Hamiltonian, and it is then possible to obtain values of E below the experimental value for the system being modelled. This does not mean that such results are not accurate or useful but only that one cannot rely (even in principle) exclusively on the variational theorem to improve the wave function.

Table 1. QUANTUM CHEMICAL METHODS

1. Semi-empirical methods
   - Model Hamiltonians
   - Adjustable molecular parameters

   e.g. Hückel-simple tight binding (16)
   Extended Hückel (17)
   CNDO (18)
2. "Ab Initio" methods
   - Choose a form for Ψ (1 or many determinants)
   - Choose a set of basis functions
   - No further (fundamental) approximations

   e.g. LCAO-MO-SCF (Hartree-Fock) (19)
   Configuration Interaction (20)
   Generalized Valence Bond
3. Pseudopotential methods
   - Like "ab initio" (or local density) but core electrons replaced by pseudopotentials

   e.g. SCF (21-23)
   CI (24)
   GVB (25)
4. Local density methods
   - Approximate local exchange-correlation potential (e.g. Xα, LSD) $v_{xc} \sim \rho^{1/3}$
   - Roots in density functional theory and theory of electron gas
   - Various ways of solving 1-electron Schrödinger equation

   e.g. LCAO (26, 55)
   Discrete Variational (LCAO) method (27)
   Scattered-Wave method (4-12, 55)

With these general principles in mind we can now turn to an overview of the methods used in quantum chemistry. In Table 1 I have divided the methods into four broad classes and have summarized the principal characteristics of each. References to a few examples of cluster applications are also given. I would now like to discuss some of these methods in more detail in order to bring out the basic features of the correlation problem.

2.1 Semi-empirical Methods

Since Dr. Baetzold (17) will be talking about semi-empirical methods and their applications to metal clusters I will limit myself to a few general comments. These methods involve model Hamiltonians which are often not even written down in explicit form but are instead defined by their matrix elements over a one electron basis set. In the simplest method (Hückel theory) not even the basis set is defined. The matrix elements are hence the central quantities and they are kept as adjustable parameters whose values are determined in order to reproduce a subset of the experimental data of interest. The idea is that a small number of experiments are used to calibrate the method, which is then used to treat further (more complicated) systems. Such methods are usually most reliable if applied to systems which are closely related to those used in the determination of the parameters. The value of such methods in providing a language and interpreting qualitative trends is undeniable; one has only to mention the Woodward-Hoffmann rules of organic chemistry which grew out of Extended Hückel calculations to justify the existence of this class of methods. There are, however, clear dangers inherent to this type of approach. The quantitative results are not always satisfactory, great care must be taken in the choice of parameters and one must also be on guard against using a particular method beyond its range of validity, i.e. for molecules or properties that are too different from those for which the method has been shown to work.

If we define correlation effects as those which are present in the exact wave function and absent in the best single determinantal function (the Hartree-Fock limit) then clearly, no explicit account of correlation is taken in the simple semi-empirical methods. However, since the parameters may be adjusted to yield experimental (i.e. fully correlated) properties one might argue that the methods take some account of correlation in a more restricted sense; the energy (and other properties) are "correlation-corrected" even though the wave function is not. The methods try, and are at least partially successful at "cheating" the correlation problem through their reliance on parameters. I will come back to this notion a little later in the discussion of local density methods.

2.2 "Ab Initio" Methods - The Role of Correlation - The Example of H_2

As I have indicated in Table 1 there are in the usual cases only two major choices to make if one wants to do an "ab initio" calculation. One has to decide on whether a single determinant is appropriate or whether many determinants will be required (and if so, which determinants to use). Then one typically has to choose a set of functions in which to expand the molecular orbitals that are used to construct the determinants. A further choice which may be made, especially for systems containing heavy atoms with a large number of "unimportant" core electrons, is to replace these with a pseudopotential and thereby reduce the problem to essentially that of the valence electrons. All this sounds deceptively simple. In practice, however, these choices are far from obvious and the value of the results depends crucially on them. The words "ab initio" by themselves are no guarantee of accuracy or appropriateness of a calculation but simply denote a general approach to quantum chemistry characterized by the above choices and no other fundamental ones.

We will see the practical consequences of some of these choices in sections 2.3 and 3 but first I would like to treat a simple example, the hydrogen molecule in its ground state, in considerable detail since this example more than any other I know can yield, once it is understood, a good intuitive feeling for the nature of electronic correlation. One's thinking about more complicated systems can often be helped by seeking analogies with H_2 (at some internuclear distance). I do not mean to imply by this that the bond in H_2 is typical; from many points of view it is not. For one thing, there are only two electrons and they have opposite spin, so there is no exchange in the usual sense. Nevertheless, in order to have a starting point let us dissect the electronic structure of H_2 as given by various quantum chemical methods. We will discuss only two quantities, the energy E, and the wave function Ψ as a function of internuclear distance, R.

2.2.1 MO - The molecular orbital method. The non- relativistic electronic Hamiltonian for H_2 is

$$\hat{H}_{el} = -\tfrac{1}{2}\nabla_1^2 - \tfrac{1}{2}\nabla_2^2 - \frac{1}{r_{1A}} - \frac{1}{r_{2A}} - \frac{1}{r_{1B}} - \frac{1}{r_{2B}} + \frac{1}{r_{12}} + \frac{1}{R_{AB}} \qquad (2.10)$$

where 1 and 2 represent electronic coordinates and A and B those of the nuclei, defined in Figure 2.

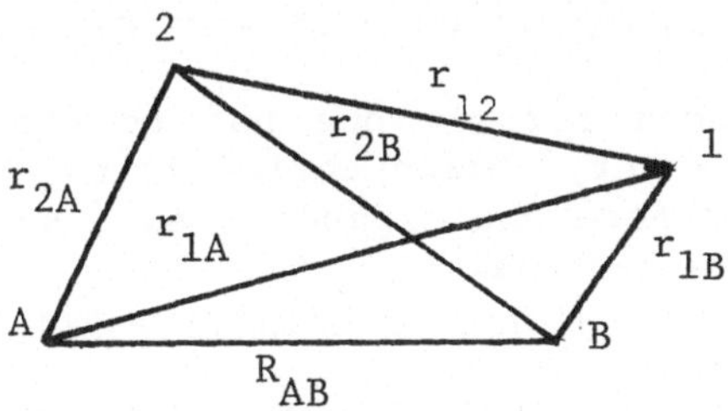

Figure 2. The hydrogen molecule

Because of the interelectronic repulsion term, $1/r_{12}$, in the Hamiltonian, the problem is not separable and so approximate solutions must be sought.

Among the simplest approximate solutions is that of the LCAO-MO method (Linear Combinations of Atomic Orbitals to form Molecular Orbitals). One writes the 2-electron wave function as a single Slater determinant constructed from two molecular spin-orbitals which have a common spatial part and α and β spin respectively.

$$\Phi_{MO}(\vec{r}_1, \sigma_1, \vec{r}_2, \sigma_2) \equiv \Phi_{MO}(1,2) = \frac{1}{\sqrt{2}} \begin{vmatrix} \phi(1)\alpha(1)\phi(2)\alpha(2) \\ \\ \phi(1)\beta(1)\phi(2)\beta(2) \end{vmatrix} \quad (2.11)$$

which may be expanded

$$\Phi_{MO}(1,2) = \frac{1}{\sqrt{2}} \{\phi(1)\alpha(1)\phi(2)\beta(2) - \phi(1)\beta(1)\phi(2)\alpha(2)\} \quad (2.12)$$

$$= \phi(1)\phi(2) \{\frac{1}{\sqrt{2}} \ (\alpha(1)\beta(2) - \beta(1)\alpha(2))\} \quad (2.13)$$

We have obtained the total wave function as a product of two factors, a simple product of spatial orbitals times a singlet spin function. (For more than two electrons this cannot be done so H_2 is an especially simple case). Since the Hamiltonian is spin-independent the energy depends only on the spatial part and we will drop the spin part from here on. The energy expectation value is given by

$$\langle E\rangle = \langle\Phi_{MO} \ |\hat{H}|\ \Phi_{MO}\rangle = \int\int\phi^*(1)\phi^*(2) \ \hat{H}\phi(1)\phi(2) \ dv_1 dv_2 \quad (2.14)$$

where dv_1 and dv_2 are integrations over spatial coordinates only.

In order to proceed further one has to choose a concrete representation of ϕ. A very common choice is the LCAO approximation. In its simplest form one expands ϕ as a sum of hydrogenic 1s functions on the two centers.

$$\phi(1) = N\{1s_A + 1s_B\ (1)\} \equiv N\ \{a(1) + b(1)\} \qquad (2.15)$$

where N is a normalization constant. A profile of ϕ is shown in the upper panels of Figure 3 for the equilibrium distance and also for a much longer distance.

We can now expand the (spatial part of) the 2-electron wave function

$$\begin{aligned}\Phi_{MO}(1,2) &= N\{a(1) + b(1)\}\ \{a(2) + b(2)\} \\ &= N\{a(1)b(2) + b(1)a(2) + a(1)a(2) + b(1)b(2)\} \qquad (2.16)\end{aligned}$$

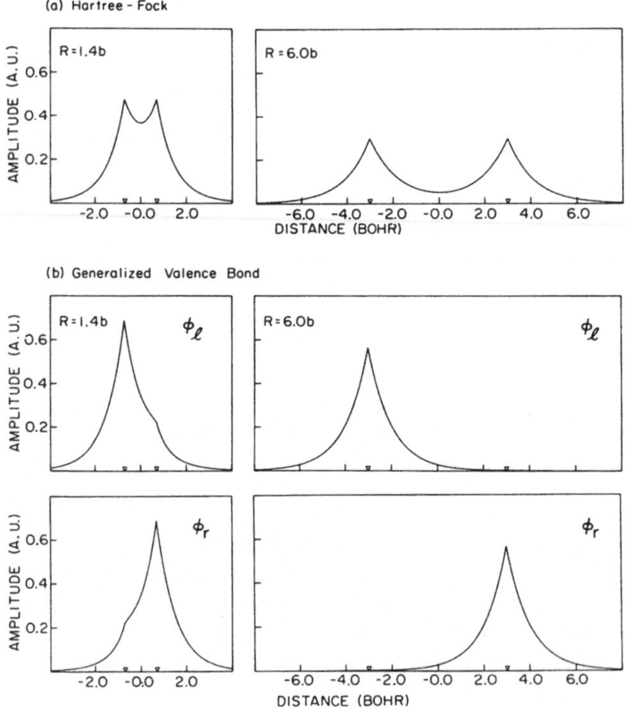

Figure 3. Hartree Fock and GVB orbitals of H_2 - Reproduced from ref. 28

The energy corresponding to this wave function is plotted as the curve labelled MO in Figure 4. If one completely optimizes the spatial form of ϕ then the Hartree-Fock (HF) limit is attained.

The MO approach yields a reasonable value for the equilibrium distance and about three quarters of the observed binding energy. The correlation energy is usually defined as the difference between the exact energy and the HF limit, so that in H_2 at equilibrium about one quarter of the binding energy is due to correlation effects.

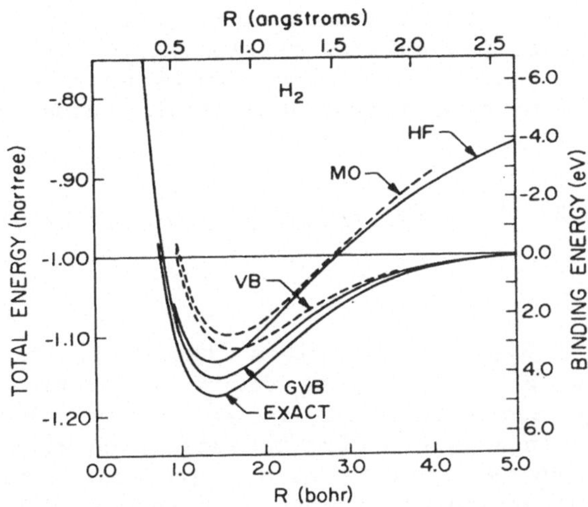

Figure 4. The MO, VB, HF, GVB and exact energies for H_2. Reproduced from ref. 28.

At long distances, however, disaster strikes. Instead of dissociating to two neutral H atoms, the MO method predicts dissociation to a state which is about 7 eV too high. The reasons for this are clear upon examination of the wave function. The first two terms in equation (2.16) represent covalent bonding where one electron is associated with each nucleus. The last two terms however are ionic; both electrons are on the same center. At large distances these latter terms should not contribute but because of the form chosen for the MO wave function, they are forced to - and with equal weight to the covalent terms. Thus in MO theory H_2 dissociates to a fictitious state which is a 50:50 mixture of covalent (H+H) and ionic (H^++H^-) states. The motions of the electrons are not properly correlated.

2.2.2 VB - The valence bond method. Since the ionic terms are causing so much trouble, one possibility is to simply leave them out. In fact this is what is done in the simplest version of valence bond theory. Here one works directly with atomic orbitals and the wave function is written as

$$\Phi_{VB}(1,2) = N\{a(1)b(2) + b(1)a(2)\} \tag{2.17}$$

where the second term is needed to keep the electrons indistinguishable. This wave function leads to the proper dissociation asymptote (curve VB of Figure 4).

To improve this function one can realize that at finite distances there is some probability of finding both electrons on one center and hence the ionic terms should make a contribution. This is the valence bond method with ionic terms for which the wave function is

$$\Phi_{VB,I} = N\{(a(1)b(2)+ b(1)a(2))+ \mu(R_{AB})(a(1)a(2)+ b(1)b(2))\} \tag{2.18}$$

The parameter $\mu(R_{AB})$ which gives the relative weights of covalent and ionic terms may be determined variationally as a function of R_{AB} and this leads to an energy curve similar to that labeled GVB in Figure 4 (for reasons which will become clear in a moment) which is in quite close agreement with the experimental curve. Unfortunately the straightforward extension of the valence bond method to many-electron polyatomic molecules is very complicated and is rarely used in practice. Instead, most of the effort in quantum chemistry has gone into methods for improving the molecular orbital wave function, principally by means of configuration interaction.

2.2.3 CI - Configuration Interaction. Since we have used two basis function $1s_a$ and $1s_b$, to construct $\Phi_{MO}(1,2)$ it is possible to form two molecular orbitals. So far we have only used one:

$$\phi(1) = N\{a(1) + b(1)\} \tag{2.19}$$

The other possibility is

$$\phi'(1) = N\{a(1) - b(1)\} \tag{2.20}$$

We also can form a determinant, $\Phi'(1,2)$ using ϕ', which corresponds to another electronic configuration, $(\phi')^2$ instead of $(\phi)^2$. We can regard Φ and Φ' as approximations to the ground state and to a doubly excited state respectively:

$$\begin{array}{ccc} - - - & \phi' & -x-x- \\ -x-x- & \phi & - - - \end{array}$$

Since Φ and Φ' are not true eigenstates of the Hamiltonian they can mix and the mixture will in general be a better approximation. Let's expand Φ' and Φ (dropping the normalization factors for clarity).

$$\Phi'_{MO} \propto \{a(1) - b(1)\}\{a(2) - b(2)\}$$

$$= a(1)\,a(2) + b(1)\,b(2) - a(1)\,b(2) - b(1)\,a(2) \tag{2.21}$$

$$\Phi_{MO} \propto a(1)\,a(2) + b(1)\,b(2) + a(1)\,b(2) + b(1)\,a(2) \tag{2.22}$$

Since the ionic terms enter Φ and Φ' with the same sign and the covalent terms enter Φ and Φ' with different signs, addition of Φ and Φ' will cancel the covalent terms while subtraction will cancel the ionic ones. An optimum linear combination can be found by appeal to the variational principle. One can write

$$\Phi_{CI} = N\{\Phi_{MO} + k(R_{AB})\,\Phi'_{MO}\} \tag{2.23}$$

This is clearly just another way of writing $\Phi_{VB,I}$ (equation 2.18) the valence bond function with ionic terms:

VB + ionic terms = MO + CI

In contrast to valence bond theory, CI is a widely used method for taking account of correlation effects. Unfortunately a CI wave function is usually highly complicated and it is often difficult to extract any simple physical picture from the list of determinants and their coefficients. Even for our simple example of H_2 it only became clear how correlation was entering Φ_{CI} once we had expanded it into its simplest terms and saw how the ionic terms were cancelled. The simplicity of an orbital picture is often lost in CI. However, there is a further method which yields a correlated wave function and yet still allows one to think in terms of orbitals.

This method was first proposed for H_2 by Coulson and Fischer (29) and has since been developed into the Generalized Valence Bond method by Goddard and his coworkers (28).

2.2.4 GVB - Generalized Valence Bond method. The Coulson-Fischer function for H_2 may be constructed by using a single covalent VB structure but instead of using pure atomic orbitals, allowing the AO's to be "contaminated" by contributions from functions on the other center. That is, the basis orbitals are:

$$\tilde{a}(1) \sim (a(1) + \lambda(R_{AB})\, b(1)) \qquad \tilde{b}(1) \sim (b(1) + \lambda(R_{AB})\, a(1)) \tag{2.24}$$

where, once again $\lambda(R_{AB})$ is a parameter to be determined variationally. We can see even at this stage that $\tilde{a}$ and $\tilde{b}$ are somewhere in between the basis functions used in MO and VB theory. If $\lambda = 1$ a delocalized molecular orbital is obtained whereas if $\lambda = 0$, localized atomic orbitals result. Intermediate values of λ yield partially localized functions, profiles of which are shown in the lower panels of Figure 3. The Coulson-Fischer function is written in normal valence bond style:

$$\begin{aligned} \Phi_{CF}(1,2) &= N[\tilde{a}(1)\, \tilde{b}(2) + \tilde{b}(1)\, \tilde{a}(2)] \\ &= N\{[a(1) + \lambda b(1)]\, [b(2) + \lambda a(2)] \\ &\quad + [b(1) + \lambda a(1)]\, [a(2) + \lambda b(2)]\} \\ &= N\{(1+\lambda^2)[a(1)\, b(2) + b(1)\, a(2)] \\ &\qquad + 2\,\lambda[a(1)\, a(2) + b(1)\, b(2)]\} \end{aligned} \tag{2.25}$$

But this looks familiar! It is just another way of writing Φ_{CI} or $\Phi_{VB,I}$. The difference is that now we can think in terms of orbitals and their degree of localization rather than mixtures of covalent and ionic terms or superpositions of configurations. The CF or GVB orbitals change smoothly from being quite (though not fully) delocalized molecular orbitals near R_e to being pure atomic orbitals at long distances. Hence a bridge between MO and VB pictures is afforded by the GVB method. It also underlines the link between the correlation of electronic motions and the localization of orbitals. In MO theory the orbitals are often forced to be completely delocalized by the symmetry of the problem and this is a prime factor in the failures of the method. Such failures are typically most dramatic for cases of small overlap (such as H_2 at large R_{AB}). In these cases localized orbitals are more appropriate.

2.3 Many electron systems

Unfortunately if one is interested in large molecules or clusters with many tens or even hundreds of electrons, then life gets very difficult indeed. Even in this age of super-computers, a correlated ab initio calculation must be approached with the courage and cunning of a battle-wise general. I do not have the space here (nor the expertise) to detail all of the considerations that go into an "ab initio battle plan" but will instead mention just a few aspects of the Hartree-Fock method to give the reader some feeling for the complexity of the problem (e.g. where things can go wrong) and also to serve as background for the discussion of local density methods in the next section. I would suggest that the reader who is interested in more details would benefit from consulting the series Modern Theoretical Chemistry (New York, Plenum, 1977, 8 volumes).

Let us see roughly what is involved in a Hartree-Fock calculation. (Remember, what follows does not take into account any correlation effects). We will restrict our attention to the case of a closed-shell molecule having 2n electrons in n doubly-occupied orbitals. The energy expectation value is given by

$$\langle E\rangle = 2\sum_{i=1}^{n} \langle\phi_i| -\tfrac{1}{2}\nabla_i^2 - \sum_A \frac{Z_A}{r_{iA}} |\phi_i\rangle + \sum_{i=1}^{n}\sum_{j=1}^{n} (2J_{ij} - K_{ij}) \tag{2.26}$$

where the 2-electron coulomb (J) and exchange (K) integrals are defined as

$$J_{ij} = \int\int \phi_i^*(1)\phi_i^*(2) \frac{1}{r_{12}} \phi_i(1)\phi_j(2)\, dv_1 dv_2 \tag{2.27}$$

$$K_{ij} = \int\int \phi_i^*(1)\phi_j^*(2) \frac{1}{r_{12}} \phi_j(1)\phi_i(2)\, dv_1 dv_2 \tag{2.28}$$

Variation of the ϕ_i to minimize $\langle E\rangle$ yields the (one-electron) Hartree-Fock equations.

$$\hat{F}(1)\, \phi_i(1) = \varepsilon_i \phi_i(1) \tag{2.29}$$

where the Fock operator $\hat{F}$ is

$$\hat{F}(1) = -\tfrac{1}{2}\nabla_1^2 - \sum_A \frac{Z_A}{r_{1A}} + \sum_{j=1}^{n} (2\hat{J}_j(1) - \hat{K}_j(1)) \tag{2.30}$$

Equation (2.30) involves coulomb and exchange operators

$$\hat{J}_j(1) = \int \phi_j^*(2)\phi_j(2) \frac{1}{r_{12}} dv_2 \quad (2.31)$$

$$\hat{K}_j(1)\ \phi_i(1) = \int \phi_j^*(2)\phi_i(2) \frac{1}{r_{12}} dv_2\phi_j(1) \quad (2.32)$$

The exchange operator is non-local and is only defined when it acts on a particular function. Since the optimum molecular orbitals ϕ_i which we are seeking appear in the definitions of $\hat{J}_j$ and $\hat{K}_j$ and hence in $\hat{F}$, the HF equations must be solved iteratively.

The HF equations are of the integro-differential type and are not easily amenable to solution. A great step forward was taken when Roothaan (30) worked out how to incorporate the LCAO approximation, which changes the equations into a matrix form that can then be solved by the powerful techniques of matrix algebra. In Roothaan's approach one expands the MO's (ϕ_i) in a basis set(χ_p)

$$\phi_i = \sum_{p=1}^{m} C_{pi}\chi_p \quad (2.33)$$

and the HF equations become

$$\mathbf{FC} = \mathbf{SC}\boldsymbol{\varepsilon} \quad (2.34)$$

where the bold characters represent the Fock matrix **F**, the coefficient matrix **C**, the overlap matrix **S** and the diagonal values of $\boldsymbol{\varepsilon}$ are the orbital eigenvalues.

The Fock matrix elements now involve 1-and 2-electron integrals over the basis functions:

$$\begin{aligned} F_{pq} &= < \chi_p|\hat{F}|\chi_q > \\ &= < \chi_p|\hat{h}^c|\chi_q > + < \chi_p|\sum_i^{occ} (2\hat{J}_i-\hat{K}_i)|\chi_q > \\ &= H_{pq} + \sum_{rs=1}^{m} P_{rs} (<pr|qs> -\tfrac{1}{2}<pr|sq>) \end{aligned} \quad (2.35)$$

where

$$P_{rs} = 2 \sum_{i}^{occ} C_{ri} C_{si} \qquad (2.36)$$

In (2.35) $\hat{h}^c$ represents the 1-electron part of the Fock operator and $\langle pq|rs\rangle$ is a 2-electron integral:

$$\langle pq|rs\rangle = \iint \chi_p^*(1)\, \chi_q^*(2)\, \frac{1}{r_{12}}\, \chi_r(1)\, \chi_s(2)\, dv_1 dv_2 \qquad (2.37)$$

The evaluation of the 2-electron integrals represents a major bottleneck to the application of the HF method to large systems. If gaussian functions are used as a basis then each integral can be calculated quite rapidly; however, in general a large number of gaussians have to be used so that a huge number of integrals has to be evaluated. This number increases as m^4, where m is the number of functions in the basis. This rapid increase is illustrated in Table 2 which gives the number of 2-electron integrals for various basis set sizes (for large m the number is about 1/8 m^4). The largest basis set so far used for a metal cluster is that of Demuynck <u>et al</u> (19) who performed a gargantuan calculation involving 819 Gaussians in a study of Cu_{13}. This calculation involved about 6 x 10^{11} integrals! It is a sobering thought that this basis corresponds roughly to the so-called double-ζ level and that calculations on smaller transition metal systems (32,33) indicate that even more basis functions are needed to approach the Hartree-Fock limit, to say nothing of what is required to include correlation.

As a few examples of the dependence of calculated properties on the quality of the basis set I have gathered in Table 3 some results from atomic HF-type calculations (34). The quantity shown is the 3d orbital energy which corresponds via Koopmanns' theorem (35) to an approximate ionization potential.

Table 2: Number of two-electron Integrals Associated with Various Size Basis Sets (31)

<u>m</u>	Number of 2-electron integrals
10	4,540
20	22,155
50	814,725
100	12,751,250
200	202,015,050
300	1,019,261,250

Table 3: 3d orbital energies from atomic Hartree-Fock type calculations (in hartrees 1h = 27.21 eV) (34)

Atom (state)		ε(3d)	
	single-ζ	double-ζ	HF limit
Sc(^{2}D)	-0.18988		-0.34357
Mn(^{6}S)	-0.25159		-0.63884
Ni(^{3}F)	-0.00054	-0.66185	-0.70993
Cu(^{2}S)	+0.50637		-0.49074

The single-ζ level is completely useless (e.g. for Cu the 3d level is unbound, the error with respect to the HF limit being about 25eV!) Going to the double-ζ level improves matters greatly. For Ni the error decreases from 19eV to 1.4 eV; much better, but still an error which might be highly significant in a molecular context. It appears after the work of Hay (32) and others that for molecular calculations involving transition metal atoms one must work at least at the triple-ζ level an atomic double-ζ set plus an additional diffuse d function which is vital for a correct description of the binding. Several authors have also pointed out the crucial influence of correlation for the transition metal atoms (36-39) and molecules involving them (20,-24,25).

While some inroads have been made into the "ab initio" treatment of metal clusters it is fair to say that the field is in its early developmental stages. Progress will come as experience is gained and as new, more powerful computers are brought on line; however, at the present time highly accurate calculations which are thought to include at least the most important correlation effects are only starting to become available for diatomic (25) or small polyatomic clusters (40,41).

Given this situation, there is a need for less onerous approximate methods which can be applied to larger systems. I believe that the local density methods described in the next section do much to fill the gap. They are much more rapid than "ab initio" approaches and we will see, primarily by way of example, that they appear to be of sufficient accuracy to provide accurate predictions and useful concepts for the study of quite complicated many-electron systems.

2.4 Local (spin) density methods

I will provide here just a brief sketch of the local density methods in order to put them in perspective with the "ab initio" methods. Slater's Volume 4 (13) is a good source for further details.

Let us rewrite the expression for the energy in Hartree-Fock theory (for a closed shell system with doubly occupied orbitals).

$$\langle E_{HF}\rangle = \sum_{i}^{occ} 2 \int \phi_i^*(1)\{-\tfrac{1}{2}\nabla_i^2 - \sum_A \frac{Z_A}{r_{iA}}\}\,\phi_i(1)\,dv_1$$

$$+ U_C + U_{XHF} + \sum_{A>B} \frac{Z_A Z_B}{R_{AB}} \qquad (2.38)$$

The coulomb energy, U_C is given by

$$U_C = \sum_i^{occ}\sum_j^{occ} 2J_{ij} = \tfrac{1}{2}\sum_i^{occ}\sum_j^{occ} \iint 2\phi_i^*(1)\,\times$$

$$\times\,\phi_i(1)\,\frac{1}{r_{12}}\,2\phi_j^*(2)\phi_j(2)\,dv_1 dv_2 \qquad (2.39)$$

$$= \tfrac{1}{2}\iint \sum_i^{occ} 2\phi_i^*(1)\phi_i(1)\,\frac{1}{r_{12}}\sum_j^{occ} 2\phi_j^*(2)\phi_j(2)\,dv_1 dv_2 \qquad (2.40)$$

$$= \tfrac{1}{2}\iint \rho(1)\,\frac{1}{r_{12}}\,\rho(2)\,dv_1 dv_2 \qquad (2.41)$$

where ρ is the electron density

$$\rho(1) = \sum_i^{occ} 2\phi_i^*(1)\phi_i(1) \qquad (2.42)$$

Hence the coulomb energy may be calculated once the (local) electron density is known. On the other hand, as is usual in quantum mechanics, our miseries start with the exchange term:

$$U_{XHF} = -\sum_i^{occ}\sum_j^{occ} K_{ij} = \sum_i^{occ}\sum_j^{occ}\iint \phi_i^*(1)\phi_j(1)\,\frac{1}{r_{12}}\,\times$$

$$\times\,\phi_j^*(2)\phi_i(2)\,dv_1 dv_2 \qquad (2.43)$$

There is no way to write this in terms of local quantities (Try it!). Remember, the exchange operators in (2.30) are different for the different orbitals. However in 1951, Slater reasoned, by considering some general features of the Fermi hole (a fictitious "hole" that follows each electron around and keeps out electrons of opposite spin) that the exchange operators for different orbitals shouldn't be radically different and that one might therefore construct a reasonable approximation by using an average behaviour. This average behaviour can be evaluated for the case of a homogeneous electron gas and the results depend only on the local density. The Xα method is based on the assumption that the most important features of the exchange in a (inhomogeneous) molecule or solid can be treated by using the electron gas results for the appropriate value of the local density. This yields the following exchange energy

$$U_{X\alpha} = c \int \rho(1)^{4/3} \, dv_1 \tag{2.44}$$

where

$$c = -\alpha 3/4 \ (81/8\pi)^{1/3} \tag{2.45}$$

In Slater's original treatment α was equal to one; however Gaspar (42) and later Kohn and Sham (43) showed that a value of 2/3 was more appropriate. In current use α is treated as a variable parameter. Schwarz (44) has determined those values which reproduce atomic Hartree Fock total energies and these are the most often used values.

When the variational theorem is applied to (2.38) with U_{XHF} replaced by $U_{X\alpha}$ the following one-electron Schrödinger equation results:

$$\hat{h} \, \phi_i(1) = \{- \tfrac{1}{2}\nabla_1^2 - \sum_A \frac{Z_A}{r_{1A}} + V_C(1) + V_{X\alpha}(1)\} \, \phi_i(1)$$

$$= \varepsilon_i \phi_i(1) \tag{2.46}$$

where

$$V_C(1) = \int \frac{\rho(2)}{r_{12}} \, dv_2 \tag{2.47}$$

and

$$V_{X\alpha}(1) = \frac{4}{3} \, c \, \rho(1)^{1/3} \tag{2.48}$$

More recently a number of authors (45-49) have suggested improved exchange-correlation potentials based on treatments of the electron gas which include correlation to various degrees. These are known collectively as local density (LD) or local spin density (LSD) potentials. It would take us too far afield to compare and contrast all of the approaches so that I must be content with referring the interested reader to the literature. In general the Xα potential has proven to be highly accurate under a wide variety of circumstances. It can however be improved, particularly for some spin-dependent problems, so that it is gradually being phased out in favor of the other potentials.

The above description of the Xα method underlines its relationship to HF theory and from this point of view one can regard Xα as an approximation to HF. However, in practice the Xα (or LD or LSD) method often gives results which are in better agreement with experiment than those of HF theory (see below). A possible path out of this logical dilemma (an approximate theory being more accurate than the theory it is meant to be approximating) is to consider density functional theory. Hohenberg and Kohn (50) proved that the non-degenerate ground state energy of an N electron system can in principle be calculated using an (unfortunately unknown) functional of the electron density, E[ρ]. The central quantity is ρ rather than the wave function. Since the Xα method has a certain density functional "flavor" (the potential energy is fully determined by the density) it may be more appropriate to view it as an approximation within this framework rather than the "ab initio" one. In this way it might be easier to accept that Xα might contain certain correlation corrections, a topic we will return to in section 3.

Indeed, a closer comparison of the HF and Xα methods reveals some fundamental differences, particularly concerning the interpretation of the eigenvalue spectrum. In HF theory the orbital energies represent unrelaxed (frozen orbital) ionization potentials

$$- \varepsilon_i^{HF} = IP_i^k \tag{2.48a}$$

where the superscript k signifies Koopmanns' theorem. To include relaxation effects one can perform a separate SCF calculation on each ionic state (the so called ΔSCF approach). As an example, the effects of relaxation are illustrated for the 3d and 4s orbitals of atomic copper on the left side of Figure 5.For the 4s orbital which is diffuse, the shift from the Koopmanns' theorem value is small, whereas for the localised 3d state it is substantial.

In the Xα method the orbital energies do not represent ionization potentials but rather orbital electronegativities

$$\varepsilon_i^{X\alpha} = \frac{\delta\langle E\rangle}{\delta n_i} \tag{2.48b}$$

where n_i is the occupation number of the ith orbital. In order to obtain an unrelaxed (frozen orbital or "Koopmanns") IP in Xα theory one has to subtract out a self-interaction term (a coulomb integral).

$$IP_i^{k,X\alpha} = -\varepsilon_i^{X\alpha} + \tfrac{1}{2}J_{ii} \tag{2.48c}$$

As can be seen from the two right-most colums of Figure 5 (HFS-Hartree-Fock-Slater is synonymous with Xα) these self-interaction terms are large especially for the localized 3d state. Once the self-interaction terms have been taken care of, the effects of physical relaxation are similar in the two theories. In the Xα method the two effects are of opposite sign and this provides some cancellation so that very often the approximate relative positions of ionic states can be judged from the orbital eigenvalues themselves. For example, for copper clusters, which we will discuss in detail in sec 3.6, this has been shown to hold to a very good approximation by explicit calculation of the IP's (8, 27). In HF theory on the other hand the effects of relaxation are dominant in situations like this where orbitals of very different spatial character are involved. We note from Figure 5 that the relaxed IP's from Xα theory are in better agreement with experiment than are those from HF theory.

So far we have only succeeded in writing down a one-electron Schrödinger equation (2.46). Now we have to solve it. A number of methods have been proposed and in Table 4 I have listed the salient characteristics of some of those which have seen service for metal clusters.

Since these methods have been well documented (and since I am rapidly running out of space) I will limit myself to a few comments on those I know best, Xα-SW and LCAO-Xα. In many respects the LCAO-Xα method is operationally similar to LCAO-HF. For instance, one is faced with the same problem of choosing a basis set. In addition it is necessary to choose further sets of auxiliary functions (gaussians) to fit ρ and $\rho^{1/3}$.

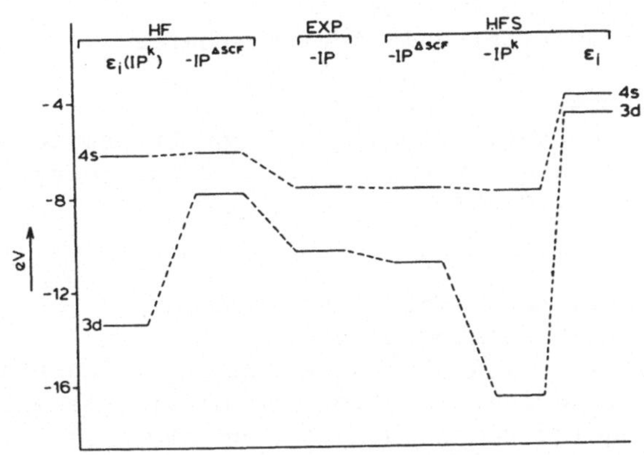

Figure 5: 3d and 4s orbital energies, Koopmans' ionization potentials (IP^k) and SCF IP in HF (19) and HFS ($X\alpha$) for the Cu atom (experimental values from ref. (91). Reproduced from reference 27.

$$\rho = \sum_r C_r^{coul} \chi_r^{coul} \tag{2.49}$$

$$\rho^{1/3} = \sum_r C_r^{xc} \chi_r^{xc} \tag{2.50}$$

the coefficients C^{coul} and C^{xc} being determined by a least squares procedure. Once this is done the two electron integrals to be evaluated have the form

$$<\chi_p(1) \mid \frac{\chi_r(2)}{r_{12}} \mid \chi_q(1)> \tag{2.51}$$

where χ_p and χ_q are orbital basis functions and χ_r is one of the fitting functions. If there are about the same number m of functions in the various basis sets then one has to evaluate the order of m^3 integrals rather than the m^4 of HF theory. Hence for large systems the advantage in computational time associated with LCAO-$X\alpha$ will be considerable. Much work remains to be done with this method especially in terms of developing operational rules for the choice of auxiliary functions and in improving the efficiency of the computer codes; however as we shall see in the next section the quality of the results so far obtained indicates that this developmental work will likely be well rewarded.

Table 4 Characteristics of Various Xα(LSD) Methods

Xα-SW (14)
(MS-Xα)
- scattered wave
- multiple scattering
- muffin-tin potential
- numerical solution in spheres
- partial waves
- most rapid method
- accurate orbitals and spectroscopic quantities
- does not yield accurate total energy curves
- e.g. Cu_{19}(8)

DVM-Xα (51,52)
- discrete variational method
- LCAO (Slater orbitals)
- fit ρ and $\rho^{1/3}$ with auxiliary functions
- numerical sampling for matrix elements
- quite rapid-work goes up about as m^2
- very large number of sampling points needed for total energy
- e.g. Cu_9 (27)

LCAO-Xα (53,54)
- LCAO (Gaussian orbitals)
- fit ρ and $\rho^{1/3}$ with auxiliary functions
- analytical integrals
- slowest of Xα methods -work goes up about as m^3
- accurate total energies
- e.g. Ni_4 (55)

The method which we have used in most of our work on metal clusters is the Xα-SW method developed by Johnson which is quite unlike the other quantum chemical methods. It has its roots in the Korringa-Kohn-Rostoker method of band theory (56,57). Rather than expanding the orbitals in some set of basis functions which are defined for all of space, one first breaks up space into spherical regions around the various nuclei. If the potential which enters Schrödinger's equation is approximated within these spheres by its spherical average

$$V_{sphere,j}(\vec{r}) \rightarrow \bar{V}_{sphere,j}(r) \tag{2.52}$$

then within the spheres Schrödinger's equation is simply that of a central potential which can be readily solved by numerical integration in a single variable, r. The wave function is given by

$$\psi_{sphere,j}(\vec{r}) = \sum_L C_L^j R_\ell^j(\varepsilon;r) Y_L(\hat{r}_j) \tag{2.53}$$

where $L \equiv (\ell,m)$, Y_L is a spherical harmonic and the radial wave function R is an implicit function of the energy. To complete the definition of the so-called muffin-tin potential one assumes a volume averaged (constant) potential for the regions between the spheres (and inside an outer sphere surrounding the whole molecule).

$$V_{intersphere}(\vec{r}) \rightarrow \bar{V}_{intersphere} \qquad (2.54)$$

The solution of Schrödinger's equation is also well known for the case of a constant potential and can be written in several forms. That used here is a multicenter expansion in partial waves. For example if the energy is below the constant potential the wave function is:

$$\psi_{intersphere}(\vec{r}) = \sum_j \sum_L A_L^j \, k_\ell^{(1)} (\kappa r_j) \, Y_L(\hat{r}_j) + \sum_L A_L^o \, i_\ell (\kappa r_o) Y_L(\hat{r}_o) \qquad (2.55)$$

where $\kappa = (V-\varepsilon)^{\frac{1}{2}}$, $i_\ell{}^{(1)}$ is a modified spherical Bessel function and $k_{\ell_1}{}^{(1)}$ is a modified spherical Hankel function. The first term in (2.55) corresponds to waves "scattered" by the atomic spheres and the second term to scattering by the outer sphere. The unknown coefficients (the C's and A's) are determined by requiring continuity of the wave function and of its first derivatives across the various sphere boundaries. This yields a set of secular equations which, despite the complicated appearance of equations (2.53) and (2.55), can be rapidly formed and solved. Further details may be found in Johnson's paper (14) and Rösch's NATO lecture notes (58) as well as papers and reviews on various applications (4-15).

A prime reason for the success of the scattered-wave method is the fact that the radial functions are found numerically and are hence entirely flexible; there is no problem of basis set choice. A major drawback of the method is that the total energy (and hence molecular geometries) cannot be accurately evaluated in an efficient manner.

3. COMPARISONS OF HARTREE-FOCK WITH Xα(LSD) RESULTS

I would now like to turn to a few examples to show how the local density methods perform in practice. I have chosen to do this via comparisons with the HF method since both approaches involve orbitals and one-electron equations. In this way we

Table 5. Advantages and Disadvantages of Hartree-Fock and Local Density Methods

Hartree-Fock

PROS:
- well defined
- can be systematically corrected
- "technology" well developed

CONS:
- computationally demanding
- often requires extensive correlation corrections
- impracticable for large system
- fundamental inadequacy for metals: $N(\varepsilon_F) = 0$

Local Density Methods

PROS:
- rapid (especially SW)
- accurate (often appear to mimic correlated calculations)

CONS:
- no procedure for going beyond one-electron approximation
- not strictly variational (approximate Hamiltonian)
- difficulties with total energy (SW)
- relatively young "technology"

will start to get an idea as to the circumstances in which the local density methods appear to be incorporating correlation corrections. I will also make a few comments as to how this is coming about, although it must be stated at the outset that there is at the present time no clear-cut formal justification for this contention.

In Table 5 I have summarized some of the most important relative advantages and disadvantages of each approach. The only new element in this table, beyond what we have discussed, at least by inference, in the preceding sections, is specific to metals. It is well known for the electron gas and has recently been proven for metals in general (59) that, within the HF approximation, the density of states at the Fermi level is rigorously equal to zero. To say the very least this hinders the interpretation of all the typical metallic properties and underscores the important role of correlation. If one has in mind a series of clusters of increasing size leading to the bulk metal, then it is certain that the infinite limit is poorly treated in HF theory.

3.1 H_2

Even though it is rather far removed from our main interest, metal clusters, let us start with H_2 since we can make detailed comparisons with the treatment of section 2.2. The potential curves calculated with local density (LD) and local spin density (LSD) methods (47) are compared in Figure 6 with those of HF, valence bond (HL) and a highly accurate (exact) calculation. It is clear that the LSD method gives a very accurate curve indeed, which is all the more impressive given that the calculations involve no parameters beyond those needed to fit the electron gas exchange-correlation data.

Like the HF method, the LD method does not dissociate properly, owing to the forced delocalization of the molecular orbital over the two centers, in order to satisfy the symmetry constraints. These constraints can be removed in a spin-polarized (LSD) calculation in which a symmetry subgroup ($C_{\infty v}$ instead of $D_{\infty h}$) is used. At long distance (beyond 3.2 a_o) the orbitals (one of α spin, one of β) localize (like the GVB orbitals of section 2) on one center or the other to form a "spin density wave" state. The two orbitals are like

$$\tilde{a}(1)\ \alpha(1) \sim \{a(1) + \lambda b(1)\}\ \alpha(1) \tag{3.1}$$

and

$$\tilde{b}(2)\ \beta(2) \sim \{b(2) + \lambda a(2)\}\ \beta(2) \tag{3.2}$$

This leads to the correct dissociation asymptote. However, unlike the GVB wave function, a determinant constructed from these "unrestricted" LSD orbitals is not a proper eigenfunction of total spin. Such a proper eigenfunction could be formed by mixing two determinants formed from the orbitals (3.1 and 3.2) and their mirror image spin-flipped counterparts. The good agreement found between LSD and exact calculations indicates that the effects on the energy of such mixing would be slight.

At shorter distances, and in particular, at the minimum (R_e) the LSD and LD curves coincide and both correspond to completely delocalized symmetry-adapted orbitals. This behaviour may be contrasted with that of the GVB orbitals in Figure 3 which, at R_e, are still localized more on one center than the other (as a result of left-right correlation). At R_e the LD orbital is likely very similar to the HF orbital (for a minimum basis set they are identical, $\phi = 1s_a + 1s_b$) and yet the LD energy is very nearly exact whereas the HF energy is not. Hence the source of the accurate LD potential curve lies not in any drastic improvement of the orbitals vis-à-vis HF but rather in the use of a local density functional for the energy. This is

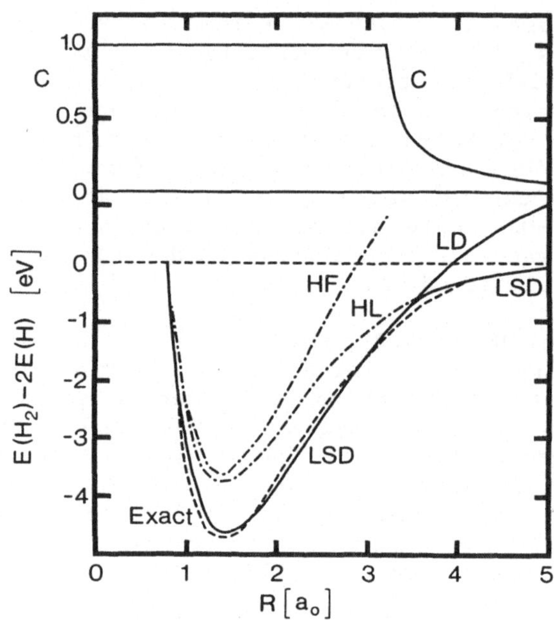

Figure 6. Energy of the hydrogen molecule as a function of the internuclear distance. The curves show the results in the LD and LSD approximations, which differ for R > 3.2. The dashed curve gives the accurate result of reference 60. The upper part of the figure shows the quantity c, which gives the degree of spin polarization. For c=1 the system is unpolarized. Reproduced from reference 47.

somewhat akin to the "correlation corrections" inherent in the parameters of semi-empirical methods mentioned earlier. However, in the present case the correlation corrections are not being introduced by any ad hoc adjustment of parameters to match experiment but are instead inherent to the functional and there is at least the hope that further work on the N-electron problem (from both the "ab initio" and density functional viewpoints) will shed some light on how this comes about.

3.2 First Row Diatomics

The success of the local density method for the potential curve of H_2 is not an isolated incident. In Table 6 I have compared experimental equilibrium distances (R_e) and dissociation energies (D_e) for some first row diatomics with values

Table 6. Equilibrium Distance (a_o) and Dissociation Energies (eV) for Diatomic Molecules from experiment and from Hartree-Fock and Xα Calculations

	exp.(61)	numerical Xα (62)	LCAO Xα (54)	DVM Xα (63)	HF(64,65)
			$R_e(a_o)$		
B_2	3.00	3.04	3.04		
C_2	2.35	2.36	2.35		2.37
N_2	2.07	2.06	2.08	2.13	2.01
CO	2.13	2.12	2.13	2.15	2.08
O_2	2.28	2.26	2.28	2.36	2.18
F_2	2.68	2.61	2.61	2.67	2.50
			D_e(eV)		
B_2	3.0	3.8	3.9		0.9
C_2	6.3	6.2	6.0		0.8
N_2	9.9	9.6	9.2	8.4	5.2
CO	11.2	12.1	12.0	11.5	7.8
O_2	5.2	7.1	7.0	6.6	1.3
F_2	1.7	3.3	3.2	2.9	-1.4

calculated by the HF method and the Xα method in three realizations (LCAO, DVM and some recent completely numerical calculations which may be regarded as representing the Xα limit). The Xα results are uniformly better, particularly for D_e. For the homonuclear molecules the full symmetry ($D_{\infty h}$) has been used so that like the case of H_2, the correlation corrections are coming in through the energy functional rather than having "correlated" (localized à la GVB) orbitals. (In passing one should note that by comparison with the numerical calculations the LCAO-Xα calculations are more accurate than the DVM-Xα which may indicate some problems with the sampling grid in the latter method).

3.3 Na_2

LD and LSD (pseudopotential) calculations for Na_2 (66) (Figure 7) show features entirely analogous to those for H_2. The agreement with experiment for the potential curve is very good indeed. In contrast to this, Hartree-Fock calculations using a pseudopotential (40,41) predict Na_2 to be unbound and an all-electron calculation (40) yields only a small fraction of the binding energy (1.45 kcal/mol vs 8.40 (exp.)).

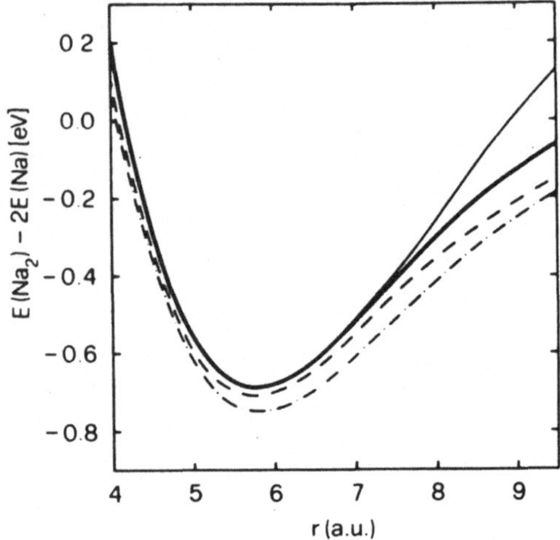

Figure 7. Binding energy curve of Na_2. Thick solid line: LSD pseudopotential results; thin solid line: LD pseudopotential results; dashed line: CI-SCF pseudopotential results (Ref. 67); dash-dotted line: experimental RKR curve (Ref. 68). Reproduced from reference 66.

3.4 Ozone

The ozone molecule provides an interesting comparison between HF and Xα theories (69-70). Here HF(71-74) theory predicts the wrong ground state [a triplet ($^3B_2$) rather than the observed singlet ($^1A_1$)] and also the wrong order for the states corresponding to the lower ionization potentials (IPs). On the other hand Xα theory(70,75) yields a $^1A_1$ ground state and also the correct order for the IPs.

In order to shed further light on the situation we have carried out (69) LCAO-Xα and HF calculations on ozone at its equilibrium geometry. The lowest three vertical IPs calculated by various methods are compared with the experimental values in Table 7. The LCAO-Xα IPs are in excellent agreement both with correlated (GVB CI) calculations and with experiment. The HF results, on the other hand do not agree with experiment, an important discrepancy being that the $^2A_2$ state (ionization of an $a_2(\pi)$ electron) has far too low an IP relative to the $^2A_1$ and $^2B_2$ states which correspond to removal of σ (lone pair) electrons. Hence, in HF theory it appears that the $a_2(\pi)$ level is not stable enough. The source of this discrepancy is well understood following the work of Hay et al (71). An acceptable qualitative description of the ozone ground state requires the mixing of at least two determinants. The HF configuration for the $^1A_1$ state, Φ_1, has two

Table 7. Ionization Potentials for Ozone (eV)

State	Exp.[a)]	Xα SW [b)]	LCAO-Xα[c)]	GVB-CI[d)]	HF(ΔE_{scf})[e)]
$^2A_1(\sigma)$	12.75	13.00	12.68	12.91	14.20
$^2B_2(\sigma)$	13.06	12.96	13.01	13.03	14.33
$^2A_2(\pi)$	13.57	13.43	13.77	13.59	12.19

a) Refs. (76-78) b) Ref. (70) c) Calculated with the 9s/5p basis of ref. (79) d) Ref. (72)

electrons in the $1a_2$ MO which may be represented approximately as $\phi(a_2) \approx N(\pi_\ell - \pi_r)$ where π_ℓ and π_r represent $p\pi$ functions on the left and right end atoms. However there exists a second important configuration, Φ_2, which involves a double excitation from $1a_2$ to a virtual b_1 orbital that is approximately $\phi(2b_1) \approx N[\lambda(\pi_\ell + \pi_r) + \mu\pi_c]$ (π_c is on the central atom), where λ and μ are MO coefficients. A mixture of Φ_1 and Φ_2 serves to reduce ionic contributions (e.g. π_ℓ^2) which are overestimated in Φ_1. The generalized valence bond orbitals corresponding to the mixture of Φ_1 and Φ_2 are highly localized on either end atom so that ozone has a high biradical character. Hence, the wave function of the ozone ground state cannot be correctly described by a single determinant of symmetry-adapted functions.

Similar to the H_2 and Na_2 examples, it is therefore clear that the accurate values for the IPs in Xα theory do not arise from taking account of correlation in the detailed sense just out-lined. There remain two possibilities. It is conceivable that the Xα molecular orbitals could be in some manner more appropriate than HF orbitals for the calculation of ionization potentials, even though the HF orbitals yield the better total energy, 2). In order to test this possibility we have calculated "Koopmans' theorem" IPs using the Xα orbitals and the HF eigenvalue expression. These are compared in Table 8 with the eigenvalues from a HF calculation with the same basis set. It is clear from Table 8 that the Xα MOs must be quite similar to those from HF theory since the same erroneous order of IPs is obtained for both sets of orbitals when the HF hamiltonian is used. Similar conclusions concerning the similarity of HF and Xα orbitals

---

2) It should be re-emphasized that our calculations used the full C_{2v} symmetry of O_3 so that no localization of the orbitals was allowed. Indeed we have carried out some tests and so far have not been able to obtain broken-symmetry solutions for O_3 at its equilibrium geometry. So, like H_2 and Na_2, "correlation-corrected" energies are being obtained with orbitals that are not "correlated" (localized).

Table 8. Hartree-Fock Orbital Eigenvalues for Ozone Calculated with LCAO Xα and with Hartree-Fock orbitals (eV)[a]

Orbital	$-\epsilon_{HF}$	
	Xα	HF
$4a_1(\sigma)$	13.88	15.10
$3b_2(\sigma)$	14.45	15.56
$1a_2(\pi)$	12.19	13.28

a) Calculated with the 8s/4p basis of ref. (80)

have previously been reached for atoms, for example, by Wood (81) and by Tseng et al. (82).

The only remaining possibility is that the essential differences between the two methods arise from the different treatment of exchange in the hamiltonians or equivalently in the energy or eigenvalue expressions. The individual orbital terms entering U_{XHF} and $U_{X\alpha}$ are compared in Table 9 . The same orbitals (from an LCAO-Xα calculation) have been used in both cases so the differences reflect differences in the energy expressions. The Xα potential has an overall smoothing effect on the exchange terms and most importantly, that for the $1a_2(\pi)$ orbital, which in HF theory is very small compared with those for near-lying levels, takes a value in Xα theory which is quite similar to those of the other orbitals. This leads to a stabilization of the $1a_2$ level by 3eV relative to the nearby $3b_2$ to reverse the erroneous order of the levels found in HF theory and also to bring the Xα IPs into agreement with those from correlated calculations and also with experiment.

On the basis of these results it is tempting to speculate that the success of the Xα method vis-à-vis HF in other cases may arise for similar reasons. One expects HF theory to encounter difficulties in cases were localized orbitals (and valence bond theory) would be more appropriate. Such cases typically involve in single determinantal theory, a high-lying occupied MO and a low-lying vacant MO with appropriate phases of the AO components to yield localization if a mixture of determinants is formed. Since this occupied MO in HF theory is at high energy one would expect it to have a relatively small exchange term. In a rough manner of speaking, configurational mixing serves to stabilize this orbital (in the sense that one might label a peak in a photoelectron spectrum with a single orbital label, even though orbital language is not rigorously appropriate) an effect which is also accomplished by the Xα method through the smoothing effect of the local exchange potential. However, further calculations will be necessary in order to examine the validity of the foregoing hypothesis.

Table 9. Comparison of Hartree-Fock and Xα Exchange Terms (eV)[a)]

Orbital (i)	$\sum_{i}^{occ} K_{ij}$	$\langle \phi_i \| v_{X\alpha} \| \phi_i \rangle$	Δ
$4a_1$	28.09	32.65	4.56
$3b_2$	28.36	32.47	4.11
$1a_2$	23.83	30.90	7.07
$3a_1$	28.43	31.63	3.20
$2b_2$	28.60	33.31	4.71
$1b_1$	25.82	30.19	4.37
$2a_1$	33.02	34.55	1.53
$1b_2$	33.18	34.70	1.52
$1a_1$	34.85	34.51	-0.34
$1s(a_1)$	132.49	127.81	-4.68
$1s(b_2)$	132.49	127.83	-4.66
$1s(a_1)$	132.47	127.83	-4.64

a) Orbitals from an LCAO-Xα calculation using the 8s/4p basis set of ref. (80).

3.5 Al_{19}, Al_{19} + O

A few years ago we carried out an extensive Xα-SW study of aluminum clusters including some chosen as models for the various surfaces of aluminum (7,84-86). These clusters were then used to study the early stages of oxidation of aluminum by chemisorbing oxygen atoms at various sites above, in and beneath the surfaces. These calculations proved highly useful in interpreting ultraviolet and X-ray photoemission results. In fact, a peak due to oxygen non-bonding electrons was predicted from the calculations (85) before it was observed experimentally (87). We were also able to deduce geometric information, in an indirect manner, through comparisons of experimental photoemission data with calculations for various geometries. A particularly interesting situation arose for the case of Al(111) + O. Our calculations for Al_{19} + O (86) indicated that the oxygen atom should be adsorbed at a perpendicular distance ($d\perp$)between 0.5 and 1.0 Å above a three-fold hollow site. Soon thereafter a LEED analysis (88) found $d\perp$ = 1.33 Å, much larger than our value. More recently surface EXAFS (89) and a reanalysis of the LEED data (90) have yielded $d\perp$ = 0.7 Å in good agreement with our predictions. Overall, the results for the aluminum-oxygen system indicate that the cluster model (with well-chosen clusters of sufficient size) coupled with the Xα-SW method provides a powerful interpretive and predictive tool for the study of chemisorption.

Recently some approximate Hartree-Fock pseudopotential calculations have been carried out for Al_{19} and Al_{19} + O (23). These provide interesting comparisons with our previous Xα work and help to put the two methods in perspective in the context of metal clusters. One quantity we can compare is the density of states (DOS) which is obtained from the cluster eigenvalues by replacing each discrete level with a gaussian in order to generate a continuous curve. Such DOS curves from Xα-SW calculations on aluminum clusters of various sizes are shown in Figure 8.

As the cluster size increases, so does the bandwidth and the DOS becomes smoother. The largest clusters provide a very good approximation to the DOS for bulk aluminum as obtained from band theory or from experiment. For example the occupied bandwidth for Al_{19} is about 85% of that of the bulk (see also(84) for further comparisons).

Of course the DOS is not a direct observable and comparisons such as the above must be made with caution and with a particular experiment in mind. For example, the interpretation of photoelectron or X-ray emission spectroscopy results (84,85) requires consideration of final-state relaxation, any possible localization of the positive hole, intensity matrix elements and also any correlation effects which are beyond those inherent in the local density approach. The fact that the DOS for Al_{19}, which is based an ground state unrelaxed eigenvalues, yields a reasonable, straightforward interpretation of these experiments has several implications. First, it implies that the basic electronic structure of Al_{19}, as probed by these experiments, resembles that of bulk aluminum. Second, it implies that, similar to the case of the Cu atom discussed in section 3.4, there are no large differential effects of the self-interaction and relaxation corrections. This has been verified by explicit calculations for a few levels. Third, it implies that the treatment of correlation provided by the calculations is adequate for these purposes.Finally, a general conclusion is that the DOS and for the case of chemisorption, the partial DOS for the adsorbate (7,85,86)) from a local-density cluster calculation is a useful quantity and can be employed directly in the interpretation of this type of experiment.

The situation is quite different in HF theory. The DOS generated from the ground state eigenvalues of a HF calculation for Al_{19} are shown in Figure 9. A straightforward comparison with experiment, such as that given above for the local density approach, is quite unsatisfactory. The band is about 5 eV too wide and there is a deep valley at the Fermi energy. This is the forerunner of the $N(\varepsilon_F)=0$ artefact of HF band theory which we mentionned earlier. It arises essentially from the fact that the

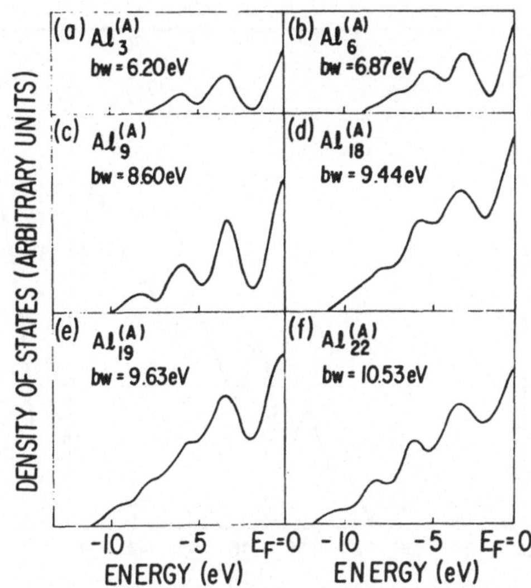

Figure 8. Density of states (arbitrary units) of the occupied valence band for representative clusters of various sizes vs orbital energy (eV) relative to the Fermi level. The curves were generated by replacing each discrete cluster eigenvalue by a Gaussian of width parameter 0.68 eV. bw is the width (in eV) of the occupied valence band. Reproduced from reference 86.

occupied and vacant orbitals in HF theory are treated differently. The eigenvalues of occupied levels correspond to ionization potentials (a field of N-1 electrons) whereas the energies of vacant levels are determined by a field of N electrons and are hence destabilized by the "extra" electron. The Hartree-Fock DOS is not particularly useful.

It is clear that in this "ab initio" approach, substantial (and arduous) relaxation and correlation corrections would have to be incorporated to bring the calculations in line with experiment, assuming (after the Xα work) that the relevant properties of Al_{19} are sufficiently bulk-like that such comparisons are appropriate.

Interestingly, despite the inadequacies of the HF description of Al_{19}, a calculation for Al_{19} + O yields an equilibrium geometry in good agreement with experiment (and with the Xα calculations) for Al(111) + O. This simply serves to underline the fact that different properties have different sensitivities to the details of the computational method.

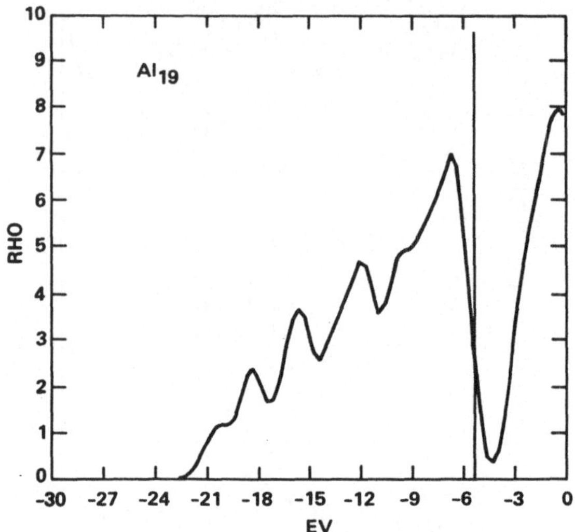

Figure 9. The continuous cluster density of states for Al_{19} formed by fitting Gaussian functions of width 1eV to the Hartree-Fock eigenspectrum. Reproduced from reference 23.

3.6 Cu_n

Considerable controversy has surrounded the application of Xα and HF theories to small copper clusters. While I do not have the space to detail every point of contention I will try to outline the major issues. Messmer et al. (8) found that the distribution of energy levels from an Xα-SW calculation for Cu_{13} showed many of the characteristics of the band structure of bulk copper. In particular, a relatively dense region of levels having high d character was overlapped by a more sparsely spaced set of s type levels, analogous to the s band overlapping the d band in bulk copper. HF orbital eigenvalues (19) however are distributed quite differently, the s levels lie at lower binding energies and the "d band" is beneath them. Hence the two groups come to opposite conclusions. The Xα calculations indicate that the electronic structure of Cu_{13} is in many respects similar to that of bulk copper whereas the HF workers conclude that the clusters are fundamentally different (although it should be pointed out that calculations within this approach for the infinite limit, i.e. HF band calculations have not been carried out).

As we have just seen for the case of aluminum, great care is needed in drawing conclusions from such comparisons. If one is interested only in ground state properties and if one accepts the justified premise that (local density) band theory yields a reasonable interpretation of these for bulk copper, then the main conclusion of Messmer et al is straightforward and fully justified. The s and d features of the DOS for ground state Cu_{13} overlap as they do for bulk copper. If one then wishes to consider experiments involving a hole in the valence band, then self-interaction, relaxation, localization (as we shall see in a moment) and correlation effects should be considered for both the cluster and the infinite solid. For the Xα calculations on Cu_{13}, it was found that the self-interaction-relaxation effects were similar for different orbitals so that a DOS based an ionic states, which can be compared with photoemission results, would be similar to that based on the ground-state eigenvalues. Again, this results from approximate cancellation of the self-interaction and physical relaxation corrections which are of opposite sign and large for the more localized d levels and small for the delocalized s levels. No explicit account of relaxation (or correlation) effects was taken in the HF calculations, although it was argued that reasonable relaxation corrections would still not yield overlap of the s and d bands.

Now we come to a somewhat subtle point. If one wants to use Cu_{13} as a model for bulk copper as given by band theory then these self-interaction and relaxation terms should be turned off since, in band theory, the positive hole is delocalized over the whole crystal. Hence there is justification, in addition to the empirical observation that the net relative effects are small for Cu_{13}, for a straight comparison of orbital eigenvalues if one is modeling the bulk. Of course, the band theory is not an exact one and one might wonder whether the effects of correlation or localization of the orbitals might be important, especially for the more compact d orbitals. This turns out to be the case for the ionic states of Cu_2. In an interesting turn of events, two recent series of calculations have been performed in order to resolve the Xα-HF controversy. Cox et al. (20) have performed HF calculations on Cu_2^+ using reduced $(C_{\infty v})$ symmetry. They find that the 4s hole remains delocalized but that the 3d hole localizes on one center. This changes the relative positions of s and d levels substantially so that the authors conclude that on the basis of ionic states "the d band appears as totally overlapped for clusters as small as Cu_5, the two bands overlapping appreciably for Cu_3 and Cu_4". They also found through CI calculations for Cu_2^+ in $D_{\infty h}$ and $C_{\infty v}$ symmetries that the symmetry breaking accounts for 2eV of the 2.5eV correlation energy (for their set of configurations). Equivalent CI calculations for neutral Cu_2 would be necessary to ascertain the correlation corrections to the IP's.

Table 10. Ionization Potentials (eV) of Cu_2 from Hartree-Fock and Xα Calculations.

		HF(ΔSCF)(20)	Xα (27)
"d"	$D_{\infty h}$	9.4	8.8-9.7
	$C_{\infty v}$	6.6	10.7
"s"	$D_{\infty h}$ (or $C_{\infty v}$)	5.7	7.7

DVM-Xα calculations have been performed by Post and Baerends (27) for Cu_2 and Cu_2^+. They find that the s and d "bands" overlap for the ground state calculation but that removal of an electron causes rehybridization of the orbitals so that for the ion only one state can be considered as arising from removal of an s electron and its IP is less than those for removal of d electrons. When the symmetry constraints are relaxed it is found that, like in HF theory, the 3d (and not the 4s) orbitals localize; however, contrary to the HF results this leads to an increased separation between "d" and "s" levels (Table 10) (see (27) for a discussion). The authors conclude that "based on the ionic states, the Xα calculations do not show the overlapping of the 3d band by the 4s band". Hence, the "controversy" has not been resolved but only the direction has changed if one considers ionic states. It would be interesting to know if the localization of the d electrons persists for ionic states of larger clusters and ultimately of bulk copper.

It is clear that more work is needed on these systems and that great care is needed in making comparisons. For example different quantities are appropriate if one is interested in ionization potentials of isolated molecules than if one seeks a cluster model of the ground state properties of bulk materials. An extensive correlated "ab initio" treatment of both neutral and ionic Cu_2 would be of great value. If a photoelectron spectrum could be obtained for a beam of Cu_2 molecules it would obviously help to clarify the situation.

4. ELECTRONIC AND MAGNETIC STRUCTURE OF TRANSITION METAL CLUSTERS

As the final subject of these lectures I would like to present a brief survey of some of our own work with the Xα-SW method applied to transition metal and "alloy" clusters. The clusters in which we have been interested typically contain a dozen or more transition metal atoms and hence it will not be possible to make comparisons with correlated "ab initio" treatments nor to make any detailed analysis of any "correlation" contributions. We must simply have faith, based on past experience, that the method is appropriate and judge the results by their utility in interpreting the relevant experimental data. (Most of this data is for bulk materials so that comparisons sometimes become somewhat clouded by the effects of the finite size of the clusters. Remember, the second sub-theme of these lectures is that there is a great need for more experimental data on isolated clusters.)

4.1 The 3d Elements

The transition metals of the 3d series are of paramount interest for a number of reasons, principally i) Iron and its alloys are at the heart of metallurgy. ii) Transition metals in supported particulate form are essential catalysts in many industrially important processes. iii) Several of the transition metals (Cr, Mn, Fe, Co, Ni and some of their alloys) possess magnetic moments which can be put to use in technology. The unique properties which make the transition metals so useful all stem from the presence of unfilled d shells and the extreme sensitivity of the properties to the details of the electronic structure (d-band filling, s-d hybridization, the detailed shape of the density of states, specific chemical interactions, etc.). As a result, metallurgical, catalytic and magnetic fine-tuning can be achieved by varying such parameters of the metal as the composition (alloying or "promoting" a catalyst), the surrounding medium (support effects) or the particle size in addition to the usual variables of temperature and pressure.

From a theoretical point of view the presence of partially occupied and interacting d shells presents a definite challenge. As we have seen, adequate "ab initio" treatment of transition metal systems requires an enormously laborious, highly correlated calculation using an extremely extended basis set. Such calculations can presently only be performed for diatomics or very small polyatomic clusters. On the other hand, band theoretical methods (13) using the $X\alpha$ (or local density (LD) or local spin density (LSD) potentials), while they do not treat the spatial and spin correlations with the same detail as the "ab initio" quantum chemical methods, do yield a useful level of agreement with experiment for a wide variety of properties (13, 91). In a molecular or cluster context the analogue of these band structure techniques is the SCF-$X\alpha$ (or LD, or LSD)-SW molecular orbital method.

$X\alpha$-SW calculations have now been performed for clusters containing about 15 atoms (a central atom and its nearest (and, for the bcc structure, second nearest neighbors) for all of the 3d metals. One of the goals of this series of calculations was to examine to what extent the properties of clusters in this size range reflect those of the bulk metal. The presence of magnetic moments in these materials is of prime importance, not only for the study of magnetism in its own right but also since the energy level shifts induced by the spin-polarization are often large and can have profound effects on the other (e.g. metallurgical and catalytic) properties. Hence, a natural starting point for the discussion of the cluster properties is to examine the magnetism. The calculations are spin-polarized and during the self-consistent cycle electrons can pass from one spin manifold to the other in order to lower the energy. At the end of the calculation, the difference between the number of up (↑) spin and the number of down (↓) spin electrons is equal to the magnetic moment of the cluster. (Orbital angular momentum is neglected since for the 3d metals it is almost entirely quenched by the crystal field). In Table 11 we compare the calculated average moments for the clusters with their experimental bulk counterparts.

The overall tendencies towards magnetization in the solids are faithfully reproduced by the clusters. The clusters of non-magnetic metals (Sc,Ti,V,Cu) all converge either to a closed-shell configuration ($n^{\uparrow}=n^{\downarrow}$) or, for the clusters having an odd number of electrons, to a configuration with a single unpaired spin. For the ferromagnetic elements (Fe,Co,Ni) the clusters are indeed ferromagnetic with average moments in reasonable agreement with the bulk values. The cases of Cr and Mn are special (antiferromagnetic order) and will be discussed in more detail below.

A further comparison with the bulk limit is afforded by generating a cluster density of states (DOS) which may be compared with that from a band calculation. Such a comparison is shown in Figure 10 for the Fe_{15} cluster (92).

Table 11. Comparison of Calculated (SCF-Xα-SW) Average Magnetic Moments for Transition Metal Clusters with their Experimental Bulk Counterparts (μ_B).

Cluster	Structure	$(n^{\uparrow}-n^{\downarrow})$	$(n^{\uparrow}-n^{\downarrow})/N$	μ (exp)
Sc_{13}	hcp	1	0.1	0
Ti_{13}	hcp	0	0	0
V_{15}	bcc	1	0.1	0
Cr_{15}	bcc	12	0.8 (AF)	0.6 (AF)
Mn_{13}	fcc	11	0.8 (AF)	
Fe_{15}	bcc	40	2.7	2.2
Co_{13}	hcp	21	1.6	1.7
Ni_{13}	fcc	8	0.6	0.6
Cu_{13}	fcc	1	0.1	0

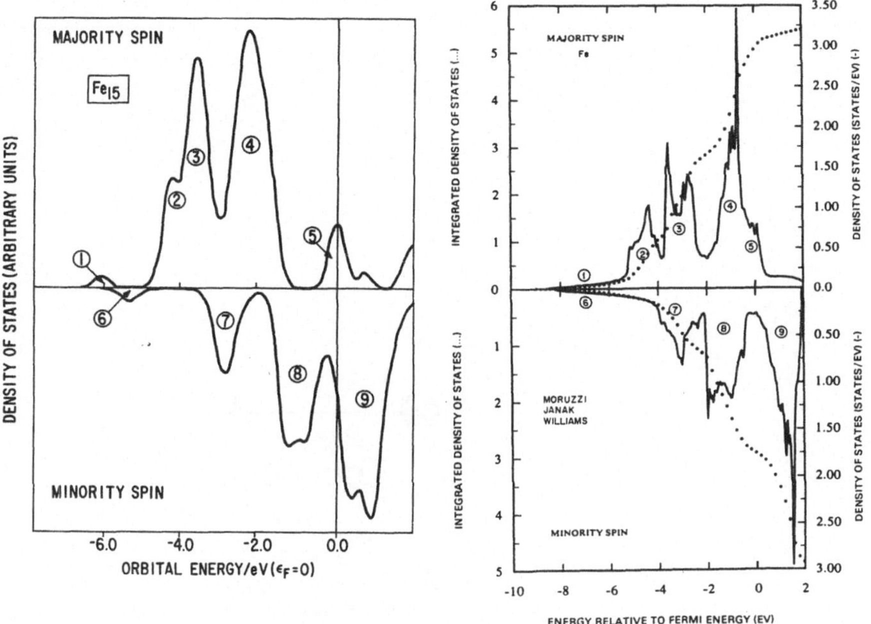

Figure 10. Comparison of cluster (92) and bulk (91) DOS for spin-polarized iron. Reproduced from reference 92.

There is a reasonable correspondence, the large exchange splitting being a dominant aspect. The major features of the band DOS are present in the cluster DOS at similar energies. The spin-density of Fe_{15} has also been generated and this is compared in Figure 11 with the experimental magnetization density deduced from neutron diffraction (93).

The distribution is characterized by nearly (but not quite) spherical regions of positive magnetization density centered at the nuclei. In the experimental figure these regions have been reduced in size in order to emphasize the oppositely polarized areas further from the nuclei which have the shape of interlocking tori. The degree of correspondence between the bulk experimental spin density and that calculated for the 15-atom cluster is remarkable.

These and other results indicate that quite small clusters are capable of embodying at least the gross features of the electronic and magnetic structures of the bulk metals. However, this should not be taken to imply that the clusters are simply miniaturized versions of the bulk solids. Indeed, as stated above

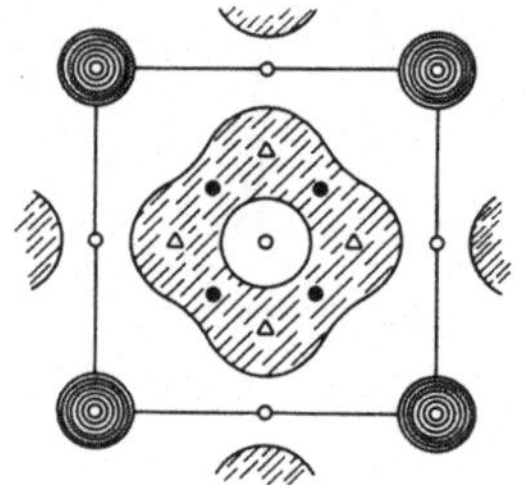

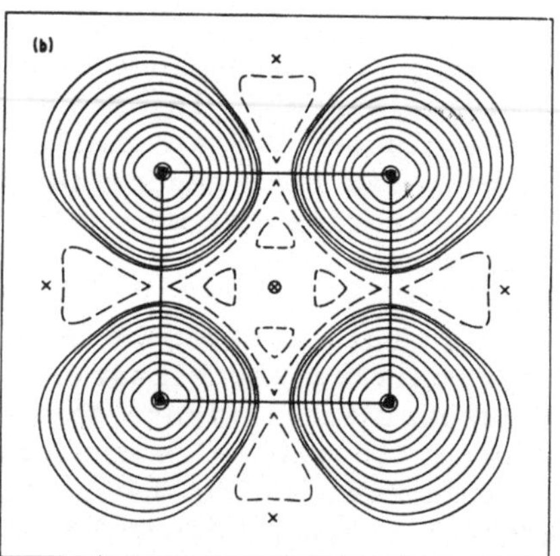

Figure 11 Comparison of spin density in a 100 plane of Fe_{15} with the experimental magnetization density from neutron scattering experiments. Reproduced from reference 92.

many of the technological applications depend on the variability of the properties upon changes in the chemical and structural environment. In fact, if the cluster results are examined at a finer level it is found that the magnetic (and other) properties which may be assigned to the various atoms do depend on their local environment. The results for Cr_{15} will illustrate this point.

The magnetic structure of bulk chromium has been extensively studied (13). It is characterized by a spin-density wave which is similar to that of an antiferromagnetic compound but of a more complex periodicity. For our purposes it is sufficient to consider chromium as a perfect antiferromagnet with alternating signs of the moments on two spin sub-lattices. The bulk moments have a maximum value of about $0.6\mu_B$. A calculation on Cr_{15} (central atom, Cr(1); 8 first neighbors, Cr(2); and 6 second neighbors, Cr(3)) yields an antiferromagnetic arrangement of local moments. In this respect Cr_{15} is similar to the bulk. The magnitudes of the moments on the three types of atom are however quite different; $-0.7\mu_B$ for Cr(1), $+ 4.1\mu_B$ for Cr(2) and $-3.4\mu_B$ for Cr(3). It seems that the effects of the finite size of the cluster are quantitatively more serious for Cr than for the other non-magnetic or ferro-magnetic clusters. In fact the chromium cluster shows a behaviour similar to that found in recent calculations (94,95) for slabs of chromium chosen to study the surface. Allan (94) finds that a non-magnetic calculation exhibits a high DOS at the Fermi level owing to states localized on the surface layer. This leads, through Stoner arguments,(92) to large surface moments. Similar statements apply to the clusters.

Nowhere in the 3d series is the need to take some account of correlation (and localization) more apparent than for chromium (and its neighbour manganese). There is a delicate balance between magnetic and chemical bonding effects and much is to be learned through a study of the sequence of clusters, starting from the dimer, Cr_2, and ending with bulk chromium. Roughly speaking it seems at the present time that small clusters (and surfaces) are dominated by the magnetic aspects. Large, antiferromagnetically aligned magnetic moments develop to the detriment of the interatomic binding. As more and more neighbours are added the situation changes until in bulk Cr one has only small magnetic moments and a highly cohesive solid. The complications of the electronic structure in this series are only beginning to be unraveled and the challenges to theory are both awesome and exciting.

The antiferromagnetic ground state is intrinsically more complex than either non-magnetic or ferromagnetic cases. This is perhaps best illustrated by recent calculations (25) on the diatoms Cr_2 and Mo_2 and on binuclear complexes involving Cr and Mo. At the very least, to allow a spin arrangement resembling antiferromagnetism, one has to consider the spin- and space-unrestricted versions of molecular orbital theory. In this type of calculation, as we have seen, one does not have proper eigenfunctions of the spin operator, $\hat{S}^2$ and furthermore the orbitals do not necessarily reflect the full spatial symmetry of the molecule. Nevertheless, for Cr_2 the unrestricted Hartree-Fock (25) or unrestricted Xα (96) picture of two large, highly localized moments, antiferromagnetically coupled, is reasonably similar in many respects to the picture which emerges from more accurate, highly correlated calculations (25). The principal difference is that in the more accurate calculations, the wave function consists of a linear combination of many determinants, each of which is "antiferromagnetic", and the proper space and spin symmetries are obtained, with the result that for Cr_2 there is no net spin density on the atoms. However, some experiments (notably neutron diffraction) can be interpreted in terms of magnetic moments on the atoms, so that in this context the unrestricted calculations have merit. Indeed band theory has successfully interpreted the neutron results for solid chromium, in terms of a spin-density wave (SDW) state, and even the incommensurability of the SDW has been explained in terms of nested Fermi surfaces (13). So while we are aware that a full many-body treatment of even the simplest chromium cluster, Cr_2, is extremely intricate and that it may be some time before a completely satisfactory picture of the electronic structure of larger chromium clusters emerges, we believe that useful results can be obtained within the context of local spin-density MO theory if proper caution is exercised and if appropriate attention is paid to the role of symmetry constraints.

We have completed (96) Xα-SW calculations for Cr_2, Cr_8, Cr_9 and Cr_{14} as well as Cr_{15}. This series underlines the important role of the symmetry constraints. All of the calculation show a basically antiferromagnetic arrangement of atomic moments. For Cr_9 and Cr_{15} which are built around a central atom, this occurs "naturally" i.e. when the full point-group symmetry of the cluster is employed. (Viewed as models for the infinite solid, the periodic symmetry of the lattice has been broken, allowing the formation of spin sub-lattices). For Cr_2, Cr_8 and Cr_{14} it is necessary to break the point group symmetry in order to allow the localization of the orbitals necessary for antiferromagnetism. All of these clusters develop moments which are much larger than that observed for bulk chromium and this can be traced to a high density of (surface) states at the Fermi level of the unpolarized cluster.

To summarize, clusters containing about 15 atoms yield a useful representation of the electronic structure of the non-magnetic metals (Sc,Ti,V) to the left of the periodic table as well as for the noble metal Cu. The ferromagnetic metals (Fe, Co, Ni) are also well represented as far as their DOS, and the approximate magnitudes of the magnetic moments are concerned. Among other matters, since most of the cluster atoms are on the surface this indicates that the magnetic properties of the surfaces of Fe, Co and Ni are not radically different from those of the bulk. The metals in the middle of the periodic table (Cr, Mn) show an intermediate behaviour. Both clusters and the bulk are dominated by the presence of antiferromagnetism which is a manifestation of important correlation effects. The magnitudes of the moments are however very different for the clusters as compared with the bulk. There are bound to be many surprises in this part of the periodic table as the theoretical, computational and, hopefully, experimental techniques are improved.

4.2. Magnetic Alloys: Fe-Ni, Fe-Co

We have recently started a series of both cluster and band calculations on ferromagnetic transition metal alloys. Alloying can have dramatic effects on the electronic, magnetic, structural and mechanical properties of these materials. Perhaps the most startling examples are found in the Fe-Ni Invar alloys (97) which show marked anomalies in a number of properties including the magnetization, the volume and the thermal expansion coefficient. It was these anomalies which initially sparked our interest in these alloys and we have proposed (9) an explanation for them in terms of a cluster model. Further work (98) has made us doubt whether the model is correct in some respects. However, what is perhaps more important, and what I wish to discuss here, is that the calculations have allowed us to isolate and understand in quite simple terms, some important interactions between chemical bonding and magnetic effects.

Let us illustrate this for the case of the bcc Fe-Co alloys. Addition of Co to Fe causes an increase in the magnetization (99) and neutron diffraction studies (100) indicate that this is mainly due to an increase in the local moment on Fe. Band calculations (101) have accounted for this effect, at least qualitatively. To get a more detailed picture of the reasons for this one can examine the local densities of states for Fe and Co in the alloy. In a simple rigid band approximation, which is often used for alloys, these would be identical. Our results (Figure 12) indicate that this approach is completely inadequate. For minority spin the shape of both the Fe and Co DOS is drastically modified from the "canonical" bcc DOS. There is a dramatic attenuation of the low energy Fe states (below ε_F) and an increase in the number of states above ε_F. The Fe moment increases in the alloy owing to the decrease in the number of minority spin states. For Co the reverse happens. Similar effects are present in the DOS from cluster calculations (9).

The key to interpreting the local DOS comes from chemical considerations. One can account for the general shape of the spin-polarized local DOS for the clusters in a schematic fashion (Figure 13) by considering the relative spin-orbital electronegativities of Fe and Co. Co has a higher electronegativity than Fe and this is reflected by its lower non-spin-polarized energy level in Figure 13, where a single line is used to represent an average energy level for the atoms in a bulk environment (the center of gravity of the local DOS). The next outermost columns of Figure 13 represent the effects of spin polarization on the Fe and Co levels. Fe moments and exchange splittings are larger than those for Co and for the clusters this brings the spin-up Fe level below that for Co, i.e. the spin-orbital electronegativities are reversed. For minority spin the opposite holds and the electronegativity difference between Fe and Co is enhanced. Finally in the central column the Fe-Co interactions are taken into account leading to the situation described above. The difference between Figure 13 and a similar diagram applicable to the band calculations is that for the latter, the Fe↑ and Co↑ levels should be nearly degenerate, leading to non-polar interactions and very similar majority spin DOS for Fe and Co (Figure 12).

These are fundamental and general concepts and should prove useful in the study of other magnetic alloys, both in the bulk or as clusters. In fact, to my knowledge, no one has so far made an isolated transition metal alloy cluster. If nothing else, our calculations indicate that here too, there will undoubtedly be some interesting, and perhaps even useful, surprises.

5. CONCLUDING REMARKS - NEED FOR EXPERIMENTS ON ISOLATED CLUSTERS

The main impressions I would like to leave you with are the following

1. Electronic correlation, defined as those effects which are not accounted for at the Hartree-Fock limit, is often a crucial feature of the electronic structure of metal clusters. From the "ab initio" point of view it is encouraging that extensions of the GVB method are now being applied to transition metal systems. These calculations provide correlated wave functions and yet retain to the maximum possible extent, a simple orbital picture. To a large degree, one of the drawbacks of the straightforward CI

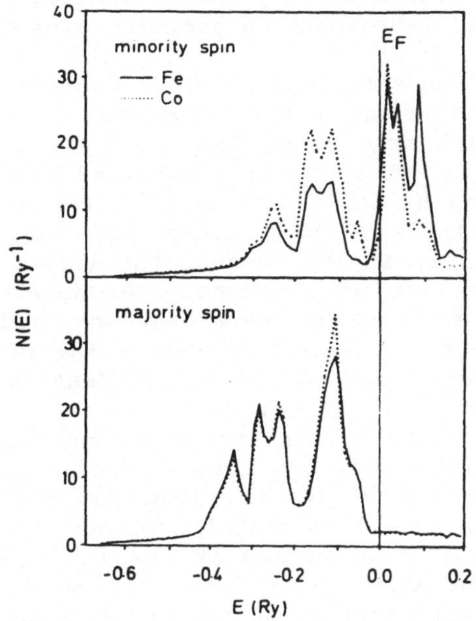

Figure 12. Local density of states (DOS) per atom and spin of FeCo. Reproduced from reference 101.

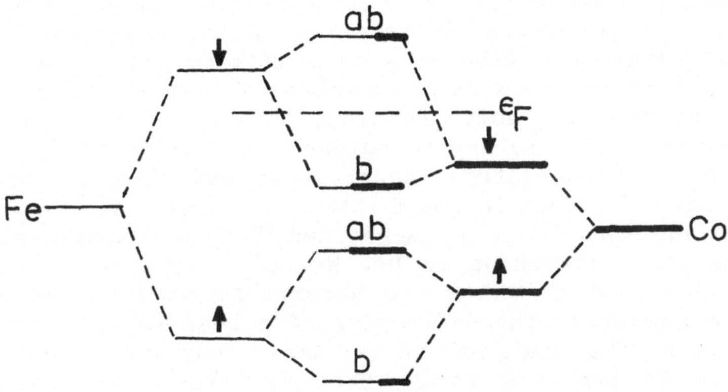

Figure 13. Schematic construction of spin-polarized MOs from non-spin-polarised AO's representing the local DOS for an Fe-Co cluster (see reference 9).

approach, the physics being lost in the complicated expansion of the wavefunction, is avoided. For the moment, this type of calculation is restricted to very small clusters and their extension to larger systems represents an exciting area for future development.

2. The local density methods often provide one-electron energies, total energies and other properties which are in much better agreement with experiment than the Hartree-Fock values. They are, in a manner of speaking, "correlation-corrected". Since these methods can be applied to quite large clusters, which show some of the features of bulk materials, and appear to be quite accurate, they are, at the moment, the method of preference for this important class of problems. To develop improvements on these methods for the cases where they are inadequate, it will be necessary to obtain a better understanding of the exchange-correlation approximation, and indeed, of density functional theory itself.

3. The importance of the symmetry constraints imposed on the orbitals is becoming more and more evident. It is often advantageous, and sometimes vital, to allow the molecular orbitals to localize. A single determinantal approach does not always allow this to be done with retention of the full space and spin symmetries of the problem. More work will be required to delineate the regime of applicability of the local density methods in this respect. "Antiferromagnetic" systems are particularly interesting (and problematic) and I believe that the series of chromium clusters along with that of the 4d congener, molybdenum will provide a very demanding but fruitful testing ground for the various theoretical approaches.

4. Finally, I would like to reiterate that there is a crying need for more experimental information on the properties of isolated metal clusters. (There is nothing more effective in keeping theorists honest than hard experimental data.) Those which I would suggest as being most pressing, at least for my own particular interests, are 1) any technique which will yield geometric information 2) photoelectron spectroscopy and 3) magnetic moment measurements. In fact I would like to close, at the risk of public humiliation, with a suggestion for an experiment which came to me after listening to Dr. Wöste's lecture. The experiments he described in which the absorption spectrum of Na_2 was obtained by measuring the deflection of a beam caused by momentum transfer from the photons of a laser was very intriguing. Wouldn't it be possible to do this experiment with the laser replaced by a synchrotron and hence to measure the X-ray absorption spectrum of a beam of clusters, subsequently obtaining geometric parameters through analysis of the Extended X-ray Absorption Fine Structure?

ACKNOWLEDGEMENTS

I am grateful to the Natural Sciences and Engineering Research Council of Canada and to the General Electric Company for support of some of the works described in this review and to NATO for supporting my participation in the ASI. I am especially thankful to Drs. A.N. Baykara, J. Kaspar and B.N. McMaster for their comments on the manuscript.

REFERENCES

1. Entre l'atome et le cristal: les agrégats, F. Cyrot-Lackmann, ed., (Les Ulis, Les Editions de Physique, 1981).
2. Proceedings of the second international meeting on small particles and inorganic clusters, Surf. Sci. 106 (1981).
3. Conférence internationale sur les petites particules et amas inorganiques, J. Physique Colloq. 38C2 (1977).
4. K.H. Johnson in The New World of Quantum Chemistry, B. Pullman and R. Parr (eds.) (Dordrecht, Reidel, (1976)) p.317.
5. R.P. Messmer, Phys. Rev. B23 (1981) 1616.
6. E.L. Anderson, T.P. Fehlner, A.E. Foti and D.R. Salahub, J. Am. Chem. Soc. 100 (1980) 7422.
7. D.R. Salahub, L,-C. Niem and M. Roche, Surf. Sci 100, (1980), 199 and refs. therein.
8. R.P. Messmer, S.K. Knudson, K.H. Johnson, J.B. Diamond and C.Y. Yang, Phys. Rev. B13 (1976) 1396.
9. J. Kaspar and D.R. Salahub, Phys. Rev. Lett. 47 (1981) 54; J. Phys. F; Metal Physics (in press).
10. G.G. DeLeo, G.D. Watkins and W.B. Fowler, Phys. Rev. B23 (1981) 1851.
11. D.R. Salahub and R.P. Messmer, Surf. Sci. 106 (1981) 415.
12. C.L. Briant and R.P. Messmer, Phil. Mag. B42 (1980) 569.
13. J.C. Slater, Adv. Quantum Chem. 6 (1972) 1; The Self-Consistent Field for Molecules and Solids (McGraw-Hill, New York, 1974) Vol. 4.
14. K.H. Johnson, Adv. Quantum Chem. 7 (1973) 143.
15. For reviews of cluster applications see e.g. R.P. Messmer, Surf. Sci. 106 (1981) 225; K.H. Johnson, Crit. Rev. Solid State Mater. Sci. 7 (1978) 101; R.P. Messmer in Nature of the Surface Chemical Bond T.N. Rhodin and G. Ertl (eds) (Amsterdam, North-Holland, 1978); A. B. Kunz in Theory of Chemisroption J.R. Smith (ed.) (Berlin, Springer, 1980); D.R. Salahub in Entre l'atome et le cristal: les agrégats F. Cyrot-Lackmann (ed.) (les Ulis, les Éditions de Physique, 1981).
16. R.P. Messmer, Phys. Rev B15 (1977) 1811.
17. R. Baetzold, this volume.

18. J.D. Head and K.A.R. Mitchell, Mol. Phys. 35 (1978) 1681.
19. J. Demuynck, M.-M. Rohmer, A. Strich and A. Veillard, J. Chem. Phys. 75 (1981) 3443.
20. P.A. Cox, M. Bénard and A. Veillard, Chem. Phys. Lett. 87 (1982) 159.
21. P. Durand in ref. 1 p. 37.
22. H. Basch, M.D. Newton and J.W. Moskowitz, J. Chem. Phys. 73 (1980) 4492.
23. B.N. Cox and C.W. Bauschlicher Jr., Surf. Sci. 115 (1982) 15
24. H. Basch, M.D. Newton and J.W. Moskowitz, J. Chem. Phys. 69 (1978) 584.
25. M.M. Goodgame and W.A. Goddard III, Phys. Rev. Lett. 48 (1982) 135.
26. B.I. Dunlap and H.L. Yu, Chem. Phys. Lett. 73 (1980) 525.
27. D. Post and E.J. Baerends, Chem. Phys. Lett. 86 (1982) 176; L. Noodleman, D. Post, and E.J. Baerends, Chem. Phys. 64 (1982) 159.
28. W.A. Goddard III, T.H. Dunning Jr., W.J. Hunt and P.J. Hay, Acc. Chem. Research 6 (1973) 368.
29. C. A. Coulson and I. Fischer, Phil. Mag. 40 (1949) 386.
30. C.C.J. Roothaan Rev. Mod. Phys. 23 (1951) 69.
31. J.W. Moskowitz and L.C. Snyder, in Modern Theoretical chemistry, H.F. Schaefer III (ed) (New York, Plenum 1977) vol. 3, p. 387.
32. P.J. Hay, J. Chem. Phys. 66 (1977) 4377.
33. P.J. Hay, J. Am. Chem. Soc. 100 (1978) 2411.
34. A.N.Tavouktsoglou and S. Huzinaga, J. Chem. Phys. 72 (1980) 1385.
35. T. Koopmanns, Physica 1 (1933) 104.
36. R.L. Martin, Chem. Phys Lett. 75 (1980) 290.
37. B.H. Bloch, T.H. Dunning Jr. and J.F. Harrison, J. Chem. Phys. 75 (1981) 3466.
38. C.W. Bauschlicher Jr. S.P. Walch and H. Partridge, J. Chem. Phys. 76 (1982) 1033.
39. C. Froese Fischer, J. Chem. Phys. 76 (1982) 1934.
40. G. Pacchioni, H.-O. Beckmann and J. Koutecky, Chem. Phys. Lett. 87 (1982) 151.
41. J. Flad, H. Stoll and H. Preuss, J. Chem. Phys. 71 (1979) 3042.
42. R. Gaspar, Acta Phys. Acad. Sci. Hung, 3 (1954) 263.
43. W. Kohn and L.J. Sham, Phys. Rev. 140 (1965) A1133.
44. K. Schwarz, Phys. Rev. B5 (1972) 2466.
45. L. Hedin and B.I. Lundqvist, J. Phys. C. Solid State Phys. 4 (1971) 2064.
46. U. von Barth and L. Hedin, J. Phys. C. Solid State Phys. 5 (1972) 1629.
47. O. Gunnarsson and B.I. Lundqvist, Phys. Rev. B13 (1976) 4274.
48. S.H. Vosko, L. Wilk and M. Nusair, Can. J. Phys. 58 (1980) 1200.

49. R.O. Jones, J. Chem. Phys. 76 (1982) 3098.
50. P. Hohenberg and W. Kohn, Phys. Rev. 136 (1964) B864.
51. G.S. Painter and D.E. Ellis, Phys. Rev. B1 (1970), 4747.
52. D.E. Ellis and G.S Painter, Phys. Rev. B2 (1970) 2887.
53. H. Sambe and R.H. Felton, J. Chem. Phys. 61 (1974) 3862; 62 (1975) 1122.
54. B.I. Dunlap, J.W. D. Connolly and J.R. Sabin, J. Chem. Phys. 71 (1979) 3396,4993.
55. R.P. Messmer and S.H. Lamson, Chem. Phys. Lett. 90 (1982) 31.
56. J. Korringa, Physica 13 (1947) 392.
57. W. Kohn and N. Rostoker, Phys. Rev. 94 (1954) 1111.
58. N. Rösch in Electrons in Finite and Infinite Structures, P. Phariseau and L. Scheire (eds) (New York, Plenum 1977) p. 1.
59. H.J. Monkhorst, Phys. Rev. B20 (1979) 1504.
60. W. Kolos and C.C.J. Roothaan, Rev. Mod. Phys. 32 (1960) 219.
61. K.P. Huber, in American Institute of Physcis Handbook, D.E. Gray (ed), (New York, McGraw-Hill, 1972), sec. 7g.
62. A.D. Becke, J. Chem. Phys. 76 (1982) 6037.
63. E.J. Baerends and P. Ros, Int. J. Quantum Chem. 12S (1978) 169.
64. P.E. Cade and A.C. Wahl, At. Data Nucl. Data 13 (1974) 339; P.E. Cade and W.M. Huo, ibid 15 (1975) 1; E. Clementi and C. Roetti, ibid 14 (1974) 177.
65. See Ref. 54 and references therein.
66. R. Car, R.A. Meuli and J. Buttet, J. Chem. Phys. 73 (1980) 4511.
67. J.N. Bardsley, B.L. Junker and D.W. Norcross, Chem. Phys. Lett. 37 (1976) 502.
68. P. Kusch and M.M. Hessel, J. Chem. Phys. 68 (1978) 2591.
69. D.R. Salahub, S.H. Lamson and R.P. Messmer, Chem. Phys. Lett. 85 (1982) 430.
70. R.P. Messmer and D.R. Salahub, J. Chem. Phys. 56 (1976) 779.
71. P.J. Hay, T.H. Dunning Jr. and W.A. Gooddard III, J. Chem. Phys. 62 (1975) 3912.
72. P.J. Hay and T.H. Dunning Jr., J. Chem. Phys. 67 (1977) 2290.
73. C.W. Wilson Jr. and D.G. Hopper, J. Chem. Phys. 74 (1981) 595.
74. W.D. Laidig and H.F. Schaefer III, J. Chem. Phys. 74 (1981) 3411.
75. H. Sambe and R.H. Felton, J. Chem. Phys. 61 (1974) 3862.
76. D.C. Frost, S.T. Lee and C.A. McDowell, Chem. Phys. Letters 24 (1974) 149.
77. C.R. Brundle, Chem. Phys. Letters 26 (1974) 25.
78. J.M. Dyke, L. Goolob, N. Jonathan, A. Morris and M. Okuda, J. Chem. Soc. Faraday Trans II (1974) 1828.
79. S. Huzinaga, J. Chem. Phys. 42 (1965) 1293.
80. F.B. van Duijneveldt, IBM Research report, RJ945 (1971).
81. J.H. Wood, Chem. Phys. Letters 51 (1977) 582.

82. T.J. Tseng, S.H. Hong and M.A. Whitehead, J. Phys. B13 83.
83. D.W. Smith and O.W. Day, J. Chem. Phys. 62 (1975) 113; O.W. Day, D.W. Smith and R.C. Morrison, J. Chem. Phys. 62 (1975) 115; M. Morrell, R.G. Parr and M. Levy, J. Chem. Phys. 62 (1975) 549.
84. D.R. Salahub and R.P. Messmer Phys. Rev. B16 (1977) 2526.
85. R.P. Messmer and D.R. Salahub, Phys Rev. B16 (1977) 3415.
86. D.R. Salahub, M. Roche and R.P. Messmer, Phys. Rev. B18 (1978).
87. W. Eberhardt and C. Kunz, Surf. Sci. 75 (1978) 709.
88. C.W. Martinson, S.A. Flodström, J. Rundgren and P. Westrin, Surf. Sci. 89 (1979) 102.
89. L.I. Johansson, J. Stöhr and S. Brennan, Applic. of Surf. Sci. 6 (1980) 419.
90. J. Rundgren, quoted in ref. 89.
91. V.L. Moruzzi, J.F. Janak and A.R. Williams, Calculated Electronic Properties of Metal (New York, Pergamon, 1978).
92. C.Y. Yang, K.H. Johnson, D.R. Salahub, J. Kaspar and R.P. Messmer, Phys. Rev. B24 (1981) 5673.
93. C.G. Shull and H.A. Mook, Phys. Rev. 103 (1966) 516.
94. G. Allan, Surf. Sci. Reports 1 (1981) 121; Surf. Sci. 74 (1978) 79.
95. N. Hirashita, G. Yokoyama, T. Kambara and K.I. Gondaira, J. Phys.F: Metal Phys. 11 (1981) 2371.
96. D.R. Salahub, B.N. McMaster, J. Kaspar, R.P. Messmer and F.J. Pinski, in preparation.
97. Y. Nakamura, IEEE Trans. Magn 12 (1976) 278.
98. J. Kaspar and D.R. Salahub, unpublished.
99. M. Shiga in Physics of Transition Metals-1980, P. Rhodes (ed.) Inst. Phys. Conf. Ser. No55 (London, IOP, 1980) p. 241.
100. M.F. Collins and J.B. Forsyth, Phil. Mag. 8 (1963) 401.
101. K. Schwarz and D.R. Salahub, Phys. Rev. B25 (1982) 3427. (1980) 4101.

ELECTRONIC STRUCTURE OF METAL CLUSTERS

R. C. Baetzold

Research Laboratories, Eastman Kodak Company, Rochester, New York 14650

ABSTRACT

The understanding of the electronic structure of transition- and noble-metal clusters has been advanced recently by complementary experimental and theoretical research. This lecture will reflect both aspects. On the theoretical side we will examine some of the techniques of calculation including extended Hückel, CNDO, and Hartree-Fock as well as perturbation theory. These methods will be applied to problems of cluster geometry, charge distribution, ionization potential, and density of states in order to compare with results for infinite and semi-infinite analogues. Several quite different approaches have shown encouraging agreement in the properties of small clusters, which are significantly different from those of the bulk. We will also consider in a simplified fashion how a model substrate can alter the properties of metal clusters. On the experimental side we will discuss photoemission (UPS and XPS) data that can be compared with theoretical data. In the evaporated-particle model systems considered, small-cluster data are at significant variation with bulk data. In addition, some UPS experiments attempting to understand the range of metal particle/adsorbate interactions will be discussed.

1. INTRODUCTION

We consider a free cluster consisting of atoms intermediate in number between the well-defined extremes of single atom and infinite solid. In theoretical investigations of such clusters we are free to choose arbitrary shapes and bonding configurations that are useful for particular questions at hand and may eliminate some of the constraints present in experimental studies. This situation is desirable since it allows one-to-one relationships to be established, clearly aiding in our understanding. The simplifications also mean that there will frequently be some gap in correspondence between the properties calculated in theory and those measured in experiment, no matter how carefully the investigations are designed. Thus, it is well to remember that most models used in theory have limitations that govern the context in which they should be applied to experiment.

Theory must be useful. Thus, the above discussion should not be understood to imply that the interaction of theory and experiment is fruitless. Several areas come to mind. First, theory should form part of the basis in planning experiment. At a second level, theory can provide an understanding of the cause-and-effect relationships that determine the outcome of an experiment. Finally, at a much more demanding level is the attempt to determine accurately with theory some particular property that can also be measured experimentally. I believe that few, if any, cases in the area of transition-metal clusters fall into this last category at the present levels of development. Thus, I prefer to use theory in some qualitative sense in attempting to provide a useful basis for planning and interpreting experiment.

Let us consider the present state of development of theory to reinforce the point of view above. The Schrödinger equation is not solved exactly beyond the one-electron case. Thus, no matter how sophisticated the particular technique, there will be some approximations. These may involve the replacement of a series of integrals by potential functions or the use of incomplete basis sets to represent the wave function. Of course, the development of "better and more reliable approximations" within some particular context is a useful and important goal of theory. It is important to compare the results obtained from different theoretical starting points because of this present state of affairs.

2. RESULTS

2.1 General Considerations

Considerations from the simplest standpoint suggest that the properties of small clusters will be different from those of the single atom or the infinite solid. Consider the small spherical face-centered-cubic (fcc) and body-centered-cubic (bcc) clusters in Figure 1. Most atoms in these clusters are on the surface and lack the full coordination number of the bulk atom, which is 12 for fcc and 8 for bcc. The computed average coordination numbers for each cluster in Figure 1 show a strong unsaturation, which should be manifested in the physical and chemical properties of the

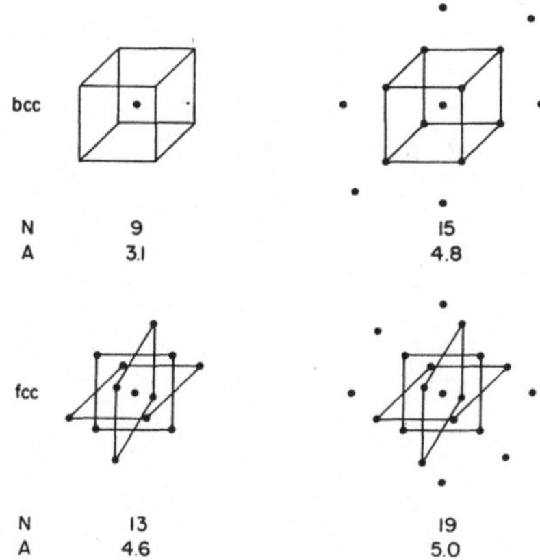

Figure 1. Sketches of the cluster geometry for small bcc and fcc structures. Average coordination numbers (A) are shown for particular sizes (N).

cluster. One expects that the cohesive energy, the ionization potential, and the density of states of the cluster should be affected.

It is now appropriate to consider the electronic structure changes resulting from the assembly of atoms in aggregate form. We may classify the atomic orbitals in three broad categories. First, there are the core orbitals, which show minor effects of interaction between adjacent atoms and so remain "frozen." We postpone a discussion of core level shifts until a later section. Second are valence s and p orbitals, which are very diffuse and overlap strongly with like orbitals on adjacent atoms. Finally, there are the d valence orbitals, which are localized on the atomic center but also have a "tail" that provides some overlap between orbitals on adjacent centers. Figure 2 sketches the situation resulting under aggregate formation from these orbital interactions.

Valence orbitals interact and lead to molecular orbitals split in energy in the dimer and larger species. The most diffuse orbitals (typically s,p type), which are typically located closest to the vacuum level, split the most in energy, as depicted for the dimer. We have labelled the highest occupied molecular orbital (HOMO) and lowest unoccupied molecular orbital (LUMO) of the dimer.

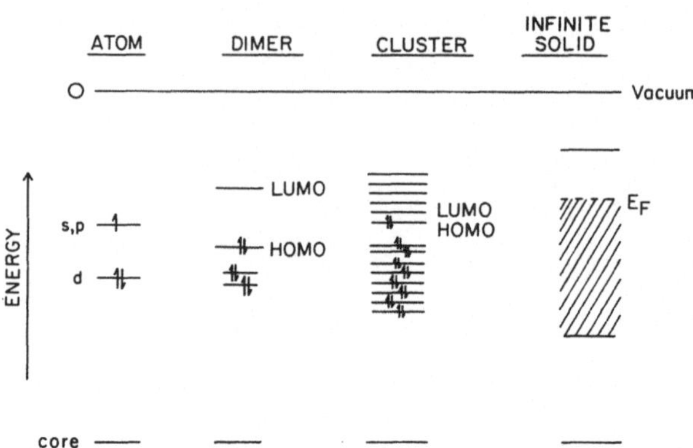

Figure 2. Sketch of the development of electronic structure (energy levels) with cluster size. Double arrows indicate occupied cluster energy levels; the shaded area corresponds to occupied states of the infinite solid.

As more atoms are added to the cluster, the splitting increases and mixing of valence orbitals of different types, called hybridization, takes place. We note that the HOMO has shifted closer to the vacuum in the cluster and the gap between HOMO and LUMO decreases. The HOMO corresponds to the first ionization potential in Koopmans' theorem, and the HOMO-LUMO gap to the energy of the first electronic transition. Finally, in the infinite solid a band of occupied levels form, which are spread apart more in energy than the spread of occupied levels in the cluster or the dimer. The highest filled level is termed the Fermi energy, E_F. In the infinite solid the s,p orbitals have formed a band across a broad range of energy and have lost their identity, with any particular atom becoming "free-electron-like". The d orbitals have formed a band but retain their local atomic character. Thus, the "jellium" approximation is appropriate for the s,p orbitals, whereas the "tight-binding" approximation becomes appropriate for the d orbitals of an infinite solid.

The mechanism of band formation may be understood in terms of the covalent mixing of orbitals familar in quantum mechanics (1). This situation is depicted in Figure 3 for two orbitals of unequal energy, where the perturbation formulas relating interactions of several orbitals are also shown. A splitting of orbitals occurs whose magnitude is proportional to the interaction matrix element H'_{ij} and is inversely proportional to their energy separation. The bonding combination $\psi_1 \propto \psi_1^0 + \lambda\psi_2^0$ and the antibonding combination

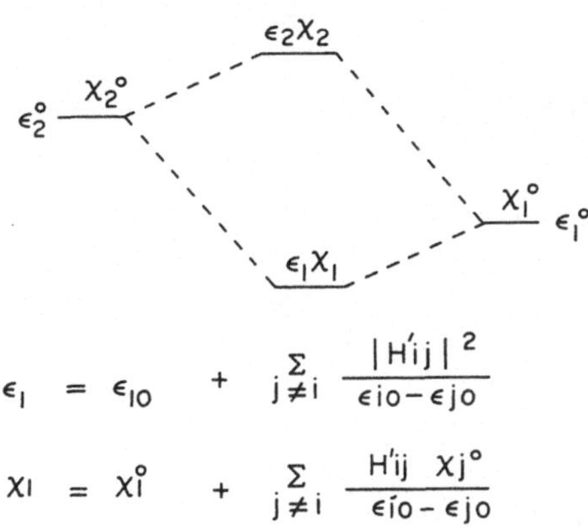

Figure 3. Perturbation relations governing the mixing of atomic orbitals (ψ_i^0) of energy (ϵ_i^0) are sketched, with the appropriate formulas.

$\psi_2 \propto \psi_2^0 - \lambda\psi_1^0 \quad \lambda > 0$ form from this type of interaction. The interaction matrix element is larger between adjacent s orbitals than adjacent d orbitals, leading to the greater splitting (broader bands) in the former case.

2.2 Computational Approaches

A number of texts describe quantum theories in mathematical detail, so only the nature of the types of calculations will be given here. We will be concerned with extended Hückel (EH) (2), CNDO (3), $X\alpha$ (4), and Hartree-Fock (5) methods of calculation, because they are the most common approaches. The nature of the approximations is quite different in each of the different approaches.

Extended Hückel theory is the simplest method of calculation applied to metal-cluster problems. First the wave function (ψ) is written as a product of molecular orbitals (ψ_i), which are a linear combination (LCAO) of the valence atomic orbitals (χ_i).

$$\psi_i = \sum_j c_{ji}\chi_j \qquad (1)$$

The Hamiltonian of the system $\hat{H}$ is taken as a sum of effective independent one-electron operators ($\hat{h}_i$).

$$\hat{H} = \sum_i \hat{h}_i \qquad (2)$$

Thus the energy of the system becomes

$$E = \frac{\langle\psi|H|\psi\rangle}{\langle\psi|\psi\rangle} = \sum_i n_i \epsilon_i \qquad (3)$$

where n_i is the occupancy of energy level ϵ_i filled in the order of lowest first. The energy levels are found by minimizing the energy through variation of the coefficients C_{ji}, leading to the secular equation

$$|H_{ij} - S_{ij}E| = 0 \ . \qquad (4)$$

where

$$H_{ij} = \langle \chi_i|\hat{H}|\chi_j\rangle$$
$$S_{ij} = \langle \chi_i|\chi_j\rangle \qquad (5)$$

and E is the matrix containing ϵ_i elements on the diagonal. In this theory the χ_i are taken as Slater atomic orbitals with exponents determined from Hartree-Fock calculations (6). This permits calculation of S_{ij} straightforwardly from Equation (5). The H_{ii} terms are identified as the atomic ionization potential with appropriate allowance for charge and configuration of the valence orbitals. The H_{ij} elements are typically computed from the Wolfsberg-Helmholtz formula

$$H_{ij} = \frac{1}{2}KS_{ij}(H_{ii} + H_{jj}) \ . \qquad (6)$$

In this method the occupied valence orbitals as well as virtual orbitals are included.

In the Hartree-Fock theory, the wave function is written as a determinant of the molecular orbitals rather than the product form as in extended Hückel theory. The allowance for electron-electron interactions is retained by not invoking an independent electron approximation as in Equation (2).

$$H = \sum_i \left[\frac{-\hbar^2}{2m_i}\nabla_i^2 - \sum_n \frac{Z_n}{r_{in}}\right] + \frac{1}{2}\sum_{i\neq j}\sum e^2/r_{ij} \qquad (7)$$

The molecular orbitals are expanded as a linear combination of atomic orbitals, and energy is minimized to yield a secular equation similar in form to Equation (4) as necessary for solution of the energy levels. The energy is given by Equation (8)

$$E = 2\sum_i \epsilon_i + \sum_{i\neq j}\sum (2J_{ij} - K_{ij}) \qquad (8)$$

for a closed-shell system. Here we see an energy form different from Equation (3) in the extended Hückel theory because of electron-electron interactions. The coulomb integral J_{ij}

$$J_{ij} = \langle \psi_i(1)\psi_j(2) | \frac{e^2}{r_{12}} | \psi_i(1)\psi_j(2) \rangle \qquad (9)$$

and exchange integral K_{ij}

$$K_{ij} = \langle \psi_i(1)\psi_j(2) | \frac{e^2}{r_{12}} | \psi_j(1)\psi_i(2) \rangle \qquad (10)$$

are expanded through Equation (1) to lead to a procedure necessitating the calculation of many integrals, which may contain up to four centers. There is one additional approximation in this method resulting from taking interactions of one electron with the average field of the remaining electrons. This approximation neglects instantaneous electron-electron interactions and leads to the so-called correlation energy, which requires several determinants in the wave function for calculation.

The Xα method eliminates some of the multicenter integrals in the Hartree-Fock method by replacing exchange integrals with a local potential of the form

$$\alpha\left(\frac{3}{4\pi}\rho\right)^{1/3} \qquad (11)$$

where α is a parameter and ρ is the electron density. This approximation is believed to include some correlation energy effects by its proponents. In place of the LCAO approximation, spheres are drawn around each nucleus in the calculation, dividing regions of different functional form of the potential. The radius of these spheres is another parameter of the method. So far calculation of the energy of transition metal systems has proven rather difficult.

The CNDO (complete neglect of differential overlap) method is a severe approximation of Hartree-Fock theory. In this procedure the number of two-electron integrals (J_{ij} and K_{ij}) is reduced by neglect of differential overlap (i.e., $\chi_i\chi_j \xrightarrow{ij} \chi_i\chi_i\delta_{ij}$), and some of the one-electron integrals are replaced by experimental atomic information. This greatly reduces the number of integrals to be evaluated and introduces experimental parameters such as ionization potential, electron affinity, and a resonance parameter.

2.3 Geometry

There are several indications that the geometry of small bare metal clusters will be different from that of the same size unit in the bulk crystalline lattice. Very compact structures in the small cluster force antibonding molecular orbitals to become particularly destabilizing, and depending upon the occupancy, this leads to unexpected geometries for the cluster. The bonding molecular orbitals are stabilized by most compact structures, so there is always a balance to be considered between the two effects. Extended Hückel theory shows these effects clearly, and the same effects have been noted in ab initio calculations (7).

Extended Hückel and CNDO calculations (8-10) for Ag and Cu have shown a preference for linear neutral clusters versus other structures. In these clusters the s valence orbitals create the dominant bonding interactions. For three-atom clusters, one bonding molecular orbital is doubly occupied and one antibonding molecular orbital is half occupied. The antibonding interaction is dominant, leading to the preferred linear structure.

An analysis of the effect of orbital occupancy for three- and four-atom clusters is shown for s, p, and d orbitals in Figures 4 and 5. We consider variable occupation scaled so that s^2, p^6, or d^{10} corresponds to 100% filled molecular orbitals in these figures. The linear, equilateral triangle, and 135° isosceles triangle geometries are considered for triatomic clusters. For s occupancy greater than 0.5, linear is favored. For p occupation, the equilateral triangle is favored up to $p^{3.0}$, with linear more stable beyond. Except for d^9, the equilateral triangle is preferred for d occupation. Of course, any particular structure contains a combination whereby the s, p, and d orbital effects must be summed. For the four-atom clusters shown in Figure 5, the effects are similar, although some of the s and p cases show a particular preference for the square. Note the parabolic structure of the binding energy resulting from the d contribution to the energy for both size clusters. This behavior closely parallels the bulk cohesive energy.

The preference for linear or nearly linear geometric structures has been reported recently in ab initio pseudopotential calculations for small Cu, Ag, and Ni clusters. Basch et al. (11) found linear Ni_3 more stable than triangular, whereas for Ni_4 the linear and square geometry are roughly equal in energy with tetrahedral much less stable. Veillard and co-workers (7) have found linear the preferred geometry for Cu_3 and Cu_4, but at larger sizes other geometries prevail. Recent calculations by Basch (12) for Ag_3 have shown a slightly bent structure, but with a very shallow potential minimum. In considering these results we note the strong similarity to EH predictions. There are, however, no

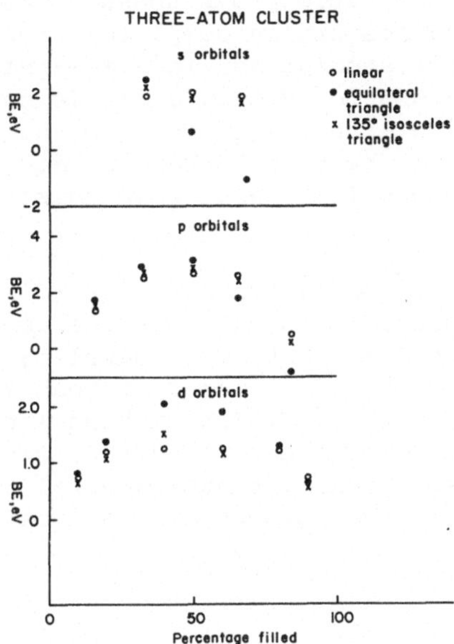

Figure 4. Computed total binding energy for three-atom clusters of different geometry for different occupation of s, p, and d orbitals. Here 100% occupation corresponds to s^2, p^6, or d^{10} configurations.

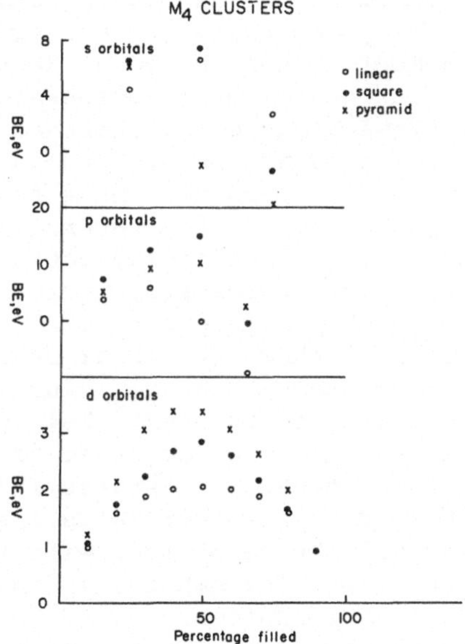

Figure 5. Four-atom clusters considered as in Figure 4.

conclusive experiments that could prove these results, although several experimental findings are in complete accord with the idea. Schulze et al. (13) studied Ag_3 in a low-temperature matrix using Raman spectroscopy and found only a single band. This finding is consistent with linear Ag_3, because the equilateral-triangular structure should have more than one active Raman mode. However, intensity arguments (14) raise some questions on this linear assignment.

Clusters of more than four atoms had three-dimensional structures as their most stable configuration in the ab initio calculations of Veillard's group (7). Extended Hückel calculations deviate from this behavior, predicting linear as most stable when computations are performed at the equilibrium bond length. This is undoubtedly due to a flaw in the EH method that gives incorrect equilibrium bond lengths. When bulk experimental bond lengths are used, the EH method gives three-dimensional as most stable for the larger clusters.

2.4 Charge Separation

Small clusters of atoms have a charge separation due to the geometric inequality of various atoms (15). The origin of this effect may be traced to the shape of the local density of states, by use of arguments based upon those developed for semi-infinite surfaces by Cyrot-Lackmann et al. (16). For the semi-infinite surfaces, the second moment of the density of states (which is a measure of band width) is proportional to the square root of the average coordination number. Thus the local density of states will be narrower for a surface atom with fewer nearest neighbors than for a bulk atom. This behavior is sketched in Figure 6. Thus, the population will be different for the surface atom versus the bulk atom. If the band is less than half full, the bulk atom will have a greater population, and the reverse is true if the band is more than half full. These considerations for semi-infinite surfaces also apply to clusters, where now the coordination number correlates with a more surfacelike or bulklike character of the atom. Figure 7 shows the local density of states computed by EH for various atoms on a Ni_{14} cluster whose structure is derived by adding one atom to the next shell of the fcc thirteen-atom structure in Figure 1. We see that as the coordination number decreases, the local density of states is more concentrated near the center of the band. Thus, for bands less than half full, the greatest population will be associated with the highest coordination number and vice versa for bands less than half full. These considerations have been put on a sound theoretical footing by use of perturbation theory (15). They have been verified computationally, within the framework of extended Hückel, for bands more than half filled, as shown by the example in Table 1. We

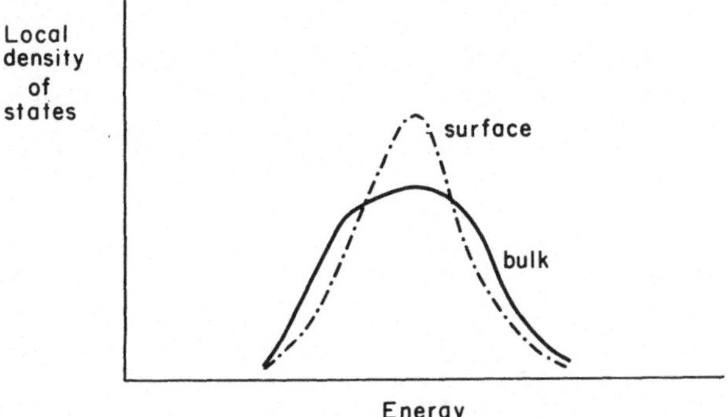

Figure 6. Sketch of the local density of states for an atom on the surface or in the interior of a semi-infinite solid.

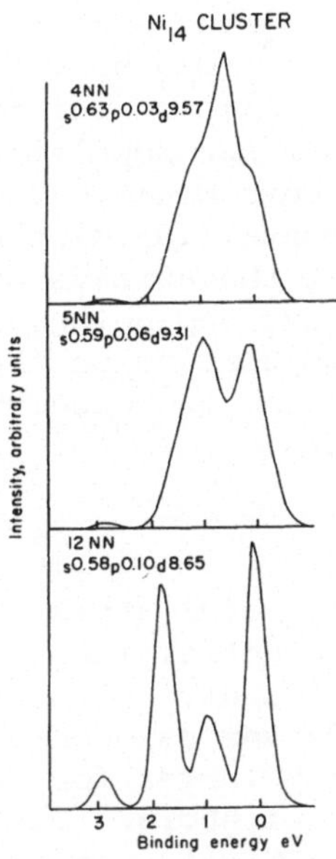

Figure 7. Computed (EH) local density of states for a 14-atom fcc cluster of Ni containing one atom beyond the 13-atom spherical cluster shown in Figure 1. Plots are for atoms with different numbers of nearest neighbors (NN). Energy is measured relative to the highest filled level.

Table 1

Cluster	Center Atom with Largest Coordination		Outer Atom with Smallest Coordination	
	sp	d	sp	d
Ag_{13}	0.71	9.83	1.12	9.92
Pt_{13}	0.80	5.95	1.23	9.04
Rh_{13}	0.89	5.17	0.60	8.64
W_{13}	0.81	4.11	0.66	5.43

observe that the d population is greatest for the cluster atom with smallest coordination.

A good analogy of the cluster excess surface d population (for population greater than d^5) can be found from theoretical (17) and experimental (18) studies on infinite surfaces. Theory has shown that an excess d surface population relative to bulk is manifested in a decrease in core binding energy of surface atoms. Recent experimental measurements (18) of the 4f core levels in Pt, Au, Ir, and W have shown that the surface atom has a decreased binding energy of up to 0.6 eV, depending on the system. Thus, the indication of excess d population on the surface atoms of clusters has a logical extension in connection with observed phenomena on infinite surfaces.

2.5 Odd-Even Effects

Small clusters of noble-metal atoms show strong odd-even oscillations in ionization potential (IP), electron affinity (EA), and bonding energy per atom (BE/n). These calculations show greater stability for closed- versus open-shell configurations, so that IP and BE/n are generally larger for even-size neutral clusters versus their odd-sized neighbors. The opposite is true for EA. This effect is found in EH calculations for Ag_n, CNDO calculations for Ag_n and Cu_n (8), and ab initio calculations for Ag_n (12). There is some experimental support for these effects, in that BE/n measurements (19) for noble-metal trimers show a marked decrease in value compared to the dimer. Also, measurements in photographic systems (20), where redox solutions are used to oxidize Ag_n clusters, have shown definite ranges where larger clusters are more easily oxidized than smaller clusters. These results are interpreted in accord with the odd-even effects.

The overall trend in BE/n versus cluster size shows an increase in most calculations. A good example of this behavior is exemplified in the work of Veillard and co-workers (7), where a roughly linear behavior in BE/n vs. n is found up to Cu_{13}. These authors conclude that "the binding energies of these small clusters appear rather different from the value for the bulk metal."

2.6 Ionization Potential

Melius et al. (21) have made an interesting set of Hartree-Fock calculations to represent a series of Ni clusters of up to 87 atoms. In this work a localized d^9 configuration is taken on each Ni atom, allowing the use of an effective potential to represent the d electrons. This reduces the conduction-band problem to one electron per Ni atom. This procedure allows bonding between adjacent atoms via s orbitals, in a manner similar to the bonding generally accepted for noble metals such as Cu and Ag. The adequacy of this approximation for Ni depends upon how closely the configuration remains at d^9 for each atom for clusters of different sizes.

Let us compare the gross trends in ionization potential computed above for fcc spherical clusters with data that have been computed by using EH for the same structures for Ag clusters (22). This analogy compares the trends in a size-induced property computed by the two methods, since both use primarily s electron representations to determine the ionization potential. Figure 8 shows this comparison derived by using Koopmans' theorem to determine IP. Although both procedures have different starting points for the IP at small cluster sizes, the trends are remarkedly

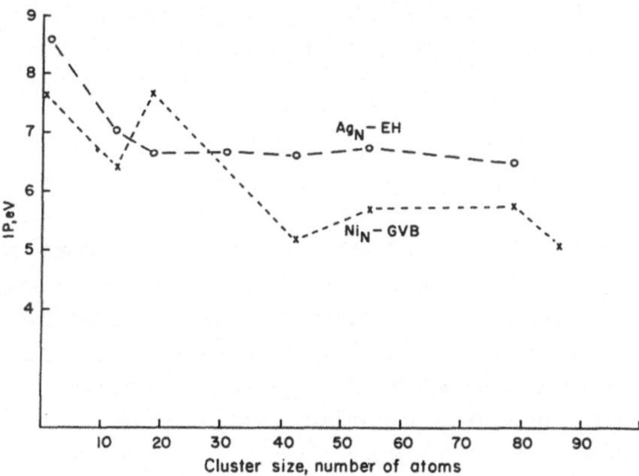

Figure 8. Ionization potential (IP) versus cluster size for spherical fcc structures. Calculations are performed by EH for Ag_n (ref. 22) and generalized valence bond (GVB) for Ni_n (ref. 21).

similar. There is a degree of nonmonotonic behavior in each curve, but after 20-40 atoms the IP has become only a mild function of size. The computed value for the infinite bulk EH calculation is shown, but no corresponding value is available for the Hartree-Fock calculation.

2.7 Density of States

The DOS of metal clusters has been examined by several techniques. There is good agreement among EH, CNDO, and ab initio pseudopotential calculations in arriving at DOS for the cluster that is rather different from that for the bulk. Consider the plots in Figure 7. Here the Ni_{14} cluster contains mostly surface atoms, giving a total DOS peaked near the center of the d band, with less intensity at the edges. Thus, the total DOS for Ni_{14} would have a rather small value at the Fermi energy, in contrast to the large value for the bulk. This effect is also seen in CNDO calculations (23) for Pd_n, and it has also been shown (11) recently in ab initio calculations for Ni_6. The latter authors have noted a corresponding effect in experimental photoemission spectra, which will be discussed later.

2.8 Substrate Effects

A metal cluster can interact with a substrate at various levels. First, there is the charge exchange required to equalize the chemical potential of electrons in the cluster and the support. The chemical potential of the support is well defined through the Fermi energy characteristic of the infinite solid. It is less well defined for the cluster, where an energy gap separates the ionization potential and the electron affinity. Nevertheless, net electron transfer could change the population of the cluster, resulting in a change in geometry in accord with the behavior sketched in Figures 4 and 5. In addition, a small cluster charge would affect the position of core levels, which should be detected by photoemission spectroscopy as discussed in the next section.

Mixing of substrate and cluster orbitals takes place when the two interact. This mixing broadens the DOS of the cluster and should be directly observed by valence photoemission spectroscopy. Depending upon the degree of mixing, cluster molecular orbital positions could be changed significantly, leading to significantly altered patterns of interaction with adsorbate.

3. EXPERIMENTAL RESULTS

Photoemission is a powerful technique for the investigation of properties of small metal clusters. The technique measures, in effect, the energy levels of a core or valence electron in the cluster or adsorbed species. This information can be used to deduce how properties such as charge or valence density of states change with cluster size. It is now generally accepted with these techniques that a cluster of ~150 atoms or more is required for a spectrum similar to that of the corresponding bulk metal.

In the photoemission process, a photon in the x-ray region (XPS) or the ultraviolet region (UPS) of energy $h\nu$ excites bound electrons. A matrix element governs the probability of excitation,

$$\langle i | H' | f \rangle$$

where $\langle i|$ is the initial wave function, $\langle f|$ is the final state wave function, and H' is the interaction operator. The selection rule provides modulation, which must be considered in deducing the DOS from a photoemission spectrum. The photoemission spectrum is a measure of intensity of photoelectrons of kinetic energy (KE) at a given photon energy ($h\nu$)

$$h\nu = KE + BE + \emptyset$$

where $\emptyset$ is the spectrometer work function and BE is the binding energy of the ground-state electron.

We have found evaporated metal particles to be a useful sample form for study. This preparation technique, described in detail by Hamilton and Logel (24), involves deposition onto a carbon or metal-oxide support. Then electron microscopy is used to determine a size-frequency distribution of particles for a given amount of metal evaporated. The structure of the particles was determined to be hemispherical, allowing a count of the number of atoms to be deduced for a given particle size.

Several authors have used UPS or XPS to show the buildup of the density of states of metal clusters with size. A common feature in the photoemission spectra of transition metals is the slow buildup of the density of states at the Fermi energy, as first noted by Mason et al. (25) in XPS studies of Pt and Pd deposited on C support. This phenomenon is also observed in difference UPS spectra, as shown in Figure 9 for Pd deposited onto carbon.

The photoemission spectrum of carbon is subtracted from the photoemission spectrum of carbon plus evaporated metal to give the spectra in Figure 9. Note that the low-binding-energy threshold shifts towards E_F and that there is a general broadening of the spectra as cluster size increases. Marked changes in the spectra take place up to 150 atoms, and even at 220 atoms the spectra are somewhat different from the thick-film spectra. This behavior with size is also characteristic of Ni and Pt clusters evaporated onto carbon.

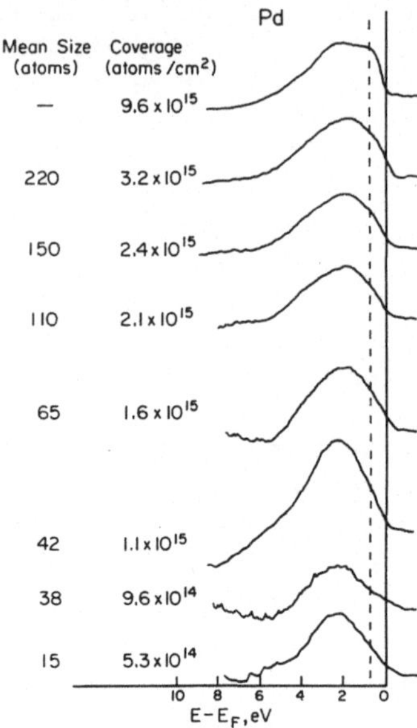

Figure 9. Difference spectra for Pd clusters evaporated onto carbon for various amounts of Pd evaporated (along with corresponding mean size). Energy is measured relative to the carbon Fermi edge.

Several manifestations of the effect of particle size on photoemission spectra have been noted. One concerns the valence state of Sm clusters deposited on carbon (26). At small sizes, divalent ($4f^6$) Sm is detected, whereas at larger sizes trivalent ($4f^5$) Sm is found. The transition region occurs at 30 ± 10 Å particle size. Another important indicator of bulk behavior is the spin-orbit splitting of gold 5d valence band for clusters deposited on carbon and measured by XPS (27). At small cluster sizes, the splitting characteristic of the free atom (1.4 eV) is found. The splitting continuously increases with cluster size, reaching the bulk value (2.7 eV) for clusters of 100 atoms or more.

The core level of small particles is shifted to greater binding energy than the corresponding level in the bulk. Table 2 shows some values of the shift for single atoms or very small clusters versus the bulk. Although sample treatment or the substrate can lead to different values for the core shift, the direction is always the same. The explanation of this effect has involved two phenomena: relaxation energy and configuration changes. Extraatomic relaxation should increase with cluster size, because of the additional electrons available for screening. The latest results (28), however, point to configuration changes as providing the dominant mechanism of core shifts.

Table 2

Atom vs. Bulk Core-Level Shift

Metal	Substract	Level	Core Shift (eV)	Reference
Ag	C	$3d_{5/2}$	0.2	29
Pd	C	$3d_{5/2}$	2.1	25
Pt	C	$4d_{5/2}$	1.6	25
Pd	SiO_2	$3d_{5/2}$	1.6	30
Au	Al_2O_3	$4f_{7/2}$	0.55	31
Pd	C	$3d_{5/2}$	0.6	32

UPS can be used to measure adsorbate-induced changes in the photoemission spectrum as a function of particle size. We have considered halogen adsorbates on a variety of different metals because of the relatively large adsorbate cross section. Chlorine and iodine were studied on substrates including small particles of Pt (33), Ni (34), Ag (35), and Cu (35). Silver has been a particularly interesting cluster for study because of the proximity in energy of the Ag d orbitals and halogen p orbitals. Figure 10

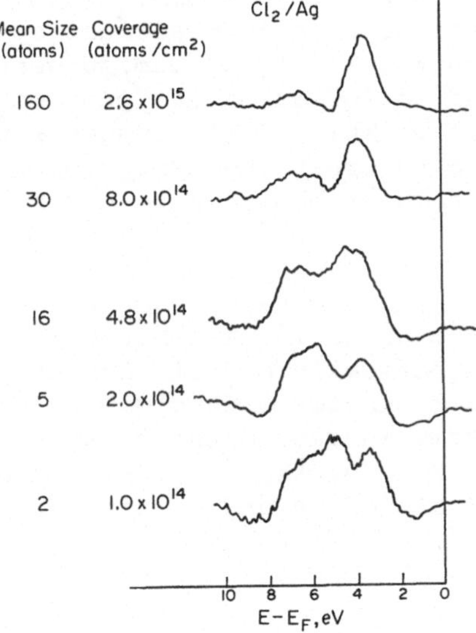

Figure 10. Difference spectra for Cl_2 exposure of silver clusters evaporated at different coverages onto carbon. The mean Ag cluster size is indicated for each spectrum.

shows the effect of chemisorbing Cl_2 on different size clusters of Ag. The increased emission contains two components, at the top and the bottom of the d emission (4-8 eV), and there is some attenuation of d emission, particularly on the larger clusters. The relative intensities of the two adsorbate emission peaks change with Ag particle size. At low particle size the high-binding-energy peak dominates in intensity, with the reverse being true for the larger particles. This same effect is found for halogen deposited after $CHCl_3$ or I_2 exposure of the Ag clusters.

The two peaks containing enhanced adsorbate emission in Figure 10 arise from bonding (p + d) and antibonding (p - d) combinations of the halogen (p) and metal (d) orbitals. The degree of orbital mixing increases when the two types of levels are closely spaced in energy. We assume this situation for the small Ag clusters. As cluster size increases, the IP decreases (18), and consequently the ability to donate electrons to the halogen improves. This leads to greater electron density on the halogen, destabilizing its p levels, which are thus shifted towards the vacuum and away from the metal d orbitals. Thus the orbital mixing decreases, giving a greater halogen component in the low-binding-energy peak, which causes an increase in relative intensity. These qualitative arguments have been verified by self-consistent EH calculations (35).

The changes in the Cl-induced spectra in Figure 10 persist at least up to a cluster size of 30 atoms. This size shows the range over which significant bonding interactions operate in this system. Other measurements in the I_2/Ag_n system suggest that clusters of 40-50 atoms are required for adsorbate photoemission spectra characteristic of the bulk. This effect suggests that "surface-molecule" approaches used to treat chemisorption using a small number of metal substrate atoms probably would not be successful in this system.

With clusters of other metals, such as Cu and Pt, less dramatic size effects are noted in the photoemission spectra than with Ag. For Cu there is basically one Cl-induced peak at all sizes, which shows only minor changes with cluster size. The lack of an effect in these systems is due to the greater separation of p,d levels than for Ag. For Cu, the Cl levels have greater binding energy than the d levels of Cu and remain fixed in position with size.

ACKNOWLEDGEMENTS

I am grateful to my colleagues for interactions that helped to formulate the views expressed here. Particular thanks are due to G. Apai, J. Hamilton, G. Mason, and E. Shustorovich.

REFERENCES

1. Hoffmann, R. Acc. Chem. Res. 4 (1971) 1.
2. Hoffmann, R. J. Chem. Phys. 39 (1963) 1397.
3. Pople, J. A. and G. A. Segal. J. Chem. Phys. 43 (1965) 5136.
4. Slater, J. C. Quantum Theory of Molecules and Solids, vol. IV (McGraw-Hill, New York, 1974).
5. Schaefer, H. F. The Electronic Structure of Atoms and Molecules (Addison-Wesley, Reading, Mass., 1972).
6. Clementi, E. and D. L. Raimondi. J. Chem. Phys. 38 (1963) 2686.
7. (a) Bachmann, C., J. Demuynck, and A. Veillard. Gazz. Chim. Ital. 108 (1978) 389; Faraday Symp. Chem. Soc. 14 (1980) 170; (b) Demuynck, J., M. Rohmer, A. Strich, and A. Veillard. J. Chem. Phys. 75 (1981) 3443.
8. Baetzold, R. C. J. Chem. Phys. 55 (1971) 4363.
9. Anderson, A. B. J. Chem. Phys. 66 (1977) 5108.
10. Baetzold, R. C. and R. E. Mack. J. Chem. Phys. 62 (1975) 1513.
11. Basch, H., M. D. Newton, and J. W. Moskowitz. J. Chem. Phys. 73 (1980) 4492.
12. Basch, H. J. Am. Chem. Soc. 103 (1981) 4657.
13. Schulze, W., H. U. Becker, R. Minkwitz, and K. Manzel. Chem. Phys. Lett. 55 (1978) 59.
14. Moskowitz, M. and D. P. Dilella. J. Chem. Phys. 72 (1980) 2267.
15. Shustorovich, E. and R. C. Baetzold. J. Am. Chem. Soc. 102 (1980) 5989.
16. Desjonqueres, M. C. and F. Cyrot-Lackmann. J. Phys. F 5 (1975) 1368.
17. Desjonqueres, M. C., D. Spanjaard, Y. Lassailly, and C. Guillot. Solid State Commun. 34 (1980) 807.
18. (a) VanderVeen, J. F., P. Heimann, F. J. Himpsel, and D. E. Eastman. Solid State Commun. 37 (1981) 555; (b) Duc, T. M. G. Guillot, Y. Lassailly, J. Lecante, Y. Jugnet, and J. C. Vedrine. Phys. Rev. Lett. 43 (1979) 789.
19. Hilpert, K. and K. A. Gingerich. Ber. Bunsenges. Phys. Chem. 84 (1980) 739.
20. Platzer, S.J.W. and G. R. Bird. 1978 International Congress of Photographic Science, Rochester, New York, paper II-7, p. 75.
21. Melius, C. F., T. H. Upton, and W. A. Goddard. Solid State Commun. 28 (1978) 501.
22. Baetzold, R. C. Inorg. Chem. 20 (1981) 118.
23. Baetzold, R. C. in Catalysis in Chemistry and Biochemistry. Theory and Experiment, ed. by B. Pullman, (D. Reidel, Boston, Mass., 1979), pp. 191-206.
24. Hamilton, J. F. and P. C. Logel. Thin Solid Films 23 (1974) 89; ibid. 16 (1973) 49.
25. Mason, M. G., L. J. Gerenser, and S.-T. Lee. Phys. Rev. Lett. 39 (1977) 288.

26. Mason, M. G., S.-T. Lee, G. Apai, R. F. Davis, D. A. Shirley, A. Franciosi, and J. H. Weaver. Phys. Rev. Lett. 47 (1981) 730.
27. Lee, S.-T., G. Apai, M. G. Mason, R. Benbow, and Z. Hurych. Phys. Rev. B 23 (1981) 505.
28. Mason, M. G. private communication.
29. Mason, M. G. and R. C. Baetzold. J. Chem. Phys. 64 (1976) 271.
30. Takasu, Y., R. Unwin, B. Tesche, A. M. Bradshaw, and M. Grunze. Surf. Sci. 77 (1978) 219.
31. Liang, K. S., W. R. Salaneck, and I. A. Aksay. Solid State Commun. 19 (1976) 329.
32. Egelhoff, W. F. and G. C. Tibbetts, Phys. Rev. B 19 (1979) 5028.
33. Baetzold, R. C. Inorg. Chem., 21 (1982) 2189.
34. Baetzold, R. C. J. Phys. (Paris) 78 (1981) 933.
35. Baetzold, R. C. J. Am. Chem. Soc. 103 (1981) 6116.

CHARACTERIZATION OF SUPPORTED METAL PARTICLES IN HETEROGENEOUS CATALYSTS: PART I. STUDIES OF HIGH SURFACE AREA MATERIALS

J. A. Dumesic and G. Connell

Department of Chemical Engineering
University of Wisconsin
Madison, Wisconsin 53706

1. INTRODUCTION

Supported metal particles play an important role in heterogeneous catalysis. Not only do they have high surface to volume ratios, but they may also possess unique electronic and geometric properties. However, to understand how these supported metals function as catalysts, and thereby to predict how new generations of heterogeneous catalysts can be prepared, it is necessary to characterize the structure and surface properties of the metal particles. Ideally, this characterization should be carried out with the sample under reaction conditions.

Unfortunately, no single spectroscopic technique is capable of providing a complete characterization of a supported metal particle. For example, consider the use of X-rays in catalyst characterization. X-ray diffraction is the traditional technique for structural studies, but it is insensitive to the chemical and surface properties of a metal particle. Extended X-ray absorption fine structure measurements probe both the structural and electronic properties of metal particles, but this information represents contributions from both surface metal atoms and atoms within the particle. X-ray photoelectron spectroscopy can be used to investigate the elemental composition and chemical bonding near the surface of a metal particle, but this technique is insensitive to the crystal structure of the particle. Having reached the conclusion that it is necessary to combine several techniques when characterizing supported metal particles, it is important to note that different techniques are best suited for studying different types of samples under different sample environments. Returning to the use of X-rays in catalyst characterization, X-ray diffraction and extended X-ray absorption fine structure measurements can be employed with the sample under reaction conditions, while studies utilizing X-ray photoelectron spectroscopy must be carried out with the sample under vacuum (less than 10^{-4} Pa). Furthermore, in the characterization of metal particles on high surface area (porous) supports, the first two techniques probe all of the particles in the sample while X-ray photoelectron spectroscopy monitors only those metal particles on the external surface of the support (i.e., particles within the pores of the support cannot be observed).

It is useful to generalize the above example involving X-rays, leading to a separation of spectroscopic techniques into two classes: (I) those techniques which probe metal particles in high surface area supports under reaction conditions, and (II) those techniques which monitor the surfaces of metal particles on the external surfaces of supports under vacuum conditions. Examples of techniques in Class I include X-ray diffraction, extended X-ray absorption fine structure measurements, Mössbauer spectroscopy, magnetic susceptibility, ferromagnetic resonance, infrared spectroscopy, and laser Raman spectroscopy. These techniques involve the use of electromagnetic radiation (e.g., microwave, infrared, visible, X rays, γ rays) that penetrates through solids and high pressure gases. Hence measurements can be made at reaction conditions. However, such measurements are not specific to the surface because the whole sample contributes to the spectra. Class II techniques include X-ray photoelectron spectroscopy (XPS), ultraviolet photoelectron spectroscopy (UPS), Auger electron spectroscopy (AES), and electron energy loss spectroscopy (EELS). These techniques involve the detection of low energy electrons which have short penetration depths through solids (e.g. nanometers) and gases at pressures greater than 10^{-4} Pa (e.g. meters at 10^{-4} Pa, nanometers at 10^{5} Pa). Hence they are surface sensitive but cannot be used under catalytic reaction conditions. As a consequence of the surface sensitivity, only metal particles on the external surface of the support are monitored. These external particles may have different properties than the unseen particles inside the pores. Hence both classes of techniques have their own limitations.

The present discussion of the characterization of supported metal particles is divided into two parts. The present paper deals with the characterization of metal particles using the techniques listed above in Class I. In addition, indirect characterizations using chemisorption methods are included in this paper. The adjoining paper involves the characterization of supported metal particles using the techniques listed above in Class II. The techniques of transmission electron microscopy, scanning transmission electron microscopy and temperature programmed desorption are also included in Part II. For convenience, the present paper is limited to metal particles on high surface area supports while Part II deals with metal particles on low surface area supports even though this restriction is not necessary.

2. STRATEGIES FOR CHARACTERIZATION OF METAL PARTICLES ON HIGH SURFACE AREA SUPPORTS

Characterization of a supported metal particle involves the determination of crystal structure, electronic properties, particle size, particle shape, and surface properties. In this respect it is important to remember that techniques with the ability to probe metal particles within the pores of high surface area supports under reaction conditions are sensitive to both surface and bulk atoms of the particles. For the measurement of crystal structure and electronic properties, this presents no problem. For the determination of particle size, shape and surface properties, a significant fraction of the total number of metal atoms in the particle must be at the surface. Because the characterization of small particles (less than ca. 5 nm in size) is the primary focus of the present paper, this does not pose a serious problem. Consider first the determination of particle size and shape. For some techniques it is possible to resolve the spectral contributions from surface and bulk metal atoms, while for other techniques the spectral signal represents a weighted-average over the surface and bulk metal atoms. In either case, correlations of measured (or theoretically calculated) spectra from particles of known geometry allow the size and/or shape of unknown particles to be determined. The above characterization procedure may be modified by comparing spectra collected with and without the presence of surface adsorbed species. If it is then assumed that adsorption involves the surface metal atoms only, then the particle size/shape can be estimated by determining spectroscopically the number of metal atoms in the cluster that are perturbed by the adsorption process.

Compared to particle size and shape determination, the study of surface properties using techniques of Class I is a more difficult task. Specifically, particle size/shape determination merely requires a correlation between particle geometry and an observable spectral feature, while the understanding of surface properties requires that this correlation be interpreted quantitatively. In addition, it is possible that the spectral feature used to monitor particle size or shape may not contain any information about the surface (as is the case for magnetic relaxation phenomena in magnetic susceptibility).

Due to the above difficulties in probing the surface properties of a metal particle using in situ (Class I) spectroscopic techniques alone, it is often desirable to combine these techniques

Table 1. Contributions of *in situ* techniques to the characterization of supported metal particles

	Determine Crystal Structure	Measure Electronic Properties	Determine Particle Size	Study Particle Shape	Specific to Surface Properties
X-ray Diffraction and Scattering	X		X	X	
Extended X-ray Absorption Fine Structure	X	X	X		
Mössbauer Spectroscopy		X	X	X	
Magnetic Susceptibility		X	X		
Ferromagnetic Resonance		X	X	X	
Vibrational Spectroscopy					X

with studies of chemisorption phenomena, which probe only the surface atoms of the cluster. In fact, surface studies of chemisorption are often accompanied by spectroscopic studies of the adsorbed molecules (using, for example, infrared or laser Raman spectroscopies). In this way, extents of adsorption are combined with spectroscopic peak shifts (in vibrational frequencies, for example) to provide information about the number and nature of the various adsorption sites on the metal particle. To supplement chemisorption studies, it should be mentioned that it is possible to probe the surfaces of metal clusters using a catalytic reaction for which the mechanism is understood. The advantage of this approach is that the catalytic activity for the probe reaction may vary by more than an order of magnitude with changes in the structure or electronic properties of the metal particle, while the chemisorption of probe molecules (e.g., H_2, CO) may not be significantly altered. Unfortunately, it may be difficult to relate quantitatively the observed changes in catalytic behavior to physical properties of the metal particle.

Table 1 lists the primary techniques that will be discussed in this paper according to their contributions to metal particle characterization. It can be seen therein that a judicious combination of techniques allows the crystal structure, electronic properties, size, shape and surface properties of metal particles to be studied. In the following sections, the use of these techniques will be introduced by means of examples from the recent literature. This discussion is meant to be illustrative, not comprehensive.

The use of electron microscopy and techniques of Class II will also be briefly mentioned where appropriate, even though Part II of this series focuses on these techniques in detail.

3. X-RAY DIFFRACTION AND SCATTERING

The scattering and diffraction of X rays by crystals has been an analytical technique for decades. A number of books can be referenced for details (e.g., 1,2). In short, X rays impinge upon a sample and are scattered by the electrons of the atoms. If these electrons are at periodic distances, a diffraction pattern arises because of the interference of the scattered waves.

Most supported particles are small crystallites on a randomly oriented support. Hence, an X-ray diffraction powder pattern is obtained, i.e., diffracted X rays are observed at all angles for which Bragg's law is satisfied. The sample is placed in an X-ray beam and the intensities of the diffracted beams are measured as a function of the diffraction angle. In general, the positions, intensities and shapes of these diffraction peaks are sensitive to the crystal structure, lattice parameter, mean-square vibrational amplitude of the atoms, crystallite size, crystallite shape, and the presence of strains and defects in the crystallites. The crystal structure and lattice parameter are found by direct measurement of the angles of diffraction. The mean-square vibrational amplitude is obtained by measurement of the intensity of the diffracted radiation, preferably at several different temperatures. Information about crystallite size, shape, strain, and defects is obtained through analysis of the diffraction peak shapes. This is discussed below.

X-ray diffraction data can be analyzed with different degrees of quantitative detail to obtain information about the size, shape and structure of supported metal particles. In the most qualitative analysis, the breadth, β, of a given diffraction peak can be used to estimate the length, L, of the crystallite in the direction normal to the diffracting planes. This is done using the Scherrer equation (e.g., see ref. 1):

$$\frac{\langle L^2 \rangle}{\langle L \rangle} = \frac{K\lambda}{\beta\cos\theta} \tag{1}$$

where the angular brackets represent average values, λ is the X-ray wavelength, θ is the Bragg angle of the diffraction peak and K is a constant approximately equal to unity. The measured value of β must be corrected for instrumental broadening, and it must be assumed that the clusters are free of strain and defects. (This assumption is valid for a given sample if the measured value of L is the same for different orders of the same diffraction condition, e.g., (111), (222) diffraction peaks.) For supported metal crystallites with characteristic lengths greater than ca. 5 nm, analyses using the Scherrer equation are easily carried out using conventional X-ray sources and analog records of diffracted X-ray intensity versus diffraction angle. Additional information can be obtained from X-ray diffraction through Fourier analysis of the diffraction peak shapes (2,3). This requires a digital record of diffracted X-ray intensity versus angle at various intervals (e.g., 0.2°) across the diffraction peaks. For small crystallites (e.g., 2 nm in diameter) the use of a high intensity X-ray source (e.g., rotating anode) may be desirable to collect diffraction patterns with high signal to noise ratios in reasonable periods of time. The Fourier series used to describe the diffraction peak profile is in terms of sines and cosines of $2\pi nh$, where n is an integer, h equals $\frac{2}{\lambda}(\sin\theta - \sin\theta_0)$, 2θ is the diffracted angle, and $2\theta_0$ is the position of the centroid of the peak. Each Fourier cosine coefficient A(n) is the product of two terms, $A^S(n)$ and $A^D(n)$, which are related to the particle size and the presence of microstrains, respectively. For small values of n, h_0^2, and $\langle E_\ell^2 \rangle$, this relation can be written as

$$\ell nA(n) = \ell nA^S(n) - \frac{2\pi^2 \langle E_\ell^2 \rangle \ell^2 h_0^2}{\alpha^2} \tag{2}$$

where $\langle E_\ell^2 \rangle$ is the mean-square strain in a column of length ℓ normal to the diffracting plane, α is the lattice parameter and h_0^2 is the sum of the squared Miller indices of the diffraction peak $(h^2+k^2+l^2)$. By analyzing two or more orders of a given diffraction peak (e.g., (200), (400)), one can plot $\ell nA(n)$ versus h_0^2 and determine the size coefficient $A^S(n)$ by extrapolation to $h_0^2 = 0$. Having thereby separated the effects of strain and particle size in the X-ray diffraction pattern, it is possible to determine the average particle length in a particular direction by plotting $A^S(n)$ versus n. For a given diffraction peak,

$$\left[\frac{dA^S(n)}{dn}\right]_{n\to 0} = \frac{1}{\langle L_{eff} \rangle_{hkl}} \tag{3}$$

where $\langle L_{eff} \rangle_{hkl}$ is the effective length of the particle in the direction normal to the diffracting planes (hkl). This effective length has contributions from the presence of stacking faults and microtwins in the crystallites. The effect of these planar defects, however, can be determined by analysis of shifts in the positions of the diffraction peaks, and this allows the true length of the crystallite in a particular direction, $\langle L \rangle_{hkl}$, to be determined from the value of $\langle L_{eff} \rangle_{hkl}$ (2,3). Finally, the distribution of crystallite lengths in a given direction can be determined if the crystallites are unstrained or if enough diffraction orders are measured to allow the effects of strain to be accurately subtracted from the effects of crystallite size (i.e., allow A(n) to be separated into $A^S(n)$ and $A^D(n)$). This is given by

$$\frac{d^2A^S(n)}{dn^2} = \text{constant} \cdot p(n) \tag{4}$$

where p(n) is the fraction of the crystallites having a length equal to $n\alpha_3$, and α_3 is the period of the Fourier analysis. The latter is determined by measuring the diffraction angles corresponding to the high-angle and low-angle extremes of the peak:

$$\alpha_3 = \frac{\lambda}{2(\sin\theta_{\text{high angle}} - \sin\theta_{\text{low angle}})} \tag{5}$$

The most extensive use of the above procedures in X-ray diffraction to characterize supported metal particles is the work of Sashital <u>et al</u>. (3) in their study of Pt supported on silica. Nitrogen adsorption measurements and small angle X-ray scattering showed that the average pore size of the silica support was 15 nm. One sample contained large Pt particles which were approximately the same size as the pore diameter. These particles showed significant strain. This was revealed by the difference between plots of A(n) versus $n\alpha_3$ for the (111) and the (222) diffraction peaks. In contrast, for another sample with smaller Pt particles, such plots superimposed, as shown in Figure 1. The absence of strain in these particles allowed the particle length distribution to be determined from the second derivative of this plot. Figure 2 presents the corresponding distributions for the <111>, <100>, and <311> directions. By comparing particle lengths in various directions it is possible to study the shape of the crystallite. For the case of Pt/SiO_2, these authors found that the average

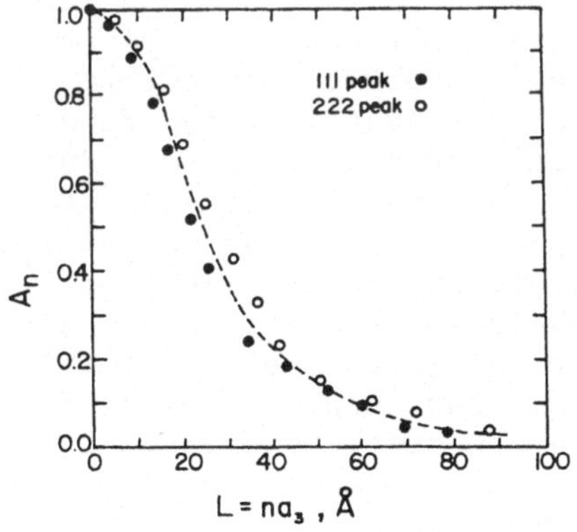

Fig. 1. Stokes'-corrected Fourier coefficients of the (111) and (222) peaks. Reprinted with permission from ref. 3.

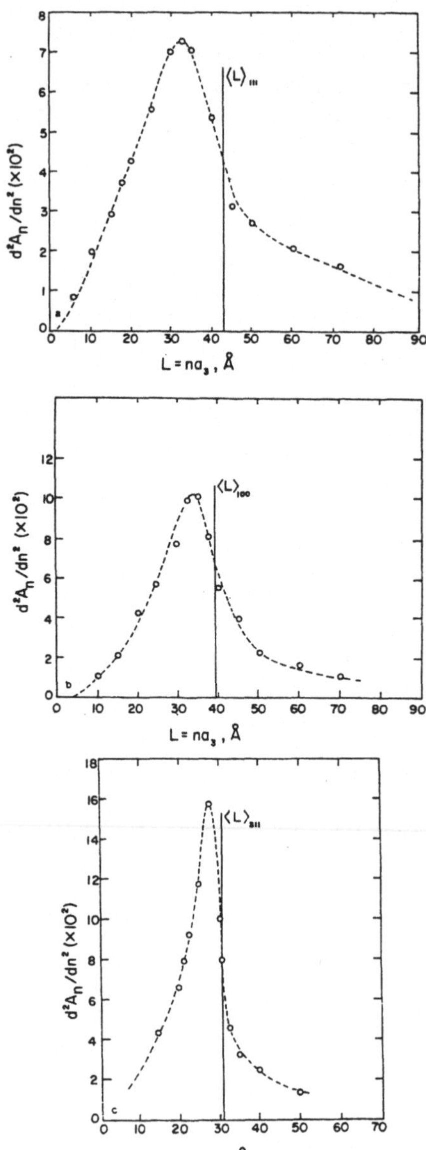

Fig. 2. Particle size distributions in the <111>, <100>, and <311> directions. Reprinted with permission from ref. 3.

lengths of the Pt crystallites were essentially the same in the <111>, <100>, <110> and <311> directions, and they therefore concluded that the particles were equiaxed (e.g., spherical). In addition, the lattice parameter of the Pt particles was equal to the value for bulk, metallic Pt. However, from the integrated intensity of the diffraction peaks, the mean-square vibrational amplitude (related to the Debye-Waller factor) was ca. 30% larger for the Pt atoms in the small Pt particles of Figure 2 than bulk Pt.

Since the breadth of the X-ray diffraction peaks increases with decreasing crystallite size, it becomes difficult to characterize particles smaller than ca. 3 nm because of difficulties in resolving the diffraction peak from the base line. In addition, broad diffraction peaks from the support may interfere with the diffraction peaks from the small metal particles. Such is the case for Pt supported on alumina. In these cases it may be advantageous to employ small angle X-ray scattering, SAXS. This technique is based on X-ray scattering from heterogeneities in the electron density of the sample. For supported metal particles, this scattering could be due to the metal particles or to the pore structure of the support. If the objective of the study is to characterize the metal particles, then the scattering from the support must be minimized. This can be done, for example, by filling the pores of the support with a fluid that has a similar electron density as the support. For Al_2O_3, a fluid effective for this purpose is CH_2I_2 (4). The scattered X-ray intensity, I(s), is a function of the scattering angle, $s = 2\theta/\lambda$. This function is a sum of the contributions from all of the metal atoms in the sample, and it is sensitive to the size and shape of the metal particles. For example, considering a distribution of spherical particles of radius R, a slope of a plot of $\ell n\ I(s)$ versus s^2 allows the Guinier radius, R_G, to be determined, where $R_G^2 = \langle R^7\rangle/\langle R^5\rangle$. In addition, if the area under a plot of sI(s) versus s is divided by the limit of $s^3 I(s)$ as s becomes large, then a quantity related to the Porod radius, R_P, is obtained, where $R_P = \langle R^3\rangle/\langle R^2\rangle$. Knowing two moments of the particle size distribution allows one to calculate the mean and variance of the distribution. This procedure has been outlined by Whyte (5). For the case of 0.62 wt. % Pt/Al_2O_3 (reduced in H_2 at ca. 800 K), Whyte et al. (4) assumed that the Pt particles were present in a log-normal distribution. The values of $2R_G$ and $2R_P$ were found to be 57 and 37 nm, respectively, corresponding to a mean particle diameter of 28 nm.

By Fourier analysis of the scattered X-ray intensity over a wide range of scattering angles (i.e., including data at both low and high angles) it is possible to determine the radial electron distribution function for a sample (6). This function gives the electron density (related to the number of atoms) at a distance r from a central atom. Metal clusters of small size (e.g., 1 nm) do not possess the long-range order characteristic of bulk materials; however, these clusters may well show short range order. For example, the distance from a central metal atom to its nearest neighbors may be nearly the same in a small cluster as in a bulk metal. It is this short range order that the radial electron distribution measures. The application of this technique to "amorphous catalysts" (i.e., catalysts lacking long range order) has been reviewed by Ratnasamy and Léonard (6). Of relevance to the characterization of supported metal particles using radial electron distribution measurements is the study by Gallezot and Bergeret (7) of 1 nm Pt particles in Y zeolite. A sample (containing 15 wt. % Pt) was reduced in hydrogen at 580 K, evacuated at 700 K and then transferred in a glove box to a controlled atmosphere cell. Radial electron distribution functions were subsequently measured at room temperature with the sample in argon, hydrogen, benzene, and various hydrogen/benzene gas mixtures. As many as seven coordination shells of Pt could be resolved. The observed Pt-Pt distances of the Pt particles in Y zeolite under vacuum were generally smaller than those in bulk Pt, and this contraction was not uniform for the different coordination shells. It was thereby concluded that the structure of these "naked" Pt particles is distorted relative to the normal fcc structure of Pt.

Upon addition of hydrogen, however, the Pt-Pt distances in the small Pt particles relaxed to those values of bulk Pt. Hence, chemisorption of hydrogen on surface Pt atoms eliminated the distortion of the Pt particles, apparently because the coordination of the Pt surface atoms is completed by the chemisorption of hydrogen. Addition of benzene to the naked Pt particles also relaxed the distortion of the particles but not as much as observed for hydrogen. With mixtures of benzene and hydrogen (i.e, benzene hydrogenation reaction conditions) the extent of relaxation was correlated with the ratio of the two gaseous constituents. This led to the conclusion that both hydrogen and benzene were adsorbed on the surface. The hydrogen, however, was adsorbed irreversibly on some sites, and could not be displaced by benzene. This clearly illustrates the use of in situ RED measurements to monitor the structure of metal particles in controlled gaseous environments. Ratnasamy et al. (8) have also shown that supported Pt particles with dispersions as high as 70% (i.e., 1.5 nm clusters) can be studied using the above radial electron distribution techniques.

4. EXTENDED X-RAY ABSORPTION FINE STRUCTURE

With the availability of synchrotron radiation sources which provide high X-ray fluxes, extended X-ray absorption fine structure (EXAFS) has become a powerful probe of supported metal particles. For details the reader is referred to the review by Lytle et al. (9). The absorption of X rays by a sample is monitored beyond an absorption edge, corresponding to the excitation of a core electron to a valence level. Such measurements provide information about the valence state of the absorbing atom as well as information about the coordination number and the distances to neighboring atoms. The advantage of EXAFS in the study of small metal particles is that the metal need not possess long range order. That is, a radial structure function is generated which is related to the number of atoms surrounding the central absorbing atom as a function of the distance from the central atom. In addition, by studying X-ray absorption edges of different elements it is possible to determine the radial structure functions for the various components of the sample under investigation. For example, the structure surrounding each of the metals in a bimetallic cluster may be studied separately, as will be shown later in this section.

To collect EXAFS information, a sample is placed in an X-ray beam and the absorption is measured as a function of photon energy. Once the energy of the absorption edge is exceeded, a photoelectron is ejected which is scattered by the neighbors of the central atom. The scattered electron wave can interfere constructively or destructively with the ejected photoelectron wave, producing oscillations in the X-ray absorption coefficient. This is shown

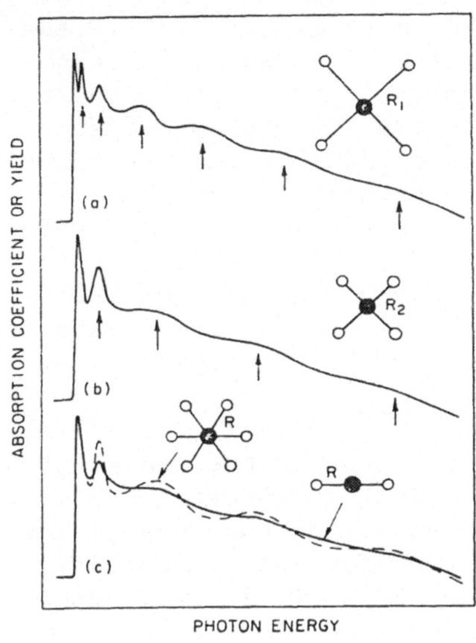

Fig. 3. Hypothetical absorption spectra for some model compounds. Reprinted with permission from ref. 10.

schematically in Figure 3. The absorption coefficient rises sharply at the absorption edge and modulation of this coefficient is observed beyond the edge. The frequency of the oscillations is inversely proportional to the distance of neighboring atoms, and the amplitude is proportional to the number of neighbors. Furthermore, the area under the curve near the edge (called the white line area) and the position of the edge may provide information about the chemical state of the absorbing atom.

The first step in EXAFS studies is to determine the EXAFS function, $\chi(k)$, as a function of the wave vector, k, of the ejected electron:

$$\chi(k) = (\mu - \mu_0)/\mu_0 \tag{6}$$

where μ and μ_0 are the X-ray absorption coefficients for an atom in the sample and in the free state, respectively. The wave vector k is defined in terms of the kinetic energy, E, of the ejected electron:

$$k = (2mE)^{\frac{1}{2}}/\hbar \tag{7}$$

where m is the mass of an electron and $\hbar$ is Planck's constant divided by 2π. The value of μ is related to $\ell n(I_0/I)$, where I and I_0 are the detected X-ray intensities with and without the sample in the X-ray beam, respectively. Thus, from plots of $\ell n(I_0/I)$ versus k it is possible to determine $(\mu - \mu_0)$ by measuring the amplitudes of the fluctuations in $\ell n(I_0/I)$ relative to a smooth curve (without fluctuations) fit through the data. The value of μ_0, which also depends on k, is related to the height of the absorption edge.

Fourier transformation of the EXAFS data yields the radial structure function, $\phi(r')$ where r' is the distance from the absorbing atom minus a phase shift. Thus, from plots of $\phi(r')$ versus r' it is possible to determine the interatomic distances between a central atom and its neighbors. As noted above, a phase shift must be added to r' in order to obtain true interatomic distances. This phase shift for a given peak in the radial structure function is sensitive to the identities of the central atom and the neighbors contributing to that peak. For a pair of atoms, the phase shift can be determined from EXAFS of known compounds, and this can be used to calculate interatomic distances from EXAFS of unknown structures containing this pair of atoms. Bond distances can typically be determined in this manner with an accuracy of 0.002 nm (9).

To interpret quantitatively a given peak in the radial structure function, it has been found convenient to invert the Fourier transform $\phi(r')$ over a range of r' encompassing that peak. The result of this procedure is an EXAFS function due to scattering from those atoms responsible for the peak in $\phi(r')$, i.e., atoms in a particular coordination shell. For example, if a given coordination shell, i, is used to generate the EXAFS function through the inverse Fourier transform, then

$$\chi_1(k) = \left(\frac{N_i}{kr_i^2}\right) F_i(k) \exp(-2k^2\sigma_i^2) \sin[2kr_i + 2\delta_i(k)] \tag{8}$$

where the subscript i refers to the coordination shell chosen, N_i is the number of atoms in the shell, r_i is the distance to the shell from the central atom, σ_i^2 is the mean square deviation of the interatomic distance r_i, $\delta_i(k)$ is the phase shift and $F_i(k)$ is a scattering function. By studying known compounds, $F_i(k)\exp(-2k^2\sigma_i^2)$ and $\delta_i(k)$ can be determined for specific pairs of atoms. This allows N_i, r_i and $\Delta\sigma_i^2$ to be determined for unknown structures through least squares fitting of inverse Fourier transforms to the above equation, where $\Delta\sigma_i^2$ is the disorder of the unknown structure relative to the known compound. For example, if the value of r_i is the same for a particular pair of atoms in a standard and an unknown sample, then comparison of the Fourier-filtered EXAFS functions for these two samples allows differences in the coordination and disorder to be determined using the equation:

$$\ln\left[\frac{\chi_{i,s}(k)}{\chi_{i,u}(k)}\right] = 2(\sigma_{i,u}^2 - \sigma_{i,s}^2)k^2 + \ln\left(\frac{N_{i,s}}{N_{i,u}}\right) \tag{9}$$

where the subscripts s and u refer to the standard and unknown samples, respectively.

Sinfelt et al. (11) studied the structure of Ru-Cu bimetallic clusters supported on SiO_2. EXAFS measurements were made on the Ru-Cu/SiO_2 catalyst, as well as Ru/SiO_2 and Cu/SiO_2 reference materials having the same Ru and Cu loadings. The radial structure functions and the inverse Fourier transforms, for Ru as the central absorbing atom, were similar for Ru-Cu/SiO_2 and Ru/SiO_2. This was not the case for Cu as the central absorbing atom, especially when the inverse Fourier transforms over the first coordination shell were compared for Ru-Cu/SiO_2 and Cu/SiO_2. It was postulated by the authors that Cu is present primarily at the surface of Ru-Cu clusters, with the cores of these clusters containing primarily Ru. Accordingly, the Ru atoms are bonded primarily to other Ru atoms, and they have nearly the full coordination of bulk Ru. In contrast, Cu atoms at the surface of Ru-Cu clusters are bonded to both Cu and Ru, and the coordination number of these Cu atoms is lower than that in bulk Cu.

To test this hypothesis further, Sinfelt et al. exposed the samples to oxygen at room temperature. The metal atoms near the surface would be expected to react with the oxygen, thereby changing the nature of their coordination shell. The radial structure functions for Ru-Cu/SiO_2, Ru/SiO_2 and Cu/SiO_2 are shown in Figure 4, before and after exposure to oxygen. The coordination of Ru (obtained from EXAFS of the K-edge of Ru) is only slightly altered by oxygen for both Ru-Cu/SiO_2 and Ru/SiO_2, as shown in the upper half of the figure. In addition, the coordination of Cu (obtained from EXAFS of the K-edge of Cu) is perturbed little by oxygen for Cu/SiO_2; however, the radial structure function of Cu is significantly changed by exposure of Ru-Cu/SiO_2 to oxygen. This can be seen in the lower half of the figure. Such behavior illustrates that Cu is present at the surface of these Ru-Cu bimetallic clusters.

Clausen et al. (12) studied EXAFS of the K-edge of Mo in Co-Mo/Al_2O_3 hydrodesulfurization catalysts. As with the above study of Ru-Cu/SiO_2, these EXAFS data were compared to studies of related materials: Mo/Al_2O_3 and bulk MoS_2. The Fourier transforms (i.e., radial structure functions) of the EXAFS functions for calcined Co-Mo/Al_2O_3 and Mo/Al_2O_3 were essentially the same. Both of these transforms showed a peak due to Mo-O pairs and the absence of peaks due to next nearest neighbors. This indicated that a well-defined (crystalline) phase of molybdenum does not exist for these calcined samples. After sulfiding the Co-Mo/Al_2O_3 and Mo/Al_2O_3 samples by reaction with H_2S/H_2, EXAFS data were again collected. The first coordination shell of Mo in these two samples, which corresponds to Mo-S pairs, was essentially the same as that in bulk MoS_2. Accordingly, it was concluded that the Mo in sulfided Co-Mo/Al_2O_3 and Mo/Al_2O_3 is present as a phase which resembles MoS_2. Because no Mo-O pairs were found, this cannot be an oxysulfide phase.

Analysis of the second coordination shell of Mo in sulfided Co-Mo/Al_2O_3 and Mo/Al_2O_3, which corresponds to Mo-Mo pairs, showed a difference from the structure of bulk MoS_2. The Fourier-filtered

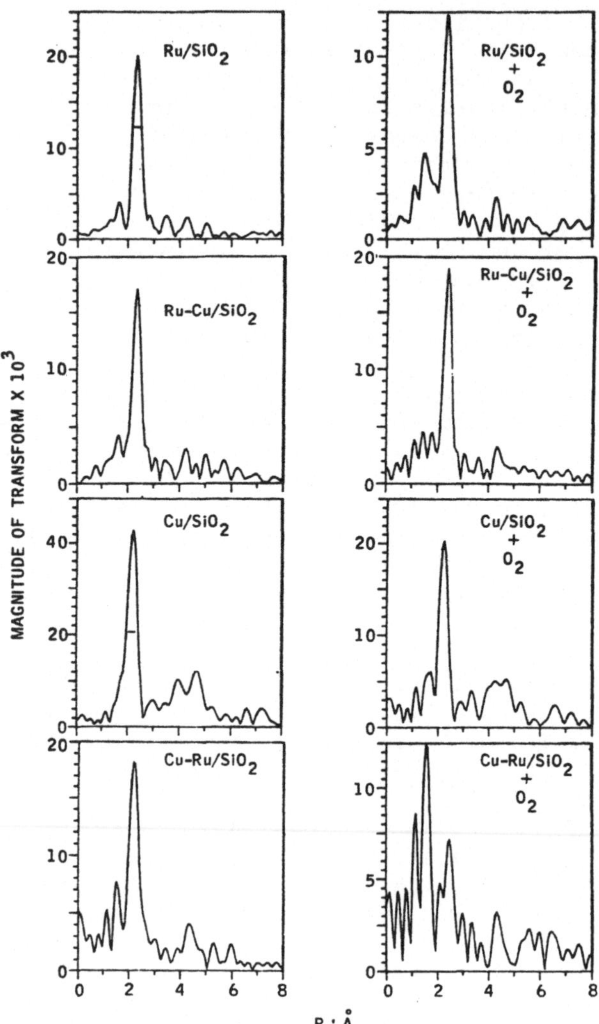

Fig. 4. The radial structure function for Ru/SiO_2, Cu/SiO_2 and $Ru\text{-}Cu/SiO_2$ before and after exposure to oxygen. Reprinted with permission from ref. 11, Copyright 1980 American Institute of Physics.

EXAFS functions had the same shape and frequency, indicating that the second coordination shell of Mo has the same constituents (i.e., Mo) and interatomic distance in all samples. The intensity of the oscillations in the EXAFS function for bulk MoS_2, however, was greater than that for sulfided $Co\text{-}Mo/Al_2O_3$ or Mo/Al_2O_3. Thus, the number of atoms in the second coordination shell of Mo is greater in bulk MoS_2 than in the two alumina-supported samples. While it was concluded above that the Mo in sulfided Co-Mo/Al_2O_3 and Mo/Al_2O_3 is present as a phase resembling MoS_2, it must be concluded further that this phase is present as small domains (ca. 1 nm in size) on Al_2O_3. Furthermore the EXAFS data for Co-Mo/Al_2O_3 and Mo/Al_2O_3 were very similar in both the calcined and sulfided states. Cobalt apparently does not disrupt the Mo phases formed on the samples, but it may be present on or within these phases. In this respect, the authors suggested that Co may substitute for Mo at surface positions of a MoS_2-like phase on Al_2O_3. It is interesting to note that similar conclusions were obtained using ^{57}Co Mössbauer emission spectroscopy to monitor the state of Co in Co-Mo/Al_2O_3 catalysts (13).

X-ray absorption measurements can be used to probe the chemical state of the absorbing atom. This is done by measuring the extent of X-ray absorption near the edge. Consider the L_{III}-edge of Re which was studied by Short et al. (14) in their investigation of Re-Pt/Al_2O_3 catalysts. This edge is due to electronic transitions from $2p_{3/2}$ states to unfilled $5d_{5/2}$ states. The number of these unfilled valence states increases as the oxidation state of Re increases, and the intensity of the L_{III} absorption edge increases accordingly. Figure 5 shows the L_{III}-edge of Re for a Re-Pt/Al_2O_3 catalyst (containing 0.9 wt. % of each metal and after hydrogen treatment at 785 K) and for metallic Re physically mixed with zeolite powder. The increased area under the absorption curve for Re-Pt/Al_2O_3 indicates that the Re in this sample is electron deficient relative to metallic Re. Quantitative comparison of these areas suggested that the Re in Re-Pt/Al_2O_3 was present as Re^{4+}. Confirmation that the Re was not in the metallic state for this sample was obtained from Fourier transforms of EXAFS data for the L_{III}-edge. In particular, the primary peak in the radial structure function could not be attributed to Re-Re or Re-Pt distances, and it was instead interpreted in terms of Re-O (or Re-Cl) pairs. The absence of peaks due to higher coordination shells of Re suggested that the phase containing Re^{4+} was highly dispersed on the Al_2O_3 support.

In a similar manner, Gallezot et al. (15) measured the area under the L_{III}-edge of Pt in a series of Y zeolite-supported Pt samples. These authors also carefully measured shifts in the edge position for the various samples. It was found that the white line area increased and the edge position shifted to higher energies when (i) the particle size was decreased from 3 to 1 nm, (ii) the charge of the ion-exchange cations in the zeolite was increased (from Na^+ to Ce^{3+}), and (iii) the samples were exposed to oxygen at 298 K. Thus, Pt clusters in Y zeolite are electron deficient relative to bulk metallic Pt, and this electron deficiency is increased by increasing the charge of the ion-exchange cations in the zeolite or by exposure of the sample to

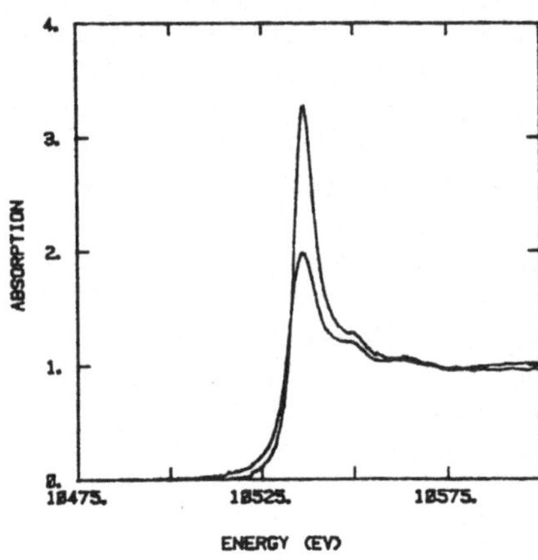

Fig. 5. Comparison of normalized Re L_{III}-edge intensities for 0.9-0.9 wt.% $Pt-Re/Al_2O_3$ (upper curve) and Re powder mixed with zeolite (lower curve). Reprinted with permission from ref. 14.

oxygen. It should be noted that discussions have recently been published regarding the best procedures for analyzing X-ray absorption edges in order to correlate changes in the chemical state of Pt samples (16,17).

5. MÖSSBAUER SPECTROSCOPY

Mössbauer spectroscopy is based on the recoil-free emission of γ rays by a radioactive source and the resonant absorption of these γ rays by Mössbauer nuclei in the sample to be studied. The physical bases for this technique and the applications of Mössbauer spectroscopy in heterogeneous catalysis have been reviewed elsewhere (18,19). In brief, resonance between the source and absorber is normally achieved by modulating the energy of the emitted γ rays using the Doppler effect. A typical Mössbauer spectrum is a plot of the γ-ray intensity transmitted through the sample as a function of Doppler velocity (e.g., units of mm/s). For certain elements, this nuclear excitation process in the sample can be used to obtain information about the electron density, local symmetry, and magnetic structure of atoms in solids or in viscous liquids. The spectral parameters derived from Mössbauer spectra are the recoil-free fraction f, isomer shift δ, quadrupole splitting ΔE_Q, and the magnetic hyperfine field H. These Mössbauer parameters are related to the strength of bonding of the atom in the solid or viscous liquid, the electron density at the nucleus, the electric field gradient at the nucleus, and the internal magnetic field at the nucleus, respectively. Depending on which particular radioactive source is used to generate the γ rays, only one element may absorb the γ ray resonantly. This can be a distinct advantage of the technique. For example, the study of iron present at low loadings (ca. 1 wt.%) on porous supports presents no problem. In addition, since the Mössbauer spectrum is a sum of the contributions from all of the atoms of that element being probed, this technique has the ability to resolve surface and bulk atoms.

The primary limitation in the application of Mössbauer spectroscopy to the characterization of metal clusters is that only a small number of elements are suited for chemical studies using this technique. For example, the nuclear energy levels must be sufficiently narrow in energy that the effects of chemical perturbations on the nuclear levels can be resolved, since it is the measurement of these chemical perturbations that allows the Mössbauer parameters δ, ΔE_Q and H to be determined. It is also desirable that a radioactive source exist with a long lifetime (greater than ca. 1 month) for convenient Mössbauer spectroscopy. With these two restrictions, the following isotopes may be useful for chemical studies using Mössbauer spectroscopy (the isotopes in parentheses are the radioactive nuclides used to generate the γ rays):

$^{57}Fe(^{57}Co)$, $^{83}Kr(^{83}Rb)$, $^{119}Sn(^{119m}Sn)$, $^{121}Sb(^{121m}Sn)$,
$^{125}Te(^{125}I)$, $^{127}I(^{127m}Te)$, $^{129}I(^{129m}Te)$, $^{149}Sm(^{149}Eu)$,
$^{151}Eu(^{151}Gd)$, $^{155}Gd(^{155}Eu)$, $^{170}Yb(^{170}Tm)$, $^{182}W(^{182}Ta)$,
$^{237}Np(^{241}Am)$, $^{145}Nd(^{145}Pm)$, $^{160}Dy(^{160}Tb)$, $^{168}Er(^{168}Tm)$,
$^{181}Ta(^{181}W)$

It is, in fact, important to note that the applicability of Mössbauer spectroscopy to the characterization of metal particles may be extended to include the study of metals containing these radioactive nuclides. Specifically, a sample containing one of these nuclides can be used as a radioactive source (e.g., ^{57}Co) and a standard absorber containing the corresponding Mössbauer isotope (e.g., ^{57}Fe) in a well-defined chemical environment is used to generate the Mössbauer spectrum. It should be noted that these so-called source experiments may be difficult to interpret for non-metallic systems due to electronic after-effects caused by nuclear decay in the sample. Finally, if the supported metal particle of interest does not contain a suitable Mössbauer isotope or radioactive nuclide, it may still be possible to use Mössbauer spectroscopy for sample characterization by doping a small amount of a Mössbauer isotope or radioactive nuclide into the metal particle. In this way, the added isotope or radioactive nuclide is used as a probe of the surrounding structure.

An example of the use of Mössbauer spectroscopy to characterize supported metallic iron particles is the work of Phillips et al. (20) and Phillips and Dumesic (21). Metallic iron particles were prepared by the thermal decomposition of $Fe(CO)_5$ at 380 K on a Grafoil support (sheets of oriented graphite crystallites). Figure 6 shows the Mössbauer spectra that were collected at 77 K after various $Fe(CO)_5$ decomposition times. The doublet present in these

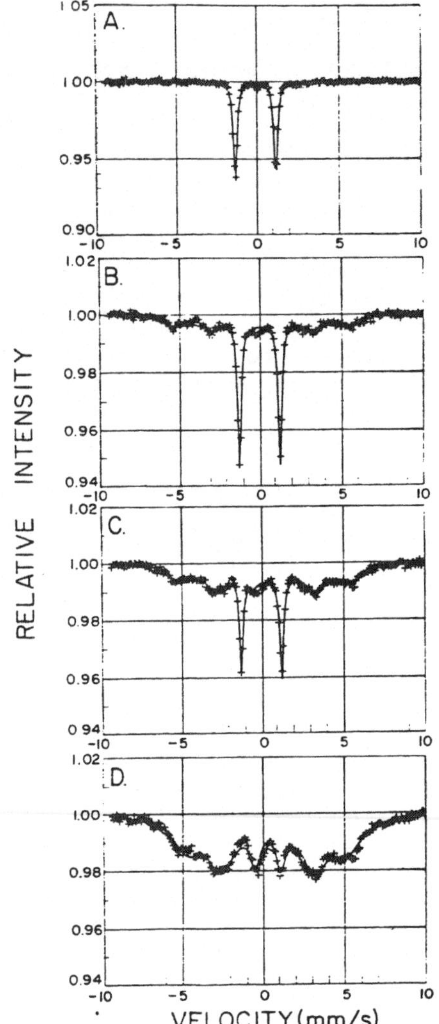

Fig. 6. Iron Mössbauer spectra taken at 77 K after decomposition of $Fe(CO)_5$ on Grafoil outgassed at 295 K: (A) after 1 h at 383 K; (B) after 4.5 h at 383 K; (C) after 14 h at 383 K; (D) after 25 h at 383 K. Reprinted with permission from Ref. 21.

spectra after short times of treatment at 380 K is due to $Fe(CO)_5$ physically adsorbed (or frozen) on the Grafoil surface. The six-peak spectral component which grows in intensity with increasing time of treatment in $Fe(CO)_5$ is due to metallic iron. Visual inspection of this series of Mössbauer spectra allows the rate of $Fe(CO)_5$ decomposition to metallic iron to be determined.

Information about the structure of the metallic iron particles on Grafoil can be obtained from a quantitative analysis of the six-peak, metallic iron Mössbauer spectrum. This is demonstrated with reference to Figure 7. The Mössbauer spectrum at 295 K of the metallic iron formed after $Fe(CO)_5$ decomposition at 380 K is shown in Figure 7A. The sample was then heated to 450 K, and a Mössbauer spectrum was collected at 295 K, as shown in Figure 7B. The spectrum of Figure 7B exhibits six narrow peaks, with widths similar to those peaks of bulk metallic iron. The splitting between peaks 1 and 6 (numbering peaks consecutively, starting from negative Doppler velocities) is a measure of the magnetic hyperfine field, and the value determined from Figure 7B is in excellent agreement with the value for bulk metallic iron at 295 K. In contrast, the Mössbauer spectrum of Figure 7A shows six broad peaks, and the splitting between peaks 1

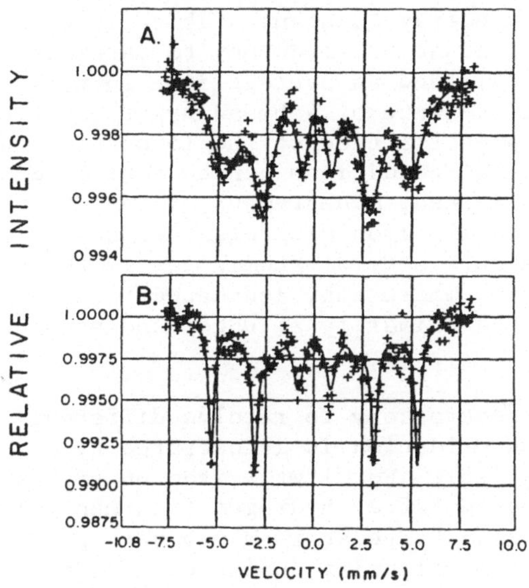

Fig. 7. Iron Mössbauer spectra at 295 K on Grafoil: (A) after $Fe(CO)_5$ decomposition at 378 K; (B) after heat treatment at 450 K. Reprinted with permission from ref. 20, Copyright 1980 American Chemical Society.

and 6 is smaller than that of Figure 7B. It is also apparent that the six peaks of Figure 7A are superimposed on a broad, curved base line. In short, the quantitative similarities of Figure 7B with the Mössbauer spectrum of bulk metallic iron suggests that large metallic iron particles (larger than ca. 10 nm) are formed after heating Fe/Grafoil to 450 K. (The spectrum of Figure 7B also shows traces of iron carbide since the sample was heated in the presence of gaseous CO liberated during $Fe(CO)_5$ decomposition.) However, the broad peaks, small magnetic hyperfine field, and curved base line of Figure 7A all suggest that smaller metallic iron particles are formed by $Fe(CO)_5$ decomposition at 380 K.

To interpret further the spectrum of Figure 7A, computer simulations are necessary. Due to the small size of the metallic iron particles, the magnetic hyperfine field is not pinned in one direction, but it may possess sufficient thermal energy to fluctuate in space (22). The movement of the magnetization of the particle from one low energy direction to another is called superparamagnetism, and it leads to a collapse of the six-peak Mössbauer spectrum. The curved base line is a result of this magnetic relaxation phenomenon. The fluctuation of the particle magnetization around a single low energy direction is termed collective magnetic excitation, and it causes a decrease in the observed magnetic hyperfine field. The small splitting between peaks 1 and 6 in Figure 7A stems from this mode of magnetic relaxation. Finally, the presence of surface atoms leads to a distribution of magnetic hyperfine fields, and this causes the broad peak shapes observed in Figure 7A. With the aid of a computer, the above three effects can be used to simulate Mössbauer spectra of metallic iron clusters. These spectra are sensitive to the choice of particle size distribution, magnetic hyperfine field distribution and the magnetic anisotropy energy constant. This latter constant relates the magnetic fluctuation frequency to particle size. Such computer simulation yielded an average iron particle size of ca. 6 nm. However, the presence of a broad hyperfine field distribution suggested that a large fraction of the iron atoms must be near the surface. Both of these conditions can be satisfied by thin, raft-like, particles. Indeed, measurement of the extent of CO chemisorption by these metallic iron clusters suggests that these raft-like clusters may be only several atomic layers thick. In addition, transmission electron microscopy indicated that these metallic iron clusters are located primarily at edges and steps on the Grafoil surface.

The ability of Mössbauer spectroscopy to resolve different environments for iron in a given oxidation is illustrated by a study of iron supported on SiO_2 (23). For 1 wt.% iron on silica, Fe^{2+} is the predominant iron species after hydrogen treatment at temperatures below ca. 650 K. Room temperature Mössbauer spectra of this Fe^{2+} show two, partially overlapping quadrupole-split doublets. This is shown in Figure 8B for 1% Fe/SiO_2 after hydrogen treatment at 498 K. The positive velocity peaks of each doublet are resolvable, and the negative-most peaks of each doublet overlap. (For comparison, Figure 8A shows the Mössbauer spectrum of Fe^{3+} after treatment of the sample in oxygen at 470 K.) The doublet with the larger isomer shift and quadrupole splitting is denoted as the outer doublet, while the doublet with the smaller isomer shift and quadrupole splitting is denoted as the inner doublet. The quadrupole splitting of high spin Fe^{2+} has two contributions. The oxygen anions surrounding the Fe^{2+} and the sixth d-electron of Fe^{2+} give rise to electric field gradients at the nucleus (while the other five d-electrons are in a half-filled, spherically symmetric shell). These two contributions are opposite in sign, and it has been found experimentally that the quadrupole splitting for high spin Fe^{2+} normally increases with increasing coordination number. Thus, the outer doublet is due to Fe^{2+} in sites of higher coordination compared to the sites for Fe^{2+} of the inner doublet. As the reduction of Fe/SiO_2 becomes more severe, the average coordination of the Fe^{2+} cations would be expected to decrease. This is reflected in the Mössbauer spectra of Figures 8B-8D. Outer doublet (high coordination) Fe^{2+} is seen therein to be progressively converted to inner doublet (low coordination) Fe^{2+} as the severity of reduction is increased from H_2 at 498 K, to CO/CO_2 (15:85) at 653 K, and to H_2 at 653 K.

The surface properties of Fe^{2+} on SiO_2 can be probed by combining Mössbauer spectroscopy with adsorption measurements. It has been found that nitric oxide adsorbs strongly at room temperature on Fe^{2+} cations (24), while it adsorbs only weakly on SiO_2. Room temperature Mössbauer spectra were collected before and after NO was adsorbed on Fe/SiO_2 (reduced in CO/CO_2 at 653 K) and after

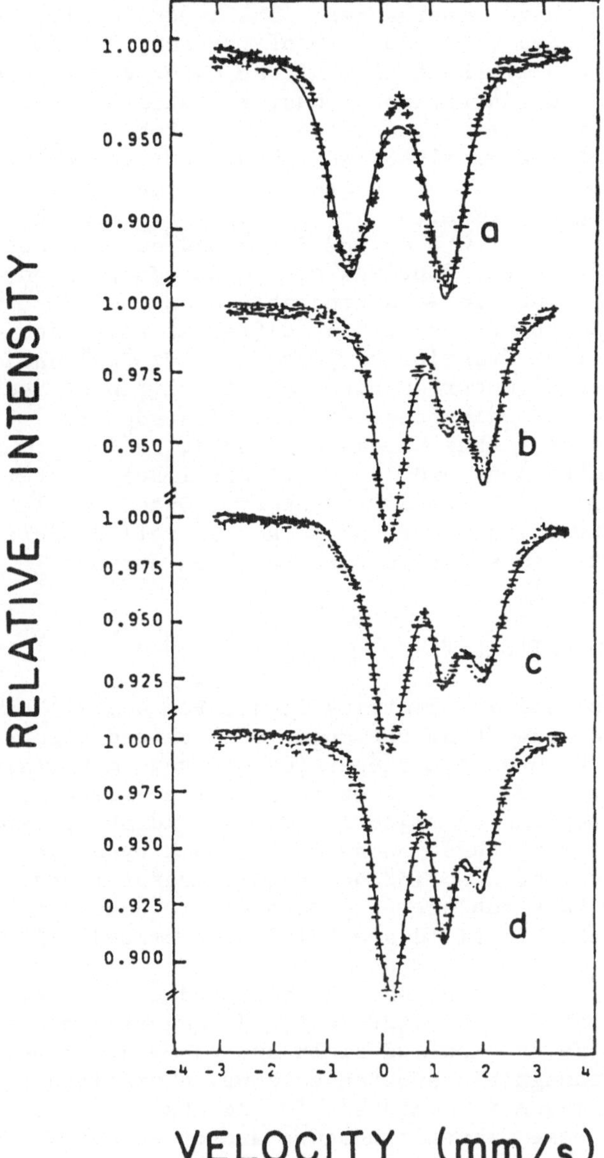

Fig. 8. Iron Mössbauer spectra taken at room temperature for 1 wt. % Fe/SiO_2: (A) after O_2 treatment at 470 K; (B) after H_2 treatment at 498 K; (C) after CO/CO_2 (15:85) treatment at 653 K; (D) after H_2 treatment at 653 K. Reprinted with permission from ref. 23, Copyright 1982 American Chemical Society.

evacuations at higher temperatures. The inner doublet Fe^{2+} was completely eliminated upon exposure of the sample to NO at room temperature while only about half of the outer doublet Fe^{2+} cations were affected by NO adsorption. Thus, essentially all of the inner doublet Fe^{2+} cations were located on the surface of the silica support, while at least half of the outer doublet Fe^{2+} cations were located within small iron oxide clusters. Upon evacuation at progressively higher temperatures, NO desorbed from the sample as was evidenced by the growth of the inner and outer doublets and the disappearance of the two doublets due to NO adsorbed on Fe^{2+} cations. At no point did an appreciable amount of Fe^{3+} form. Measurement of Mössbauer spectra at different temperatures showed that the recoil-free fraction of inner doublet Fe^{2+} cations was greater than that of outer doublet Fe^{2+} (i.e., inner doublet Fe^{2+} cations are more strongly bonded to the SiO_2 support). All of the above results suggest that the inner doublet Fe^{2+} cations are present as isolated ions and/or thin (raft-like) clusters in strong interaction with the SiO_2 support. Outer doublet Fe^{2+} cations, in contrast, are present in not as well dispersed clusters in weaker interaction with the SiO_2 support.

6. MAGNETIC SUSCEPTIBILITY

Some metals used as catalysts (e.g., Fe, Co, Ni) have magnetic properties which arise from the spins of unpaired electrons. Measurement of the magnetic properties of these materials leads to information about the chemical state of these elements, the particle size distributions of magnetic phases, and the interaction of adsorbed species with these phases. The book by Selwood (25) provides an excellent description of such measurements. As an introduction to this subject, the possible responses of a sample in an applied magnetic field are briefly described below.

Diamagnetic substances have no unpaired electrons. When these materials are placed in a magnetic field, magnetic moments are induced which oppose the applied field and hence are repelled by it. In contrast, paramagnetic substances contain unpaired electrons with associated magnetic moments. In the absence of an applied magnetic field these magnetic moments are oriented randomly; however, when an external field is applied the magnetic moments tend to align with the field, and the sample is thereby attracted to the external field. Paramagnetic materials are characterized by their susceptibility, κ, defined by:

$$\kappa = \frac{M}{H} \tag{10}$$

where M is the magnetization (i.e., magnetic moment per unit volume of material) and H is the external field. The magnetization generally follows the Langevin equation:

$$\frac{M}{M_s} = \coth\left(\frac{m_p H}{k_B T}\right) - \left(\frac{k_B T}{m_p H}\right) \tag{11}$$

where M_s is the saturation magnetization (corresponding to the complete alignment of all the magnetic moments in the sample), m_p is the magnetic moment of the individual spins in the sample, k_B is the Boltzmann constant, and T is the temperature. At low fields ($M << M_s$) this equation reduces to

$$M = \frac{N_p m_p^2 H}{3k_B T} \tag{12}$$

where N_p is the number of spins per unit volume of sample. According to this low-field approximation, the susceptibility of paramagnetic materials is independent of the applied field and varies inversely with temperature. At high fields ($\frac{M}{M_s} \approx 1$) the Langevin equation has the following asymptotic form:

$$\frac{M}{M_s} \approx 1 - \frac{k_B T}{m_p H} \tag{13}$$

Below a certain temperature the magnetic moments may become aligned with respect to each other, even in the absence of an applied magnetic field. The magnetization which the sample possesses at zero applied field is called the spontaneous magnetization. If the magnetic moments are oriented in a parallel fashion, then ferromagnetism results and the temperature below which magnetic ordering takes place is called the Curie temperature. An antiparallel orientation between neighboring spins results in antiferromagnetism or ferrimagnetism, and the ordering temperature is called the Néel temperature. (In an antiferromagnetic material, the antiparallel magnetic moments completely cancel each other, while this cancellation is not complete in a ferrimagnetic material.)

The above concepts form the basis for determining the chemical state of a magnetic material. For example the extent of reduction of NiO to metallic nickel can be easily monitored by measuring the limiting value of the sample magnetization at high applied fields (i.e., the saturation magnetization). Metallic nickel is ferromagnetic while NiO is antiferromagnetic, and the conversion of NiO to Ni during reduction treatments therefore leads to significant increases in the magnetization of the sample. In addition, different magnetic phases (e.g., α-Fe_2O_3, Fe_3O_4, FeO, Fe) usually have different Néel or Curie temperatures, thereby providing another means of phase identification using magnetic methods. These ideas will be made more quantitative by means of examples presented later in this section.

The ordering of magnetic moments which takes place below the Curie or Néel temperatures leads to the formation of magnetic domains in massive materials. Within each domain the magnetic moments are ordered, but the direction in which the moments are oriented is different for different domains. However, particles less than ca. 20 nm in size are composed of single domains. Thus, the magnetic moments in these particles are all ordered with respect to each other, and the particle behaves paramagnetically with a net magnetic moment equal to the vector sum of the individual moments within the particle. This phenomenon is called superparamagnetism. It should be noted that the net magnetic moment of the particle may be pinned in certain directions due to magnetic anisotropies, as discussed in the section on Mössbauer spectroscopy. At temperatures below the so-called blocking temperature, T_B, the particle behaves ferromagnetically (or ferrimagnetically) instead of superparamagnetically. In general, T_B decreases with decreasing particle size.

When a collection of superparamagnetic particles is placed in an applied magnetic field, the measured magnetization is expressed by the Langevin equation (equation 11). The magnetic moment of an individual spin, m_p, which appeared in the Langevin equation for paramagnetism is replaced by the net magnetic moment of the particle, M_p, for superparamagnetism:

$$M_p = M_{sp}V \tag{14}$$

where M_{sp} is the spontaneous magnetization of the material and V is the volume of the particle. (The value of M_{sp} has been found to be approximately independent of particle size.) According to the Langevin equation, plots of M/M_s versus H/T should superimpose at different temperatures if the sample behaves superparamagnetically . Furthermore, the shapes of these curves are related to the particle size. Assuming that the particles are spherical, it is possible to calculate the radius, R, of the particles from the limiting forms of the Langevin equation at high and low fields. In particular, at low fields:

$$R_{LF}^3 = \frac{9k_B T(M/H)_{LF}}{4\pi M_{sp} M_s} \tag{15}$$

where the subscript LF refers to values determined from data at low fields. At high fields (HF), the following relation holds:

$$R_{HF}^3 = \frac{3k_B T}{4\pi M_{sp}\left(1-\frac{M}{M_s}\right)_{HF}} \tag{16}$$

If all of the particles were of the same size, then R_{LF} would equal R_{HF}. As the breadth of the particle size distribution increases, however, the value of R_{LF} becomes larger than R_{HF}. In fact, the difference between R_{LF} and R_{HF} has been used as a size heterogeneity index, Δ, where

$$\Delta = \frac{(R_{LF} - R_{HF})}{(R_{LF} + R_{HF})/2} \tag{17}$$

The surface average particle radius, R_s, which would be determined from chemisorption measurements of the metallic surface area, is between R_{LF} and R_{HF}:

$$R_{HF} < R_s < R_{LF} \tag{18}$$

It should be noted that detailed analysis of the shape of the M/M_s versus H/T curve allows the particle size distribution to be probed, as reviewed by Richardson (26).

For the characterization of supported metal particles, magnetic susceptibility is also sensitive to chemisorption phenomena. For example, it has been proposed by a number of investigators (as reviewed by Richardson (26)) that the chemisorption of hydrogen, carbon monoxide and a variety of simple hydrocarbons (e.g., small olefins and alkanes) leads to a decrease in the magnetic moment of the surface metal atoms involved in the adsorption. By measuring the surface coverage by adsorbed species and the change in saturation magnetization which accompanies this adsorption the change, α, in magnetic moment caused by one adsorbed molecule can be determined. (This change is typically expressed in Bohr magnetons, μ_{BM}, the approximate value of the magnetic moment of one unpaired electron in the direction of the applied field.) If the value of α is divided by the magnetic moment of a surface atom prior to adsorption, then the magnetic bond number, N, is obtained. This corresponds to the number of surface metal atoms that are demagnetized by each adsorbed molecule. For example, one could imagine that for CO adsorption on nickel, the value of N would be higher for CO molecules adsorbed on two surface atoms (i.e., bridge bonding) compared to CO molecules adsorbed on a single surface atom (i.e., linear bonding). In this way, N would be sensitive to surface geometry, as discussed later in this section. Furthermore, the extent to which a molecule (e.g., a hydrocarbon) dissociates into fragments, each of which interacts with a surface metal atom, can be probed by measuring N.

The use of magnetic susceptibility to characterize the bulk and surface properties of silica-supported nickel particles has been reported by Primet et al. (27). To probe the chemical state of Ni on silica, the saturation magnetization was measured as a function of the reduction temperature. The reduction of nickel oxide to metallic nickel was complete for reduction temperatures higher than ca. 900 K. The nickel particles formed at various extents of reduction behaved superparamagnetically (at temperatures above 77 K), allowing their size to be determined using the high and low field approximations of the Langevin equation. The average between R_{LF} and R_{HF} was taken to represent approximately the surface average particle size, R_s. This latter value was then compared to surface average particle sizes determined from hydrogen chemisorption, R_{H_2}. For completely reduced samples, R_s and R_{H_2} were nearly equal, indicating that all of the nickel surface was accessible to hydrogen. In contrast, R_{H_2} was larger than R_s for incompletely reduced samples, and from this it was suggested that the surface of metallic nickel in incompletely reduced samples was partially covered by unreduced nickel oxide species.

The surface properties of the silica-supported nickel particles were probed by studying the adsorption of CO. The change in saturation magnetization as a function of CO coverage is shown in Figure 9 at three different temperatures (195, 293, 373 K). From the slopes of these curves, the magnetic bond number for CO adsorption on metallic nickel was calculated. Because the slopes at 293 and 373 K were similar, it was proposed that the mode of CO adsorption was the same at these temperatures. For this fully reduced Ni/SiO_2 sample, the corresponding magnetic bond number was calculated to be 1.85. The mode of CO adsorption at 195 K, however, was different from that at 293 K, as evidenced by a smaller magnetic bond number. Similar behavior was observed for a variety of fully-reduced samples having different Ni particle sizes (from 2.5 to 10 nm), and the magnetic bond number for CO adsorption (at temperatures between 293 and 373 K) was indépendent of particle size. On partially reduced samples, however, the magnetic bond number was smaller, a typical value being 1.45. To interpret these data it was proposed that two forms of adsorbed CO were present on the nickel surface: (i) linearly-bonded CO, associated with one Ni atom and (ii) multiply-bonded CO, associated with more than one Ni atom. The magnetic bond number would be expected to be greater for the latter mode of CO adsorption. For fully-reduced samples, the relative amounts of these species were independent of Ni particle size. The presence of nickel oxide species on the surface of partially-reduced samples, however, favored linearly-bonded CO at the expense of multiply-bonded CO. Evidently, the nickel oxide species dilute the metallic nickel atoms on the surface, thereby decreasing the number of Ni ensembles (containing two or more Ni atoms) on which CO can be multiply-bonded.

The above model of the Ni surface was tested by collecting infrared spectra of adsorbed CO. Bands were observed at 2075-2045 cm^{-1} and at 1935-1800 cm^{-1}. The bands in the former range,

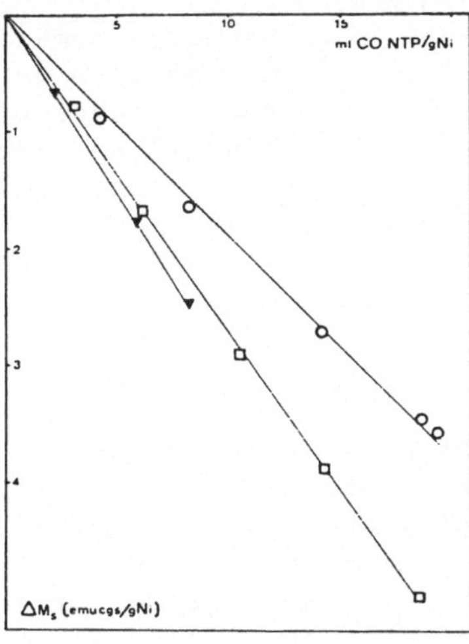

Fig. 9. Variations of the saturation magnetization versus adsorbed quantity of CO at adsorption temperatures of (o) 195 K, (□) 293 K, and (▼) 373 K. Reprinted with permission from ref. 27.

denoted by A_A, were attributed to linearly-bonded CO, and the bands in the latter range, denoted by A_B, were interpreted in terms of multiply-bonded CO. Indeed, the intensity of the A_A bands decreased relative to the A_B bands as the reduction of Ni became more complete, in agreement with the observed changes in magnetic bond number described above. For further details about the preparation, activation and reduction of silica-supported Ni catalysts, as studied by magnetic methods, the reader is referred to the recent paper by Martin et al. (28).

To illustrate the use of magnetic susceptibility in the characterization of supported bimetallic clusters, the study by Dalmon (29) of Ni-Cu on SiO_2 will be introduced. Figure 10 shows the saturation magnetization for a Ni-Cu/SiO_2 sample (4.5 wt.% Cu, 8.5 wt.% Ni) as a function of reduction temperature in hydrogen. The conversion of oxidized nickel to metallic nickel is complete for reduction temperatures higher than ca. 900 K. For a series of completely-reduced Ni-Cu/SiO_2 samples with different Cu-contents,

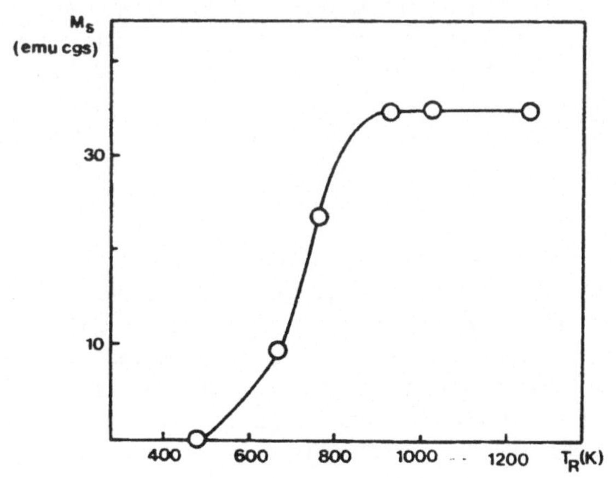

Fig. 10. Saturation magnetization versus reduction temperature. Reprinted with permission from ref. 29.

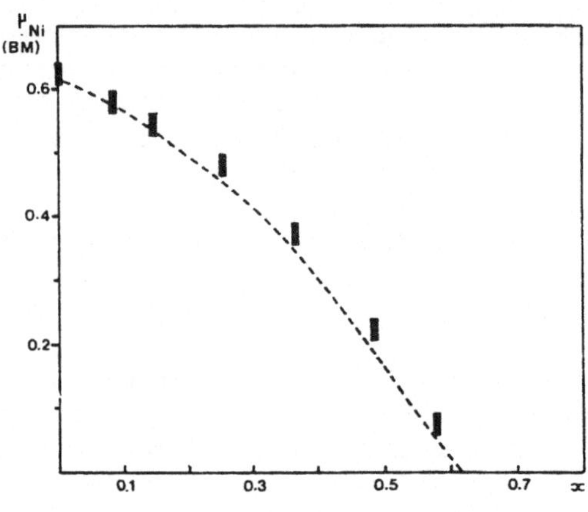

Fig. 11. Magnetic moment per Ni atom versus the Cu content x. The dashed line corresponds to bulk alloys. Reprinted with permission from ref. 29.

the saturation magnetization was determined. The plot in Figure 11 was thereby constructed, where the filled rectangles represent the magnetic moment per nickel atom (expressed in μ_{BM}) as a function of the Cu-content of the Ni-Cu/SiO_2 samples. The dashed line is the curve obtained for bulk Ni-Cu alloys of known composition. The good agreement between the experimental points for Ni-Cu/SiO_2 and bulk Ni-Cu alloys is a verification that bimetallic Ni-Cu particles were, in fact, formed on SiO_2. In addition, the Curie temperatures for the Ni-Cu/SiO_2 samples were determined, and these agreed well with the corresponding Curie temperatures for bulk Ni-Cu alloys having the same Ni:Cu ratios. This was further evidence for the formation of bimetallic Ni-Cu particles on SiO_2.

Dalmon used the superparamagnetic properties of the Ni-Cu particles to determine their size. This was done by first plotting M/M_s versus log (H/T) at different temperatures, as shown in Figure 12 for a sample containing 4.5 wt.% Cu and 8.5 wt.% Ni on SiO_2. The superposition of these plots indicated that these Ni-Cu particles behaved superparamagnetically, and the low and high field approximations of the Langevin equation were used to determine R_{LF} and R_{HF}, respectively. The particle size was observed to increase with increasing reduction temperature, from a value of ca. 6 nm at 920 K to ca. 13 nm at 1270 K.

7. FERROMAGNETIC RESONANCE

Ferromagnetic resonance (FMR) is a technique which can be used to determine the Curie temperature, the magnitude of magnetic anisotropies and the quantity of ferromagnetic or ferrimagnetic materials in samples. As such, this technique is complementary to magnetic susceptibility. Ferromagnetic resonance spectra are collected with the sample in a microwave cavity (i.e., in an esr spectrometer working in the X-band). A magnetic field perpendicular to the microwave field is varied and the absorption of the microwave radiation by the sample is monitored. The absorption is

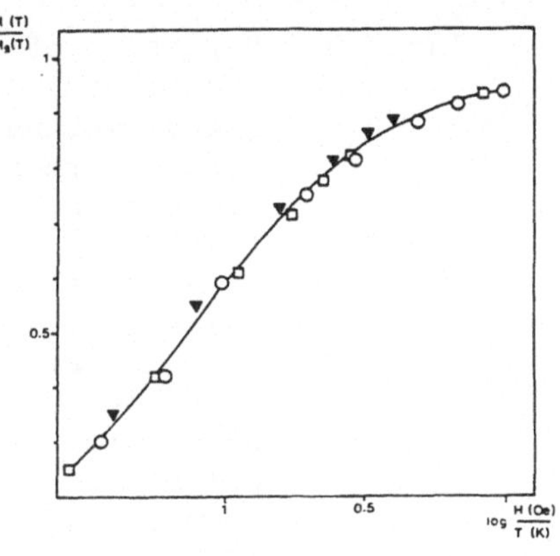

Fig. 12. Plot of M/M_S (magnetization/saturation magnetization) versus log(H/T) (magnetic field/temperature) at (o) 77 K, (▼) 200 K, and (□) 285 K. Reprinted with permission from ref. 29.

a result of a resonance corresponding to the precession of the magnetic moments in the external magnetic field (30). It is common to display the first derivative of the absorption versus magnetic field to separate the signal from the background.

The use of FMR in the characterization of magnetic particles has been reviewed by Simoens and Derouane (31). The intensity of the ferromagnetic resonance signal is proportional to the magnetization, M, of the sample. Thus in a manner analogous to that used in magnetic susceptibility, it is possible to use FMR to determine the saturation magnetization of the sample. This value is related to the amount of ferromagnetic or ferrimagnetic material present in the sample. By measuring the magnetization of the sample as a function of temperature it is possible to determine the Curie or Néel temperature of the magnetic phase. This is done by plotting M^2 versus T and extrapolating to the temperature at which M^2 equals zero. As discussed with respect to magnetic susceptibility, the measurement of the magnetic ordering temperature provides valuable information for identifying the chemical nature of the magnetic phase. The shape of the ferromagnetic resonance signal is sensitive to the existence of magnetic anisotropies. In general, these anisotropies may be the result of particle shape, crystal structure, stresses and interfacial effects; and, they cause the magnetic energy of the particle to be a function of the direction of the magnetization. Accordingly, certain low-energy directions exist for the orientation of the particle magnetization. These preferred orientations may be described in terms of effective magnetic fields (called magnetic anisotropy fields) which are different in different directions. When the particle is placed in an external magnetic field, the magnetic anisotropy fields interact with the applied field, leading to a broadening of the FMR signal. Perhaps the most commonly encountered form of magnetic anisotropy is due to particle shape. A spherical particle does not possess magnetic shape anisotropy, and a narrow, Lorentzian FMR lineshape results. As

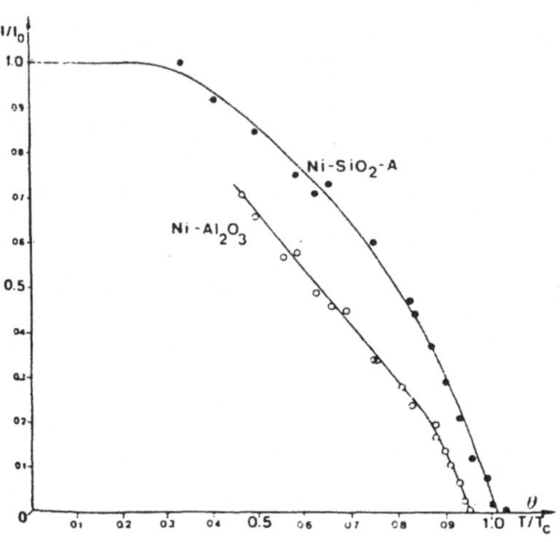

Fig. 13. Variation of the reduced FMR signal intensity (I/I_o) versus reduced temperature (T/T_c, T_c = 631 K) for (o) Al_2O_3 supported, and (•) SiO_2 supported catalysts. Reprinted with permission from ref. 32.

deviations from spherical symmetry become greater (e.g., ellipsoidal, plate-like, and rod-like particles), the magnetic shape anisotropy increases and the FMR signal broadens. It is this sensitivity to particle shape that distinguishes FMR from magnetic susceptibility. It should be noted that the magnitude of the magnetic shape anisotropy is proportional to the spontaneous magnetization of the particle, for a given particle shape.

Derouane et al. (32) used FMR to study Ni supported on $\eta-Al_2O_3$ and SiO_2. The FMR signal was measured for reduced samples at various temperatures, and plots of the FMR signal intensity, I, versus temperature were used to determine the Curie temperatures for these samples. This is shown in Figure 13, as a plot of I/I_o versus T/T_c, where I_o is the FMR intensity at the lowest measurement temperature and T_c is the Curie temperature of bulk metallic nickel. The measured Curie temperature for silica-supported Ni was equal to that of bulk metallic Ni while that for alumina-supported Ni was about 35 K lower. This was interpreted in terms of unreduced nickel species (e.g., $NiAl_2O_4$) present as small inclusions in the metallic Ni particles. Indeed, it was shown earlier by Derouane et al. (33) that partially-reduced metallic nickel particles have a lower Curie temperature than fully-reduced nickel particles. Also, the Curie temperature for fully-reduced nickel particles is independent of particle size, down to about 2.5 nm.

The Ni/SiO_2 sample studied by Derouane et al. (32) was pressed into a pellet to obtain partially oriented silica sheets. After reduction at 1273 K, the FMR signal was monitored as the sample was rotated in the external magnetic field. The FMR signal was markedly different for each orientation. The sensitivity of the signal to the orientation of the sample indicates that the metallic Ni particles possess magnetic shape anisotropy, and furthermore, that these particles are oriented with respect to the sheets of silica. Thus, FMR can be used in this manner to probe the existence of structural (e.g., epitaxial) relationships between metal clusters and the supports on which these clusters are deposited.

Jacobs et al. (34) used ferromagnetic resonance to study the chemical state of metallic nickel particles in Y zeolite. It was postulated that a bidispersion of metallic nickel existed: small nickel particles were present in the supercages of the zeolite, Ni_i, and larger nickel particles were present on the external surfaces of the zeolite, Ni_e. Ferromagnetic resonance studies were carried out on Ni-Y zeolite samples which had been reduced at different temperatures (from 673 to 873 K). The FMR signal was fit by two components: a narrow Lorentzian line and a broader Gaussian component. The intensity of the Gaussian component increased relative to the Lorentzian line as the reduction temperature increased. From temperature programmed oxidation and reduction studies it was suggested that the amount of metallic nickel on the external surface of Y zeolite increased with increasing reduction temperature. Thus, the Gaussian component was attributed to Ni_e and the Lorentzian line was interpreted in terms of Ni_i. The narrow width of the Lorentzian line indicated that the small nickel particles in the supercages of Y zeolite possess weak magnetic anisotropy, i.e., they must be nearly spherical. In contrast, the nickel particles on the external surface of Y zeolite showed strong magnetic anisotropy (as indicated by the breadth of the Gaussian FMR signal). If these particles are assumed to be ellipsoidal in shape, then the ratio of the lengths of the major and minor axes must be approximately 3:1 to account for the observed FMR lineshape.

The Curie temperatures for the various Ni-Y zeolite samples were determined from plots of M^2 versus temperature. As mentioned earlier, extrapolation of these plots to M^2 equal to zero gives the Curie temperature. Near the Curie temperature, however, such plots should be linear with slopes equal to M_0^2/T_c, where M_0 is the saturation magnetization at absolute zero and T_c is the Curie temperature. Such plots for the Ni-Y zeolite samples showed two linear portions having different slopes and intercepts with the temperature axis. The component with the Curie temperature nearly equal to that of bulk metallic nickel (i.e., 632 K) was attributed to Ni_e while the component with a lower Curie temperature (ca. 600 K) was attributed to Ni_i. From the relative slopes of these two linear portions of the M^2 versus T plots, the relative amounts of Ni_e and Ni_i were estimated (since the amount of each phase is proportional to the value of M_0 determined from the slope, M_0^2/T_c). It was thereby shown that the amount of Ni_e increased relative to the amount of Ni_i as the reduction temperature was increased, consistent with the behavior of nickel described above. The lower Curie temperature of Ni_i compared to Ni_e was explained in terms of a metal-support interaction for metallic nickel particles located inside the supercages or to incomplete reduction of these nickel particles.

8. VIBRATIONAL SPECTROSCOPY

Infrared spectroscopy is a powerful technique for the study of adsorbed species, and it is in this respect that this technique can be used to probe the surface structure and composition of metal particles. The reader is referred elsewhere (35,36) for detailed discussions. For illustration, consider the adsorption of CO. The CO stretching frequency is dependent on the mode in which CO is bonded to the surface. In general, this stretching frequency decreases as the CO becomes more strongly adsorbed or coordinated to a greater number of surface metal atoms. Thus ideally, the infrared spectrum of adsorbed CO (i.e., a plot of infrared absorbance versus energy of the ir radiation) can be used to probe the nature and distribution of surface sites on a metal cluster. It must be noted, however, that even for a uniform surface, the CO stretching frequency may be dependent on surface coverage due to dipole-dipole interactions between adsorbed species. The surface composition of multimetallic clusters can also be studied by infrared spectroscopy if the CO stretching frequency is different for CO adsorbed on the different metal components of the cluster. As a limiting case, the CO may adsorb specifically on only one metal of the multimetallic cluster, and then nonspectroscopic measurements of the extent of CO adsorption can be used to determine the surface concentration of that metal component. Besides stretching (or bending) frequencies, the various modes of vibration of adsorbed species are also characterized by extinction coefficients. These coefficients are equal to the effective cross sections of the adsorbed species for vibrational excitation due to absorption of infrared radiation. For a given surface concentration of adsorbed species, a large extinction coefficient leads to an intense peak in the ir spectrum while a small extinction coefficient leads to a weak peak. It has been shown recently that the extinction coefficient for the stretching of adsorbed CO is sensitive to the metal on which the CO is adsorbed (37). Thus, for a bimetallic cluster, the measured extinction coefficient for vibration of an adsorbed species can be used to determine the surface composition if the extinction coefficients for this adsorbed species on the individual metals are known. In both of the above schemes for surface composition measurement using infrared spectroscopy, it must be noted that surface composition changes induced by adsorption are assumed to be negligible. If the adsorbed species interact specifically and strongly with a particular metal component of the cluster, then this assumption may, in fact, be invalid.

An example of surface composition analysis using infrared spectroscopy is the work of Ramamoorthy and Gonzalez in their study of Pt-Ru bimetallic clusters on silica (38). The procedures employed were based on the differences in adsorption behavior of CO and NO on Pt and Ru at room temperature. Specifically, CO adsorbed on Ru shows a strong band at 2030 cm^{-1} which is rapidly displaced from the surface by NO. This adsorbed NO gives rise to ir bands at 1860, 1820 and 1630 cm^{-1}, and the adsorbed NO is not displaced by CO. On Pt, adsorbed CO shows a strong band at 2070 cm^{-1}, and this adsorbed CO is not rapidly displaced by NO. Instead, a slow-reaction takes place between adsorbed CO and gaseous NO, forming CO_2 and N_2O. If NO is first adsorbed on Pt, a single band at 1760 cm^{-1} is observed, and this adsorbed NO can be completely displaced by CO. Thus, the strategy used to probe the surface composition of supported Pt-Ru clusters was to (1) expose the sample to CO, (2) expose the sample to NO, and (3) expose the sample to CO once again. During the first step, CO adsorbs on Ru and Pt, leading to an asymmetric ir band. During the second step, CO is displaced by NO from Ru, and some of the CO on Pt may slowly react with gaseous NO. The third step is used to insure that the Pt surface is saturated with adsorbed CO. As a result of this three-step procedure, CO is adsorbed only on Pt surface atoms and NO is adsorbed only on Ru surface atoms; and, the quantities of these adsorbed CO and NO species can be determined from the intensities of the ir bands at 2070 and 1810 cm^{-1}, respectively. In this way, Ramamoorthy and Gonzalez showed that, for a series of different Pt-Ru clusters, the surface and bulk compositions were very similar.

When utilizing a procedure such as that outlined above for surface composition analysis, it is important to verify that bimetallic clusters are in fact present on the support. Without additional information, the infrared spectra from adsorbed species on bimetallic clusters or on a physical mixture of the two pure metals could not be distinguished. In this respect, the following information was used to show that Pt and Ru formed Pt-Ru clusters on the silica-supported samples studied by Ramamoorthy and Gonzalez. X-ray diffraction showed a peak at an angle between that of the Pt (111) and the Ru (002) peaks. In addition, the vibrational frequencies of the CO and NO adsorbed on Pt and Ru, respectively, were observed to shift as the Pt:Ru ratio in the sample was varied. This was attributed to electron transfer from Ru to Pt during the adsorption of NO and CO. The propensity for NO to dissociate on pure Ru, leading to a strong band at 1860 cm^{-1} due to NO adsorbed on a Ru site perturbed by oxygen, was diminished when both Pt and Ru were present on SiO_2. Finally, the formation of isocyanate species from CO and NO on Ru was not observed when Pt was present in high concentrations, and the formation of these species requires a large number of adjacent Ru surface sites. Thus, not only can infrared spectroscopy be used to probe the surface composition of bimetallic clusters, but also it is useful in verifying that bimetallic clusters (in contrast to a mixture of monometallic clusters) are, in fact, present on the support.

The use of infrared spectroscopy and adsorption techniques to probe surface structures of metal particles is not as well-developed as are the procedures for surface composition analysis. It may be possible to use infrared spectroscopy to monitor probe molecules which adsorb only on specific types of surface sites. Such is the case for the adsorption of N_2 on ensembles of surface metal atoms which allow the N_2 to penetrate into the surface (39). (The N_2 molecule becomes infrared-active when adsorbed in this manner.) As mentioned earlier, it is also possible to relate shifts in the stretching frequencies of adsorbed species to changes in the coordination of these species on the surface. Consider, for example, the adsorption of NO on silica-supported Fe^{2+} (23). The characterization of Fe^{2+}-containing clusters on silica, using Mössbauer spectroscopy, was discussed previously in this paper. Four bands were observed in the infrared spectrum of NO adsorbed on Fe^{2+}/SiO_2: 1910, 1830, 1810 and 1750 cm^{-1}. By comparison with ir studies of NO adsorption on Fe^{2+} in Y zeolite (24), the pair of bands at 1910 and 1810 cm^{-1} could be ascribed to the symmetric and asymmetric stretching frequencies, respectively, of iron dinitrosyl species. Steric considerations suggest that these dinitrosyl species would form on highly unsaturated iron cations: i.e., the low coordination Fe^{2+} cations of the Mössbauer spectroscopy inner doublet. Upon evacuation of Fe/SiO_2 after previous exposure to NO (at room temperature), the intensity of the band at 1750 cm^{-1} increased at the expense of the bands at 1910 and 1810 cm^{-1}. This suggests that the band at 1750 cm^{-1} is due to mononitrosyl species associated with Fe^{2+} cations at low coordination, formed by the removal of one NO group from each of the dinitrosyl species. Thus, the ir bands at 1910, 1810 and 1750 cm^{-1} are all due to NO associated with inner doublet Fe^{2+} cations. Indeed, Mössbauer spectroscopy showed that all of the inner doublet Fe^{2+} cations were perturbed upon exposure of the sample to NO. In addition, a significant fraction of the Fe^{2+} cations of the outer doublet were affected by the adsorption of NO, and this should be reflected in the infrared spectrum of the adsorbed NO. Accordingly, the ir band at 1830 cm^{-1} can be interpreted in terms of mononitrosyl species associated with the highly coordinated Fe^{2+} cations of the outer doublet. In summary, the lower coordination number (prior to adsorption) of inner doublet Fe^{2+} cations, compared to outer doublet Fe^{2+}, is reflected in the ability of these cations to form dinitrosyl species and in a lower stretching frequency for mononitrosyl species. This example illustrates how infrared spectroscopy of adsorbed species may be used to probe the structure of surface sites. Although the example chosen involved metal oxide clusters, it should be noted that this technique may be applied equally well to metal clusters. In this respect, the observation of dicarbonyl species on Rh/Al_2O_3 samples has been interpreted in terms of isolated Rh atoms (40) or Rh atoms at the edges of thin Rh rafts (41) on the Al_2O_3 support.

Another important application of infrared spectroscopy in probing the structure of small metal clusters is its use as a "fingerprint" technique. Consider, for example, a polynuclear metal carbonyl cluster. The infrared spectrum of the CO stretching region (near 2000 cm^{-1}) is a combination of overlapping bands due to vibrational modes involving combinations of linear and bridging CO ligands. The structure of small metal clusters on supports can thus be probed by comparing the infrared spectra of CO on these clusters with the ir spectra of known metal carbonyl clusters. The presence of CO on these clusters may be due to incomplete decarbonylation of a metal carbonyl cluster used to prepare the sample or it may be due to the adsorption of CO intentionally exposed to the sample.

The above discussion focused on the indirect characterization of supported metal particles using infrared spectroscopy to study surface adsorbed species. Analogous experiments could also be conducted using laser Raman spectroscopy (e.g., see refs. 42, 43, 44,45). It is important to note, however, that it is possible to study directly the structure of supported clusters by means of vibrational spectroscopy. For illustration, consider the case of metal oxide clusters. The infrared and Raman spectra of oxides are sensitive to the coordination of the compound. This is especially true for laser Raman spectroscopy, for which the spectral bands are sharp and intense. For example, laser Raman studies of a series of molybdenum polyanions (e.g., MoO_4^{2-}, $Mo_7O_{24}^{6-}$, $Mo_8O_{26}^{4-}$ and $PMo_{12}O_{40}^{3-}$) established that the metal-oxygen stretching vibration increases as the aggregation of Mo increases (46). This correlation was then used to monitor the degree of Mo aggregation on Al_2O_3 and SiO_2 after various sample preparations and pretreatments. For example, MoO_3 was not observed for Al_2O_3-supported samples with low Mo loadings (i.e., less than half of the support covered by Mo species), and the degree of Mo aggregation increased with increasing Mo loading. Furthermore, the Mo was more highly aggregated on SiO_2 than on Al_2O_3, due to a weaker interaction between Mo and SiO_2. The addition of Co to the samples reduced the degree of aggregation of Mo, and the formation of $CoMoO_4$ was detected.

In contrast to metal oxides, much less work has been done in the direct characterization of supported metal clusters using vibrational spectroscopy. Because absorption of infrared radiation requires the presence of an oscillating dipole, infrared spectroscopy is not sensitive to metal-metal stretching frequencies. Laser Raman spectroscopy, however, is related to the polarizability of the vibrational modes, and this technique can be used to monitor directly the metal-metal stretching frequencies of supported metal clusters. Deeba et al. (47) studied accordingly the structure of supported osmium clusters prepared by contacting partially dehydroxylated alumina with $Os_3(CO)_{12}$. The laser Raman spectrum of this sample is shown in Figure 14A. For comparison, the laser Raman spectra of $H_2Os_3(CO)_{10}$ and $Os_3(CO)_{12}$ are shown in Figures 14B and 14C. The Os-Os stretching modes appear at wavenumbers between ca. 100 and 200 cm^{-1}. The supported cluster shows a strong band at 160 cm^{-1}, and the $H_2Os_3(CO)_{10}$ and $Os_3(CO)_{12}$ clusters show two bands at 145 and 188 cm^{-1} and at 120 and 161 cm^{-1}, respectively. The presence of the Os-Os stretching frequency for the supported cluster indicates that the Os-Os bonds remain intact during sample preparation. The similarity of the three spectra of Figure 14 suggests that the supported cluster is a triosmium species. When this sample was heated in CO at ca. 700 K, the band at 160 cm^{-1} disappeared, and the triosmium cluster was converted to a mononuclear surface complex.

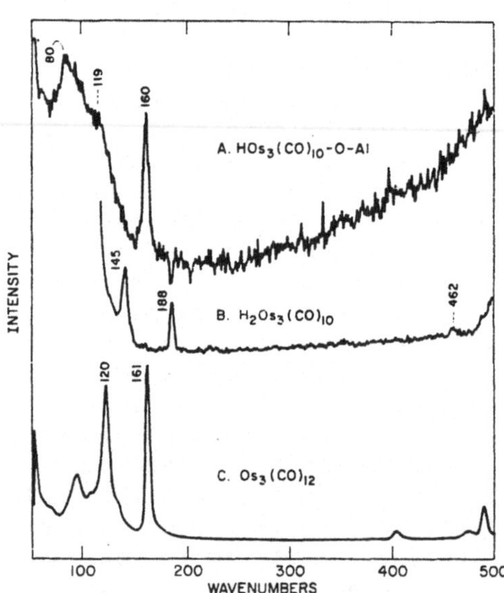

Fig. 14. Raman spectra of the solid phase of: (A) alumina-supported triosmium cluster; (B) $H_2Os_3(CO)_{10}$; (C) $Os_3(CO)_{12}$. Reprinted with permission from ref. 47.

9. CONCLUDING REMARK

A variety of techniques can be used to carry out in situ characterizations of metal particles on high surface area supports. While these techniques probe all of the atoms in the metal particles, they can be sensitive to the surface properties of small metal particles, especially when combined with studies of chemisorption phenomena. The use of spectroscopic techniques with short analysis depths (e.g., 5 nm) to characterize metal particles on high surface area supports may be complicated by the pore structure of the support. For this reason, it may be desirable to supplement the characterization of high surface area, supported metal samples (as discussed in this paper) with the study of metal particles on low surface area, geometrically flat supports (as introduced in Part II of this series).

ACKNOWLEDGEMENT

One of us (JAD) wishes to acknowledge the partial financial support from the Camille and Henry Dreyfus Foundation during the preparation of this manuscript. We also acknowledge funding from Chevron Research which aided in the preparation of this paper.

REFERENCES

1. Klung, H.P. and Alexander, L.E., "X-ray Diffraction Procedure", 2nd ed., Wiley, New York 1974
2. Warren, B.E., "X-ray Diffraction", Addison-Wesley, Reading, Mass. 1969
3. Sashital, S.R., Cohen, J.B., Burwell, R.L., Jr., and Butt, J.B., J. Catal. 50, 479 (1977)
4. Whyte, T.E., Jr., Kirklin, P.W., Gould, R.W., and Heineman, H., J. Catal. 25, 407 (1972)
5. Whyte, T.E., Jr., Catal. Rev. 8, 117 (1973)
6. Ratnasamy, P. and Léonard, A.J., Catal. Rev. 6, 293 (1972)
7. Gallezot, P. and Bergeret, G., J. Catal. 72, 294 (1981)
8. Ratnasamy, P., Léonard, A.J., Rodrique, L., and Fripiat, J.J., J. Catal. 29, 374 (1973)
9. Lytle, F.W., Via, G.H., and Sinfelt, J.H., in "Synchrotron Radiation Research" (H. Winick and S. Doniach, Eds.) p. 401, Plenum Press, New York 1980
10. Stöhr, J., in "Emission and Scattering Techniques" (P. Day, Ed.) p. 215, NATO Advanced Study Institute Series C, Vol. 73, D. Reidel Publishing Company, Holland 1981
11. Sinfelt, J.H., Via, G.H., and Lytle, F.W., J. Chem. Phys. 72, 4832 (1980)

12. Clausen, B.S., Topsφe, H., Candia, R., Villadsen, J., Lengeler, B., Als-Nielsen, J., and Christensen, F., J. Phys. Chem. 85, 3868 (1981)
13. Wivel, C., Candia, R., Clausen, B.S., Mφrup, S., and Topsφe, H., J. Catal. 68, 433 (1981)
14. Short, D.R., Khalid, S.M., Katzer, J.R., and Kelley, M.J., J. Catal. 72, 288 (1981)
15. Gallezot, P., Weber, R., Dalla Betta, R.A., and Boudart, M., Z. Naturforsch. A 34, 40 (1979)
16. Lewis, P.H., J. Catal. 69, 511 (1981)
17. Dalla Betta, R.A., Boudart, M., Gallezot, P., and Weber, R.S., J. Catal. 69, 514 (1981)
18. Dumesic, J.A. and Topsφe, H., in "Advances in Catalysis" (D.D. Eley, H. Pines, and P.B. Weisz, Eds.), Vol. 26, p. 121, Academic Press, New York 1977
19. Topsφe, H., Dumesic, J.A., and Mφrup, S., in "Applications of Mössbauer Spectroscopy" (R.L. Cohen, Ed.), Vol. 2, p. 56, Academic Press, New York 1980
20. Phillips, J., Clausen, B., and Dumesic, J.A., J. Phys. Chem. 84, 1814 (1980)
21. Phillips, J. and Dumesic, J.A., Appl. Surf. Sci. 7, 215 (1981)
22. Mφrup, S., Dumesic, J.A., and Topsφe, H., in "Applications of Mössbauer Spectroscopy" (R.L. Cohen, Ed.), Vol. 2, p. 1, Academic Press, New York 1980
23. Yuen, S., Chen, Y., Kubsh, J.E., Dumesic, J.A., Topsφe, N., and Topsφe, H., J. Phys. Chem. 86, 3022 (1982)
24. Segawa, K.-I., Chen, Y., Kubsh, J.E., Delgass, W.N., Dumesic, J.A., and Hall, W.K., J. Catal. 76, 112 (1982)
25. Selwood, P.W., "Chemisorption and Magnetization", Academic Press, New York 1975
26. Richardson, J.T., J. Appl. Phys. 49, 1781 (1978)
27. Primet, M., Dalmon, J.A., and Martin, G.A., J. Catal. 46, 25 (1977)
28. Martin, G.A., Mirodatos, C., and Praliaud, H., Appl. Catal. 1, 367 (1981)
29. Dalmon, J.A., J. Catal. 60, 325 (1979)
30. Vonsovskii, S.V., Ed., "Ferromagnetic Resonance", Pergamon Press, New York 1966
31. Simoens, A.J. and Derouane, E.G., in "Stud. Surf. Sci. Catal. Vol. 4; Proc. of the Conf. on Growth and Prop. of Small Metallic Clusters" (C. Proyanowsky and G. Dourdon, Eds.) p. 201, Elsevier, Amsterdam 1980
32. Derouane, E.G., Simoens, A.J., and Vedrine, J.C., Chem. Phys. Lett. 52, 549 (1977)
33. Derouane, E.G., Simoens, A.J., Colin, C., Martin, G.A., Dalmon, J.A., and Vedrine, J.C., J. Catal. 52, 50 (1978)
34. Jacobs, P.A., Nijs, H., Verdonck, J., Derouane, E.G., Gilson, J., and Simoens, A.J., J. Chem. Soc. Faraday Trans. 1 75, 1196 (1979)

35. Bell, A.T. and Hair, M.L., Eds. "Vibrational Spectroscopies for Adsorbed Species", ACS Symposium Series 137, American Chemical Society, Washington, D.C. 1980
36. Ferraro, J.R. and Basile, L.J., Eds. "Fourier Transform Infrared Spectroscopy", Vol. 3, Academic Press, New York 1982
37. Miura, H. and Gonzalez, R.D., J. Phys. E: Sci. Instrum. 15, 373 (1982)
38. Ramamoorthy, R. and Gonzalez, R.D., J. Catal. 59, 130 (1979)
39. Van Hardeveld, R. and van Montfoort, A., Surf. Sci. 4, 396 (1966)
40. Yates, J.T., Jr., Duncan, T.M., Worley, S.D., and Vaughan, R.W., J. Chem. Phys. 70, 1219 (1979)
41. Yates, D.J.C., Murrell, L.L., and Prestridge, E.B., J. Catal. 57, 41 (1979)
42. Cooney, R.P., Curthoys, G., and Tam, N.T., in "Advances in Catalysis" (D.D. Eley, H. Pines, and P.B. Weisz, Eds.), Vol. 24, p. 293, Academic Press, New York 1975
43. Egerton, T.A. and Hardin, A.H., Catal. Rev. 11, 71 (1975)
44. Klaassen, A.W. and Hill, C.G., Jr., J. Catal. 69, 299 (1981)
45. Schrader, G.L., Basista, M.S., and Bergman, C.B., Chem. Eng. Comm. 12, 121 (1981)
46. Cheng, C.P. and Schrader, G.L., J. Catal. 60, 276 (1979)
47. Deeba, M., Streusand, B.J., Schrader, G.L., and Gates, B.C., J. Catal. 69, 218 (1981)

CHARACTERIZATION OF SUPPORTED METAL PARTICLES IN HETEROGENEOUS CATALYSTS: PART II. STUDIES OF LOW SURFACE AREA, MODEL MATERIALS

G. B. Raupp, T. J. Udovic and J. A. Dumesic

Department of Chemical Engineering
University of Wisconsin
Madison, Wisconsin 53706

1. INTRODUCTION

As discussed in part I of this series dealing with the characterization of supported metal particles, spectroscopy techniques with short analysis depths (i.e., Class II techniques) are sensitive to only those metal particles on the external surface of a high surface area (porous) supported metal sample. To fully interpret the information that can be obtained from these techniques it is desirable to "unfold" the pore structure of the high surface area support and to thereby place each metal particle in direct line-of-sight with the spectrometer. This is the rationale for the study of "model supported metal samples", which are composed of metal particles on low surface area, geometrically flat supports. This is shown schematically in Figure 1. The nature of the oxide support is chosen to model a

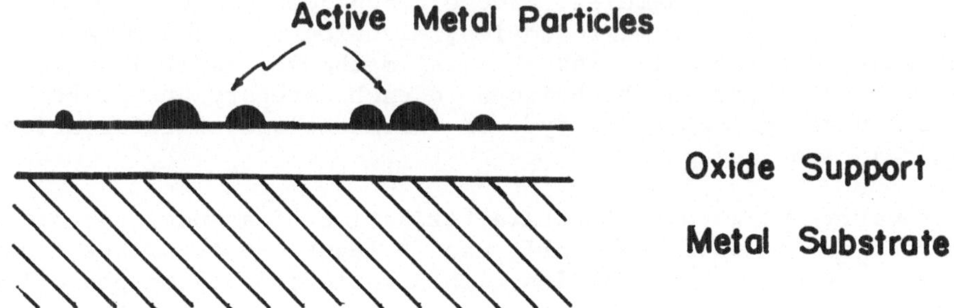

Fig. 1. Schematic of a model supported catalyst.

typical high surface area support, e.g., Al_2O_3, SiO_2, TiO_2. The metal substrate is often employed to provide mechanical support of the oxide film, to allow the sample to be heated resistively, and to reduce the effects of charging encountered when electron spectroscopic techniques are used for sample characterization. The details of the preparation of such model supports are dependent on the system of interest. For example, titanium and silicon may be oxidized in air to give thin films of TiO_2 and SiO_2 on Ti and Si, respectively. Aluminum may be anodized to give Al_2O_3 on Al. The metal particles may be deposited in a variety of ways, although vacuum evaporation has proven most common. Other preparation techniques will be presented in the applications section of this paper.

What is ultimately desired from the study of model supported metal samples is a complete morphological, physical, and chemical characterization of the metal particles. This includes the determination of particle size/shape, crystal structure, electronic properties, surface and bulk composition, and the adsorptive and catalytic properties of the metal particles. Unfortunately, no single technique can provide this complete characterization of a model supported metal sample. Accordingly, a combination of techniques is necessary, reminiscent of the strategy outlined in part I for the characterization of metal particles on high surface area supports.

2. STRATEGIES FOR CHARACTERIZATION OF METAL PARTICLES ON LOW SURFACE AREA, MODEL SUPPORTS

There exists a myriad of analytical techniques for probing solid surfaces. The information provided by a particular analytical tool depends on its specific incident probe (e.g., electrons, photons) and the way in which this probe interacts with the solid surface to give emitted (or transmitted) particles which carry the information about the sample. Luckily for the unindoctrinated, only a handful of techniques are commonly used in the study of model supported metal samples. These techniques can be categorized as electron microscopies, electron spectroscopies, vibrational spectroscopies and chemical methods. Table 1 summarizes the important techniques in each category and gives a general description of the type of characterization information obtained in each category.

To carry out a thorough characterization of a model supported metal sample, one would typically use at least one technique from each of the first two categories. This would provide morphological, structural, compositional and chemical information about the supported metal particles. This could be complemented by

Table 1. Categories of techniques used to characterize metal particles on model supports

Category	Information Obtained
A. Electron Microscopy 1. Transmission Electron Microscopy (TEM) 2. Scanning Transmission Electron Microscopy (STEM) 3. Controlled Atmosphere Electron Microscopy (CAEM)	morphology, crystal structure, elemental composition, elemental mapping
B. Electron Spectroscopy 1. Auger Electron Spectroscopy (AES) 2. X-ray Photoelectron Spectroscopy (XPS) 3. Ultraviolet Photoelectron Spectroscopy (UPS) 4. Conversion Electron Mössbauer Spectroscopy (CEMS)	surface composition, chemical bonding, valence state and band structure, magnetic information
C. Vibrational Spectroscopy 1. Electron Energy Loss Spectroscopy (EELS) 2. Inelastic Electron Tunneling Spectroscopy (IETS) 3. Infrared Reflectance Spectroscopy (IRS)	vibrational levels for adsorbed species and solid state
D. Chemical Techniques 1. Temperature Programmed Desorption (TPD) 2. Catalytic Probe Reactions	surface coverages, adsorption energetics and reactivities of adsorbed species

spectroscopic knowledge of the nature of adsorbates on the metal particle, using XPS, UPS or one of the vibrational spectroscopies. In addition, the adsorption energetics and reactivities of these adsorbed species could be studied using chemical methods. Indeed, by a combination of techniques from each of the four categories, strategies can be developed which allow the adsorptive and catalytic properties of supported metal particles to be understood in terms of the physical and chemical properties of these particles.

The following section gives a brief description of each technique listed in Table 1. The defining relations are introduced to give the reader a basic understanding of the governing principles, specific variables measured, and limitations of the various techniques. The subsequent section describes in some detail published studies of model supported metal systems.

These represent examples of the strategies involved in characterizing such materials.

3. EXPERIMENTAL TECHNIQUES

3.1 Electron Microscopies

3.1.1 Transmission Electron Microscopy (TEM). The transmission electron microscope employs Gaussian optics to focus an electron beam which illuminates the entire object simultaneously. The high resolution of TEM is due to the fact that the de Broglie wavelength of a highly energetic electron beam is of the order of a fraction of an Angstrom. The electrons that form the image must pass through the specimen. Contrast is achieved through differential attenuation of the transmitted electron beam due to regions of different sample thickness or different atomic number (i.e., high atomic number elements have greater scattering power than low atomic number elements), and to effects of electron diffraction. The fact that the electron beam must pass through the sample places an upper limit on sample thickness of the order of several hundred nanometers. High resolution work demands even thinner samples, optimally less than several tens of nanometers thick, dependent on the atomic number of the material.

TEM can be operated in three different modes, each giving different information about the sample. These modes are bright-field, dark-field, and selected area diffraction. In bright-field imaging, regions of the sample which scatter, or diffract the electrons appear dark since diffracted electrons are blocked by the objective aperture and therefore do not reach the viewing screen. Conversely, in dark-field imaging, only a selected diffracted beam is allowed to pass through the aperture; therefore, diffracting regions of the sample appear bright and transmitting regions appear dark. Thus, the bright-field image is an enlarged view of the zero order diffraction spot, while the dark-field image is an enlarged view of a higher order spot. Important information can be obtained about crystallite orientation, particle thickness and crystalline defects by comparing bright- and dark-field images from the same sample area. Selected area electron diffraction is accomplished by placing an area-selecting aperture in the plane of the objective lens. Areas as small as 0.5 μm diameter can be selected. The diffraction pattern provides information about the crystal structures and orientations of the various phases present in the sample. Also, knowledge of the diffraction pattern is essential in the interpretation of the dark-field images. Detailed description of the principles involved in TEM, as well as the state-of-the-art instrumentation can be found in a recently published book authored by Thomas and Goringe (1).

3.1.2 Scanning Transmission Electron Microscopy (STEM). A scanning transmission electron microscope combines the versatility of a scanning electron microscope (SEM) with the high resolution power of the TEM. A scanning transmission electron microscope differs from a transmission electron microscope in that rather than illuminating the entire sample with a broad electron beam, it illuminates a small spot of the sample with an extremely fine electron beam (e.g., 1 nm in diameter) and forms an image sequentially by rastering the beam across the sample. The image is displayed on a cathode ray tube scanned synchronously with the incident electron beam. Depending on which electrons are detected and used to generate the signal to the cathode ray tube, different types of images are obtained. For example, secondary electrons are generated near the surface of the sample due to the interaction of the incident electron beam with the specimen. The image formed by detecting these electrons above the sample is identical to that of a scanning electron microscope (SEM), and this image is sensitive to the topographical features of the sample surface. If an electron detector is used to measure the intensity of the electron beam which has been transmitted through the sample, then the image formed is essentially identical to that obtained using a transmission electron microscope. Image resolution obtained in this STEM-mode is equivalent to that obtained in conventional transmission electron microscopy (e.g., 0.2 nm). In addition, by adjusting the electron detector to accept only elastically scattered electrons, one achieves a signal dependent on the local elemental composition of the sample due to the strong dependence of elastic scattering cross-section on atomic number Z (up to about $Z = 40$). Since none of the signals require focusing of an image on a detector, STEM has a greater depth of field than TEM (e.g., by a factor of 1000).

An important feature of a scanning transmission electron microscope is the use of a fine electron beam which can be stopped on a particular point of the sample. X-rays emitted from this region of the sample can be detected and used to determine the elemental composition of that region. The size of the sample region analyzed in this manner depends on a number of parameters (e.g., sample thickness, atomic number), but a typical analysis diameter could be from 5 to 10 nm. Further analysis is possible by measuring the characteristic energy loss of transmitted electrons. This allows measurement of absorption edges and valence bands of the region of the sample being irradiated by the electron beam. In addition, by studying the diffraction of the stopped electron beam (so-called micro-diffraction studies) it is possible to determine the crystal structure and orientation of individual metal particles that are as small as ca. 5 nm.

For a complete review of STEM principles, instrumentation, and experimental methodology, see the recent review article by Delannay (2).

3.1.3 Controlled Atmosphere Electron Microscopy (CAEM). The chief limitation of the electron microscopies described above is that the samples cannot be observed in a high pressure environment. Interest in direct observation of dynamic phenomena has motivated efforts to develop controlled atmosphere heated specimen stages. For a complete review of the technique see (3) or (4).

Since one is interested in dynamic measurements using CAEM, a television camera and video recorder are used to record the images (although alternate methods are possible (5)). The disadvantage of this monitoring method is that the video line scan imposes a limit on the ultimate resolution for a given magnification. Resolution is further degraded by scattering of the electron beam by gas molecules and by contamination buildup produced by various gases. A typical resolution for CAEM has been quoted to be ca. 2.5 nm. Typical upper pressure and temperature limits for commercial specimen stages are 250-760 Torr and 1300-1500 K.

CAEM has been used to investigate a wide variety of gas/solid interactions, one of which will be described in some detail in the applications section. A particularly powerful feature of this technique is that it can provide quantitative estimates of kinetic parameters involving changes in the sample electron density, morphology, diffraction contrast or motion of metal particles on a support.

3.2 Electron Spectroscopies

3.2.1 Auger Electron Spectroscopy (AES). In the Auger process, depicted schematically in Figure 2, a core hole is filled by an outer shell electron with emission of an Auger electron or a photon. The initial core hole may be formed with high energy X-rays, ions, or electrons. In AES, electrons prove most convenient. The kinetic energy of the emitted electron is characteristic of the element from which it came. For the KL_IL_{III} process shown in the figure, the kinetic energy of the Auger electron is

$$E_{KL_IL_{III}} = E_K - E_{L_I} - E_{L_{III}} \tag{1}$$

Note that this can be a cascading process, as the KL_IL_{III} transition has left the atom in an excited state due to to the L_{III} level vacancy. De-excitation can then occur through an $L_{III}MM$ process and so on. In general the strongest Auger transitions are those involving the valence shell and the core level closest to it in energy. These are also the most surface sensitive

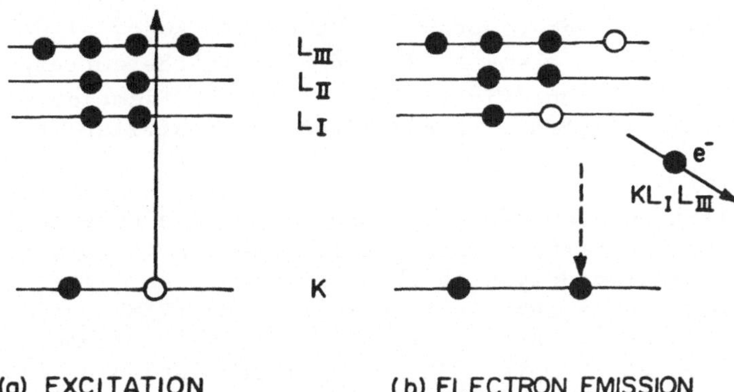

Fig. 2. Auger electron emission process. Reprinted from Gabor A. Somorjai: Chemistry in Two Dimensions: Surfaces. Copyright © 1981 by Cornell University. Used by permission of the publisher, Cornell University Press.

because of their relatively small kinetic energy and resultant shallow escape depth.

To maximize the Auger electron signal relative to the background, the number of electrons detected at a given energy (N(E)) is differentiated with respect to energy $(\frac{dN(E)}{dE})$. This suppresses the background signal since it is a slowly varying function of energy, and it accentuates the Auger signal since it varies sharply with energy at the Auger transitions. The detection limit of AES for most elements is of the order of 1000 p.p.m. of the analyzed volume (3-10 atomic layers), although lower limits can be achieved in ideal cases. It should be noted that H and He cannot be detected by AES since their lack of L shell electrons precludes Auger transitions.

The capability of AES for quantitative composition analysis is complicated by the generation of Auger electrons by electrons which have been backscattered from beneath the surface and by the fact that the multiple transition process in AES is not quantitatively understood. In practice, a semi-empirical approach using elemental standards works best. In the absence of such standards, tabulated relative sensitivities of the various Auger peaks can be used to determine relative atomic concentrations via:

$$C_X = \frac{I_X/S_X}{\sum_i I_i/S_i} \tag{2}$$

where I_X is the peak-to-peak height of the Auger signal (derivative mode) for element X and S_X is the corresponding relative elemental sensitivity factor. Quantification works best for homogeneous samples with constituents of similar atomic number.

The electron beam used to produce the core hole in AES can be focused and rastered across the surface. A cathode ray tube is scanned synchronously with this electron beam. This is the basis for scanning Auger electron microscopy (SAM), and this technique provides a spatial map of the surface elemental composition of the sample. The lateral resolution of SAM is approximately 50 nm. The electron beam can also be stopped on a particular point on the surface, thereby allowing an Auger electron spectrum to be collected at that spot. See Carlson (6) for more detailed explanations of AES.

3.2.2 X-ray Photoelectron and Ultraviolet Photoelectron Spectroscopies. X-ray photoelectron spectroscopy (XPS), also known as Electron Spectroscopy for Chemical Analysis (ESCA), and the closely related ultraviolet photoelectron spectroscopy (UPS), are based on the energy analysis of photoelectrons emitted when a solid is bombarded with monoenergetic photons. These two processes can be introduced with the aid of Figure 3. In XPS soft X-rays, either Al or Mg Kα, with energies greater than 1000 eV ionize core shell electrons. UPS on the other hand, uses low energy ultraviolet photons as incident radiation. Typically Ar or He provide discrete line sources of 11.6 and 21.2 eV, respectively, and these are of sufficient energy to ionize only valence shell electrons. It should be noted that although UPS in one respect provides less information than XPS due to its inability to excite core electrons, it provides more detailed information on the valence shell due to its inherent greater sensitivity and resolution for these electrons.

An energy balance on the photoelectric process gives the kinetic energy, KE, of the emitted electron in terms of the incident photon energy, $h\nu$, and the binding energy, BE_i, of the electron from level i:

$$KE = h\nu - BE_i \tag{3}$$

Photoelectron spectra are conventionally displayed as number of electrons detected versus decreasing binding energy.

Because XPS and UPS probe different energy levels, their

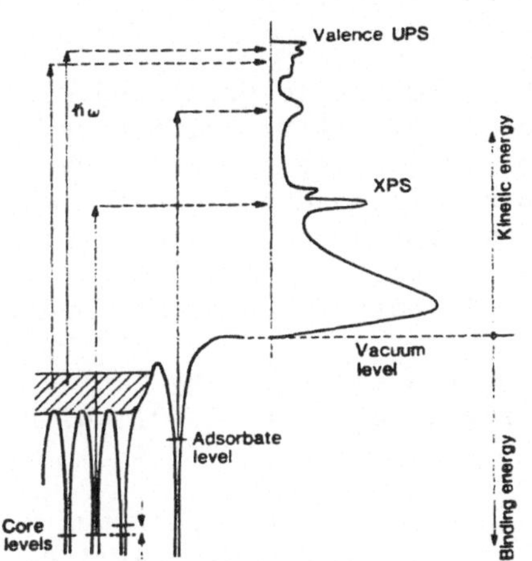

Fig. 3. Photoelectron emission in XPS and UPS. Reprinted with permission from N. V. Smith and D. P. Woodruff, Science, Vol. 216, pp. 367-372, 23 April 1982. Copyright 1982 by the American Association for the Advancement of Science.

spectra contain different types of information. In XPS, each element has its own signature of core level photoelectron emission lines facilitating qualitative elemental analysis. It should be noted that core level emission leaves the atom in an excited state. De-excitation through Auger emission is therefore also observed and the Auger peaks can aid in elemental identification. In addition, shifts in the positions of the XPS peaks provide reliable chemical information. In the simplest terms, outer shell or valence electrons are shared in forming chemical bonds, and this alters the energies of the core levels. These shifts in binding energies of the core electrons can be as large as 10 eV, although shifts of the order of 1 eV are most common. Utilization of a monochromatized source coupled with a high-resolution detector allows determination of binding energy shifts greater than ca. 0.1 eV. The capability for quantitative analysis using XPS is somewhat better than that of AES, in large part due to the fact that the single electron process is well understood. Standard sensitivity factors exist for the two commonly used sources in XPS, and chemical composition can be determined in a manner analogous to that described for AES. The detection limit of XPS (ca. 0.3-0.5% of a monolayer) is somewhat worse than AES. This is due primarily to lower source intensities for XPS. Furthermore, the surface sensitivity of XPS is poorer than AES due to the higher kinetic energies of the electrons analyzed in the former technique.

In UPS only the valence electrons are probed. Although the same valence band region is also probed by XPS, UPS is superior to XPS in this energy region due to (i) larger ionization cross sections for the emission of valence electrons using UV radiation, (ii) higher photon fluxes available with UV sources, and (iii) higher resolution obtainable with UV sources (several meV linewidth versus 1 eV for soft X-rays).

Quantitative chemical analysis using UPS is precluded for several reasons. First, the valence levels of the different elements are not unique, leading to overlapping peaks in UPS spectra. Furthermore, spectra consist of broad peaks superimposed on a scattered electron background, making deconvolution extremely difficult. The power of UPS is that it probes the valence electrons, and these are the electrons that participate in chemical bonding.

Further information on photoelectron spectroscopy can be found in Carlson (6) or in several review articles (7-9) which have appeared in the literature.

3.2.3 Conversion Electron Mössbauer Spectroscopy (CEMS). Mössbauer spectroscopy is based on the phenomenon of recoil-free resonant emission and absorption of γ-rays by nuclei of atoms tightly bound in a solid, as described in part I of this series. After a Mössbauer-active nucleus has resonantly absorbed a γ-ray, it decays back to its ground state with either re-emission of a γ-ray or with emission of an atomic s-level electron. The latter process is called internal conversion, and it leaves the atom in an excited state from which Auger de-excitation can occur emitting additional (Auger) electrons or X-rays.

Transmission Mössbauer spectroscopy, as described in part I, is performed by measuring transmitted γ-rays. One advantage of this technique is that it is based on photon detection and a vacuum environment is not required. Furthermore, model catalyst systems need not be studied since the high penetrating power of the γ-rays lets this technique probe inside the pores. For the same reason, however, transmission Mössbauer is not surface sensitive. Also, transmission Mössbauer spectroscopy does not possess the sensitivity required to detect a small quantity of a Mössbauer isotope (e.g., 10^{15} atoms of ^{57}Fe) on a model supported metal sample having an area of ca. 1 cm^2. In the back-scatter geometry, the Mössbauer spectrum can be obtained by counting either the γ-rays, the X-rays, or the internal conversion electrons emitted by the sample. For ^{57}Fe, the most commonly used Mössbauer isotope, the ratio of emitted conversion electrons to re-emitted γ-rays is approximately ten and it is therefore advantageous to detect the emitted conversion electrons. Because the electron may have high energy, CEMS probes depths up to 300 nm from the surface. Thus CEMS is only somewhat surface sensitive. The advantage of CEMS in the study of model supported metal samples is its high sensitivity. That is, by working in the back-scatter geometry, one detects electrons emitted by the sample at an angle (ca. 45°) from the incident γ-ray beam; therefore, the incident γ-rays do not interfere with the detection of the emitted electrons, and a

Mössbauer spectrum with a high signal to noise ratio results.

3.3 Vibrational Spectroscopies

The vibrational structure of species adsorbed on model surfaces can be probed by three techniques: inelastic electron tunnelling spectroscopy (IETS), electron energy loss spectroscopy (EELS), and infrared reflection spectroscopy (IRS).

3.3.1 Inelastic Tunnelling Spectroscopy (IETS). IETS allows detection of the vibrational modes of adsorbate molecules on barrier oxide substrates. Several good reviews of the technique are presented by Kirtley (10), Weinberg (11), and Hansma (12). The sample consists of a base-metal/metal-oxide/counter-metal sandwich with adsorbate molecules situated at the metal-oxide/counter-metal interface. Sample synthesis involves growing a thin (ca. 2 nm) barrier oxide layer atop the base electrode, adsorbing the desired molecules on this oxide surface, and covering with the counter-metal (usually Pb) by vacuum evaporation. Depositing metal particles onto the barrier oxide surface prior to adsorbate gas exposure makes IETS a viable technique for probing the adsorbate vibrational structure on model supported metals.

Applying a bias voltage V across this matrix separates the Fermi levels of the two metals by an energy, eV, resulting in quantum-mechanical tunnelling of electrons from the base electrode to the counter electrode through the insulating oxide barrier. The overall tunnelling current I arises from electron tunnelling processes occurring either elastically (i.e., without energy loss) or inelastically (i.e., with energy loss). Important inelastic processes are those due to excitation of vibrational modes of the adsorbate molecules present. The elastic tunnelling current increases in a smooth fashion with increasing bias voltage. In contrast, increasing V past the threshold for each new inelastic process results in additional pathways for tunnelling electrons as evidenced by a series of steps in a dI/dV vs. V plot. Plotting d^2I/dV^2 vs. V sufficiently suppresses elastic contributions while enhancing the inelastic contributions by producing peaks at each threshold energy.

Advantages of IETS are (i) its high sensitivity ($\sim 5 \times 10^{-3}$ monolayers), (ii) its capability to detect both infrared- and Raman-active vibrational modes, and (iii) its broad spectral range (230-500 meV $\cong$ 250-4000 cm^{-1}). Two disadvantages are (i) the presence of the counter-metal electrode which can shift the energy of vibrational modes and (ii) the need for cryogenic sample temperatures to attain acceptable peak resolution and tunnelling current.

3.3.2 Electron Energy Loss Spectroscopy (EELS). EELS and its applications to the study of adsorbed species are described by

Ibach <u>et al</u>. (13), Dubois and Somorjai (14), and Thiel and Weinberg (15). In EELS, the model surface is subjected to a monoenergetic beam of low-energy electrons (ca. 10 eV) after exposure to the desired adsorbate molecules. These incident electrons can scatter from the surface either elastically or inelastically, in the latter case losing a quantum of energy to a vibrational mode of an adsorbed species. Energy-analysis of these scattered electrons yields a spectrum of scattered intensity versus energy loss. This spectrum contains a series of energy-loss peaks corresponding to the different vibrational modes detected.

Typically, the scattered electrons are analyzed in the specular direction since, at this angle, the intensity of the inelastic electron current is maximized. For specular detection, only admolecule vibrational modes possessing a dipole moment with a nonzero component perpendicular to the surface are EELS-active. This dipole-normal selection rule does not apply to off-specular (i.e., impact) scattering, however, and all normal mode vibrations are observed in this case.

Like IETS, EELS has a high sensitivity ($<1 \times 10^{-3}$ monolayers for strong scatterers) and a broad spectral range (300-4000 cm^{-1} scanned within 20 min.). In addition, EELS can be performed on model surfaces in conjunction with other complementary techniques (e.g. XPS, UPS, AES), and there is no need for a top (counter-metal) electrode, which can complicate the interpretation of spectra from IETS. The main disadvantages of EELS are its limited resolution ($\sim$60-80 cm^{-1} = $\sim$7-10 meV) and the need for low operating pressures (less than ca. 10^{-3}Pa) to minimize electron-gas collisions.

3.3.3 Infrared Reflection Spectroscopy (IRS). For conventional catalysts, transmission infrared spectroscopy has been the most commonly used technique to characterize adsorbed species. Unfortunately, in spite of its advantages, including absence of pressure limitations, transmission ir lacks sufficient sensitivity for detecting monolayer adsorbates on flat films. The external reflection mode of ir, termed infrared reflection spectroscopy, has sufficient sensitivity for detecting adsorbates on reflective films when configured to give reflections of p-polarized light at grazing incidence. Adsorption of CO on polycrystalline metal foils has been the most popular application of IRS. Very few studies of monolayer organic adsorbates on metal or metal oxide films have been published; however, adsorption of acetic acid and 2,4-pentadiene on alumina films has recently been reported by Allara (16). The spectra were quite similar to those obtained using IETS. Although sensitivity in IRS is inferior to that in IETS, the overall greater simplicity of IRS should lead to its application to model supported catalysts. Because such applications have not appeared in the literature to date, a complete discussion of IRS is not given here. For a review of the basic principles involved, see Greenler (17).

3.4 Catalytic Methods

3.4.1 Temperature Programmed Desorption (TPD). TPD is a technique that can be used to provide information about the binding of adsorbed species on surfaces of model supported metal samples. Ehrlich (18), Schmidt (19), Smutek et al. (20), and King (21) have written reviews on this topic. Experimentally, the sample in a continuously-pumped, ultra-high vacuum system is dosed with the desired adsorbate molecules at a sufficiently low temperature that desorption of the admolecules is negligible. By increasing the surface temperature, T_S, at a linear rate, the admolecules desorb into the gas phase and are detected mass-spectrometrically. The desorption rate, r_d (molecules $m^{-2}s^{-1}$), of each species is described by:

$$r_d = \sum_i^w r_{d_i} = \sum_i^w \frac{-dn_i}{dt} = \sum_i^w k_{d_i} n_i^{m_i} = \sum_i^w \nu_i \exp(-E_{d_i}/RT_S) n_i^{m_i} \quad (4)$$

where i refers to each binding state i for the adsorbed species, w is the total number of binding states, n is the admolecule surface concentration (molecules m^{-2}), k_d is the desorption rate constant, ν is the preexponential factor, m is the order of desorption, E_d is the desorption activation energy, and R is the gas constant. For sufficiently large system pumping speeds in conjunction with sufficiently small heating rates, the partial pressure increase ΔP of each desorbed species is directly proportional to r_d, and plots of ΔP versus T_S produce a spectral peak for each binding state i initially present on the surface.

TPD spectra can be manipulated in various ways to determine the important kinetic parameters: E_d, ν, and m (18-21). For example, E_d (in kJ mol^{-1}) can be crudely estimated by dividing the desorption peak temperature (in K) by 4 (22). Also, since the area under the desorption curve is proportional to the initial surface coverage of the corresponding admolecule, knowledge of the heating rate, system pumping speeds, and mass spectrometer sensitivities permits calculation of relative as well as absolute admolecule surface concentrations. Thus TPD is a technique for determining both the supported-metal surface area and the surface chemical behavior of model supported metal samples.

3.4.2 Catalytic probe reactions. The primary limitation in conducting conventional reaction kinetics studies of model supported metal catalysts is that the surface area (per unit weight) of these samples is orders of magnitude lower relative to a conventional powder catalyst. This translates into a problem of gas-phase detection sensitivity for the catalytic reaction products, which can be circumvented at high pressure (e.g., atmospheric pressures) using a recycle reactor, or at low pressure by using temperature programmed reaction spectroscopy (TPRS) (23). In high pressure reaction kinetics measurements using recycle of the reactants, the gases contact the catalyst many times over the course of the experiment, thereby increasing the overall conversion of reactants to products. In such experiments, one must confirm that reaction on the surfaces of the reactor and recycle loop is negligible. At low pressures, TPRS can give quantitative reaction behavior in much the same way that TPD is used to study the energetics of adsorbed species. In short, reactant molecules are adsorbed on the surface at low temperatures and the evolution of reaction products is monitored mass-spectrometrically as the sample temperature is raised at a linear rate. High intensity molecular beams have also been used to study reaction behavior of single crystals or polycrystalline metal foils in ultra-high vacuum, but as of yet these techniques have not been applied to model supported metal catalysts.

4. APPLICATIONS OF STUDIES USING MODEL SUPPORTED METAL SAMPLES

In commercial applications of supported metal catalysts, the loss of metal surface area due to sintering of the metal crystallites is an important problem. Regeneration schemes have been developed to redisperse the agglomerated crystallites and subsequently restore catalytic activity. Essentially two basic models have appeared in the literature to take into account observed sintering phenomena. In the crystallite migration model, sintering occurs by migration of entire crystallites over the support surface. Particle collision and coalescence then forms larger crystallites. In the atomic migration model, atomic or molecular species are assumed to break free from edges of small crystallites and diffuse across the support, followed by recombination with another large stationary crystallite. The latter model can explain both sintering and redispersion phenomena whereas the former, in the absence of crystallite splitting, can explain sintering only.

In order to better understand the detailed steps involved in the sintering of Pt crystallites on alumina, Chu and Ruckenstein (24) observed the behavior of model supported Pt/Al_2O_3 catalysts with transmission electron microscopy as a function of treatment conditions. Thin, nonporous alumina films were prepared through anodization of high purity aluminum foil to a thickness of about 30 nm. The anodization produced amorphous alumina films, which were subsequently "floated" on a dilute mercuric chloride solution that etched away remaining unoxidized Al metal. The films were transferred to electron microscope grids, dried, and then transformed to polycrystalline γ-Al_2O_3 by heating in air at 1070 K for 72 hours, as confirmed by selected area electron diffraction. The Pt metal was deposited on the support by vacuum evaporation of a high purity Pt wire to a contiguous overlayer thickness of ca. 2 nm. Treatment in flowing H_2 at 770 K for 6 hours led to the formation of small Pt crystallites on the support.

To determine the effect of gas environment and temperature on the sintering of Pt, samples were treated either in vacuum, H_2, N_2, O_2, H_2/H_2O, or N_2/H_2O at temperatures ranging from 770-1120 K. Transmission electron micrographs of the same regions of the sample before and after treatment were used to monitor changes in particle size, shape, number and position. In short, it was found that little sintering occurred for specimens heated in vacuum, H_2, N_2, O_2, and N_2/H_2O at temperatures up to 1020 K. Specimens heated in H_2/H_2O, however, experienced severe sintering at temperatures between 770 and 920 K. This behavior is illustrated in Figure 4, which gives the particle size distributions determined from micrographs of the sample after treatment in H_2 at 770 K for 6 h (4a), and after treatment in H_2/H_2O at 770 K for 6 h (4b), 920 K for 1 h (4c) and 920 K for an additional 3 h (4d). The micrographs revealed that the average size of the Pt crystallites increased from 5 nm to 9, 11 and 16 nm during the above treatments, respectively, the number of crystallites decreased and the distributions broadened significantly.

Detailed examination of electron micrographs taken of the same region of a specimen following treatment in H_2/H_2O at 920 K for 1 and 4 h, respectively, was utilized to help elucidate the mechanism(s) involved in particle sintering. It was thereby con-

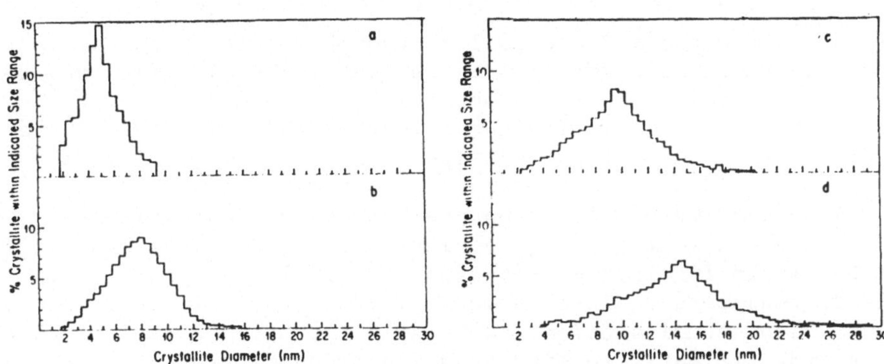

Fig. 4. Particle size distributions of Pt crystallites on Al_2O_3.
(a) after treatment in H_2 at 770 K for 6 h,
(b) after treatment in H_2/H_2O at 770 K for 6 h,
(c) after treatment in H_2/H_2O at 920 K for 1 h, and
(d) after treatment in H_2/H_2O at 920 K for an additional 3 h.
Reprinted with permission from (24).

cluded that relatively large particles, of the order of 10 nm in size, migrated considerable distances (e.g., up to 50 nm) during this treatment in H_2/H_2O. The apparent sintering mechanisms for samples treated in environments other than H_2/H_2O were similar, with the exception that crystallites larger than 5 nm in diameter were essentially immobile. In these cases, sintering occurred only through migration of crystallites smaller than this critical size. It was further shown that although sintering in O_2 or H_2 was initially slow, successive oxidation-reduction cycles led to redispersion (i.e., decrease in Pt particle size) during treatment in O_2 and rapid sintering during treatment in H_2. On the basis of these results, the authors suggested that water in the presence of H_2 weakens the attractive interaction between Pt crystallites and the Al_2O_3 support and hence promotes crystallite migration. In the case of the oxidation-reduction cycles, it was thought that water is produced by the reduction of oxidized Pt.

Rapid sintering of Pt was observed in all gas environments at 1120 K; this was attributed to the phase transformation of the support from γ-Al_2O_3 to α-Al_2O_3, as shown by selected area electron diffraction. The severity of sintering is demonstrated in Figure 5, which shows the same region of a sample before (5A) and after (5B) treatment in 10^{-4} Torr H_2 at 1120 K for 6 h. Figure 6 is the electron diffraction pattern corresponding to the sample of Figure 5B. The ring pattern is characteristic of Pt metal, while the bright spots are due to α-Al_2O_3. It was determined in similar experiments that heating model catalyst samples in vacuum at 1120 K for 6 h not only led to phase transformation to α-Al_2O_3 but also caused reaction of Pt with the support to form an intermetallic Pt_8Al_{21} compound.

In a study employing controlled atmosphere electron microscopy, McVicker, Garten, and Baker (25) investigated the sintering and redispersion of iridium on Al_2O_3 supports which were doped with oxides of Ba, Ca, and Sr. The authors suggested that sintering of Ir occurred under oxidizing conditions via atomic migration, and could be inhibited or even eliminated if chemical traps (i.e., Group IIA oxides) were present on the support surface. These traps capture atomic migrating species to form new particles before the migrating species can combine with larger particles. In effect, redispersion is observed. The above model was proposed on the basis of chemisorption and X-ray diffraction measurements of the average Ir particle size as a function of treatment conditions for high surface area Ir/Al_2O_3 catalysts. It was demonstrated that oxides of Ca, Ba, and Sr stabilized the Ir surface area under oxygen atmospheres at high temperatures. This stabilization was envisaged to occur as a result of formation of an immobile, stable surface iridate from atomic iridium and a well-dispersed Group IIA-oxide. Supporting evidence for this assertion was provided by the ineffective behavior of Mg, which

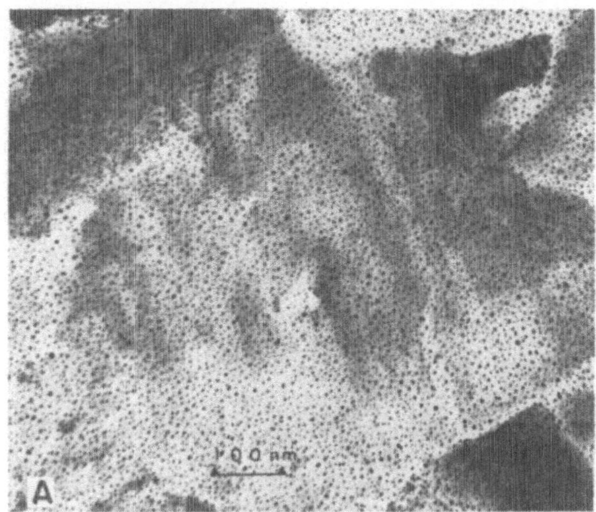

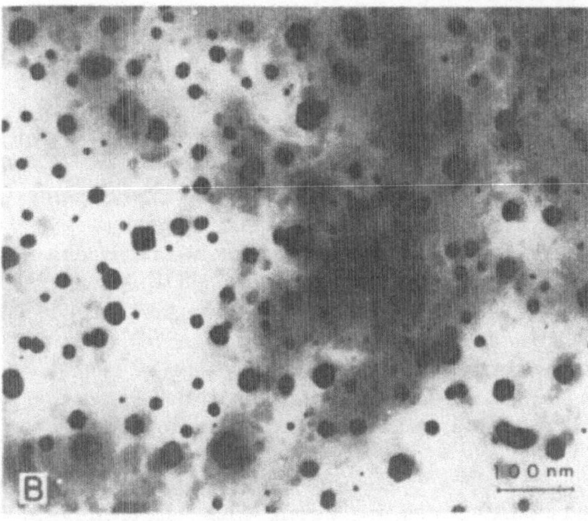

Fig. 5. Transmission electron micrographs of Pt on Al_2O_3 model samples showing the effect of phase transformation of Al_2O_3 on Pt particle size. (A) fresh sample, and (B) after treatment at 1120 K for 6 h in 10^{-4} Torr H_2. Reprinted with permission from (24).

is not known to form an iridate. Following oxidative redispersion, surface iridates were readily decomposed in hydrogen to regenerate Ir metal and the Group IIA-oxide.

The knowledge obtained on the behavior of Ir/Al_2O_3 was confirmed and extended through direct observation of model supported metal samples with TEM and CAEM. Model supports were prepared by anodization of an aluminum film to give an alumina layer ca. 50 nm thick, followed by dissolution of unconverted Al metal in 50% HCl. Iridium and barium were deposited onto the films via atomization of dilute (0.1%) aqueous solutions of chloroiridic acid and barium nitrate.

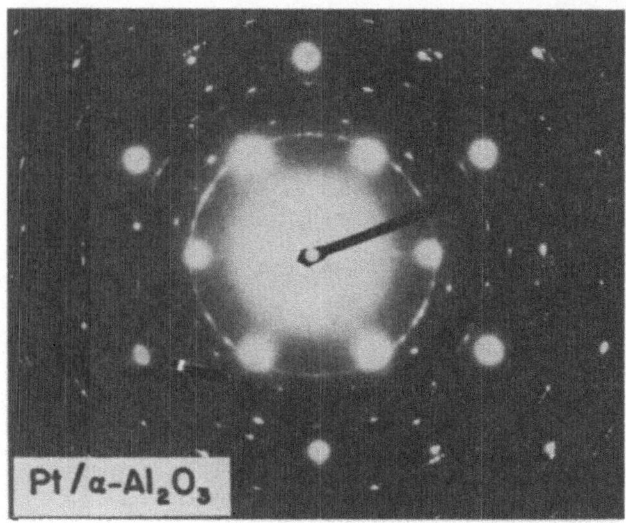

Fig. 6. Electron diffraction pattern corresponding to the specimen of Fig. 5B. The bright spots correspond to α-Al_2O_3, and the rings correspond to Pt. Reprinted with permission from (24).

Transmission electron micrographs of such Ir/Al_2O_3 specimens as a function of increasing treatment temperature in oxygen are shown in Figures 7A-7C. The micrographs clearly demonstrate that sintering of Ir takes place at temperatures above ca. 870 K, as observed previously for high surface area samples. In addition, comparison of the micrograph for the sample calcined at 870 K for 0.5 h (Figure 7B) with the micrograph for this same sample following addition of BaO, reduction, and recalcination (Figure 7D) confirms that BaO not only prevents further sintering, but also promotes redispersion.

The value of particle size measurement using TEM, in contrast to determination of average particle size only using chemisorption methods, is illustrated by analysis of the particle size distributions for samples given the treatments of Figure 7. These distributions, shown in Figure 8, are bimodal in nature. For all samples the largest fraction of crystallites was between 1-3 nm diameter. As the oxidation temperature increased, the total number of small crystallites decreased and the second peak in the distributions moved to larger diameters. These findings support the atomic migration mechanism for sintering, which predicts that large crystallites grow at the expense of smaller particles. The redispersion induced by the addition of BaO is shown by comparing distributions B and D.

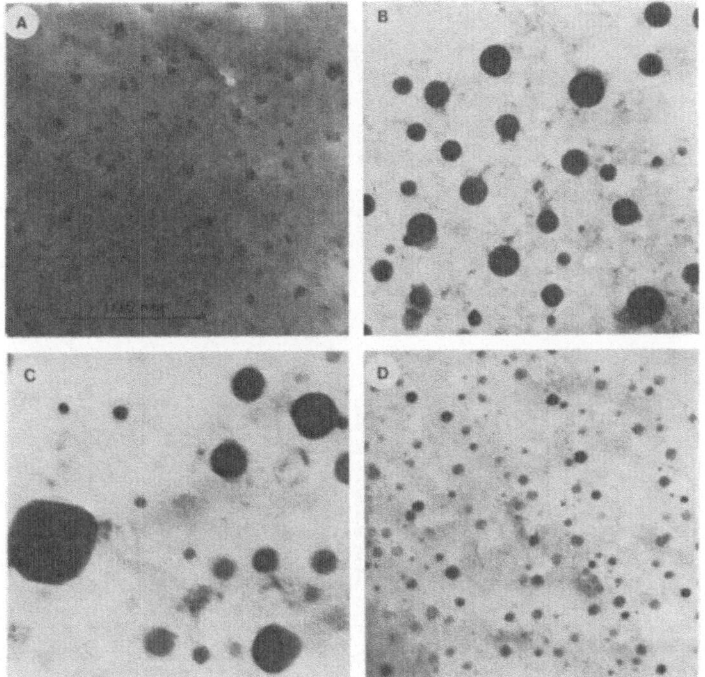

Fig. 7. Transmission electron micrographs of model Ir/Al_2O_3 and $Ir/BaO-Al_2O_3$ samples calcined under 20% oxygen. (A) Ir/Al_2O_3 770 K, 1.0 h; (B) Ir/Al_2O_3 870 K, 0.5 h; (C) Ir/Al_2O_3 970 K, 0.3 h; and (D) prepared by doping (B) with BaO, reducing, and then recalcining at 870 K for 0.5 h. Reprinted with permission from (25).

Direct observation of the sintering and redispersion processes was made using controlled-atmosphere electron microscopy. CAEM not only allowed the monitoring of morphological changes in real time, but also enabled determination of kinetic parameters for Ir growth and redispersion. These observations of Ir/Al_2O_3 model samples confirmed that sintering of Ir under oxidizing conditions takes place via migration of atomic or molecular species, since the motion of Ir particles larger than ca. 2.5 nm (the resolution of CAEM) could not be detected. For Ir on Al_2O_3, growth was first observed at 710 K, but remained slow through 870 K, and then became rapid as the temperature was increased to 970 K. The behavior of Ir on Al_2O_3-BaO was quite different. For temperatures less than ca. 970 K redispersion was observed. Optimum conditions for redispersion appeared to be near 870 K. At temperatures greater than 970 K, sintering of Ir occurred, but at a rate significantly lower than the samples which lacked BaO. The apparent activation

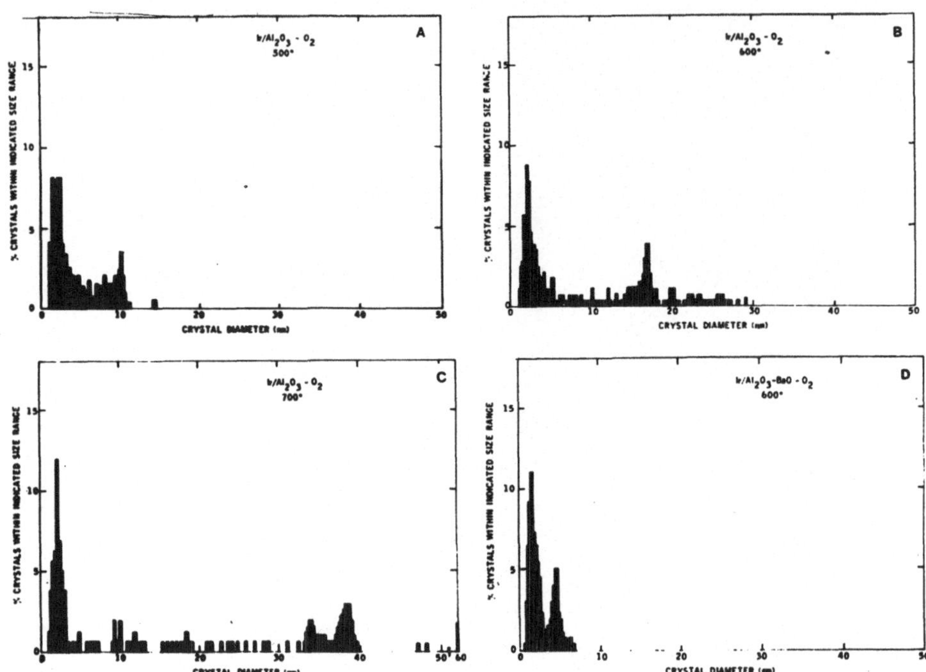

Fig. 8. (A-D) Particle size distribution curves obtained after the respective sample treatments outlined in Figure 7. Reprinted with permission from (25).

energy, for the sintering of Ir on Al_2O_3, determined from such CAEM measurements, was found to be 16 kcal/mol. From this, the authors suggested that the sintering process was controlled by formation of a volatile, mobile iridium oxide species rather than migration across the surface.

Redispersion phenomena in multi-component metal particles have been studied by Schmidt and co-workers (26-29). For example, Wang and Schmidt (29) examined the surface compositions and morphologies produced by oxidation-reduction cycles of small particles of Rh and Pt-Rh on amorphous silica and γ-alumina using XPS and high resolution STEM.

The model, silica-supported samples studied were prepared in the following manner. A high-purity Si disc was oxidized in air at 1370 K for 1 h to form a 100 nm thick layer of amorphous SiO_2. Sequential vacuum deposition of 1 to 2 nm films of Pt and Rh, followed by heat treatment in flowing hydrogen, produced dispersed metal particles on the support. Thin regions of silica suitable for analysis by STEM were prepared by placing a 50 nm

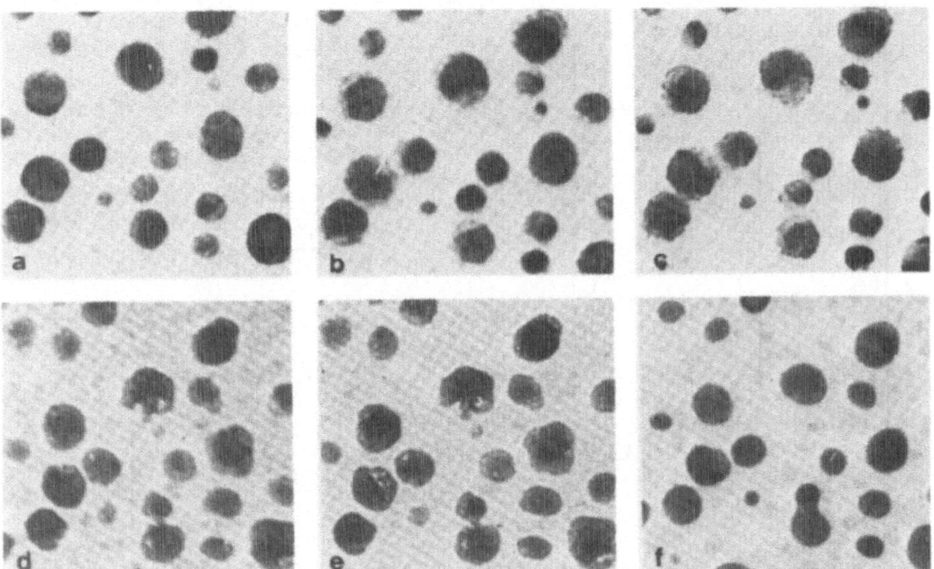

Fig. 9. Transmission electron micrographs showing the morphologies of 48% Pt-52% Rh alloy particles on SiO_2 illustrating segregation and splitting of alloy particles. The sample was heated sequentially (a) in H_2 at 920 K, (b) in air at 870 K, and then reduced in H_2 at (c) 300 K, (d) 420 K, (e) 770 K, and (f) 970 K, each for 1 h. Micrograph (c) shows tiny Rh crystallites while (d) and (e) indicate dark cores of Pt and lighter shells of Rh. Reprinted with permission from (29).

thick Si flake over a hole in the Si disk prior to oxidation and metal evaporation.

Figure 9 shows electron micrographs of the same field of view of a sample containing Pt-Rh (48% Pt-52% Rh) on amorphous SiO_2. The metal particles of Figure 9a were formed by treatment in H_2 at 920 K. After subsequent treatment in air at 870 K, thin layers of Rh_2O_3 formed at the edges of the metal particles, as can be seen in Figure 9b. (Rhodium oxidation was confirmed by electron diffraction.) Micrograph 9c shows that reduction in H_2 at room temperature produced small, metallic Rh crystallites at the edges of the original particle. Upon heating in hydrogen at temperatures up to 770 K, all of the rhodium oxide was reduced to metal and the particles coalesced into their original morphologies. However, as shown in micrographs 9d and 9e, the particles appeared to maintain a light shell of Rh around dark cores of Pt. This apparent surface enrichment in rhodium could be eliminated through

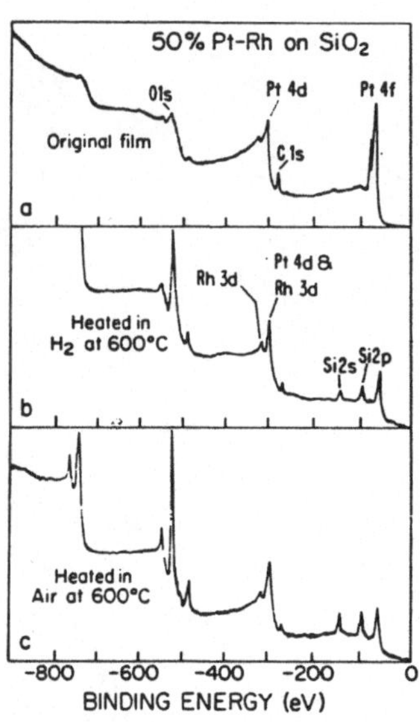

Fig. 10. XPS survey spectra of a silica-supported Pt-Rh (~50% Pt - 50% Rh) model sample (a) as prepared, (b) after treatment in H_2 at 870 K and (c) after treatment in air at 870 K. Reprinted with permission from (29).

further treatment in H_2 at 970 K, as shown in Figure 9f.

XPS measurements were directly correlated with the observed morphology changes to confirm and quantify the postulated metal phase behavior. Typical survey spectra are shown in Figure 10a for the initial film with 1 nm Pt deposited on top of 1 nm Rh on SiO_2, 10b for the same sample following hydrogen reduction at 870 K, and 10c following oxidation in air at 870 K. Several conclusions could be drawn from these spectra. First, no detectable contaminants other than a trace of C were evident. Second, Figure 10a shows that 2 nm of metal completely masked the SiO_2 substrate from analysis, and 1 nm of Pt nearly masked the Rh, illustrating that the surface sensitivity of the XPS measurements was of the order of 1 nm.

The areas under the Rh-$3p_{3/2}$ peak at 498 eV and the Pt-4f peaks at 71 and 74.5 eV, measured from high resolution scans of these peaks, were used to determine the surface composition of the Pt-Rh/SiO_2 samples after the various oxidation-reduction treatments. Figure 11 shows that when a H_2-reduced sample was treated in air at 870 K, the surface Rh/Pt ratio increased by a factor of 3, with enrichment becoming most important for treatment temperatures greater than 820 K. Even after subsequent hydrogen reduction at temperatures up to 770 K, the bimetallic particles maintained a Rh/Pt ratio as high as 2. Treatment at 970 K in hydrogen, however, homogenized the particles to their original bulk composition. Also shown in Figure 11 is the XPS composition analysis obtained following argon sputtering the samples after the different treatments. The composition was shown to be

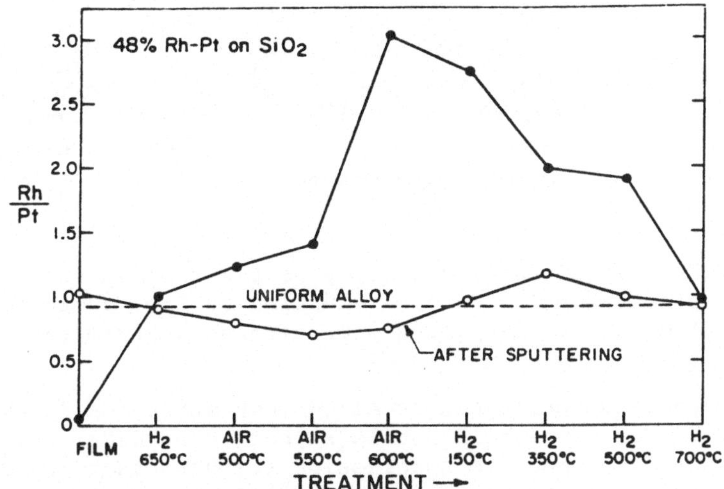

Fig. 11. Atomic ratio Pt/Rh obtained from XPS peak areas as a function of treatment for a 52% Pt - 48% Rh on silica model sample (solid circles), and ratio after 1 h sputtering (open circles). The broken line represents the surface composition if the particles were of uniform composition. Reprinted with permission from (29).

approximately equal to that of the original, homogeneous alloy particles. For air-oxidized samples, the data suggested that the cores of the particles were slightly enriched in Pt, as expected in view of the surface enrichment in Rh.

Figure 12 summarizes the model developed by the authors for the phenomena observed, based on the combination of STEM and XPS results. Oxidation produces a Rh_2O_3 layer over the bulk alloy particle and the SiO_2 support in the area around the particle. Upon heating in H_2, reduction occurs in three stages dependent on temperature. First, reduction of Rh_2O_3 to metallic Rh at 370-420 K produces small Rh crystallites on the SiO_2 in the vicinity of original Pt-Rh particles, as well as a thin shell of Rh over the Pt-Rh particles. In the second stage, the small Rh crystallites sinter and coalesce into the original particle as the sample is heated to higher temperatures, ca. 570 K. This leaves the Pt-Rh particles enriched at the surface in Rh. In the third stage, homogenization occurs, but only at temperatures sufficiently high to make bulk diffusion important. This requires temperatures in excess of 870 K.

The nature of "strong metal-support interactions" for

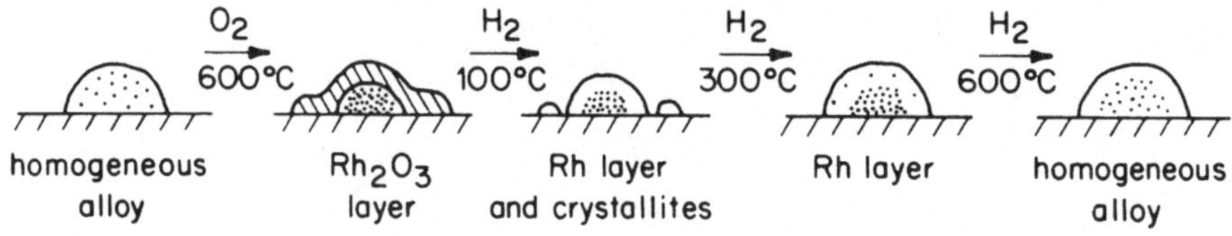

Fig. 12. Summary of idealized alloy particle cross sections produced by oxidation-reduction cycling at temperatures indicated. Reprinted with permission from (29).

titania-supported metals has received much attention in the recent literature. For Group VIII metals supported on TiO_2, this interaction results in complete suppression of hydrogen or carbon monoxide chemisorption on the metal following high temperature reduction (e.g., 770 K) (30-33). TEM results suggested that the metal crystallites spread over (or wet) the support to form thin, raft-like particles, with simultaneous partial reduction of the support. This in some way, either geometrically and/or electronically, was assumed to account for the chemisorption effects observed. Indeed, a number of systems exhibiting the above behavior appear to possess interesting catalytic properties, including Ni/TiO_2 (34), Ru/TiO_2 (35) and Rh/TiO_2 (36) for CO/H_2 hydrocarbon synthesis reactions, and Rh/TiO_2 and Pt/TiO_2 (37) in hydrogenation/dehydrogenation reactions. In a series of papers, Taturchuk and Dumesic (38-40) examined the consequences of "strong metal support interactions" for Fe/TiO_2 model supported samples. Iron was chosen as the supported metal so that electronic and magnetic information about the metal phase could be obtained using CEMS, in addition to the structural, morphological and chemical information obtainable by TEM and XPS.

The model Fe/TiO_2 samples were prepared by oxidation of chemically smoothed high purity titanium foils for 2-3 h at 570-670 K in oxygen. This produced an approximately 50 nm thick film of rutile TiO_2. Vacuum evaporation of metallic iron completed the preparation for the XPS and CEMS samples. For the TEM samples, the metal backing was etched away and the thin oxide film placed on titanium electron microscopy grids prior to deposition of iron.

Transmission electron micrographs taken after the Fe/TiO_2 samples were treated in H_2 at progressively higher temperatures could be interpreted in terms of three separate temperature regimes. At low temperatures, iron crystallites formed from the original iron-overlayer, and this process was complete at 677 K.

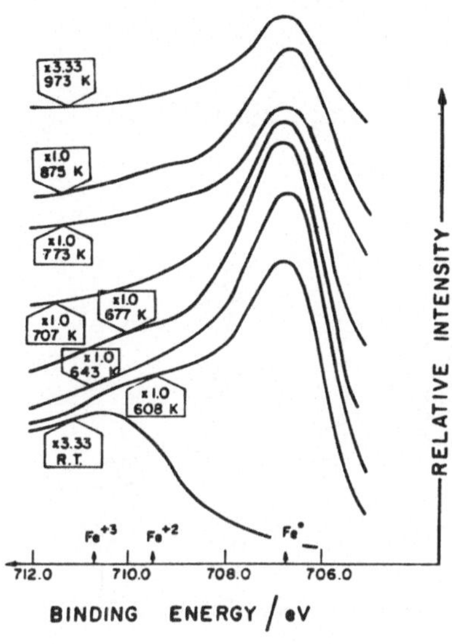

Fig. 13. Fe - $2p_{3/2}$ XPS spectra of an Fe/TiO_2 model sample after hydrogen reduction at the temperature indicated. Vertical sensitivities are given adjacent to the reduction temperature. Reprinted with permission from (39).

After continued reduction at 707 K, slight sintering of iron was observed. In the second temperature regime (ca. 770 K), the iron particles were observed to spread over (or wet) the titania support. This was based on micrographs in which many of the smaller crystallites (e.g., 5 nm) showed low contrast, consistent with a thin, raft-like morphology. A third regime was observed for reduction temperatures exceeding 875 K. At these temperatures, even larger iron particles adopted this thin-crystal morphology. In addition, the average particle size decreased while the amount of iron present as distinct crystallites also decreased. This suggested that the iron had either diffused into the support or spread diffusely over the support. TEM could not distinguish between these two possible interpretations. XPS and CEMS studies, however, showed that iron had diffused into the support as a dispersed, strongly interacting species.

The XPS scans of the Fe-$2p_{3/2}$ region and the Ti-$2p_{3/2}/2p_{1/2}$ region for progressively higher temperature reduction treatments are shown in Figures 13 and 14, respectively. From the known peak positions of Fe^{3+}, Fe^{2+}, and Fe^0, the Fe-$2p_{3/2}$ spectra showed that essentially all the iron was reduced to the metallic state during the reductions at 608 and 643 K. Of special interest was the observed decrease in spectral area following reduction at 773 K or above. Since TEM results ruled out sintering as a possible explanation for this decrease, it was concluded that iron had diffused into the support. The Ti XPS scans showed that titanium was present as Ti^{4+} after reduction at temperatures up to 773 K for both the iron-containing sample and an iron-free TiO_2 "control" sample. After reduction at 875 K, however, new components in the photoelectron spectrum of the Fe/TiO_2 sample appeared, suggesting that the support had been partially reduced. Moreover, since this behavior was not observed for the TiO_2 con-

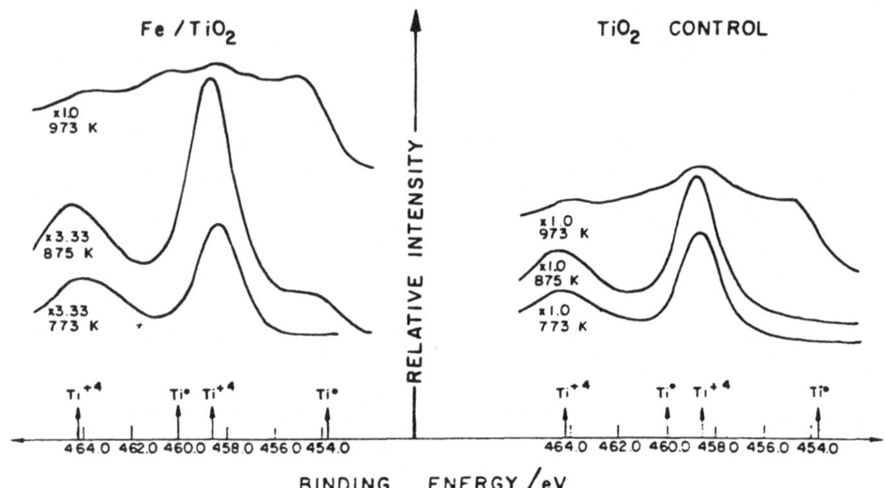

Fig. 14. Ti - $2p_{3/2}/2p_{1/2}$ XPS spectra of an Fe/TiO_2 model sample and an Fe-free TiO_2 film after hydrogen reduction at the indicated temperatures. Vertical sensitivities are given adjacent to reduction temperature. Reprinted with permission from (39).

trol sample, unless the reduction temperature exceeded 973 K, it appeared that the presence of iron facilitated the reduction of titania.

The conversion electron Mössbauer spectra collected following sequential reduction treatments of Fe/TiO_2 are shown in Figure 15. Although the spectra are complicated by the degree of overlap of the peaks, as shown by the expected positions of the Fe^{3+} singlet, Fe^{2+} doublet, and metallic iron sextuplet, computer fitting allowed detailed interpretation. Relative spectral areas (proportional to relative amounts present) of Fe^{3+}:Fe^{2+}:Fe^0 were 13:55:32% as prepared, 0:59:41% after hydrogen reduction at 608 K, and 0:0:100% after reduction at 677 K. These data show that Fe^{3+} was reduced to Fe^{2+} and Fe^0 at temperatures near 600 K, followed by further reduction of Fe^{2+} to metallic iron at 643 and 677 K. Additional hydrogen treatment at 707 K produced no further spectral changes. Hydrogen reduction at 773 K produced significant broadening of the metallic iron peaks relative to the spectra observed after treatment at 677 or 707 K, without any change in isomer shift or magnetic splitting. These results are consistent with electron micrographs which showed that the iron crystallites spread over the titania surface. Since thin iron films have been

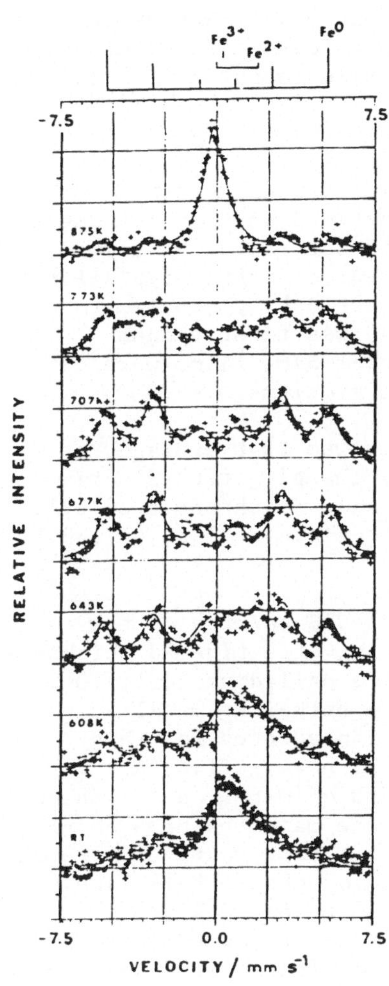

Fig. 15. CEMS spectra of an Fe/TiO_2 model sample following hydrogen reduction at the indicated temperatures. One vertical square equals 2% effect. Reprinted with permission from (39).

shown to give spectra with broad lines (41), the observed broadening of the Mössbauer spectrum is consistent with the transformation of three-dimensional iron crystallites into thinner iron particles, without changing the chemical state of iron. The CEMS spectrum after treatment at 875 K shows that a dramatic change in the chemical state of the iron has taken place. Only a trace of the metallic iron sextuplet remained, while a new peak assigned to Fe_xTi ($1 \leq x \leq 2$) or γ-Fe dominated the spectrum. Because the iron remained in the zero-valent state while lacking magnetic ordering the iron must have been in strong interaction with the support. Furthermore, total Mössbauer spectral area decreased by ca. 30%, confirming the conclusion based on XPS results that the iron had diffused into the support.

Of fundamental importance in the interpretation of the catalytic behavior of small metal particles dispersed on a support are the electronic properties of the particles as a function of particle size. Moreover, perturbation of the electronic properties of the metal crystallites may be intimately related to the support material, as illustrated in the behavior of Fe/TiO_2 discussed above. Model supported metal samples are well-suited for direct measurement of electronic properties of the metal particles as a function of particle size, since narrow particle size ranges may be prepared and characterized. In an example of such a study, Takasu et al. (42) investigated small particles of palladium supported on amorphous silica films, using electron microscopy combined with photoelectron spectroscopy. By monitoring changes in the Pd valence band using UPS as a function of increasing particle size, they showed that the Pd particles did not display bulk-like metallic properties for sizes below about 2-3 nm in diameter.

Samples with different average particle diameters were prepared by evaporation of Pd in effective thicknesses from 0.05 to 1.0 nm. ("Effective thickness" was defined as that thickness of Pd overlayer that would be obtained if the palladium formed a contiguous film. At the low loadings used, discrete crystallites were formed instead.) For the samples probed by photoelectron spectroscopy, ca. 8 nm of silicon dioxide was first vacuum deposited onto a copper substrate. Thicker SiO_2 films were not employed due to observed sample charging problems.

Transmission electron micrographs showed that average Pd particle size could be readily varied by changing the effective film thickness. For example, effective film thicknesses of 0.1, 0.2, 0.5 and 1.0 nm gave average particle sizes of 1.6, 2.1, 3.0, and 4.2 nm, respectively.

Figure 16 compares the HeI (21.2 eV) UPS spectra of the Pd/SiO_2 samples as a function of increasing effective Pd thickness with the UPS spectrum of a polycrystalline palladium foil. Only above a certain Pd particle size were the bulk-like metallic properties of Pd approached. The criterion adopted for this judgment was that metallic particles would have sufficient overlap of atomic states to exhibit a d-band width of 5-6 eV and a sharply defined Fermi level E_F. Spectrum d has the main features of the bulk valence band emission, in particular a band width of more than 4 eV. The effective Pd thickness needed to achieve bulk-like metallic properties was therefore taken to be ca. 0.3 nm, corresponding to an average Pd particle size of 2-3 nm. It should be noted that a hemispherical palladium particle of 2.5 nm diameter contains about 250 atoms. For particles smaller than ca. 2-3 nm (up to spectrum d), no intensity was observed at the eventual E_F. Furthermore, the observed d-band emission was significantly narrower. These results reflected the "atomic-like" properties of small Pd particles. Since the characteristics of the metallic state are due to the overlap of atomic energy levels to form a continuous band, the transition to metallic behavior does not occur until the Pd cluster size is such that the separation between the levels is smaller than their inherent width.

The core levels of Pd also exhibited perturbations when the particle size was decreased. These were evidenced by XPS measurements of the Pd-$3d_{5/2}$ peak. The binding energy was observed to increase by 1.6 eV and the peak width increased by almost 1 eV when the average Pd particle size was decreased from ca. 4 nm to

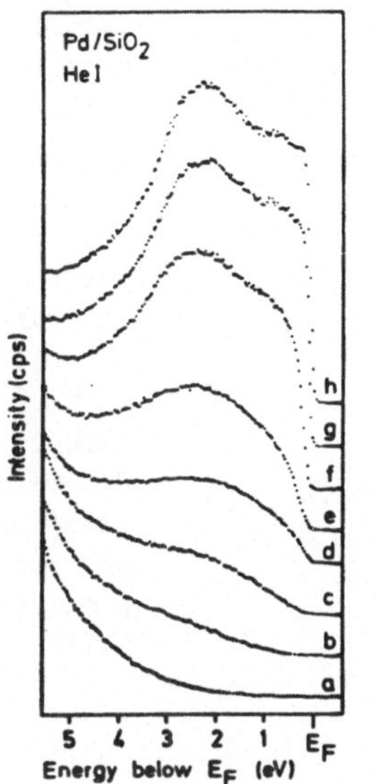

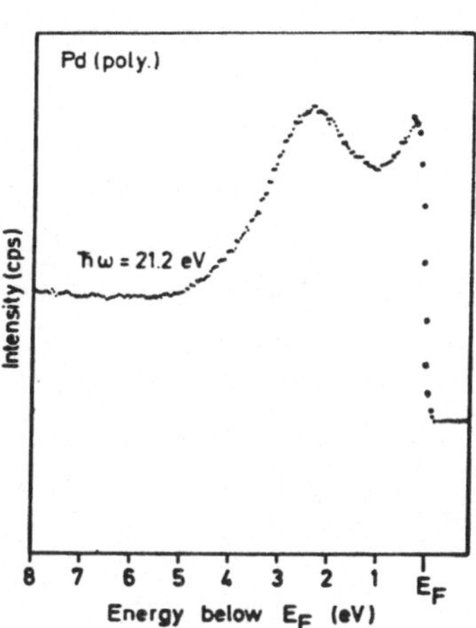

Fig. 16. UPS spectra (HeI) of Pd/SiO_2 model samples as a function of increasing particle size and of a polycrystalline Pd foil. For the model samples, spectrum (a) is of clean SiO_2, whereas (b) is of 0.04, (c) 0.1, (d) 0.3, (e) 0.5, (f) 1.0, (g), 1.4, and (h) 2.0 nm effective Pd thickness. Reprinted with permission from (42).

less than 1.5 nm. These effects were interpreted for small particles in terms of reduced screening by the valence electrons of the core hole created by the photoemission event relative to a bulk metal.

The application of vibrational spectroscopy to the study of adsorbed species on model supported metal samples is exemplified in several paper (43-47). For example, Dubois et al. (47) employed EELS to characterize chemisorption of carbon monoxide on Rh/Al_2O_3. Experimentally, the model supported samples were synthesized by (i) evaporating from 20 to 200 nm of Al onto a Pt(111) substrate, (ii) oxidizing the Al in oxygen at 300-480 K to form a 1 nm thick Al_2O_3 layer, and (iii) depositing 0.1-2 nm of Rh onto this oxide surface by evaporation in $1x10^{-5}$ Torr of CO. The result, previously determined by Kroeker et al. (44) using TEM on similar samples, was an alumina support covered with highly dispersed Rh particles (e.g., 2-3 nm Rh particles for an effective Rh thickness of 0.4 nm). As evidenced by Figure 17, AES was utilized to confirm the purity of the model supported metal sample as well as the presence of chemisorbed CO (the latter responsible for the carbon peak at ca. 270 eV).

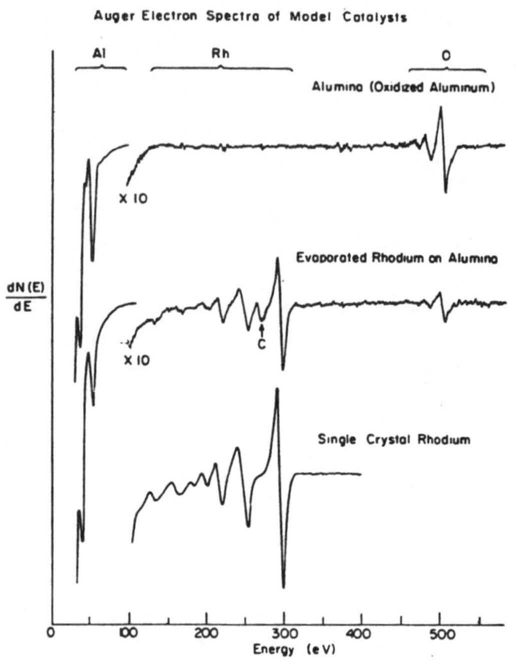

Fig. 17. AES spectra of clean, oxidized aluminum support (top), Rh/Al_2O_3 model sample as prepared by vacuum evaporation of Rh in the presence of CO (middle), and of clean, single crystal Rh surface (bottom). Reprinted with permission from (47).

Figure 18 illustrates EELS data for the clean alumina support as well as for two effective thicknesses (0.4 and 2 nm) of Rh on alumina in $1x10^{-5}$ Torr CO at 300 K. The broad peak at 860 cm^{-1}, accentuated in the clean Al_2O_3 spectrum and attenuated for the rhodium covered supports, was assigned to aluminum oxide phonons. The low frequency peak just above 400 cm^{-1} was assigned to Rh-CO stretching and bending modes, and the two peaks at 2020 and 1870 cm^{-1} were assigned to C≡O stretching modes. The spectra were not sensitive to changes in Rh particle size. Heating the Rh/Al_2O_3 samples in vacuum to 530 K followed by exposure to CO at 300 K led to a decrease in the intensities of the peaks at 2020 and 1870 cm^{-1} which was more pronounced for the latter peak. This was interpreted in terms of sintering of the Rh particles or the decomposition of CO at 530 K.

Figure 19 compares the spectra obtained from EELS, IETS (44) and infrared spectroscopy (48) of CO adsorbed on different Rh/Al_2O_3 samples. The IETS data were obtained from a model Rh/Al_2O_3 sample that was exposed to CO and covered with a Pb counter-electrode. The infrared spectra were obtained in the transmission mode using a high surface area Rh/Al_2O_3 powder sample at both low and saturation CO coverages. This figure indicates the superior resolution but the narrow spectral range of infrared spectroscopy compared to EELS and IETS, and the reasonable low-frequency resolution of IETS. In short, the peaks at 1870 and

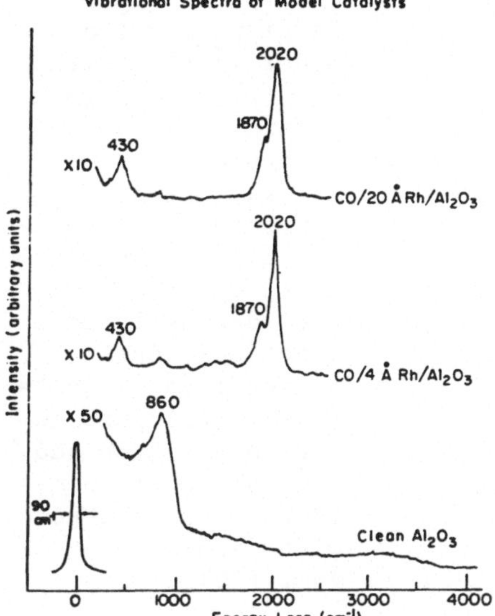

Fig. 18. High resolution EELS spectra for alumina support (bottom) and model Rh/Al_2O_3 samples at two different Rh effective thicknesses as indicated (top and middle). Reprinted with permission from (47).

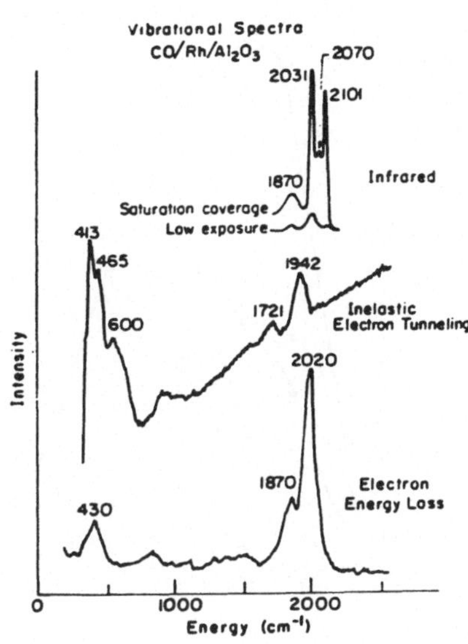

Fig. 19. Vibrational spectra of CO adsorbed on Rh/Al_2O_3 using IR (top), IETS (middle), and EELS (bottom). Reprinted with permission from (47).

near 2020 cm^{-1} can be attributed to multiply-bonded and linearly-bonded CO, respectively. For the high CO pressures and small Rh particles used in the infrared spectroscopy studies, the pair of peaks at 2030 and 2100 cm^{-1} were interpreted in terms of gem dicarbonyl species. The peaks at 400-600 cm^{-1} are due to Rh-CO stretching and bending modes, and the peak near 900 cm^{-1} arises from Al_2O_3 phonons. Note that the peaks in IETS are shifted from those in EELS, possibly due to the effects of the Pb counter-electrode in the former technique.

Ladas et al. (49) investigated the adsorption and oxidation of CO on Pd/Al_2O_3 model supported catalysts. Their work involved the combined use of TEM, TPD, and steady-state kinetic measurements. The model catalysts were prepared by vapor deposition of palladium using an electron-gun evaporator onto a thin slab of single crystal $\{\bar{1}012\}$ α-Al_2O_3. Control of the substrate temperature (in the range 723-823 K), Pd vapor flux, and time of deposi-

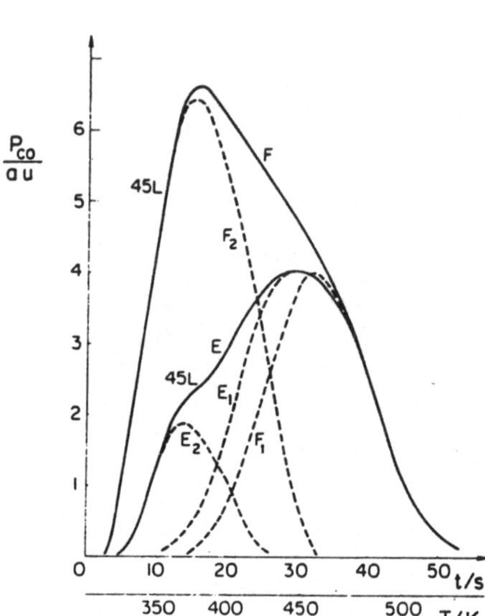

Fig. 20. TPD spectra of CO desorption from Pd/Al_2O_3 for two model samples with different average particle sizes. E represents 4.9 nm particles and F represents 1.5 nm particles. The total spectra have been deconvoluted into individual states 1 and 2 (note subscripts) pertaining to the high and low states, respectively. Reprinted with permission from (49).

tion allowed deliberate variation in the mean particle diameter of the deposited particles within reasonably narrow distributions (half-widths of ca. 2 nm). The slabs were heated resistively through a 1 μm thick film of β-Ta sputtered onto the backside. Prior to Pd deposition, the slabs were heated to 1420 K for 10-30 minutes in ultra-high vacuum. AES showed that this produced a clean surface, leaving less than a tenth monolayer of residual C and traces of Si and Ca. Samples suitable for study using TEM were made by prethinning the Al_2O_3 to slabs <10 μm in thickness followed by ion-beam etching to form a small hole. The areas around the edge of the hole proved thin enough for high resolution TEM measurements.

Four Pd/Al_2O_3 samples were prepared, having average Pd particle diameters equal to 1.5, 2.8, 4.9, and 8.0 nm, respectively. For each specimen, the transmission electron micrographs collected before and after catalytic experiments (CO oxidation) were essentially identical; therefore, sintering or redispersion during temperature cycling and chemical reaction were eliminated as possible complicating effects.

The chemisorption of CO on Pd/Al_2O_3 was found to depend on the Pd particle size, as illustrated by Figure 20 for samples with 4.9 nm and 1.5 nm average size, respectively. For both specimens, two distinct adsorption states were observed, and these have been deconvoluted in the TPD spectra shown. For low exposures, CO desorbed from a single adsorption state (state 1) giving a desorption peak centered near 450 K. As exposure was increased, this peak shifted to lower temperatures revealing that the energy for the desorption of CO decreased with increasing surface coverage. For high exposures, a second adsorption state was filled, leading to a desorption peak near 375 K. Although the desorption peaks exhibited slight upward shifts in temperature (~15 K) as the Pd particle size decreased, it was observed that the relative population of the two adsorption states was much more strongly dependent on particle size. In Figure 20, the TPD spectra for the two Pd/Al_2O_3 samples were normalized such that the height of the first state was the same. Accordingly, the lower binding energy state was the predominant peak for the 1.5 nm Pd particles, while for the 4.9 nm particles this peak was only a shoulder on the higher binding energy peak. From these data calculations of the total CO coverage, θ_T, at saturation for both states, with surface areas based on TEM measurements, revealed an almost two-fold increase in θ_T as the particle size decreased from 4.9 to 1.5 nm. The saturation surface coverage by the first adsorption state, however, was approximately independent of particle size. The increase in θ_T (with decreasing particle size) was due to an increase in the saturation coverage by the second adsorption state. This increase was rationalized by the assumption that more than one CO molecule could adsorb on highly unsaturated Pd atoms (e.g., at corners and edges of Pd particles), and smaller particles contain a larger proportion of these surface atoms than larger particles. For Pd, analogies between this latter mode of CO adsorption and the behavior of metal carbonyl clusters have been discussed by Ertl (50).

The variation of the heat of CO adsorption, ΔH_{CO}, with surface coverage, θ, was estimated from TPD spectra of 1.5 and 6 nm particles of Pd on Al_2O_3. This is shown in Figure 21 and compared with data from various bulk single-crystal Pd planes. The initial heat of adsorption (ΔH_{CO} at zero coverage) was approximately the same for both the supported Pd particles and single crystal Pd (with the exception of the {110} plane). Thus it was concluded that the energetics of CO adsorption (in the first adsorption state) are independent of the Pd particle size.

Having demonstrated the insensitivity of CO adsorption to the Pd particle size, the authors investigated the oxidation of CO over the Pd/Al_2O_3 samples. The reactants, CO and O_2, were continuously leaked into the vacuum chamber to maintain predetermined partial pressures, and the steady partial pressure of CO_2 produced was measured. This allowed determination of the number, N, of CO_2 molecules produced per surface Pd atom (as titrated by the adsorption of CO in the first adsorption state) per second. This was done as a function of temperature and fixed pairs of O_2 and CO pressures. Quantitatively, the dependence of this turnover frequency on Pd particle size at two selected reaction temperatures (445 and 518 K) and constant partial pressure conditions

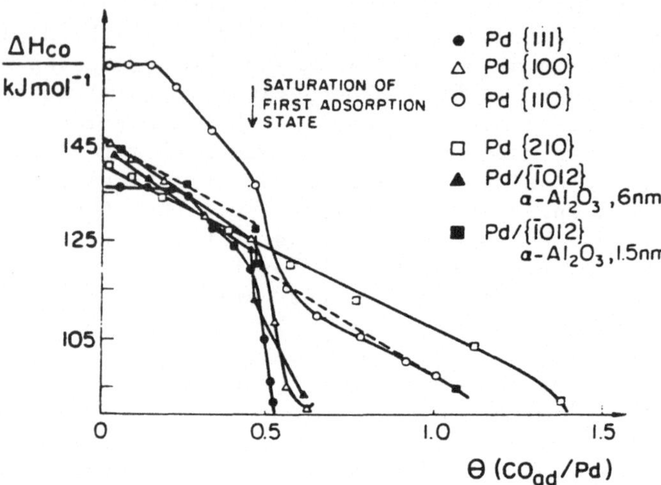

Fig. 21. Differential heat of adsorption versus CO coverage for Pd/Al_2O_3 model samples compared to data for single crystal Pd. Reprinted with permission from (49).

(P_{O_2}/P_{CO} = 1.1) is shown in Figure 22. At the lower temperature, N exhibited no dependence on particle size. At 518 K, however, N was independent of size down to ca. 5 nm, but nearly tripled for 1.5 nm particles. The interpretation of these data was that the rate determining step in the oxidation of CO over Pd at 445 K was a surface reaction between adsorbed CO and oxygen species. Evidently, the surface coverages by these species are independent of the Pd particle size (at constant temperature and pressure conditions). At higher temperatures (e.g., 518 K), the rate determining step becomes the adsorption of CO. The metal atoms at corners and edges of Pd particles were argued to be more accessible to collisions by gas phase molecules than metal atoms in flat surfaces, thereby explaining the observed increase in N with decreasing Pd particle size.

It is important to note that Ladas et al. compared their reaction kinetics data obtained over model supported catalysts at low pressure with the kinetics of CO oxidation over a high surface area 5% Pd/SiO_2 catalyst near atmospheric pressure (51). This comparison is shown in Figure 23. In spite of the near seven orders of magnitude difference in total pressure, the dependence of the rate on the CO and O_2 partial pressures exhibit distinct similarities. Indeed, the value of the turnover number over the model Pd/Al_2O_3 sample and the high surface area Pd/SiO_2 catalyst at identical P_{O_2}/P_{CO_2} is different by only a factor of 2. This

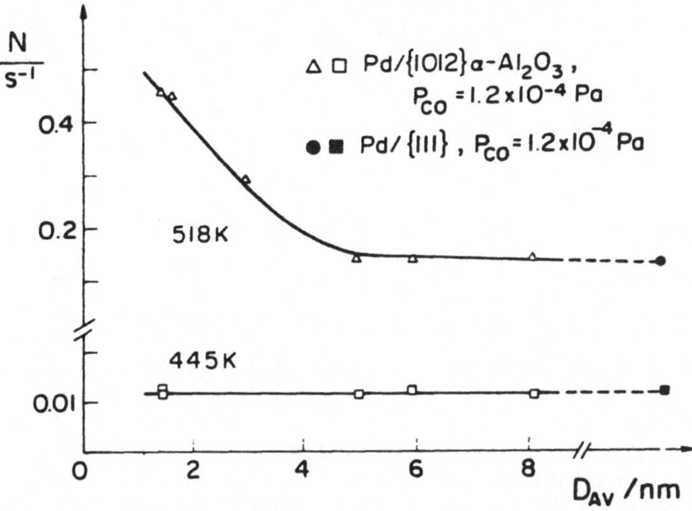

Fig. 22. CO oxidation turnover frequency versus average particle size for Pd/Al_2O_3 model catalysts at two reaction temperatures. Conditions were P_{O_2}/P_{CO} = 1.1 and P_{CO} = 1.2x10^{-4} Pa. Also included is single crystal data for Pd {111}. Reprinted with permission from (49).

demonstrates that model supported metal samples can, in fact, be used to "model" high surface area supported metal catalysts.

In another reaction study, the hydrogenolysis of methylcyclopentane (MCP) over model Pt/Al_2O_3 samples was investigated by Glassl et al. (52). The product selectivity for this reaction was found to be a function of the Pt particle size. Particle size was varied by a combination of evaporation of different Pt film thicknesses and various treatments in hydrogen and oxygen (53). Size, shape, and size distribution of the platinum crystallites were characterized by TEM. Two types of supports were prepared: "Type A" supports were prepared by reactive evaporation of Al metal in low pressure oxygen and "Type B" supports were prepared by anodization of high purity Al foil, followed by air drying at 400 K.

Reaction measurements were performed over an 80 cm^2 sample in a static reactor operating at atmospheric pressure and temperatures near 550 K. Product distributions were measured by sampling the gas phase followed by gas chromatographic analysis. In all runs 2-methylpentane (2-MP), 3-methylpentane (3-MP), and n-hexane (n-H) were the principal reaction products. The measured product

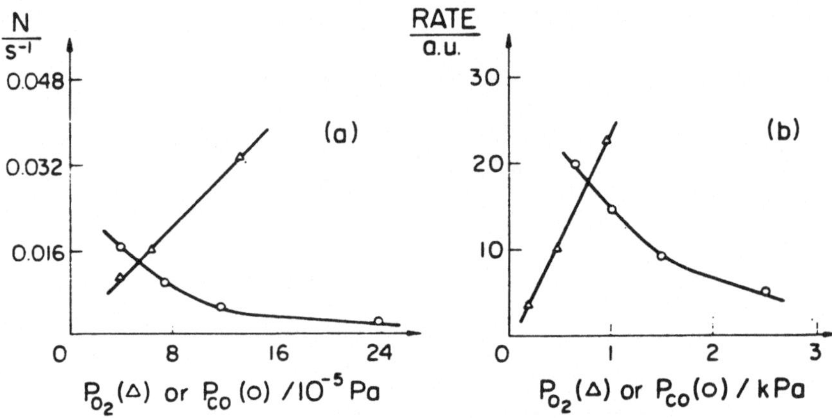

Fig. 23. Comparison of the pressure dependence of the turnover frequency (CO oxidation) (a) at low pressure over Pd/Al_2O_3 model supported catalysts with (b) high pressure over Pd/SiO_2 powder catalysts. In (a), T = 448 K, (Δ) P_{CO} = $6.7x10^{-5}$ Pa, and (o) P_{O_2} = $3.6x10^{-5}$ Pa. In (b), (Δ) T = 446 K, P_{CO} = 1 kPa, and (o) T = 427 K, P_{O_2} = 0.9 kPa. Reprinted with permission from (49).

distributions revealed a strong particle size dependence for the supported platinum particles. In particular, for Type A catalysts the proportion of 2-MP in the product gas increased from 44% at a mean Pt diameter of 1.5 nm to more than 70% at 4.4 nm. This occurred at the expense of n-hexane which decreased from 40% to 9% for the same Pt particle size range. The content of 3-MP remained essentially constant. The observed changes in product selectivity with increasing particle size were explained in terms of alternate reaction pathways dependent on the nature of the adsorbed reaction intermediate. For small particles, diadsorbed intermediates bound to atoms of low-coordination undergo statistical, or nonselective, scission of the cyclopentane ring. For larger crystallites, steric factors prevent the tertiary carbon atom of MCP from bonding to the higher-coordination sites. In this case a tetraadsorbed species was proposed to undergo selective ring opening to form 2-MP and 3-MP, i.e., n-hexane formation is inhibited. Justification for this model was obtained by dividing the Pt particle size distribution into three arbitrary size ranges and fitting the experimental product distributions for each Pt/Al_2O_3 sample to contributions from Pt particles in each size range. The results of these calculations are given in Table 2 for Type A catalysts. It was found that the decrease in n-H production with increasing particle size was most pronounced

Table 2. Product distributions for different particle size ranges calculated by fitting the experimental data

Diameter range (nm)	Percentage n-H	Percentage 3-MP	Percentage 2-MP	$\frac{2\text{-MP}}{\text{n-H}}$
<2	46.2	14.8	39.0	0.844
2-3	18.1	20.5	61.4	3.39
>3	8.6	19.5	71.9	8.36

Reprinted with permission from (52).

in the 2-3 nm size range. It is in this same size region that the fraction of edges and corner atoms to the total number of surface atoms changes sharply (54).

It was further noted that for a given Pt particle size, the selectivity to n-H was greater for samples prepared with Type-B Al_2O_3 supports. Accordingly, it was proposed that the acidity of a Type B support was greater than that of a Type A support, and that the increased production of n-H over Pt on the former support was due to a bifunctional reaction involving both Pt and Al_2O_3. Indeed, the behavior of a Type B support could be converted to that of a Type A support by exposure to NH_3 (since NH_3 would be expected to neutralize the acid sites on Al_2O_3).

5. CONCLUDING REMARK

A combination of techniques involving electron microscopy, electron spectroscopy, vibrational spectroscopy and chemical methods can be used to characterize metal particles on low surface area, geometrically flat supports. Indeed, such samples have been shown to model the properties of metal particles on high surface area (porous) supports. A powerful methodology for investigating the properties of supported metal particles is to use a variety of techniques to study the metal particles on high and low surface area supports, as described in parts I and II of this series, respectively.

ACKNOWLEDGEMENT

We wish to acknowledge the financial support of the Henry and Camille Dreyfus Foundation and of a Kodak Fellowship which allowed work on this manuscript by JAD and GBR, respectively. We also acknowledge funding from Chevron Research which aided in the preparation of this paper.

REFERENCES

1. Thomas, G., and Goringe, M.J., "Transmission Electron Microscopy of Materials", John Wiley & Sons, Inc., New York (1979).
2. Delannay, F., Catal. Rev.-Sci. Eng. 22(1), 141 (1980).
3. Baker, R.T.K., and Harris, P.S., Controlled Atmosphere Electron Microscopy, J. Phys. & Sci. Instrum. 5, 793 (1970).
4. Baker, R.T.K., CAEM of Gas-Solid Interactions, "CRC Crit. Review in Solid State Sci.", 345, Aug. (1976).
5. Ponsalle, L., Wiebel, G., Debut, D., and Matte, J.C., in "Proceedings 7th Int. Congr. on EM", Grenoble, Vol. 2, Société Francaise de Microscopie Electronique, Paris (1970), 355.
6. Carlson, T.A., "Photoelectron and Auger Spectroscopy", Plenum, New York (1975).
7. Delgass, W.N., Hughes, T.R., and Fadley, C.S., Catal. Rev. 4, 179 (1970).
8. Menzel, D., J. Vac. Sci. Technol. 12, 313 (1975).
9. Brundle, C.R., Surface Sci. 48, 99 (1975).
10. Kirtley, J., in ACS Symposium Series, Vol. 137 (Bell, A.T. and Hair, M.L., eds., Vibrational Spectroscopy of Adsorbed Species (1981)), p. 217.
11. Weinberg, W.H., in "Inelastic Electron Tunnelling Spectroscopy" (Wolfram, T., ed.), Springer-Verlag, Berlin (1978), p. 24.
12. Hansma, P.K., in "Inelastic Electron Tunnelling Spectroscopy" (Wolfram, T., ed.), Springer-Verlag, Berlin (1978), pp. 13, 186.
13. Ibach, H., Hopster, H., and Sexton, B., Appl. Surface Sci. 1, 1 (1977).
14. Dubois, L.H., and Somorjai, G.A., in ACS Symposium Series, Vol. 137, p. 163.
15. Thiel, P.A., and Weinberg, W.H., in ACS Symposium Series, Vol. 137, p. 191.
16. Allara, D.L., in ACS Symposium Series, Vol. 137, p. 37.
17. Greenler, R.G., J. Chem. Phys. 44, 310 (1966).
18. Ehrlich, G., Adv. Catal. 14, 255 (1963).
19. Schmidt, L.D., Catal. Rev. 9, 115 (1974).
20. Smutek, M., Cerny, S., and Buzek, F., Adv. Catal. 24, 343 (1975).
21. King, D.A., Surface Sci. 47, 384 (1975).
22. Madix, R.J., Adv. Catal. 29, 1 (1980).
23. Madix, R.J., CRC Critical Reviews in Solid State and Materials Sciences, 7, 143 (1978).
24. Chu, Y.F., and Ruckenstein, E., J. Catal. 55, 281 (1978).

25. McVicker, G.B., Garten, R.L., and Baker, R.T.K., J. Catal. 54, 129 (1978).
26. Chen, M., and Schmidt, L.D., J. Catal. 56, 198 (1979).
27. Chen, M., Wang, T., and Schmidt, L.D., J. Catal. 60, 356 (1979).
28. Wang, T., and Schmidt, L.D., J. Catal. 70, 187 (1981).
29. Wang, T., and Schmidt, L.D., J. Catal. 71, 411 (1981).
30. Tauster, S.J., Fung, S.C., and Garten, R.L., J. Amer. Chem. Soc. 100, 170 (1978).
31. Tauster, S.J., and Fung, S.C., J. Catal. 55, 29 (1978).
32. Baker, R.T.K., Prestridge, E.B., and Garten, R.L., J. Catal. 56, 390 (1979).
33. Baker, R.T.K., Prestridge, E.B., and Garten, R.L., J. Catal. 59, 293 (1979).
34. Vannice, M.A., and Garten, R.L., J. Catal. 56, 236 (1979).
35. Vannice, M.A., and Garten, R.L., J. Catal. 63, 255 (1980).
36. Gajardo, P., Gleason, E.F., Katzen, J.R., and Sleight, A.W., in "Proceedings 7th Int. Congr. on Catalysis, Tokyo, 1980", Paper No. E1.
37. Meriaudener, P., Ellestad, H., and Nacroche, C., in "Proceedings 7th Int. Congr. on Catalysis, Tokyo, 1980", Paper No. E2.
38. Taturchuk, B.J., and Dumesic, J.A., J. Catal. 70, 308 (1981).
39. Taturchuk, B.J., and Dumesic, J.A., J. Catal. 70, 323 (1981).
40. Taturchuk, B.J., and Dumesic, J.A., J. Catal. 70, 335 (1981).
41. Morup, S., Dumesic, J.A., and Topsoe, H., Appl. Mössbauer Spectrosc. 2, 1 (1980).
42. Takasu, Y., Unwin, R., Tesche, B., Bradshaw, A.M., and Grunze, M., Surface Sci. 77, 219 (1978).
43. Evans, H.E., Bowser, W.M., and Weinberg, W.H., Surface Sci. 85, L497 (1979).
44. Kroeker, R.M., Kaska, W.C., and Hansma, P.K., J. Catal. 57, 72 (1979).
45. Kroeker, R.M., Kaska, W.C., and Hansma, P.K., J. Catal. 61, 87 (1980).
46. Evans, H.E., and Weinberg, W.H., J. Am. Chem. Soc. 102, 872 (1980).
47. Dubois, L.H., Hansma, P.K., and Somorjai, G.A., Appl. Surface Sci. 6, 173 (1980).
48. Yates, J.T., Jr., Duncan, T.M., Worley, S.D., and Vaughan, R.W., J. Chem. Phys. 70, 1219 (1979).
49. Ladas, S., Poppa, H., and Boudart, M., Surface Sci. 102, 151 (1981).
50. Ertl, G., J. Vacuum Sci. Technol. 14, 435 (1977).
51. Cant, N.W., Hicks, P.C., and Lennon, B.S., J. Catal. 54, 372 (1978).
52. Glassl, H., Hayek, K., and Kramer, R., J. Catal. 68, 397 (1981).
53. Glassl, H., Kramer, R., and Hayek, K., J. Catal. 68, 388 (1981).
54. van Hardefeld, R., and Hartog, F., Surface Sci. 15, 189 (1966).

PECULIAR ASPECTS OF HETEROGENEOUS NUCLEATION AND GROWTH PROCESSES RELATED TO METAL SUPPORTED CATALYST

A. MASSON

Laboratoire Physico-chimie des Surfaces (CNRS)
E.N.S.C.P., 11 rue P. et M. Curie
75231 PARIS CEDEX 05
FRANCE

1. INTRODUCTION

It seems obvious now, that metal thin film obtained by condensed vapor on insulator substrate, is a good model for heterogeneous catalysis (1), (2). We don't want to expose here an exhaustive approach of heterogeneous nucleation and growth mecanisms, but only underline the new experimental facts and theoretical concepts studied in the early seventies and which could help to the understanding of the behaviour of metal support catalysts. Reader who wants to go deeply in the theoretical aspects of heterogeneous nucleation must follow the authoritative books of HIRTH and POUND (3), MATTHEWS (4) and finally the up to date one's of LEWIS and ANDERSON (5).

GENERAL FRAMEWORK. It is now classic after BAUER (6) to

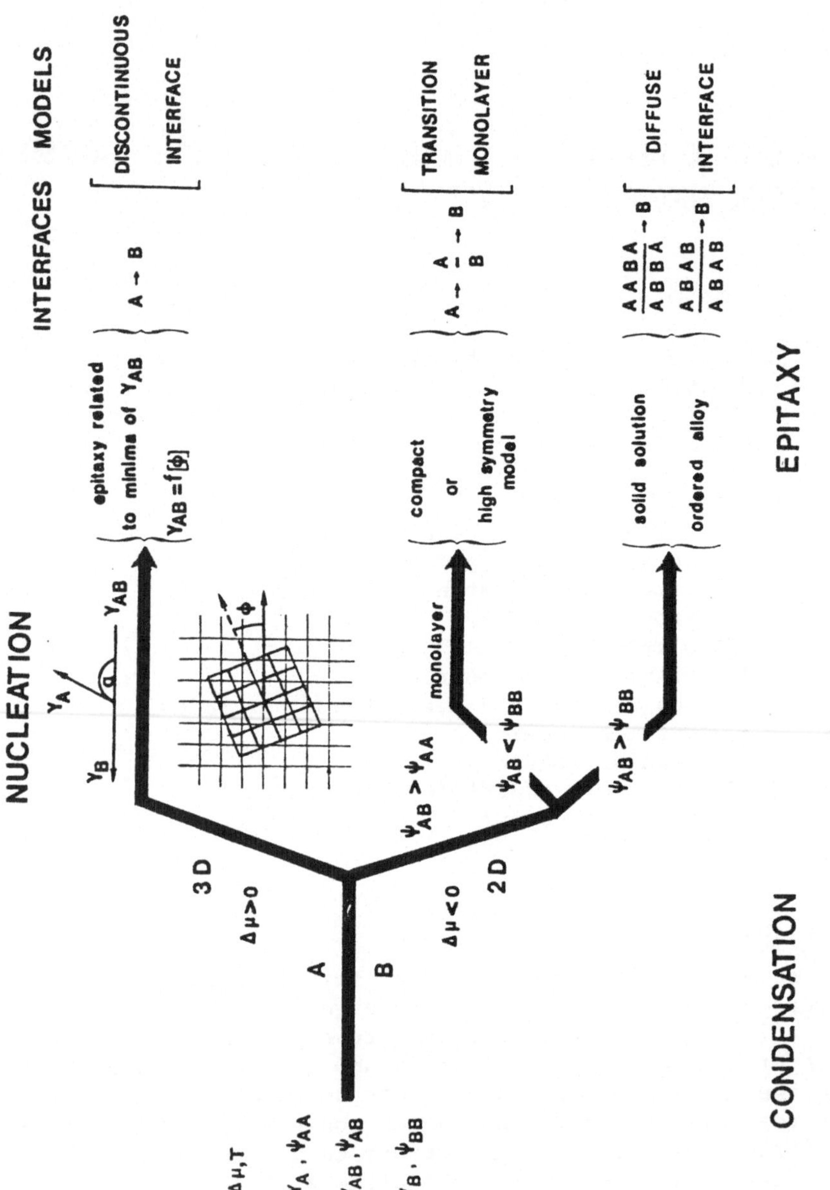
NUCLEATION
INTERFACES MODELS
EPITAXY
CONDENSATION
$\Delta\mu, T$
γ_A, ψ_{AA}
γ_{AB}, ψ_{AB}
γ_B, ψ_{BB}
A
B
3D
$\Delta\mu > 0$
2D
$\Delta\mu < 0$
γ_B
γ_A
γ_{AB}
ϕ
$\psi_{AB} > \psi_{AA}$
$\psi_{AB} < \psi_{BB}$
$\psi_{AB} > \psi_{BB}$
monolayer
epitaxy related to minima of γ_{AB}
$\gamma_{AB} = f[\phi]$
compact or high symmetry model
solid solution
ordered alloy
A → B
DISCONTINUOUS INTERFACE
A → A/B → B
TRANSITION MONOLAYER
AABA/ABBA → B
ABAB/ABAB → B
DIFFUSE INTERFACE

FIG. 1

describe the weak metal support interaction model, typical of metal insulator interaction, in a general framework, schematically represented in fig. (1) on which the different growth processes are reported.

In the case of 3 D nucleation process the "observable" generally appears as a macroscopic density ρ_o of particles and a theory of film growth should be able to describe the following phenomena

1) Condensation and nucleation on the substrate
2) Saturation of particle density
3) Coalescence of particles
4) Epitaxy : Mutual orientation of deposit and substrate.

We shall only give the "milestones" of the given A.S.I. talk, on this part, but on the contrary, emphasize on the spatio temporal variation of the density ρ_o of particles directled correlated to the mass transfer mecanism, a problem close to the sintering of catalyst. Moreover the epitaxy orientation will be exposed as a postnucleation phenomenon.

2. KEYNOTES FOR HETEROGENEOUS NUCLEATION

2.1 Thermodynamic approach

Universally described after VOLMER (7), FRENKEL (8) and HIRTH and POUND (3), the equilibrium of a population of adsorbed monomers of concentration C_1, with a collection of polymers of size i in concentration C_i is treated by the way of the action mass law of a chemical reaction $C_i \leftrightarrows iC_1$. A cluster of size i needing a work formation ΔG_i. The concept of critical nucleus of GIBBS (9) related to a maximum in ΔG_i for a minimum size i*, and the "cap shape" model of POUND enable us to explicite G_i* and i* only in terms of macroscopic parameters $\Delta\mu$, γ_i, α, respectively supersaturation of the vapor phase, surface energies and wetting angle.

Following the notation of LEWIS (5), we can write.

$$/1/ \quad G_1^* = \frac{A\,\gamma^3 f(\alpha)}{(kTLn\Delta\mu)^2} \quad \text{and} \quad i^* = \frac{B\,\gamma^3 f(\alpha)}{(kTLn\Delta\mu)^3} \quad \text{(in number of atoms)}$$

2.2 Time independant nucleation rate

The frequency of nucleation, is the rate at which critical nuclei become supercritical and give by growing the density of observable stable particles.

In the supersaturated adatoms population, it is a stochastic bimolecular reaction which is the driving force requisite to overcome the critical size i*. Since FRENKEL (8), it is transform in a formely equivalent process of diffusion in the i space (size space), according to a diffusion of particle of size i, owing to diffusion and under a force F deriving from a potential ΔG_i.

The continuity equation is then written:

$$\frac{\delta C(i,t)}{\delta t} = -\frac{\delta J}{\delta i} \qquad \boxed{2}$$

with $J = - D_i \text{ grad } C(i,t) - \mu C(i,t) F$

and $\mu = \frac{D_i}{kT}$ mobility of particle of size i

$F = \frac{\delta \Delta G_i}{\delta i}$ D_i diffusion coefficient

$C(i,t)$ number of particles of size i at time t

The steady state solution of equation /2/ being:

$$J = ZWC_i^* = A \exp \frac{(G_i^* + E_a - E_d)}{kT}$$

with Z a non equilibrium factor or Zeldovitch factor

W a probability factor for atomic impact

A a frequency factor

E_a activation energy of adsorption of adatom

E_d activation energy for diffusion of adatom

2.3 Atomistic model

We found with formula /1/ that i* is generally of subatomic size, in order that WALTON (10) has settled a statistical mechanics model.

WALTON assume that the binding energy of a cluster of size i is written as $E_i = iU_1 - U_i$ which is the decreasing in potential energy of forming the cluster from i adatoms, U_i being the total energy of particles in a particular distribution Q_i.

The partition function is then calculated and the Walton equation of equilibrium conditions for N_i clusters is written as :

$$\frac{N_i}{N_o} = \left(\frac{n}{N_o}\right)^i \exp \frac{E}{kT}$$

with N_o = number of sites on the substrate

$N = \Sigma_i N_i$ = number of clusters with $N_o >> N$

n_1 = density of single atoms.

The rate equation can be explicit as :

$$J = A \exp \frac{(E_i^* + (1+1) E_a - E_d)}{kT}$$

with E_i^* = binding energy of cluster of critical size i*

E_a = adsorption energy of adatom

E_d = activation energy of migration of adatom

A = frequency term

LEWIS (5) has shown that the two models can be finally put together.

3. MASS TRANSFER MECHANISMS

If, as we have just seen, the atomic parameters are well

suited to describe the time independant nucleation rate and if the ZINSMEISTER (11) kinetic approach for time dependant solution is still well adapted, things become more complicated when in the last sixties new experimental evidences (12, 13) rise the question of a possible new approach of mass transfer process.

New insights. The in-situ electron microscope experiment of BASSET (14) describes for the first time, the "liquid like behaviour" of a distribution of particles obtained by a condensation of vapor of gold onto a molybdenite surface. At the same period, the morphological change with time of a distribution of gold particles deposited on amorphous SiO is studied by SKOFRONICK (12). The problems open for discussion were a possible intrinsic mobility of tiny metal particles and a dynamical approach of coalescence phenomenon.

Before explaining the different models proposed, we need a good characterisation of a distribution of particles.

Such studies on distribution have taken a new impulse after the introduction of automatic image analyser, keeping in mind that generally the "observable" is an electronic micrography.

3.1 Distribution functions

A collection of particles is well described by the following parameters.

- the mean density ρ_o per surface unity on particles of size $i\rho_o = \Sigma_{i=1}^{i=\infty} n_i$
- the histogram of size i, is a statistical distribution of sizes, the moment of order 1 represents the mean value $\langle d \rangle$
- the positive square root of the moment of order 2 is the dispersion $\bar{\sigma}$ with $\bar{\sigma}^2 = \langle d^2 \rangle - \langle d \rangle^2$
- the radial distribution function P(r) is easily defined by measurements of all interparticle distances N(r, dr) so that $N(r, dr) = \rho_o P(r)\ 2\pi r dr$
- finally, the distribution of nearest neighbours distances is $W(r, dr) = 2\pi\rho_o\ e^{-\pi\rho r^2}\ dr$

SCHMEISSER (15) and ZANGHI (16) have reported exhaustive experimental investigations of all these functions leading to the following main conclusions.

3.2 Coalescence and migration of particles

a) Radial distribution function

By doing annealing a low temperature of thin film of gold particles condensed on alkali halide surfaces. ZANGHI (17) has found a variation of radial distribution functions shown on fig (2)

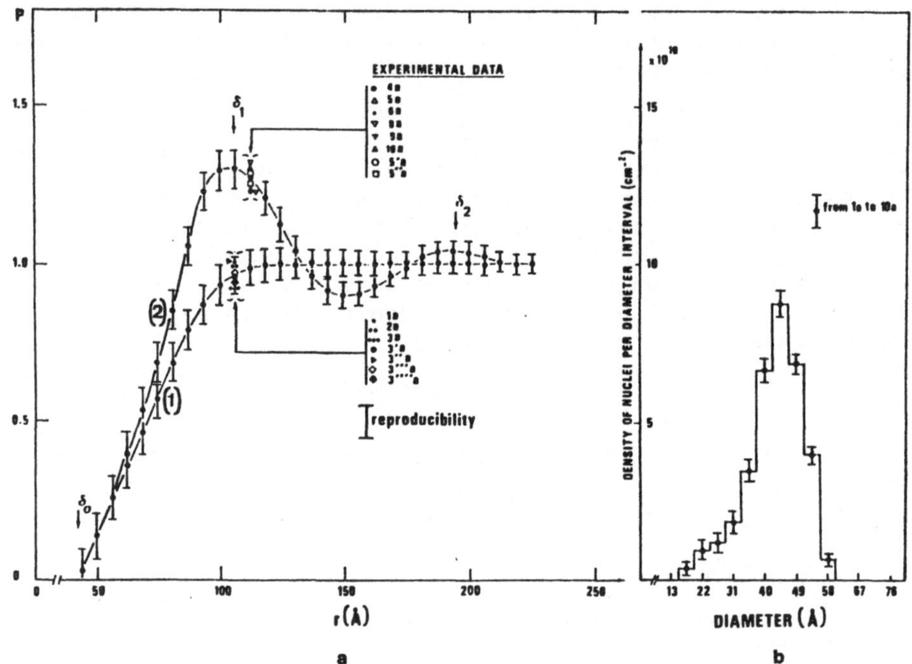

FIG. 2 Radial distribution function of gold particles, on (100) alkali halide substrate. From (17).

It is clear that, as the system is mass conservative, the oscillations on P(r) and the increase of the first maximum is only explained by an intrinsic mobility of particles. It was a "a posteriori" demonstration of the same conclusion given by MASSON et Al (18) by analysing of diffusion profiles.

b) Non-random distribution

Taken from the work of SCHMEISSER, the nearest neighbours distances distributions w(r, dr) are shown of fig. (3).

Those experimental distributions are compared with theoritical one for random distribution of particle, and it is clearly demonstrated by the systematic shift towards the great size that such a shift discloses a correlation between particles, the correlation can be related to an atomic diffusion process during nucleation as described by VENABLES (19) or attributed to a migration of small particles (15), (17).

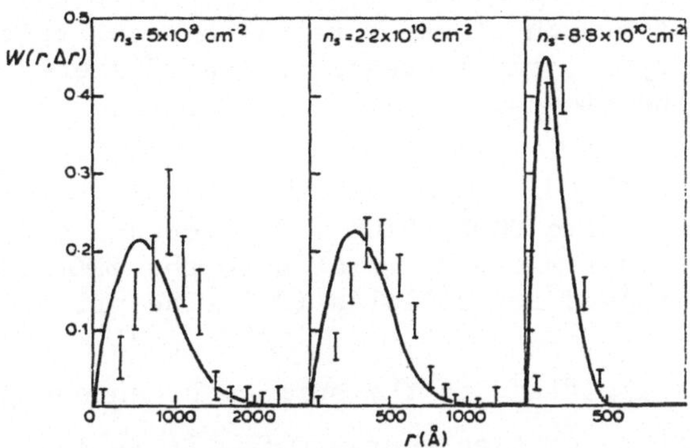

FIG. 3 Nearest neighbours distances distribution of gold particles on NaCl (100) surface. From (15).

c) Size distribution

The simple distribution to follow with time is the size distribution. Taken also from SCHMEISSER such a variation is shown on fig. (4)

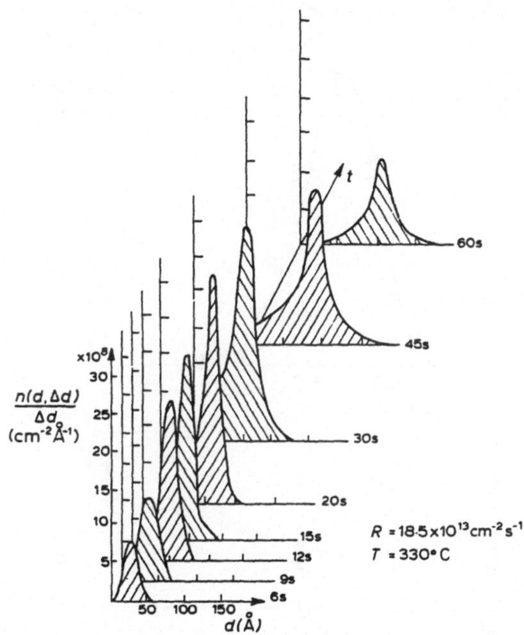

FIG. 4 Variation of cluster size-distribution with time for gold particles on Nad substrate. From (15).

This is a good approach to reveal the coalescence phenomena which is underlined by the change in shape of this kind of distribution (time equals 45 s. on fig. (4)). It is also clearly shown that theory of growth of the particle has to explain the decreasing of density ρ_o, the increase in mean diameter <d> and the increase in dispersion σ.

3.3 Dynamic models

For the first time SKOFRONICK (12) proposes a dynamic model in 1967. The formalism can be related to the SMOLUKOVSKY coagulation model developped for colloïds (20). The principal hypothesis are :

- particles are fixed on the surface by an energy E_o
- if a particle wins an energy $E>E_o$, it is able to leave the well of potential and glides on the surface with a constant speed
- all impact of particle gives a coalescence without energy of activation, the shape of the crystallite being invariant.

The description of the experimental hystogram is made by a gamma function with two parameters as seen on fig. (5)

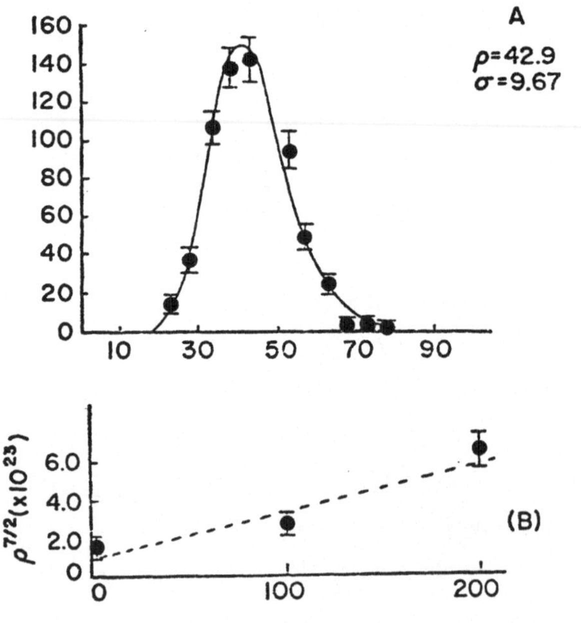

FIG. 5 Fitting of histogram of gold particle on gold on Sio amorphous substrate by two parameters gamma function and dispersion variation. From (17).

and the two solutions of the problem are given under two explicit equations.

- The first equation describes the variation of the mean radius R versus time so that

$$|R(t)^{7/2} - R(0)^{7/2}| = 2.8\ \gamma t \qquad /3/$$

with $\gamma = \frac{3M}{2\pi\rho} \left|\frac{3kT}{\rho}\right|^{1/2} \exp(-E_o/kT)$

M the mass of crystallite per cm^2, ρ the specific mass of crystallite

- The second equation describes the variation of dispersion σ versus time

$$R^2(t) - R^2(0) = 7.1|\sigma^7(t) - \sigma^2(0)|$$

Equation /3/ gives a value of E_o in the case of gold on amorphous SiO, of E_o = 1.5 eV.

METOIS (21) presents a mechanism putting together Smolukovsky model and the kinetic equations of Frenkel (11).

The main assumptions of METOIS are :

- definition of a probability factor of coalescence
- non mobility of particles after impact
- the coalescence induces a transformation between (111) oriented particle to (100), recrystallisation phenomenon is still open for discussion.

The fitting of histogram and experimental model is given in figure (6).

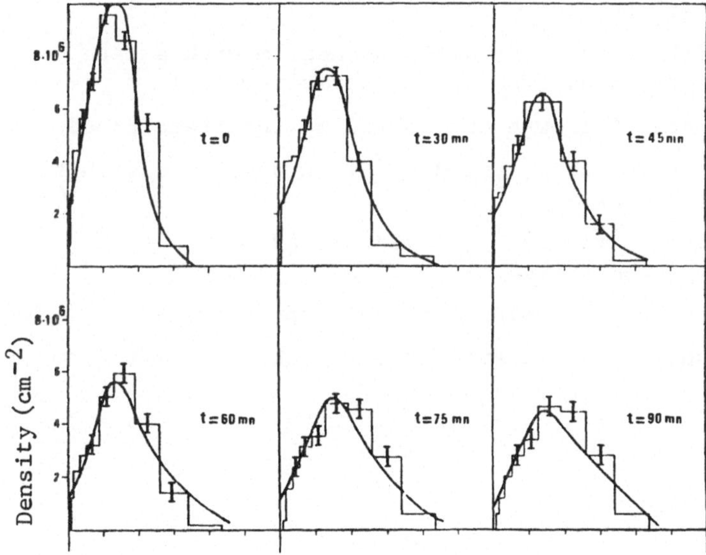

FIG. 6 Fitting of experimental and theoritical histograms. From (21)

In the opposite of Oswald ripening model which predicts a tendancy to lead to an unique crystal. METOIS and SKOFRONICK approaches lead to an enlarged distribution.

Other dynamic models have been described until now, RUCKENSTEIN and PULVERMACHER (22) for instance.

3.4 Oswald ripening

If the dynamic model is well adapted for low temperature annealing of thin film - at high temperature - the Oswald ripening is better suited.

B.K. CHAKRAVERTY (23) has given a solution for the time dependant distribution of particles, with a mass-transfer mechanism founded on atomistic process. For a given temperature an equilibrium state is taken for a density of particles and a population of adatoms given by the Gibbs-Thomson equation. The cluster are growing by atomic impact.

We explicit the master equation of LIFSHITZ (24) because it is a very general equation taken by many authors.

F(R, t) the distribution function of size for cristallite of size R at time t, and the rate of growth $\delta F/\delta t$ are related by the continuity equation :

$$\frac{\delta F(R,t)}{\delta t} + \frac{\delta}{\delta R}\left[F(R,t) . \frac{\delta F}{\delta t}\right] = 0$$

such an equation has been solved by CHAKRAVERTY, for two choices of the rate of growth :

a. a diffusion controlled mechanism with a rate of growth taken from B.C.F. formalism (25)

b. a rate of growth controlled by interface process

For the first case, the density is decreased with time as $F(R,t) \, N \, \frac{1}{t}^{3/2}$

the mean radius is increasing as $R(t) = R_o \, (t/t_s)^{1/4}$

with an increasing for the dispersion as $\sigma_t = A\, R(t)$

The results are presented for theoretical histogram on fig. (7).

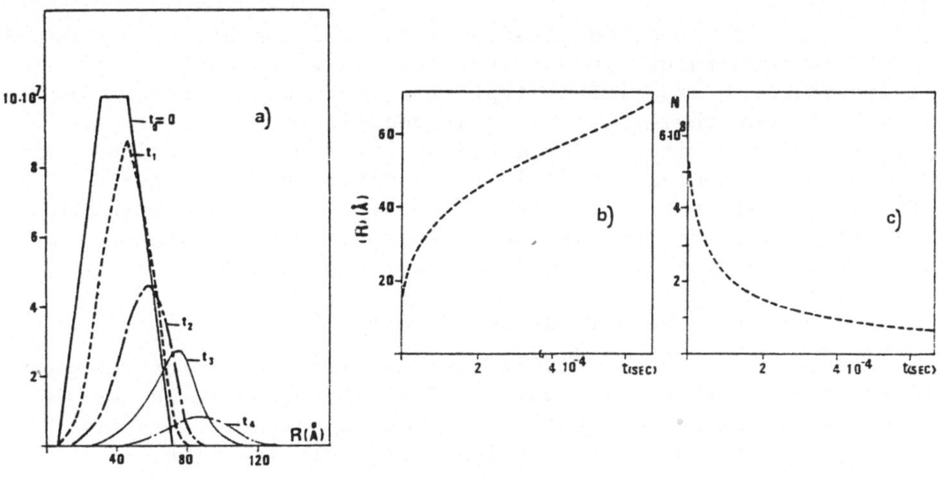

FIG. 7 Theoretical variation of distribution functions for an Oswald ripening mechanisms.
a) Histogram - b) mean radius - c) dispersion.
From (23).

4. EPITAXY POST NUCLEATION PHENOMENON

A peculiar aspect of condensed thin film is shown on fig. (8).

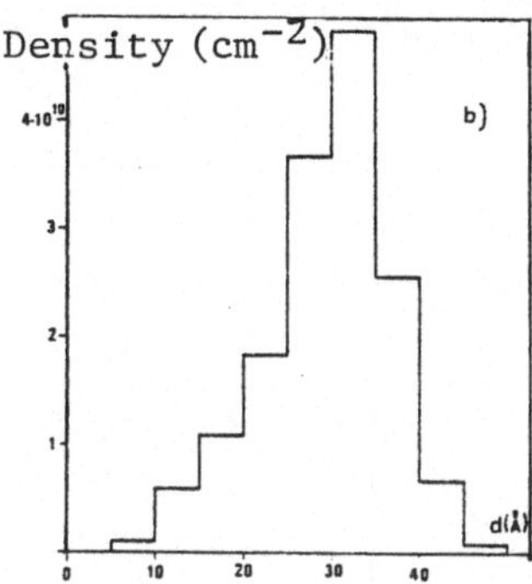

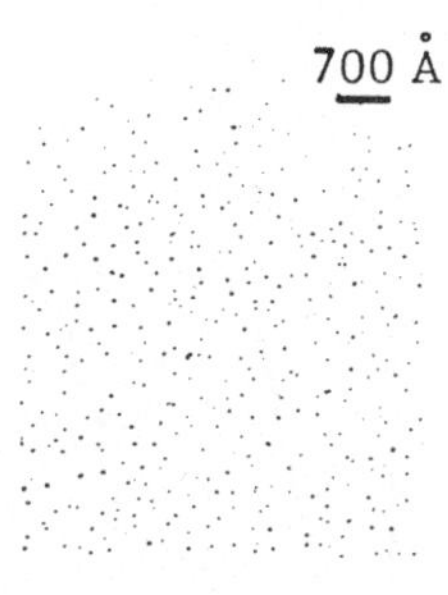

FIG. 8 Epitaxy post nucleation phenomenon.
Nucleation without epitaxy. From (18).

It is possible to obtain nucleation without orientation, an annealing at low temperature in a conservative system leads to epitaxial growth. Such experiments typical of weak metal support interaction, rule out the necessity of a specific structure for the critical nuclei as it was suggested by the Rhodin and Walton theory of epitaxial nucleation (10). It is clear that nucleation is neither necessary nor sufficient to obtain an oriented thin film. So it is became possible to imagine a progressive epitaxy mecanism, and described epitaxy as a post nucleation phenomenon.

Progressive epitaxy and interface models

After the work of BASSET (14) on electron microscope insitu experiments, REISS (26) was the first to propose the possibility of translation and rotation of a metal particle without the need of a huge amount of activation energy of migration. We underline the main hypothesis and the results are reported on fig. (9).

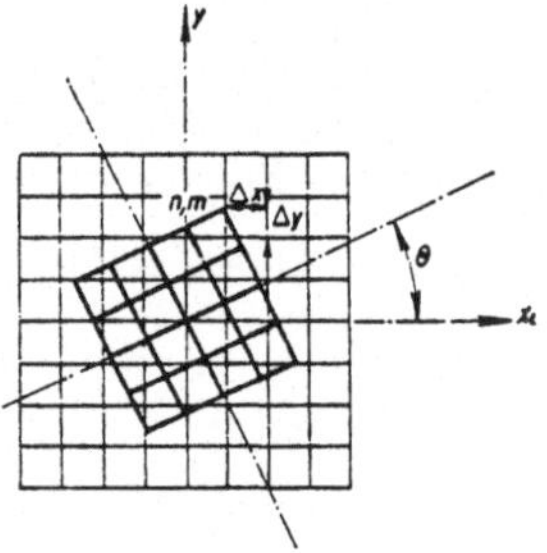

FIG. 9 Two-dimensional island on surface with a periodic potential of quadratic symmetry. From (18).

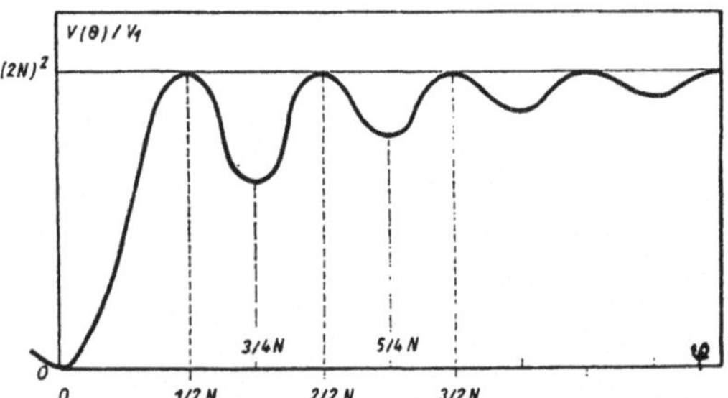

Potential energy of the island as function of misorientation. From (18).

A two dimensional quadratic cluster is laid down on a two dimensional substrate of the same symetry. The potential energy of rotation of the nucleus versus, the misfit of desorientation ϕ is written as a complementary diffraction curve, so that :

$$\frac{V(\phi)}{V_1} = (2N^2) \left[1 - \frac{\sin^2\phi}{\phi^2}\right]$$

with $(N + 1)^2$, the number of atoms in the cluster and $\phi = 2\pi N_\theta$ The translation can be also describe by the same kind of equation

The main conclusion is that the discrete orientations are possible, the largest corresponding to the minimum in the potential energy. Finally, the usual parallel epitaxy ($\theta = 0$) corresponds to the maximum in energy. A nucleus in that parallel orientation is difficult to shift.

At the same time, MASSON et Al. (18) get at this conclusion, by studying diffusion profiles, and electron diffraction patterns of gold thin film condensed on alkali halide surfaces.

A brownian migration model was settled by KERN et al. (27) on the assumption of a specific interface between the particle and the substrate.

Such a model is well described by the MOIRE imagies and related to the island model of MOTT for high angle boundary. In a more general and energetic approach, the particular problem of mutual orientation of crystals is well developped in FLETCHER (28).

Finally in 1974, POPPA et Al. (29) have observed directly in ultra-high vacuum electron microscope, the rotation of relatively large growing particles of gold on (100) oriented MgO substrate as it is shown on fig. (10).

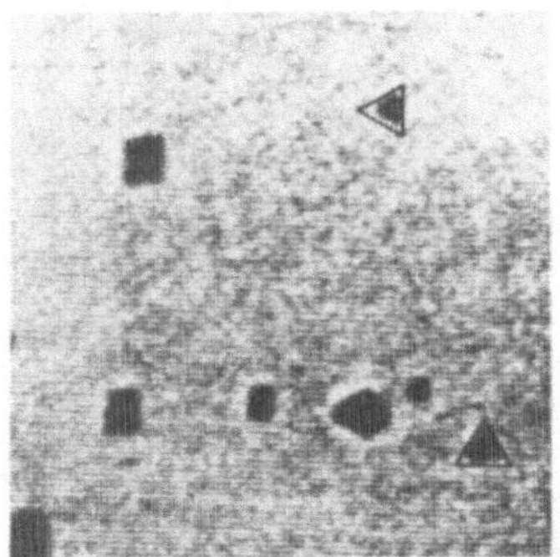

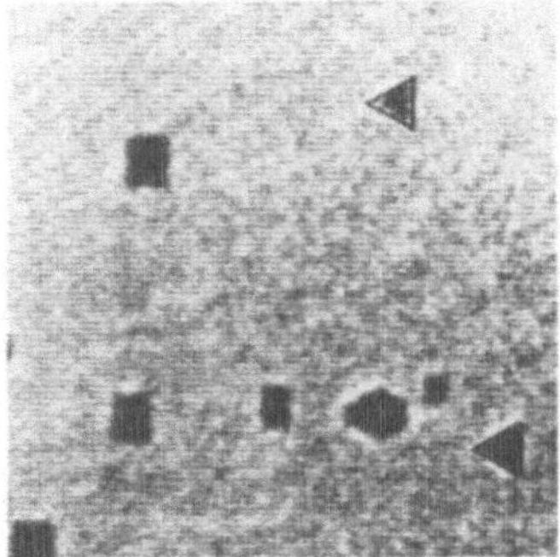

FIG. 10 In situ EM observation of the rotation of relatively large growing gold crystallites at high substrate temperature (T = 500° C). Different but equivalent epitaxial orientations on the (100) - oriented MgO substrate are assumed by the gold crystallites. From (4).

5. MISCELLANEOUS

As we have just discussed, the weak metal substrate interaction is well described by thin metal film model. We have emphasized the problem of mass transfer, because we think that it is correlated to the problem of ageing by sintering in the metal support catalysts. Moreover thin metal film have been taken for experimental model in many others physico-chemical investigations. The attractive problem of the variation of intrinsic electronic properties has been for example studied by X an UV spectroscopy by MASON (30) and TAKASU (31), a critical discussion has been recently given by CITRIN and WERTHEIM (3), on the special problem of energy shift and electron transfer correlation.

Metal clusters evaporated into reactive matrices at very low temperature seems to be a good new model for chemisorption studies (33).

The non bulk like structure is still an open problem for the high resolution electron microscopy (34), the growth mecanism and stability of such particles are not well explained. Recently, inert gaz evaporated methods have shown that those structures could be avoided (35).

Finally, the problem of equilibrium form and faceting phenomena, early studied by SUNDQUIST (36) and WITTERBOTTON (37), could found a new impulse.

REFERENCES

1. Ladas, S., Poppa, H. and M. Boudart. Surf. Science 102 (1981) 583.
2. Kao, C.C, Tsai, S.C. and Y.W Chung. J. of Catalysis 73 (1982) 136.
3. Hirth, J. and G. Pound. Prog. Mat. Sci. Mac Millan N.Y. 11 (1963).
4. Matthews, J. Epitaxial Growth, Acad. Press N.Y. (1975).
5. Lewis B. and J.C. Anderson. Nucleation and Growth of Thin Film, Acad. Press Inc. (London) (1978).
6. Bauer, E. Z. Kristallogr. 110 (1958) 372.
7. Volmer, M. Kinetik der Phasenbildung Steinkopff Dresden (1939).
8. Frenkel, J. Kinetic Theory of Liquids, Oxford Press (1946).
9. Gibbs, J.W. Scientific Papers, Dover N.Y. (1961).
10. Walton, D. J. Chem. Phys. 37 (1962) 2182.
11. Zinsmeister, G. Thin Solid Films 2 (1968) 497.
12. Skofronick, J.G. and W.B. Phillips. J. Appl. Phys. 38 (1965) 304.
13. Masson, A. and R. Kern. J. Crystal. Growth. 2 (1968) 227.
14. Basset G.A. Condensation and Evaporation of Solids (Gordon and Breach N.Y. (1962) 599.
15. Schmeisser, H. Thin Solid Films 22 (1974) 83 (a and b).

16. Zanghi, J.C. Thesis Marseille (1975).
17. Zanghi, J.C, Métois, J.J. and R. Kern. Phil Mag. 29.5 (1974) 1213.
18. Masson, A, Métois, J.J. and R. Kern. Advances in Epitaxy Endotaxy. 4-6 Schneider (Ed.) Vol. II Leipzig (1971) 103.
19. Venables, J. in Matthews (ref. 4).
20. Smolukovsky, V.M. Physik Zeitscher XVII (1916) 585.
21. Métois, J.J., Zanghi, J.C., Erre, R. and R. Kern. Thin Solid Films 22.3 (1974) 331.
22. Ruckeinstein F. and B. Pulvermacher. J. Catalysis 29 (1973) 224.
23. Chakraverty, B.K. J. Phys. Chem. Sol. 28 (1967) 2401.
24. Lifshitz, I.M. and V.V. Slyozov. J. Phys. Chem. Solids 19 (1961) 35.
25. Burton, W.K., Cabrera, N. and F.C. Frank. Phil. Trans. Roy. Soc. London A 243 (1950)
26. Reiss, H. J. Appl. Phys. 39 (1968) 5045.
27. Kern, R, Masson, A. and J.J. Métois. Surf. Science 27 (1971) 483.
28. Fletcher, N.H. in Matthews (ref. 4).
29. Poppa, A. in Matthews (ref. 4).
30. Mason, M.G., Gerenser, L.J. and S.T. Lee. Phys. Rev. Letters 39 (1977) 288
31. Takasu, Y., Unwin, R., Tesche, B., Bradshaw, A.M. and M. Grunze. Surf. Science 77 (1978) 219.
32. Citrin, P.H. and G.F. Wertheim to publish.
33. Schmeisser, D., Jacobi, K. and D.M. Kolb. Appl. Surf. Science 11/12 (1982) 164.
34. Schabes-Retchkiman, P.S. and M.J. Yacaman. Appl. Surf. Science 11/12 (1982) 149.
35. Kimato, K. and I. Nishida. J. Phys. Paris (1977) C_2 195.
36. Sundquist, B.E. Acta Met. 12 (1964) 67.
37. Winterbottom, W.L. Act. Met. 15 (1967) 303.

FORMATION, ACTION, AND PROPERTIES OF CLUSTERS IN THE PHOTOGRAPHIC PROCESS

Erik Moisar

AGFA-GEVAERT AG., D-5090 Leverkusen, and
J.W.Goethe-University, Inst. of Appl. Physics, D-6000 Frankfurt

1. INTRODUCTION

For nearly 150 years clusters in the broad sense of structured aggregates attached to or embedded into a crystalline matrix (such as silver halides) play an essential role in the photographic process, even if their presence, their action, and their nature have not been recognized in the early days of photography. To-day their presence and action are fully recognized. Their nature and their mechanisms of formation and action, however, are still not completely understood and are discussed in several and controversial ways. It is the aim of the present paper to review some facts and theories related to the formation and the properties of these clusters. This treatment will be preceded by a short outline of the basic principles of the photographic process.(For details c.f. references 1 and 2).

2. BASIC PRINCIPLES OF THE PHOTOGRAPHIC PROCESS.

2.1. General Description

The light-sensitive entities used for storage and retrievial of optical information are micro crystals of silver halides embedded in gelatin matrix. If during exposure a micro crystal has absorbed a few photons, a so-called latent image speck will be formed. This is a cluster or an aggregate consisting of a few silver atoms. Its presence greatly changes the properties of the host crystal. If brought into a reducing medium (the developer) an exposed crystal bearing a latent image speck will be immediately reduced to metallic silver. An unexposed crystal, on the other hand, remains unaffected and can be removed later from the photographic layer by a solvent (the so-called fixing agent) such as thiosulfate. A photographic image, therefore, resembles a mosaic picture: Black stones are set in regions where light has fallen upon, and white stones correspond to unexposed areas. Each mosaic stone and, hence, each micro crystal in a photographic image is a binary information element bearing either the information "sufficient light" (black) or "no light" (white). The smallest detail of the object which can be resolved in the image is obviously limited by the size of the

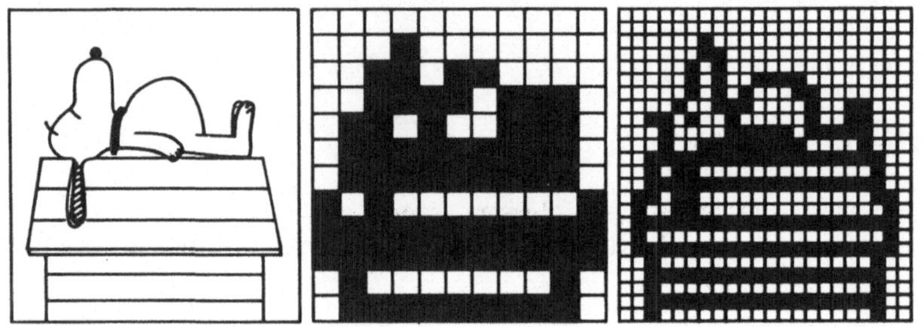

Fig.1. An object (left) is depicted with large (centre) or small (right) mosaic stones. The information capacity increases with decreasing size of the picture elements.

mosaic stones (Fig. 1). The same is true for photographic recording materials: The information storage capacity, the resolution, is ultimately limited by the size of the micro crystals.

There is another relevant effect related to crystal size. At a given light intensity and exposure time the average number of absorbed photons per crystal is proportional to the crystal volume. Hence, large crystals may collect a sufficient number of quanta in order to form latent image specks and thus to become developable, while smaller crystals do not. The latter, therefore, remain undeveloped unless the exposure intensity and/or exposure time is increased. Sensitivity, therefore, depends largely on crystal size (Fig. 2).

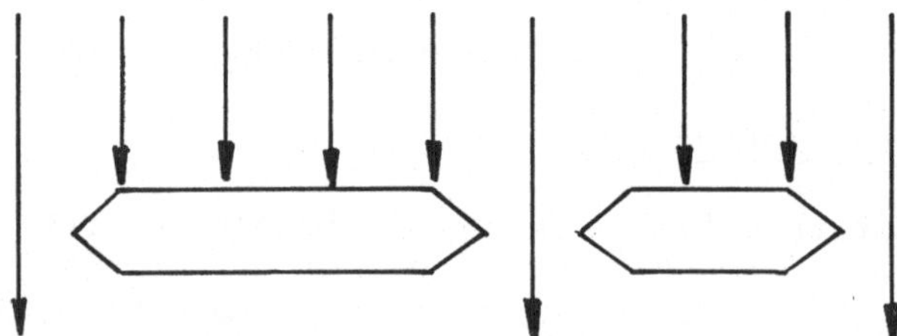

Fig.2. If e.g. four photons are required to form a latent image speck, at a given exposure level a large crystal could collect this essential number, while a small crystal absorbs less photons.

The latent image silver clusters which catalyze the rapid reduction of the whole crystal consist of a few silver atoms. Few means more than one (about four) and therefore more than one (again at least four) absorbed photons in a crystal are required in order to build up a latent image speck. This crucial process will be treated in detail later. So far, however, it then follows that at very low exposure levels none of the crystals have absorbed enough photons and none of them accordingly can be developed at all. If the number of developed crystals in a layer element is plotted versus the amount of light irradiated onto the layer (measured in photons per unit area, in lux seconds or whatever; preferentially in a logarithmic scale) a so-called characteristic curve is obtained which at lower exposures does not show any response. At a certain threshold value the ordinate value starts to raise, and at high exposure values a saturation corresponding to complete reduction of all crystals can be expected. In practice instead of the number of developed crystals the easily measurable optical density is used in the ordinate.

In a first approximation it is assumed that all micro crystals have exactly the same size. It is further assumed that each crystal becomes developable after absorption of exactly the same number (n^*) of photons. Even then the characteristic curve has a certain latitude which may surprise at first sight. This is due to the statistics of quantum incidence. If n^* is the number of absorbed photons required for the formation of a latent image silver speck and if q is the average number of photons absorbed per crystal, then the fraction P of all crystals having by chance absorbed n^* or more quanta (and thus becoming developable) is given according to Poisson's distribution law by

$$P = 1 - e^{-q} \sum_{n=0}^{n^*-1} \frac{q^n}{n!} \tag{1}$$

(fig. 3). In a second and more realistic approximation the constant number n^* is replaced by a mean value $\bar{n}^*$. There may exist crystals

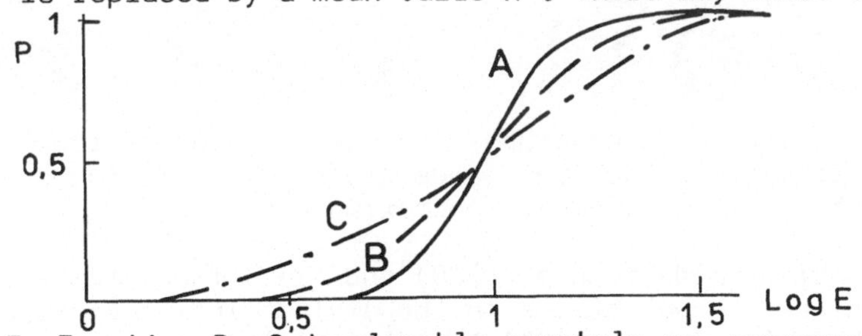

Fig.3. Fraction P of developable crystals vs. exposure.
Curve A: n^* and crystal size is constant
Curve B: distribution of n^* , size still constant
Curve C: distribution both of n^* and size.

which become developable after absorption of $n<\bar{n}^*$ photons, there may exist such which require $n>\bar{n}^*$ quanta and, of course, some of the crystals will acquire developability after absorption of just $\bar{n}^*$ photons. The respective n^* values are a measure of the sensitivity of the crystals, and thus a sensitivity spread among the individual crystals has been introduced. It is still assumed that all crystals are of uniform size, but compared to the first approximation now a characteristic curve of wider latitude is obtained. If we finally assume also a distribution of crystal sizes we have arrived at a very realistic model. As stated before, small grains absorb less, larger crystals absorb more photons and this leads then to a further increase in latitude or to a lower slope of the characteristic curve. This range describes the limits between the lowest and the highest intensities which can be recorded.

Two complex properties thus decisively influence the response of a photographic system towards exposure: crystal size and size distribution which are responsible for photon absorption, and the individual sensitivities of the crystals which determine, how many absorbed photons lead to the formation of a latent image speck.

2.2. Crystallography of the Silver Halides

The silver halides used in photographic systems are AgBr and AgCl or mixed crystals of AgBr and AgI or AgCl. AgI alone is not used. AgCl and AgBr belong to the cubic system, similar to NaCl. The equilibrium shape will be accrdingly a cube (Fig. 4). Like NaCl, however, AgBr appears also in a different shape which is an octahedron, bounded by (111) faces. This is due to a very different

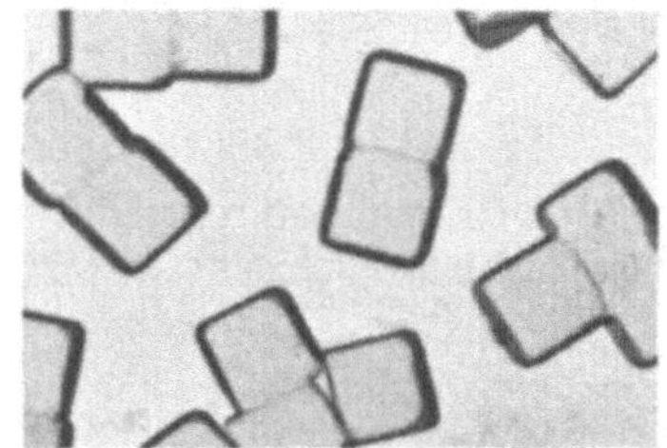
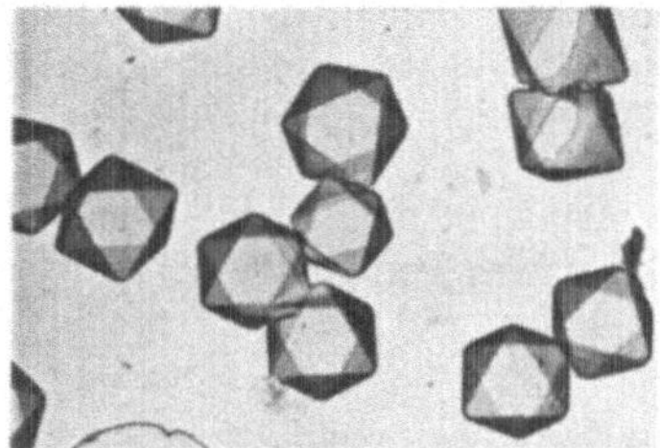

Fig.4. Electron micrographs (carbon replica) of regular AgBr crystals with (100) faces (left) or (111) faces (right). Crystal diametre approximately 0.8 μm.

adsorption of bromide ions at (100) and (111) faces respectively. At low bromide concentrations in the ambient medium bromide ions are adsorbed stronger to (100) than to (111) faces. At high bromide concentrations, however, this sequence is reversed: (111) faces adsorb more bromide than (100) planes (3). According to the Gibbs adsorption equation a higher coverage causes a larger decrease of the surface energy. Since a crystal in equilibrium tends to develop a shape where the total surface energy approaches a minimum, at very low bromide concentrations in the ambient phase AgBr displays a shape bounded by (100) faces - which is a cube - while at high bromide concentrations (111) faces predominate. The more polar AgCl develops, with some rare exceptions, only (100) planes.

In most of the commercial photographic materials the silver halide micro crystals differ in their appearance very much from the ideal cubic or octahedral shapes discussed so far (Fig. 5).

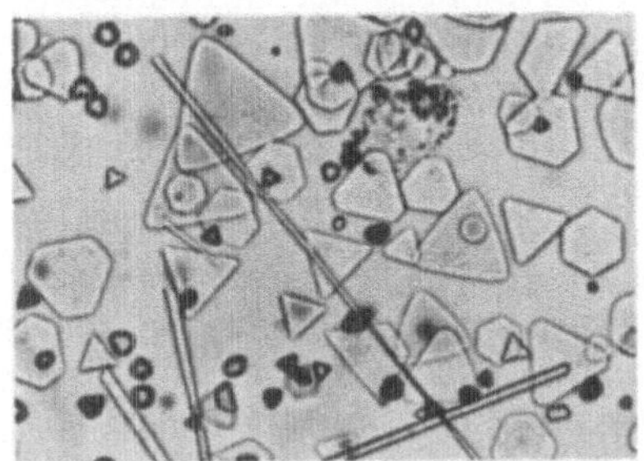

Fig.5. Electron micrograph of twinned AgBr crystals (carbon replica).

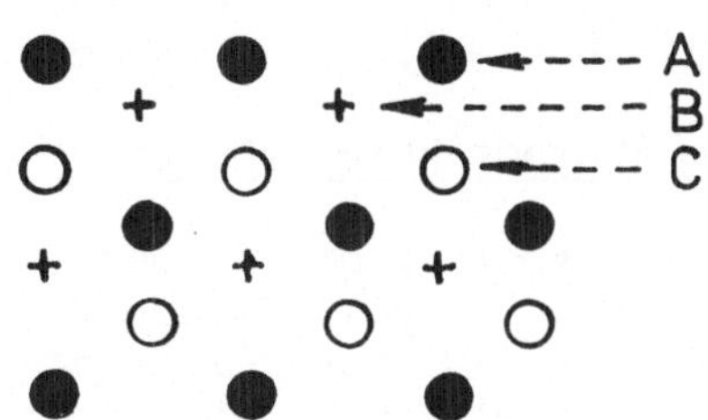

Fig. 6. Scheme of the arrangement of three consecutive (111) lattice layers.

They exhibit non-isotropic growth shapes. Obviously growth in one direction has been accelerated, while it seems to be restrained in the other directions. Such non-isotropic crystals are always observed when crystal formation starts and proceeds in the presence of a high bromide excess in the environment. Under such conditions (111) faces predominate. Growth of a crystal generally occurs in two steps: a two-dimensional nucleus is formed first on an existing crystal plane. This process requires an activation energy which is brought forth by the supersaturation in the ambient medium. To the edges of such a two-dimensional nucleus further lattice ions are attached until a new crystal plane is completed. Then again a nucleation process takes place etc. In AgBr a two-dimensional nucleus is positioned on top of the (111) lattice layer underneath according to the cubic close-packed scheme (Fig. 6). This means that the ions of the nucleus are situated above interstices of the host layer, if one looks perpendicularly at the growing (111) face. The ions of the next layer to follow are again situated above interstices of the second lattice layer and also above the remaining interstices of the first layer. The following fourth layer finally corresponds to the orientation of the first one etc.Hence, the lattice layers are oriented in the sequence ...ABCABCABC... . During very rapid nucleation processes one layer by chance and error may slip into a wrong position ...ABCABACBAC... (underlined), which locally corresponds to the hexagonal close-packed scheme ...ABABAB... . The planes formed behind this faulty layer may again grow correctly, but this so-called stacking fault has divided the crystal in two parts, resembling image and reflected image (4,5). At the intersection of the twin planes and the crystal surface grooves are formed where the activation energy for deposition of additional matter is greatly reduced. Thus crystal growth occurs preferentially at these sites which leads to the anisotropic growth shapes mentioned before. Aside from their interaction during growth structurally distorted regions like the twin planes behave differently from undistorted crystal bulk. Usually the ionic conductivity is increased at such sites, there chemical surface reactions take place easier, and there the photographic process, which will be outlined next, occurs with greater ease.

2.3. The Photographic Process

In the dark the silver halides exhibit a very low electric conductivity due to ionic disorder of Frenkel type. Depending on the temperature a few silver ions are displaced from their lattice positions and migrate by a thermally activated jump process as interstitial silver ions about the lattice. Once an interstitial silver ion has been formed it represents a site of an excess positive charge in the otherwise neutral crystal. The vacancy which is left behind bears an apparent negative charge, since the charges of the surrounding six anions are no longer compensated by the charge of the silver ion. The total formation energy of a Frenkel pair can be split into the contribution from forming an interstitial silver ion and such from forming the vacancy respectively. Since these two components of the total free energy are usually different, the defect formed easier will move farer away from the point of generation, which is usually the surface. In order to maintain electric neutrality a counter charge, located at the surface, will exist. The resulting electric field limits the diffusion of the other species towards the bulk until a dynamic equilibrium is attained (6-8). Then any observable change ceases. If, however, another charge carrier (which one, will be mentioned immediately) enters the region of the electric field, it will be repelled or attracted by the surface, depending on the sign of its charge.

The conductivity of silver halides greatly increases as soon as the crystal is exposed. As in any insulator the two electronic states assigned to the filled valence band and the empty conduction band are separated by a gap which is approximately 2.5 eV wide in AgBr and close to 3 eV in AgCl. Accordingly, AgBr absorbs only light of wavelengths shorter than 450 nm, and the absorption in AgCl terminates close to the beginning of the UV region.

Light absorption ejects an electron from the valence band, where it has been bound to a bromide ion, into the conduction band. In this state, within its lifetime of about 10^{-7} seconds, it is mobile. It has left behind a bromide ion depleted of an electron, in other words, a bromine atom. Since in the crystal the charges of lattice ions are just compensated by the charges of the surrounding ions of the other sign, each lattice ion appears to be neutral in respect to the lattice. We have seen before, that removal of a silver ion from a lattice position leaves behind a vacancy bearing an apparent negative charge. This happens also to the bromide ion which has lost one electron. The bromine atom thus formed is now the site of an apparent positive charge due to the uncompensated charges of the adjacent silver ions. For this and also a second reason we avoid to call it a bromine atom as long as it occupies a lattice site, and we prefer to call this entity a positive hole. By an exchange process the positive hole can accept an electron from another adjacent bromide ion. Thus the former becomes a bromide ion again, while the latter is now the site of a positive charge, a positive hole. By an electron jump in one direction the positive charge (the positive hole) has been displaced in the opposite direction. Thus both the photo electron and the hole are mobile within the crystal, the latter, however, due to the complicated jump process with a mobility roughly two orders of magnitude smaller than that of the electron.

The processes of absorption, creation of electron/hole pairs and their migration as free entities, as excitons, or as self-trapped species are much more complicated as indicated here. Such details, however, are beyond the scope of the present review.

The latent image speck is a cluster consiting of a few silver atoms. Its efficient formation is our main concern. Mobile silver ions (the interstitial ions) are always present in the crystal, and an electron has been just set free by exposure. Very clearly, both of them could react and a silver atom will be formed:

$$Br^- \longrightarrow h^+ + e^- \tag{2}$$

$$e^- + Ag_i^+ \longrightarrow Ag \tag{3}$$

This, however, is not yet a latent image speck. It must be enlarged up to a size of, say, Ag_4 and obviously the reaction of electrons and interstitials has to be repeated several times. This requires the absorption of several more photons, but it requires more: First of all, the subsequent reaction steps will be useless, unless they occur at the site of the silver atom formed first or unless remotely formed silver atoms could somehow get together. The chances should be extremely small from a simple statistical point of view. There are some 10^8 to 10^9 possible reaction sites in one micro crystal, and without a driving force or some other assistance one could never expect the generation of a sufficiently large silver aggregate after investment of just a few photons. Even the first reaction step, the formation of the first silver atom which later is supposed to grow to a silver cluster, has to be directed to a favourable site for the following reason: The latent image silver cluster catalyzes the development process. It obviously greatly reduces the activation energy of the initial development stages. This is not due to some action at distance, but requires the intimate contact of the three phases involved in development - reducing matter, silver cluster, and silver halide crystal. Then, however, the step leading to the first deposited silver atom should not occur randomly in the bulk, but should take place at the crystal surface. The chances, if we simply compare the geometric probabilities are roughly 1:200 which means, they are again small to occur at a useful location, unless some driving force directs the first reaction step to the surface and the consecutive growth steps to that same site.

But even if such a driving force were helpfully present, there still may remain other obstacles which could unfavourably influence the photographic process. One absorbed photon creates one photo electron which in turn could lead to the formation of one silver atom if, of course, this electron will not be spoiled uselessly. There are many such loss reactions possible, but we shall treat here only two of them which very probably are the most dangerous: Since the electron and the positive hole bear charges of opposite sign, there is the immediate danger that due to Coulombic forces they could approach each other and recombine

$$e^- + h^+ \longrightarrow Br^- \qquad (4)$$

This would restore the crystal in its original state prior to exposure. Alternatively, the positive hole could attack a silver atom or an aggregate of silver atoms previously formed

$$Ag_n + h^+ \longrightarrow Ag_{n-1} + Ag_i^+ + Br^- \qquad (5)$$

Both these reactions are loss processes which decrease the quantum yield. In order to optimize the photographic process, care should be taken to avoid them.

As has been shown previously, an electric field due to the accumulation of ion vacancies at the surface and of excess interstitial ions exists in the sub-surface region of the crystal. Charged entities of opposite sign, if brought into this field, will be drawn away in opposite directions. This could lead to a separation of photo electrons and positive holes and if this happens, the danger of pernicious recombination would decrease. While such internal fields have been proved to exist in large single crystals of silver halide, their significance in micro crystals is still uncertain. There are at most some hundred of interstitials and, hence, vacancies in a micro crystal present. Compared to about a million of lattice ions at the surface this is a negligible number. Surface effects due to polarization of surface ions, to adsorption, or solvation could easily outdo the expectedly small effects caused by a non-uniform distribution of Frenkel defects. So there still remains the problem, how a recombination of electrons and positive holes can be avoided. From these admittedly very crude statistical considerations it follows, that the quantum sensitivity of silver halide micro crystals should be very low. By quantum sensitivity is understood the number of absorbed quanta required to render developable a certain fraction (e.g. 50 per cent) of all crystals. We know, however, that in highly sophisticated modern photographic materials the quantum sensitivity according to this criterion corresponds to perhaps ten or twenty absorbed photons. In other words, their sensitivity is very high. If, however, simple and primitive silver halide crystals are used, photographic materials produced from them are extremely insensitive and would require hundreds or more absorbed photons. What causes this enormous gain in sensitivity if we use what has been called before "highly sophisticated" silver halide micro crystals, what has been done to them, and how can it be explained?

It has been found by experience that very small amounts of certain foreign matter, particularly Ag_2S (and, even better, if accompanied by Au^+ ions) incorporated into the surface of the micro crystals greatly increase their sensitivity. It has been discovered, that in the presence of such so-called sensitivity centres the deposition of silver aggregates is very strictly directed to the location of the sensitivity centres (9,10). In other words, the topography of photolytic silver aggregate formation coincides with the topography of the sensitivity centres (Fig. 7).

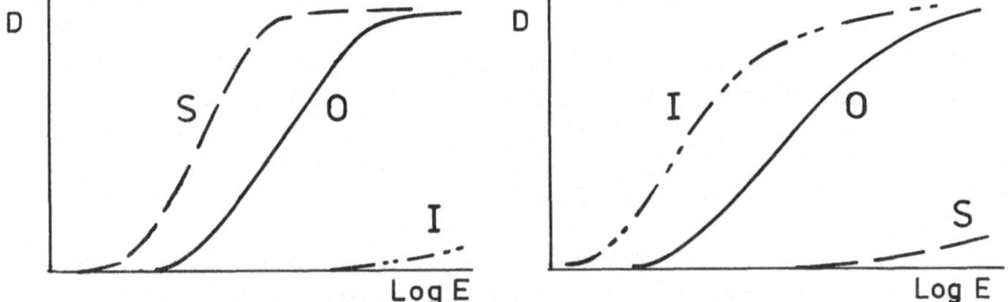

Fig.7. Characteristic curves (optical density D which is proportional to the number of developed crystals vs. exposure) after surface development (left) which develops only crystals bearing surface latent image specks and after internal development (right) which develops only crystals bearing silver specks in the crystal volume.
Crystals doped with Ag_2S at the surface (curve S) form predominantly surface silver specks and hardly such in the bulk. Internally doped crystals (curves I) form image specks in the bulk and hardly such at the surface. Undoped crystals (curves O) form both surface and internal latent image specks.

Another effect resulting from Ag_2S doping shall be mentioned: By exposing a photographic layer in a microwave field energy is absorbed by free photo electrons and this will detune the circuit. By this method photo electron lifetimes can be measured, and it has been found that Ag_2S doping essentially reduces the lifetimes (11). It then follows that in the presence of the sensitivity centres electrons are consumed faster. From the topographic relation between the sensitivity centres and the latent image specks one can conclude, that this accelerated process will be the reaction of electrons with interstitial silver ions leading to silver aggregate formation. And indeed, in the presence of sensitivity centres the sensitivity is greatly increased or, in other words, a silver halide micro crystal doped with Ag_2S centres becomes developable after absorption of much less photons than a primitive, undoped crystal. Obviously the sensitivity centres exert the driving force mentioned earlier, which causes a more efficient course of the photographic process ending up with the formation of a latent image speck. Even in an undoped, primitive crystal exposure will eventually lead to the formation of a silver cluster. The Ag_2S centres merely make this process more efficient. They therefore act like catalysts for silver formation.

2.4. Development

It is generally believed that an exposed silver halide micro crystal bearing a latent image speck will be developable, while an unexposed crystal is not developable. This is not quite correct: Even an unexposed crystal will be developed, reduced to metallic silver, if held in the developer for a sufficiently long time. The spontaneous development of unexposed crystals is called "fog". In the presence of a latent image silver aggregate, however, development starts much faster. Here we meet again an action of a small cluster which is very similar to general catalysis.

What happens during development? As stated before, the developer contains a redox system. If the redox potential is more negative than the electrochemical equilibrium potential Ag^+/Ag under the prevailing conditions, reduction of silver ions to silver atoms should start. The reducing component of the developer (hydroquinone in the simplest case) is thereby oxidized (to quinone in this case). This statement, however, is still incomplete, since it does not include the action of the latent image silver aggregates. If such are absent (as in unexposed micro crystals) either a comparatively long time would elapse until reduction starts or a much more negative redox potential would be required in order to rapidly start the reduction process. It appears as if a high activation energy separates the crystal in the initial state from a state where reduction has just started. This activation energy can be surmounted by increasing the difference of the electrochemical potentials. Otherwise the begin of reduction, induced by random processes requires a long waiting time. The action of the latent image silver cluster consists then in a decrease of this activation energy and like a catalyst it thus speeds up the initiation of development. These problems, which are closely related to the formation of a silver phase will be treated later. Here only the general electrochemistry of development shall be reviewed.

The latent image silver speck can be regarded as a small noble metal electrode (12,13) facing on one side a redox system (the developer), on the other side the crystal containing mobile interstitial silver ions (Fig. 8). The silver electrode accepts electrons from the reducing medium, and when the redox potential of the latter is sufficiently negative, the electrons will be transferred to interstitial silver ions, which will become discharged. Silver atoms thus are deposited at the rear side of the electrode which will be pushed out of the crystal by the growing silver phase (Fig. 9). The reduction of the crystal, once it has started, will

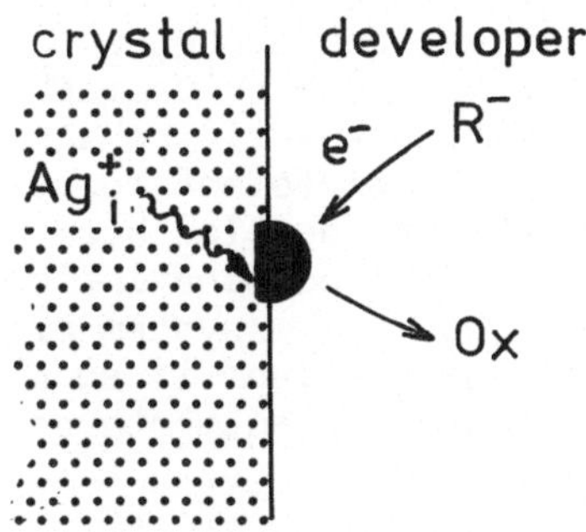

Fig.8. Model of development. Reducing molecules (R^-) transfer electrons to the latent image speck where they discharge approaching interstitial silver ions.

Fig.9. Electron micrograph (carbon replica) of a silver bromide crystal. Development has been stopped in an early stage. Silver filaments have started to grow.

be completed very soon. After the reduction has started, the rate is in the first place limited by the diffusion rates of the participating entities.

In colour processes the oxidation products of the developers (e.g. a quinone diimine, if the reducing entity has been a paraphenylene diamine derivative) react with colourless dye-precursors (couplers) and form yellow, magenta, and cyan dyes respectively. These couplers are situated in three individual layers. Each of them contains silver halide crystals which have been made sensitive either for the blue, the green, or the red part of the visible light spectrum. While silver halide is intrinsically sensitive only for blue light which it absorbs, its sensitivity can be extended to light of longer wavelengths by sensitizing dyes. These so-called spectral sensitizers are adsorbed to the crystal faces. After light absorption in their absorption range they transfer an electron from an elevated energy state to the conduction band of the silver halide crystal. There it may cause silver formation like a photo electron formed by intrinsic light absorption in the crystal.

3. THEORY OF THE PHOTOGRAPHIC PROCESS

3.1. Experimental Frame of the Theories

The basic phenomenological principles of the photographic process have been briefly reviewed in the last chapter. Our knowledge of the individual steps which commence at exposure and which terminate with the formation of a latent image silver aggregate is still fragmentary so far. These steps can neither be directly observed nor followed in detail. Supported by experimental investigations, however, theories have been developed which attempt to establish schemes of pathways of the various constructive, competing, and destructive reactions which altogether constitute the photographic elementary process. In addition to such atomistic theories which are aimed at a description of the details of the different reaction steps, a more general and thermodynamic treatment has also been developed. Its main object is not so much a detailled description how processes take place, but rather if changes in the system are allowed (and what then their probabilities would be), or if such changes are forbidden. Atomistic and thermodynamic treatments do not exclude each other.

Any theoretical treatment should agree with significant experimental facts. Some of them have been mentioned already and will be listed again; others, which are essential for following the theoretical treatment shall be added:

(1) The latent image silver speck consisting af a few silver atoms is formed by the reaction of photo electrons and interstitial silver ions which both are mobile in the crystal. For various reasons we are convinced that this build-up occurs in steps

$$Ag \longrightarrow Ag_2 \longrightarrow Ag_3 \longrightarrow Ag_4 \; \ldots\ldots \tag{6}$$

Hence, each entity - except for the first Ag atom - has a precursor and, as long as growth continues, each entity is followed by the next larger one. Should this sequence of growth steps be governed merely by chance processes, it could rarely take place. A photographic theory, therefore, should offer plausible reasons why a beginning growth process continues at the same site instead of starting new sequences somewhere else. In other words: an explanation is required for such a concentration process.

(2) In an undoped pure silver halide crystal the above sequence can start randomly everywhere. In the presence of foreign centres (sensitivity centres, e.g. Ag_2S) photolytic silver formation is directed to the site of these sensitivity centres. Thus a more economic silver deposition is achieved which results in increased sensitivity.

(3) Photo electrons and the simultaneously formed positive holes may recombine. This would reduce the number of photo electrons available for silver formation and it thus would greatly diminish sensitivity. Loss due to recombination is suppressed by sensitivity centres.

(4) Within a certain range a given photographic effect (e.g. a given number of developed crystals per unit area) is merely determined by the product of exposure intensity and time, i.e. by the total number of photons absorbed by the layer element (reciprocity law). At very low intensities, however, a deviation from the reciprocity law is observed (low intensity reciprocity failure = LIRF; Fig. 10). The onset of a LIRF starts, where the reciprocity curve deviates from a horizontal. This

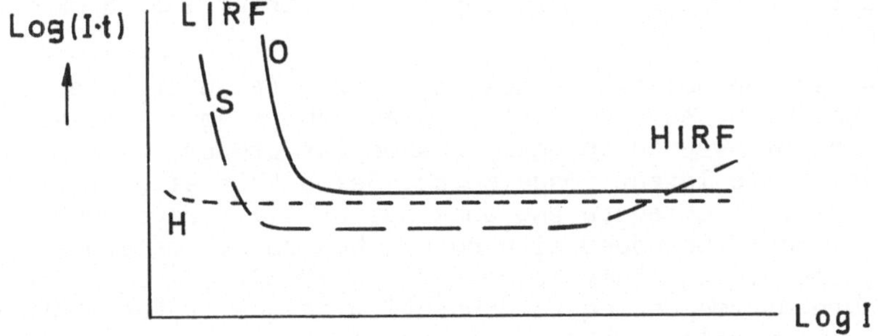

Fig.10. Reciprocity diagram. The total exposure (product of intensity and time) which leads to the same fraction of developable crystals (or to the same optical density after development) is plotted vs. intensity.
Curves are shown for a primitive, undoped crystal population (O) and for crystals doped with Ag_2S (S), and for a layer of primitive crystals which contains a halogen acceptor (H).

onset is shifted to lower intensity values if sensitivity centres such as Ag_2S are present. LIRF is very efficiently eliminated if halogen acceptors, such as nitrite or hydrazines are present in the layer. The photogenerated positive holes may diffuse to the crystal surface where they enter the ambient phase as halogen atoms:

$$h^+_{(crystal)} \rightleftharpoons \frac{1}{2}Br_2 \qquad (7)$$

There they can react with halogen acceptors and become rapidly removed. This will shift equilibrium (7) to the right side. LIRF has been attributed to instability of intermediate entities in early build-up steps of the latent image according to equation (6).

(5) At high exposure intensities again a deviation from reciprocity law is sometimes observed (HIRF, c.f. Fig. 10). It is absent if the crystals are in a relatively primitive state, and it appears strongly if the crystals have been doped with Ag_2S centres. The inefficiency of latent image formation at high exposure intensities has been attributed to competition of several sensitivity centres present in one crystal. Growth processes thus are believed to start at many sites and none of them will lead to formation of a sufficiently large silver aggregate unless much more photons have been absorbed than at moderate intensity values. The high dispersity of silver deposition at high exposure intensities has been indeed experimentally proved (14), but then immediately the question should arise why competition obviously plays a less important role at moderate exposure intensities where the reciprocitity law is much better obeyed.

(6) So far in the build-up sequence of equ. (6) a critical (magic) size has been assumed. At this size (about Ag_4) a silver aggregate is believed to abruptly change its properties and becomes a developable latent image speck. Below this size it is just a small silver aggregate and does not catalyze development. If a crystal could be doped with such sub-size silver clusters (say, one or two atoms below "critical" size; i.e. Ag_3 or Ag_2), exposure corresponding to absorption of one or two photons should then enlarge the sub-size silver aggregates to full latent image size (Fig. 11). Such experiments have been done (15,16) and the expected increase in sensitivity has been repeatedly observed, but surprisingly the latent image specks did not always appear at the site of the sub-size silver aggregates - the presumed latent image precursors - which obviously failed to grow. Even worse, exposure led to a destruction of the sub-size aggregates (17). This destruction has been attributed to oxidation by holes

$$Ag_n + h^+ \longrightarrow Ag_{n-1} + Ag_i^+ \qquad (8)$$

Such a behaviour of very small silver aggregates is not compatible with their presumed role as latent image precursors.

(7) In reversible redox systems (Fe^{2+}/Fe^{3+}) silver aggregates above a critical size can grow, while smaller aggregates become destroyed (18). This critical size is not constant. It decreases when the redox potential of the system is made more negative, and it increases, when the potential is made more positive.

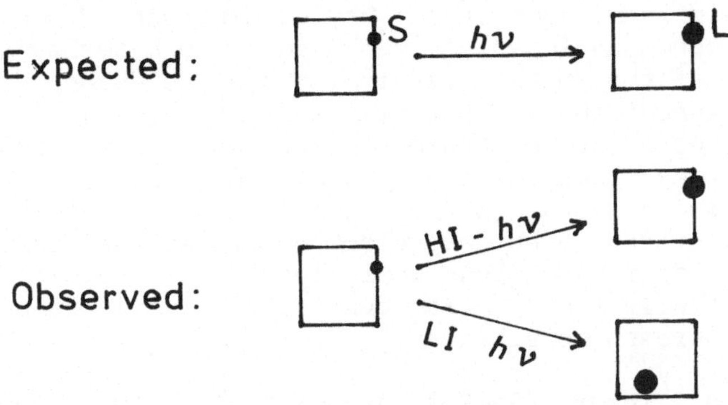

Fig.11. Top: A sub-size silver aggregate (S) is expected to grow to a latent image speck (L) upon exposure.
Bottom: in reality it sometimes grows, particularly at high exposure intensities. Sometimes, however, exposure leads to latent image speck formation in crystal regions where no sub-size silver aggregates have been present (schematic).

A surprisingly simple relation between "critical" size and redox potential has been found. It is similar to the well-known Gibbs-Thomson equantion which describes the increase in vapour pressure or solubility with decreasing particle size.

3.2. Atomistic Theories of the Photographic Process

3.2.1. Gurney-Mott Theory (19). Photolytic silver formation occurs in two steps. Photon absorption leads in an electronic step to the formation of electronic charge carriers - photo electron and hole. In an ionic step the electron is supposed to react with an interstitial silver ion. Prior to this reaction the electron is immobilized by capture in an electron trap. In the original theory this trap was believed to be uncharged. After electron trapping it acquires a negative charge. Approach of an interstitial is aided by Coulombic forces. Intrinsically present electron traps (e.g. structural defects) are believed to be comparatively shallow and therefore not very effective. Deep and very efficient electron traps are introduced by doping the crystal e.g. with Ag_2S (sensitivity centres).

It was also assumed that the photo hole is less mobile than an interstitial silver ion. Otherwise the occupied electron trap would serve as a recombination centre where the trapped electron could recombine with an approaching hole also attracted by Coulomb forces. The Gurney-Mott theory is therefore not so much concerned with the fate of the holes.

The validity of some of the key assumptions of the Gurney-Mott theory has been questioned: It has been found experimentally that the mobility of the photo holes in contrast to Gurney's and Mott's assumption exceeds the interstitial mobility by several orders of magnitude. Thus a loss of trapped electrons by recombination with holes could take place. Then, if several Ag_2S sensitivity centres are present in a crystal (a fact, which has been experimentally found) and if they act as deep electron traps, they could compete in electron trapping and thus would lead to an undesirably high dispersity of silver nucleation events; this would rather decrease instead of increase sensitivity.

3.2.2. Mitchell Theory (20-23). Arguments like those mentioned above led to the installment of an alternative theory. The presence of deep and irreversible electron traps in an unexposed crystal is denied. However, shallow traps or, rather, shallow positive potential wells are believed to be present due to lattice singularities or due to Ag_2S monolayer centres at the surface. At these sites by repeated binary processes, each of them comprising the random approach of an interstitial silver ion and a photo electron silver atoms, Ag_2 molecules, and finally Ag_3 aggregates are formed. Neither the silver atom nor the Ag_2 molecule are efficient traps for electrons or for interstitial silver ions respectively, which causes these first steps to be rather inefficient.The silver atom is comparatively unstable and, like a Ag_2 molecule, it may trap a positive hole. If this happens at a Ag_2 molecule it will be reduced in size to a silver atom which then dissociates and injects an electron to the conduction band. So absorption of one additional photon liberates in total two electrons without the appearance of a free hole:

$$\left.\begin{array}{rcl} Br^- & \xrightarrow{1h\nu} & e^- + h^+ \\ Ag_2 + h^+ & \longrightarrow & Ag + Ag_i^+ + Br^- \\ Ag & \longrightarrow & Ag^+ + e^- \\ \hline \text{Crystal} + Ag_2 & \xrightarrow{1h\nu} & 2e^- + 2Ag_i^+ \end{array}\right\} \quad (9)$$

This may be quite helpful, since it reduces the danger of electron/hole recombination and increases the number of electrons.

Starting at a size of Ag_3 adsorption of a silver ion to the aggregate is accompanied by a greater free energy change than the transfer of a silver ion to an interstitial position. Hence, a Ag_3 aggregate will immediately react with an interstitial by forming a tetrahedral Ag_4^+ cluster which now becomes a deep electron trap. It will efficiently capture a photo electron. This converts the Ag_4^+ aggregate to a neutral Ag_4 cluster, which then adsorbs an interstitial ion and thus again constitutes an electron trap etc. As soon as one Ag_3 has been formed, the inefficient and more or less randomly occurring first build-up stages are thus replaced by a directed and efficient concentration process.

In addition to serving as the sites of shallow positive potential wells Ag_2S monolayer centres (sensitivity centres) fulfill according to the Mitchell theory also a second task: They are believed to react with holes. Thus they reduce the hazard of electron/hole recombination and improve the quantum efficiency of the photographic process.

The Mitchell theory among other effects explains the occurrence of LIRF: At very low exposure intensities and, hence, very low average electron concentrations the binary first reaction steps (the simultaneous approach of an electron and an interstitial) become less probable. It furthermore explains why, at least at low and moderate exposure intensities, numerous sensitivity centres present in a crystal do not lead to a scattered silver formation which is to be expected if they were deep electron traps.

Other effects such as the occurrence of HIRF or the dependence of the "critical" size of a latent image centre on the redox potential of the developer are less convincingly covered.

3.3. The Photographic Process as an Event of Phase Formation

3.3.1. Basic Laws of Phase Formation. Without any doubt exposure of a silver halide crystal leads to the separation of a new phase (silver) in a host phase (silver halide). This new phase is formed, it grows as long as exposure lasts, and it is greatly and rapidly enlarged during subsequent development. Many of the effects listed in section 3.1. parallel those which are observed in common phase formation processes (such as condensation, crystallization etc.), and this led to an entirely different approach for describing the photographic process as an event of photo-induced phase transition (24-27). Prior to such a treatment the basic laws of phase transition events shall be summarized.

Phase formation always requires a state of supersaturation in a solvent or host phase. In a kinetic model the supersaturation is believed to fluctuate locally. There may be at a given instant regions where the actual concentration of the solute exceeds the average supersaturation value, and there may be others where the actual concentration equals the average supersaturation value or is smaller. These fluctuations lead to a transient formation of small clusters. Most of them will vanish as soon as the permanently occurring fluctuations lead to a locally lower concentration of the solute. Some of them, however, by chance can reach a size above of which they are stable. Above this critical size a decay becomes rather unprobable, and such aggregates will grow as long as a supersaturation persists.

The formation of aggregates is accompanied by a change of the free energy ΔG. For systems such as a supersaturated vapour phase in which droplets of the condensed phase of radius r are formed, the change of the free energy is given by

$$\Delta G = 4\pi r^2\sigma - \frac{4}{3}\pi r^3 \frac{RT}{V} \ln \frac{c}{c_\infty} \qquad (10)$$

(σ = specific surface energy; V = molar volume; c/c_∞ = supersaturation). Plots of ΔG vs. r exhibit a maximum at a certain size r^* which decreases with increasing supersaturation (Fig. 12) according

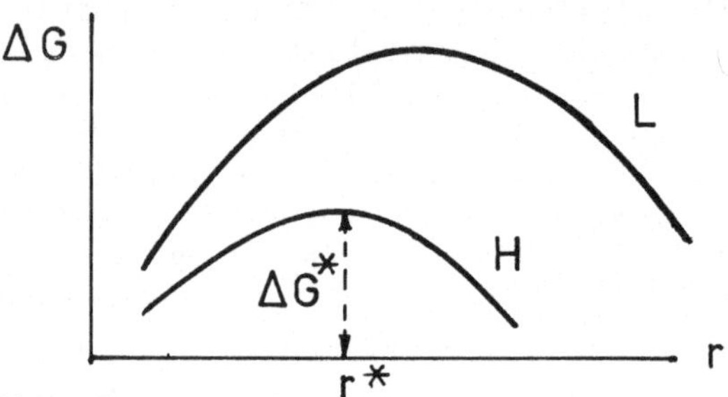

Fig.12. Gibbs free energy vs. particle size for a system of high (H) or low (L) degree of supersaturation

to the well known Gibbs-Thomson equation

$$r^* = \frac{2\sigma V}{RT \ln c/c_\infty} \qquad (11)$$

which is the derivative of equ. (10). A particle of exactly the size r^* is semi-stable. Both decay or growth will cause a decrease of the free energy and both are accordingly equally probable. A particle of that size is called a nucleus, and the energy ΔG^* required for its formation is the activation energy of nucleation (nucleation energy). Unlike activation energies e.g. of chemical reactions, however, the activation energy of nucleation is not independent on the concentration of the solute (the supersaturation). Strictly homogeneous nucleation events in pure solute/solvent systems are comparatively rare. Particles of foreign matter either deliberately introduced or present as impurities may act as condensation centres where the nucleation energy is lowered (Fig. 13). The majority of phase formation processes, hence, can be regarded as heterogeneous nucleation events. Even the surface of the solute/solvent system acts as a site of increased nucleation probability.

The driving force of nucleation is the supersaturation in the system. It not only determines the critical nucleus size r^* and the nucleation energy ΔG^*, but it influences also the rate of nucleation which very generally follows an exponential relation

$$\frac{dN}{dt} = A \exp\left[-\frac{\Delta G}{kT}\right] \qquad (12)$$

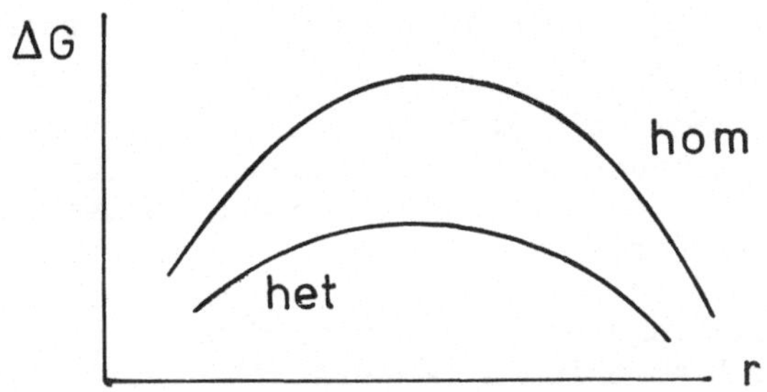

Fig.13. Gibbs free energy vs. particle size for homogeneous (hom) or heterogeneous (het) nucleation, both at the same level of supersaturation.

Hence, increasing supersaturation decreases both critical size and the value of the corresponding nucleation energy, and increases the nucleation rate (28,29).

Some of the properties of particles of the condensed phase such as particle radius r and the interfacial energy σ are well-defined for comparatively large aggregates (crystal nuclei, condensed droplets etc.). These magnitudes are physically rather questionable and meaningless if the aggregates are very small and approach atomic dimensions. For such systems quite a different theoretical treatment - the atomistic nucleation theory - is required (30,31). Instead of the above mentioned macroscopic properties atomistic parameters such as binding and separation energies are used. Such calculations lead to very unexpected results: They are almost identical to those obtained by using the classical, macroscopic treatment. The latter, therefore, appears to be a valuable tool for describing phase formation processes even in such systems to which, due to extremely small aggregate sizes, the macroscopic treatment should really not be applied.

3.3.2. Thermodynamics of the system Ag/AgBr (32-34). Phase formation has been generally treated as the separation of a condensed phase from a supersaturated solution phase. The photographic process is certainly such an event of phase separation, induced by photostimulation. If this is correct, it will imply the following statements:

- AgBr acts as a solvent for silver. There should exist then an equilibrium solubility which very probably is extremely small.
- The silver dissolved in AgBr is very likely completely dissociated

$$Ag \rightleftharpoons Ag_i^+ + e^- \qquad (13)$$

  to interstitials and electrons. Since silver interstitials due to the Frenkel equilibrium are present in large excess compared to the electrons, the nominal concentration of the latter as the minority constituent is equivalent to the concentration of dissolved silver.

- An increase of the nominal electron concentration above the equilibrium value by whatever means (including liberation of electrons due to exposure) will supersaturate the AgBr phase and could lead to separation of silver.

These key assumptions have been proved to be correct: A large silver halide crystal to which a temperature gradient has been applied, was at the side of higher temperature contacted with metallic silver. Silver atoms obviously dissolved (35) and condensed in regions of lower temperature, where the silver solubility is lower (Fig. 14). In a more recent experiment (36) again a temperature gradient was applied to a binary silver salt crystal. It was contacted to a DC source by a platinum electrode which injected electrons, and a AgI/Ag junction respectively which injected silver ions into the crystal. In cooler parts of the crystal the excess electrons combined with interstitial silver ions and led to growth of a silver filament (Fig. 15). Thus the transport of silver (as

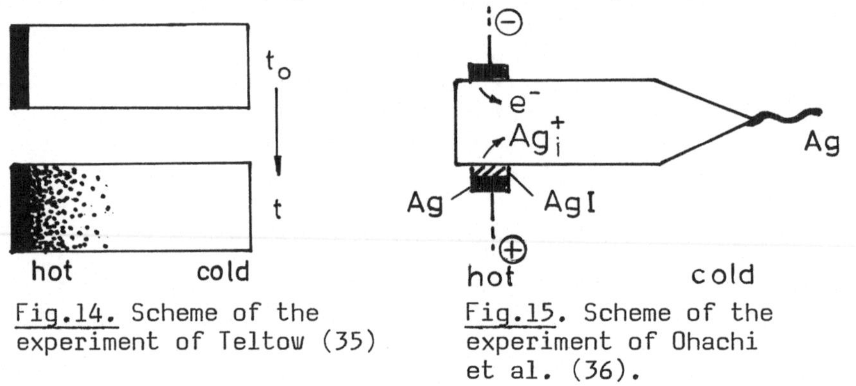

Fig.14. Scheme of the experiment of Teltow (35)

Fig.15. Scheme of the experiment of Ohachi et al. (36).

electrons and interstitials) through the crystal has been experimentally observed.

In a AgBr crystal which is in contact with a large silver phase the mass action law can be applied to the following equilibria:

(a) Frenkel equilibrium

$$[Ag_i^+]\cdot[Ag_v^-] = \text{const.} \tag{14}$$

where $[Ag_v^-]$ is the concentration of silver vacancies.

(b) Silver dissolution and dissociation

$$Ag_{solid} \rightleftharpoons Ag_{dissolved} \rightleftharpoons Ag_i^+ + e^-$$

$$[Ag_i^+]\cdot[e^-] = \text{const.} \tag{15}$$

$[Ag_i^+][e^-]$ can be regarded as the "solubility product" of silver in silver halide. Equ. (14) and (15) are connected by the same concentration of interstitials.

(c) Electron/hole equilibrium. In the dark a (very low) concentration of electrons and holes exists

$$[e^-][h^+] = \text{const.} \qquad (16)$$

In equilibrium the electron concentrations of equ. (15) and (16) are identical.

(d) Redox equilibrium. If the crystal is also in contact to an ambient phase containing a redox system, the electron concentration in the crystal corresponds to a certain redox potential in the ambient phase and vice versa:

$$\log[e^-] \propto -E_{redox} \qquad (17)$$

From equ. (16) it then follows that a similar relation connects hole concentration and redox potential

$$\log \frac{1}{[h^+]} \propto -E_{redox} \qquad (18)$$

(e) Hole/halogen equilibrium. Since holes arriving at the surface of the crystal are converted to halogen atoms (and vice versa)

$$h^+ \rightleftharpoons \tfrac{1}{2}Br_2 \qquad (19)$$

the redox potential is finally also a measure of the hole concentration or the partial halogen pressure

$$\log \frac{1}{[h^+]} \propto \frac{1}{2}\log \frac{1}{p_{Br_2}} \propto -E_{redox} \qquad (20)$$

These relations lead to a thermodynamic diagram of state of the system Ag/AgBr (Fig. 16). In equilibrium where e.g. the silver phase neither grows nor becomes destroyed the electron concentration of equ. (15) corresponds to the "solubility" of silver in silver halide; in respect to silver the silver halide phase is then just saturated. Such an equilibrium state requires according to equ. (17) a certain redox potential. If the latter is increased, the hole concentration according to equ. (20) will unavoidably exceed the equilibrium value and silver can be oxidized:

$$Ag_n + h^+ \longrightarrow Ag_{n-1} + Ag_i^+ + Br^- \qquad (21)$$

If the redox potential, in turn, is made more negative the concentration of electrons is expected to exceed the equilibrium value. The system thus becomes supersaturated with electrons and, hence, with silver and growth of the silver phase

$$Ag_n + Ag_i^+ + e^- \longrightarrow Ag_{n+1} \qquad (22)$$

may occur. The processes described in equ. (21) or (22) could always take place as soon as the concentrations of electrons or holes respectively are increased above their equilibrium values by whatever means. An example was shown in Fig. 15 where the system

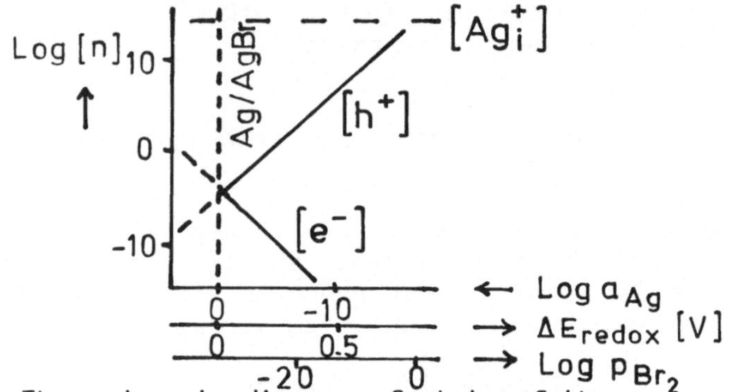

Fig.16. Thermodynamic diagram of state of the system Ag/AgBr (33). Plot of the charge carrier concentrations (ordinate; logarithmic scale) vs. silver supersaturation (a_{Ag}), redox potential in the ambient phase (E_{redox}, measured vs. the Ag^+/Ag-potential), or vs. the partial bromine pressure (abscissae).
Equilibrium prevails at the conditions given by the vertical denoted Ag/AgBr. Only under these conditions stable coexistence of metallic silver and AgBr is possible.

has been supersaturated by injecting electrons.Only under the conditions of the coexistence point given by the equilibrium concentrations the system Ag/AgBr will not change its composition.

So far a system consisting of silver and silver halide has been treated. If the system contains only silver halide without the presence of a silver phase, the above relations are still valid. The system, however, remains unchanged in composition in the whole region of electron concentrations below the equilibrium value or, according to equ. (16) in the region of hole concentrations above the hole equilibrium value up to a concentration which corresponds to a bromine partial pressure of one atmosphere. Between these two limits an elevated hole concentration or an oxidizing ambient phase will not affect AgBr. At electron concentrations above the equilibrium value, however, the system becomes unstable, too. Silver could be formed if the saturation value according to equ.(15) is exceeded, in other words, if the system becomes supersaturated.

Exposure increases according to equ, (2) both the electron and the hole concentrations. This should provoke simultaneously the onset of reactions (21) and (22). If the system is a closed one silver will be formed, but immediately it will be destroyed by attack of holes. Thus the net balance after exposure will be zero, since all reactions involved are reversible. Such a system could at best serve as a solar energy cell, but it will never be suitable for storage of optical information by forming a permanent latent image silver phase. This, however, could be easily achieved if an irreversible step - preferentially the irreversible removal of holes - is introduced. Holes which migrate to the crystal surface form there halogen atoms. The latter may react with redox systems in the ambient phase. In practical photographic systems the ambient phase always contains organic substances like gelatin, dyes etc. Most of these compounds readily react irreversibly with halogen. Thus the actual surface concentration of halogen is decreased, this will shift the partition equilibrium (7) to the right side, and this causes a removal of holes from the crystal.

A more efficient way of eliminating holes comprises the introduction of hole acceptors into the crystal. One of them is Ag_2S which, as has been mentioned before, is incorporated into the crystal surface in order to increase sensitivity. If Ag_2S is introduced to the system AgBr or Ag/AgBr the relations discussed above are still valid. However, if the hole concentration exceeds a certain value, not only silver but also Ag_2S can be attacked by holes. Thus part of the holes can be irreversibly eliminated from the system and a corresponding fraction of the initially formed silver escapes oxidation and remains unaffected in the system (Fig. 17).

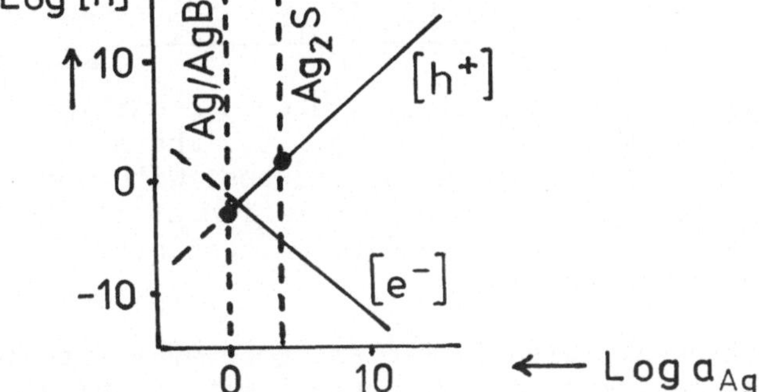

Fig. 17. Thermodynamic diagram of state of the system $Ag/AgBr/Ag_2S$ (37).
Coexistence of Ag_2S and holes is possible only at hole concentrations below a value given by the intersection of the straight $[h^+]$ and the vertical denoted Ag_2S. Between this value and the one given by the intersection of straight $[h^+]$ with the vertical denoted Ag/AgBr only silver will be attacked by holes. Above this value holes may attack also Ag_2S.

In the thermodynamic treatment contact of AgBr with a large silver phase has been assumed. Changes in size of a phase do not change the equilibrium concentrations, as long as the phase remains large. If, however, the size shrinks to the dimensions comparable to those of a few atoms, the chemical potentials of the constituents of this micro-phase will be greater than those of the macroscopic phase. The increased vapour pressure of very small droplets or the increased solubilty of very small particles are well-known examples. It is, therefore, safe to assume that e.g. the chemical potential of a silver atom in a micro-phase of silver will exceed the value it attains in a large silver phase. Hence, with decreasing size a silver aggregate in contact with a silver halide crystal will become more "soluble" in the crystal. This means that in equilibrium a higher electron concentration in the silver halide is expected to exist than in silver halide contacted with a macroscopic silver phase. In the thermodynamic diagram of state, therefore, the point of coexistence of silver and silver halide will be shifted to the left side (Fig. 18). The equilibrium

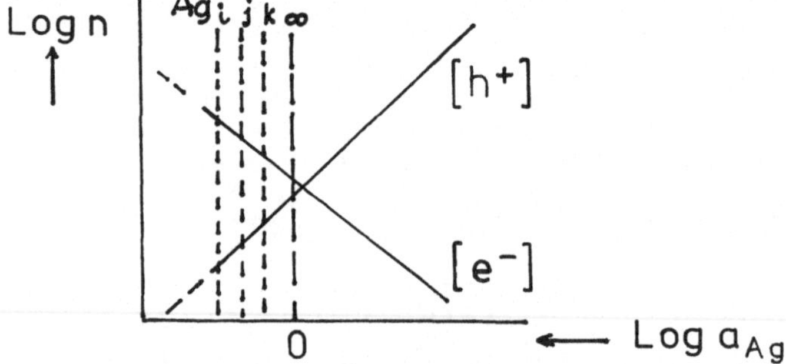

Fig.18. Dependence of coexistence conditions of the system Ag_n/AgBr (with n=i,j,k..., where $i < j < k$) on the size of the silver aggregates. With decreasing size the equilibrium concentration of electrons increases, while the equilibrium concentration of holes decreases.

concentration of holes in the presence of a very small silver phase will be lower than in contact with a macroscopic silver phase. Such considerations lead to conclusions which are important for an understanding of the photographic process: Growth of a very small silver aggregate consisting of a few atoms requires a higher electron concentration (or, a higher supersaturation) than a very large silver phase. On the other hand, oxidation of such a small aggregate starts at lower hole concentrations than that of a large silver phase.

3.3.3. Application of Phase Formation Laws to the Photographic Process. Exposure, as has been mentioned repeatedly, liberates in a silver halide crystal electrons and holes. The average electron concentration is given by the product

$$[e^-]_{exp.} \propto g\cdot\tau^- \tag{23}$$

where g stands for the generation rate, which is proportional to the exposure intensity, while τ^- is the lifetime of electrons (approximately 10^{-7} s). If $[e^-]_0$ is the equilibrium concentration (c.f. section 3.3.2.), then the system becomes supersaturated by the factor $a_{Ag} = [e^-]_{exp}/[e^-]_0$. A supersaturation can be likewise achieved by an ambient redox system in contact with the silver halide. Equilibrium exists, if $a_{Ag}=1$ or $\Delta E_{redox}=0$. The system is not in equilibrium if $\Delta E_{redox} < 0$ which will be generally the case when the crystal is in contact with a reducing medium, e.g. a developer. In conventional developers the redox potentials are approximately 200 - 300 mV below the Ag^+/Ag equilibrium potential which means, that they impose a supersaturation a_{Ag} in the range of $2\cdot10^3$ ... 10^5. These supersaturations correspond according to equ. (11) to critical sizes between four and ten silver atoms (Fig. 19). Aggregates of critical size are situated at abscissa values corresponding to the maxima of the respective ΔG vs. size (r or n) curves. Such aggregates could immediately grow without

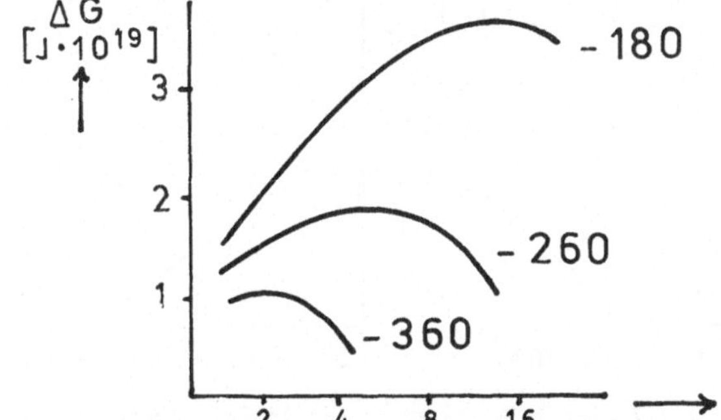

Fig. 19. Free energy vs. aggregate sizes of silver clusters on AgBr in reducing media (developers). The respective redox potentials are indicated at the curves.

any activation energy. In describing photographic systems it is preferred to use the expression that crystals, bearing such aggregates, will be immediately developed (reduced).

At a virgin crystal surface the full energy barrier corresponding to the activation energy has to be surmounted in order to form a critical size aggregate which then, when it has been produced, again will initiate immediate development. This initial process is a slow one, and the rate of nucleus formation depends according to equ. (12) exponentially on the activation energy and, hence, on the supersaturation. At a time when all the crystals bearing a critical size silver aggregate have completed development, random processes have led only at a very small fraction of virgin (unexposed) crystals to formation of a critical size silver nucleus and thus to development. This fraction of unexposed, although developed crystals leads to a certain density - the fog mentioned earlier in section 2.4. The discrimination during development between exposed crystals bearing a latent image silver cluster and such which have not been exposed is therefore due to the barrier of an activation energy. The catalytic action of the latent image silver clusters consists in reducing this activation energy nearly to zero. It can be concluded from Fig. 19 that the critical size of a latent image cluster is not a generally fixed value (a "magic" size), but that it depends on the redox potential of the developing redox system.

The activation energies of chemical reactions are generally independent on concentrations. In phase formation processes, however, the nucleation energy ΔG^* (which is equivalent to the activation energy) decreases with increasing supersaturation, as can be seen from equ. (10). This has been experimentally tested for silver nucleation in developers. From Arrhenius plots of the

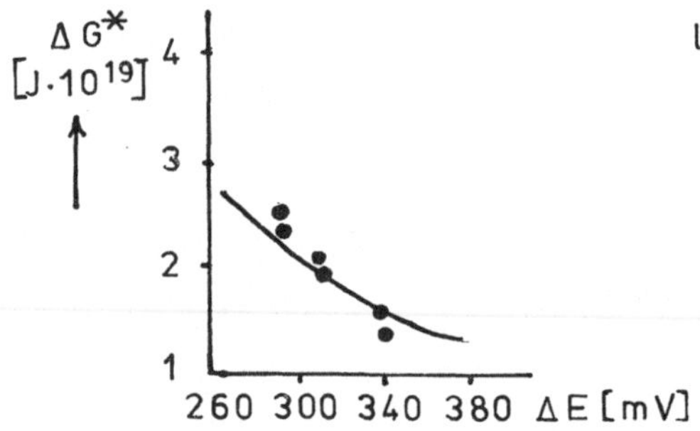

Fig.20. Nucleation energy vs. redox potential (=supersaturation) for silver aggregate formation at virgin AgBr crystals in developers.
Solid curve: Calculated from equ. (10)
Dots : Experimental values.

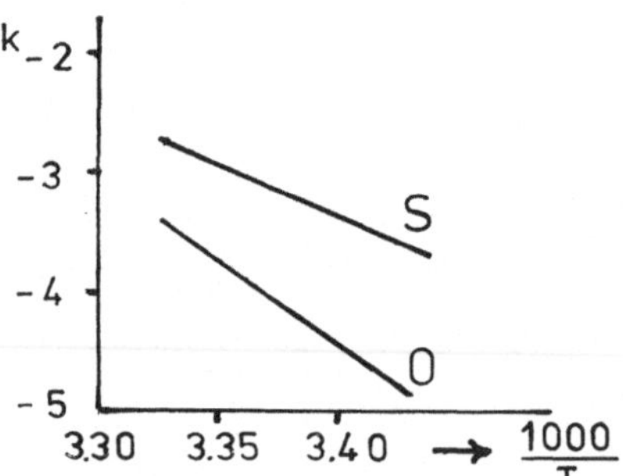

Fig.21. Arrhenius plots of the rate constants of silver aggregate formation in developers.
Curve O: Virgin AgBr crystals
Curve S: AgBr crystals doped with Ag_2S at the surface.

of the rate constants of fog development of unexposed silver halide layers the activation energies can be calculated for a wide range of redox potentials (i.e.: supersaturations, c.f. Fig. 20). Experimentally the expected dependence has been excellently confirmed (34).

At foreign condensation centres phase formation energies are reduced (Fig. 13). Very obviously the Ag_2S sensitivity centres (c.f. section 2.3.) act as such condensation centres for silver nucleation in developers. They increase the rate of silver nucleation as can be seen from Arrhenius plots (fig. 21); the slope is lower and, hence, the activation energy is smaller if the crystal surface has been doped with Ag_2S.

So far the general laws of phase formation have been applied to processes where an existing silver phase either grows or, if initially absent, is formed in the redox system of a developer. Formation or growth depend on a state of supersaturation which originates from electron transfer from the redox system to the silver halide crystal. There is no reason to exclude from such considerations processes, where by other means the electron concentration in the crystal is temporarily increased to a non-equilibrium value. This, e.g. happens also during exposure, and the supersaturated system could relax by forming a silver phase. According to equ. (23) a low exposure intensity (a low generation rate for electrons) leads to a small supersaturation level. According to equ. (12) the formation rate of silver nuclei will then be low. It will be perhaps too low in order to expect an appreciable number of nuclei formed within the total exposure time. Despite a sufficient total number of absorbed photons no silver nuclei have been formed. This is an obvious and simple explanation of the occurence of a low intensity reciprocity failure (c.f. Fig. 10, section 3.1.).

In the reducing medium of a developer foreign condensation centres (e.g. Ag_2S) reduce the activation nergy of silver nucleation. It seems reasonable to assign to them the same action in nucleation events also during exposure. This could immediately explain the topographic relation between the location of Ag_2S centres and the site of latent image cluster deposition. The lower activation energy at the sites of Ag_2S clusters greatly speeds up nucleation processes there, while nucleation at undoped, virgin regions of the surface requires a higher activation energy and proceeds, therefore, much slower. At the site of Ag_2S centres, hence, the probability of silver nucleation is greatly increased.

In addition to serving as a catalyst for silver nucleation (by reducing the activation energy) Ag_2S centres fulfill also a second task. According to Fig. 17 they can be oxidized by positive holes, which thus are irreversibly removed. This reduces the danger of electron/hole recombination and increases the average electron lifetime. An increased lifetime, in turn, increases according to equ. (23) the mean electron concentration, leads to an elevated supersaturation, and thus increases the nucleation rate. All these effects of Ag_2S should cause an increased efficiency of the photographic process - in other words, an increased sensitivity. If the exposure intensity and thus the generation rate of holes is very low, the average hole concentration will attain also a very low value. If the hole concentration drops below the value given by the intersection of the straight $[h^+]$ and the vertical denoted Ag_2S (c.f. Fig. 17), any oxidation of Ag_2S by holes will cease. Below a certain hole concentration Ag_2S therefore will not remove holes. This explains immediately, why Ag_2S shifts the onset of a low intensity reciprocity failure to lower intensities (c.f. Fig. 10) without being able to remove it entirely.

One of the most remarkable features of the photographic process is the concentration of consecutive events (forming silver atoms from interstitials and electrons) upon one site until a sufficiently large silver cluster has been built up, although in the light of merely statistical considerations this process appears to be rather unprobable (c.f. section 2.3.). This crucial phenomenon - the concentration process - has been explained by assuming certain pathways (19, 21-23). A very general explanation follows immediately from phase formation considerations: The formation of a nucleus containing n atoms proceeds in a sequence of random events. The total formation energy G can be regarded as the sum of n individual free energy increments G_1, G_2, .. G_i .. G_n which accompany the individual steps (fig. 22). The first one is

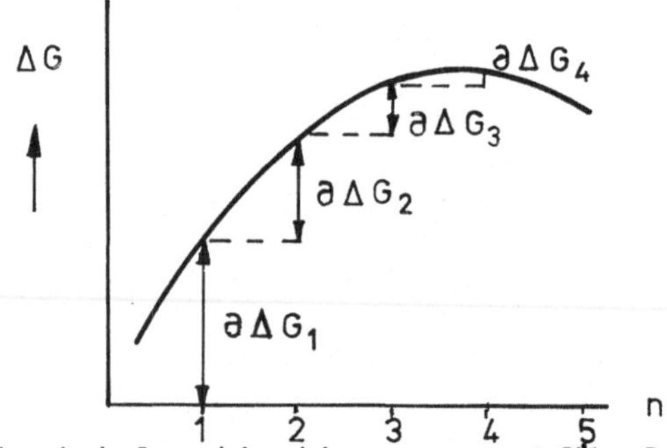

Fig.22. The total activation energy ΔG^* of nucleation is made up of energy increments $\partial\Delta G_i$ assigned to the various build-up steps $Ag_{i-1} \rightarrow Ag_i$.

obviously the largest, while the following increments decrease until at critical size $\partial\Delta G_{n^*}$ becomes nearly zero. Application of an expression similar to equ. (12) leads then to the conclusion that the rate (or the probability) of growth steps should increase with decreasing $\partial\Delta G_i$. Thus, growth of an existing, previously formed intermediate-size entity should be generally more probable than formation of a new entity. This, however, is the energetic cause of the concentration phenomenon.

Aside from the expected and experimentally observed dependence of the activation energy of silver formation on the redox potential (supersaturation) and on the presence or absence of foreign condensation centres (Ag_2S) the behaviour of small, sub-size silver aggregates is the most convincing justification for using the formalism of phase formation events in order to describe the photographic process. It has been mentioned in section 3.1.(6) (c.f. Fig. 11) that very small silver aggregates (say, Ag_2) below the critical size of developabilty do not always trap electrons during exposure and thus would grow to latent image silver aggregates, but that they react with holes as outlined in equ. (9). Such sub-size silver aggregates which do not behave as latent image precursors can be introduced into the crystal by careful chemical reduction (14-17) or by a weak sub-threshold pre-exposure with light or e.g. with γ-rays (37). At high exposure intensities, however, these sub-specks seem to grow which implies electron trapping (34); under such conditions the very same sub-specks act like latent image precursors. If the reduction treatment or the pre-exposure is carried out under slightly enforced conditions, some of the sub-size silver aggregates very probably will be slightly larger (say, Ag_3 instead of Ag_2). These slightly larger aggregates grow during exposure to latent image silver specks which then evidently means that they readily trap photo electrons (16, 37). Small silver aggregates accordingly may behave in two entirely different ways: With decreasing size and/or decreasing exposure intensity they tend to react with holes and to become reduced in size. At increased size and/or increased exposure intensity, however, they tend to trap electrons and thus will grow. This, however, corresponds to the teaching of the phase formation theory. At a given supersaturation a sub-critical size aggregate which is very close to critical size requires just a very small investment of free energy in order to reach critical size. Its growth, therefore, could take place comparatively easy. If a sub-size aggregate is farer away from critical size a much larger investment of free energy would be required for growth. Such an event, therefore, is much less probable. Instead, it will decay and thus lower the free energy of the system. As mentioned before, decay of a silver aggregate takes place by reaction with holes.

4. OUTLOOK

To-day, nearly 150 years after it has been discovered, the photographic process with light-sensitive silver halides is increasingly confronted with other competing processes for storage of optical information. A draw-back of the classical photography is the necessity of processing steps (development etc.) using wet chemical systems. Such steps are generally time consuming and require, if carried out on a commercial scale, very expensive apparatus. An adventage of silver halide photography which so far can not be beaten by modern competing systems (electrophotography, and others) is the enormous sensitivity. This is due to the amplification factor: Approximately four absorbed photons catalyze the reduction of a micro crystal consiting of 10^8 to 10^9 silver ions to the same number of silver atoms.

Silver halide photographic systems produced commercially have reached a remarkable standard of quality which is due to a highly sophisticated technology of production. It is surprising that the degree of our knowledge, how the photographic process works and what processes take place during exposure, is of much poorer level. We are unable to observe directly all the complicated processes in the micro crystals which start at exposure. We can not tell the difference between unexposed and exposed micro crystals which somewhere bear latent image silver clusters unless we subject them to development and thus irreparably destroy them. From model experiments, some of them far away from real photographic systems, we indirectly collect information on possible reaction pathways and try to fit them together into a frame which shall depict, how things happen in a micro crystal. It is a fascinating task, but we are aware that we have solved it still incompletely and not always unambiguously. Accordingly the contemporary atomistic theories which have been briefly reviewed in the present paper, of course, also suffer from incompleteness and, perhaps, from ambiguity.

The alternative way to approach the photographic process as an event of phase formation suffers likewise from incompleteness, and it may be doubted if microscopic processes such as formation of minute silver aggregates can be really described by a comparatively coarse theory which at best is valid for larger systems. It is beyond the scope of this review to discuss this question in detail. But the phase formation theory, if legally or illegally applied to photography, describes all photographic facts and effects so beautifully and self-consistently, that its application seems to be justified with just one restriction: Phase formation theory must not be regarded as a method which states how the photographic process proceeds; it should be regarded as a language for depicting how the photographic process could be described.

REFERENCES

(1) The Theory of the Photographic Process (ed. T.H.James), Macmillan Publ. Co., New York, London, 1977

(2) Grundlagen der photogr. Prozesse mit Silberhalogeniden (ed. H.Frieser, G.Haase, and E.Klein), Akad. Verlagsges., Frankfurt, 1968

(3) E.Moisar and E.Klein, Ber. Bunsenges. physikal. Chemie 67:949 (1963)

(4) R.W.Berriman and R.H.Herz, Nature 180:293 (1957)

(5) E.Klein, H.J.Metz, and E.Moisar, Photogr. Korrsp. 99:99 (1963), 100:55 (1964)

(6) J.Frenkel, Kinetik Theory of Liquids, Oxford Univ. Press, 1946

(7) K.L.Kliewer, J. Phys. Chem. Solids 27:705 (1953)

(8) K.Lehovec, J. Chem. Phys. 21:1123 (1953)

(9) E.Moisar and S.Wagner, Ber. Bunsenges. physikal. Chemie 67:356 (1963)
(10) E.Moisar, J. Photogr. Sci. 13:46 (1965)
(11) L.M.Kellogg, Photogr. Sci. Eng. 18:378 (1974)
(12) W.Jaenicke, Photogr. Sci. Eng. 6:85 (1962)
(13) R.Matejec and R.Meyer, Photogr. Sci. Eng. 6:265 (1962)
(14) H.E.Spencer and R.E.Atwell, J. Opt. Soc. Am. 54:498 (1964)
(15) E.Moisar, Photogr. Korresp. 106:149 (1970)
(16) E.Palm, F.Granzer, E.Moisat, and D.Dautrich, J. Photogr. Sci. 25:19 (1977)
(17) H.E.Spencer, Photogr. Sci. Eng. 11:352 (1967)
(18) I.Konstantinov and J.Malinowski, J. Photogr. Sci. 23:1,145 (1975)
(19) R.W.Gurney and N.F.Mott, Proc. Royal Soc. A164:15 (1938)
(20) J.W.Mitchell, J. Photogr. Sci. 6:57 (1958)
(21) J.W.Mitchell, Photogr. Sci. Eng. 22:1 (1978)
(22) J.W.Mitchell, Photogr. Sci. Eng. 23:1 (1979)
(23) J.W.Mitchell, Photogr. Sci. Eng. 26:270 (1982)
(24) E.Moisar, F.Granzer, D.Dautrich, and E.Palm, J. Photogr. Sci. 25:12 (1977)
(25) E.Moisar, F.Granzer, D.Dautrich, and E.Palm, J. Photogr. Sci. 28:71 (1980)
(26) J.Malinowski, Photogr. Sci. Eng. 23:99 (1979)
(27) E.Moisar and F.Granzer, Growth and Properties of Metal Clusters (ed. J.Bourdon), Elsevier Sci. Publ. Co., Amsterdam, 1980
(28) M.Volmer, Kinetik der Phasenbildung, Steinkopff, Dresden and Leipzig, 1935
(29) A.C.Zettlemoyer (ed.), Nucleation, M.Dekker, New York, 1969
(30) S.Stoyanov, Thin Solid Films 18:91 (1973)
(31) A.Bonnissent and B.Mutaftschiew, J. Chem. Phys. 58:3727 (1973)
(32) C.Wagner, Z. Elektrochem. 63:1927 (1959)
(33) R.Matejec, Z. Elektrochem. 66:459 (1962)
(34) E.Moisar, Photogr. Sci. Eng. 26:124 (1982)
(35) J.Teltow, Z. physik. Chem. 195:213 (1950)
(36) T.Ohachi and I.Taniguche, Crystal Growth (ed. R.A.Laudise et al.), Elsevier Publ. Co., Amsterdam, 1972
(37) D.Dautrich, F.Granzer, E.Moisar, and E.Palm, J. Photogr. Sci. 25:169 (1977)

FORMATION OF CLUSTERS IN BULK MATERIALS

A.E. Hughes

Materials Development Division,
Atomic Energy Research Establishment,
Harwell, Oxfordshire, U.K.

1. INTRODUCTION

Small precipitate particles dispersed in any solid can be regarded as local clusters of atoms forming a second phase whose properties may not be the same as those of the same compound in bulk form. For example, many metallic alloys form precipitates of special structure constrained by their small size or their coherence with the solid matrix. One example is the structure of Guinier-Preston zones found in aluminium alloys (1). The range of such systems is so wide that it is impossible to describe anything except a very small part of the subject here. We shall concentrate on systems that consist of small metallic particles ('colloids') in insulating matrices. Nearly all of the examples will be drawn from work on alkali metal colloids in the alkali halides, since such systems have been studied for a long time by very many different techniques. Some mention will also be made of closely-related systems such as silver colloids in alkali halides and metal colloids in oxide lattices. Work on all these systems, and on other examples where the properties of small particles of metals are important, has been reviewed recently (2). Some of the material covered here is therefore to be found in more detail in that article.

Alkali metal colloids in alkali halides were first studied in natural crystals such as rocksalt and in crystals in which an excess of alkali metal was produced by thermochemical treatment. The best-known example is 'blue rocksalt', in which the blue colour is caused by the optical extinction of small particles of sodium metal dispersed in the crystal. Studies of these systems actually preceded definitive studies of other colour centres and the alkali metal colloid is the first properly identified colour centre in these crystals (3,4). In many respects they are good model systems for the study of processes of cluster formation and growth since the concentration and size of the metal colloids can be deduced from their optical extinction spectrum. However, they are not without more practical interest, since metal colloids can be an important radiation damage product at high doses (5). This is relevant to situations such as the disposal of radioactive waste in salt formations (6,7) and insulating materials for future fusion reactors (8,9).

2. METAL-EXCESS ALKALI HALIDES

The best way to produce a simple stoichiometric excess of alkali metal in an alkali halide crystal is to heat it in the vapour of the alkali metal (3). This process is called additive coloration. At sufficiently high temperatures (typically $\gtrsim 400^{\circ}C$) the excess metal is incorporated into the lattice by the formation of F centres: halide ion vacancies with one trapped electron. F centres can also be introduced by irradiation, but in this case halogen interstitial species are also produced making the system more complicated. This will be discussed later in Section 6.

At lower temperatures F centres coagulate and form metal colloids. The equivalence between a cluster of F centres and a small region of metal is illustrated very schematically by Figure 1. The colloid-F centre system is effectively a one-component, two phase system; the F centre is the species forming the dispersed phase, analogous to a vapour or a species in solution in the crystal, whilst the colloid is the condensed phase. The two phases can coexist in thermodynamic equilibrium at a definite F centre concentration given by (4)

$$c_F^e = A e^{-Q_F/kT} \qquad (1)$$

where A is a constant and Q_F is the enthalpy of formation of an F centre from the colloid phase. This equilibrium can be studied from the optical extinction spectrum of a crystal. Figure 2 shows spectra of KBr:K taken after the equilibrium has been frozen-in by quenching from various high temperatures. It can be seen that the equilibrium favours F centres at high temperatures and colloids at low temperatures. From such measurements c_F^e can be found at various temperatures and the data used to find Q_F. Figure 3 shows c_F^e as a function of temperature for several alkali halides and Table 1 gives some deduced values of Q_F (10).

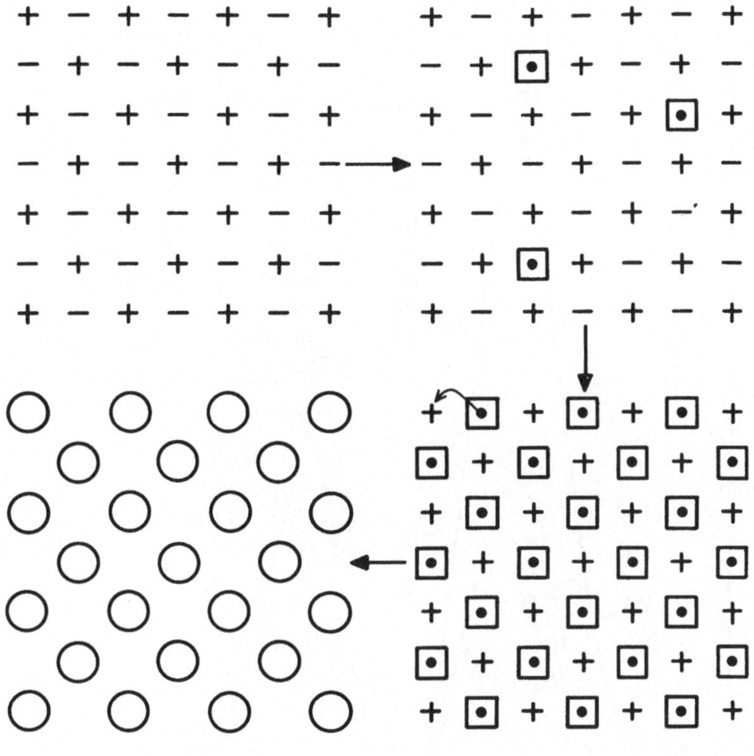

Fig. 1. This is a highly schematic illustration designed to show how the coagulation of F centres in an alkali halide is equivalent to the formation of a metal colloid. Note that if no atomic re-arrangement occurs, the clustering of F centres leaves metal atoms in a face centred cubic arrangement (adapted from Chassagne et al, ref. 15).

Table 1. Values of the enthalpy of evaporation of F centres from colloids in alkali halides (after Scott et al, ref. 10).

	NaCl	KCl	KBr	KI
Q_F(eV)	0.39 ± 0.10	0.35 ± 0.02	0.49 ± 0.02	0.39 ± 0.02

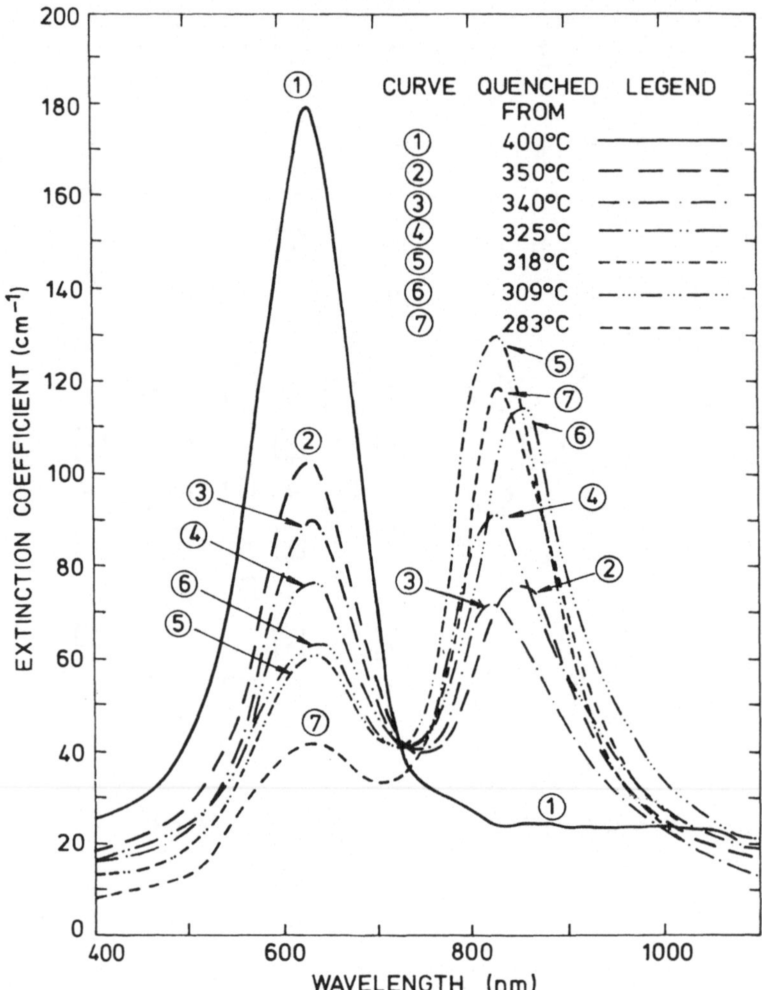

Fig. 2. Extinction spectra of potassium colloids in equilibrium with F centres in KBr at several temperatures (after Scott et al, ref. 10). The band near 600 nm is the F band and the band near 800 nm is due to the colloids.

3. EXPERIMENTAL METHODS FOR THE STUDY OF COLLOIDS

Table 2 shows the wide range of techniques that have been used to study small metal particles in ionic crystals and glasses. These are described in full by Hughes and Jain (2) and only a few examples will be given here. Probably the most powerful technique is direct observation of colloids by microscopy. Optical microscopy of large

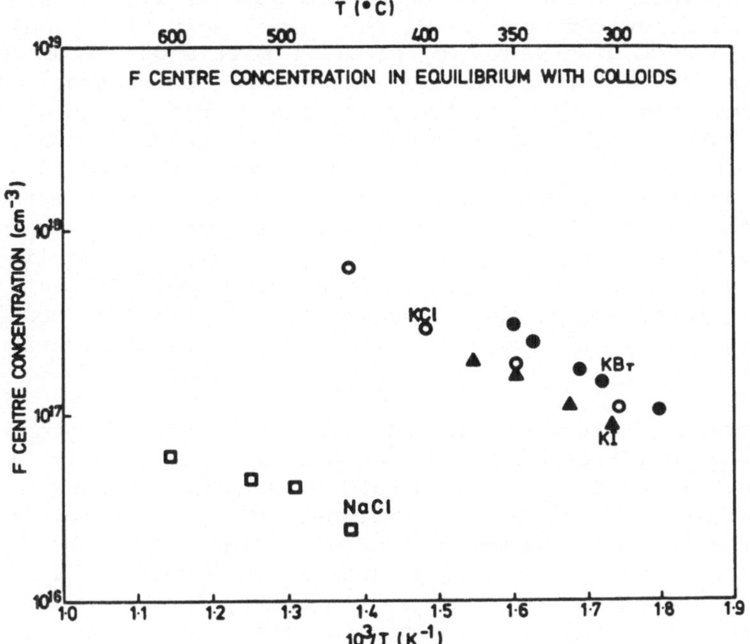

Fig. 3. Concentration of F centres in equilibrium with colloids in NaCl, KCl, KBr and KI (data taken from Scott et al, ref. 10).

Table 2. Experimental studies of metal colloids in ionic crystals and glasses.

Optical and Infrared Spectroscopy
Magnetic Susceptibility
Conduction Electron Spin Resonance (CESR)
Nuclear Magnetic Resonance (Knight shift)
Electrical Conductivity (work function)
Photoconductivity
Optical Microscopy
Electron Microscopy and Diffraction
X-ray Diffraction and Scattering (crystal structure)
Neutron Scattering (size and shape)
Raman Scattering
Differential Thermal Analysis (variation of melting point)
Hardening (interaction with dislocations)
Thermal conductivity (interaction with phonons)
Kinetics (nucleation and growth)

colloids (especially those large enough to scatter light strongly) has been carried out for many years (e.g. Amelinckx (11)), but only recently has direct observation by electron microscopy become possible through the use of sample preparation and observation techniques that minimize radiation damage to the sample by the electron beam (12). Hobbs et al (13) reported the observation of colloids in several alkali halides and these studies were extended to a thorough study of K colloids in KCl by replica and direct methods by Chassagne et al (14,15). Figure 4 shows images from colloids, where it can be seen that they form at or near dislocation lines as had also been inferred from optical ultramicroscopy. In fact colloids have long been used as a means of decorating dislocation structures in alkali halides to permit the study of dislocation networks (11).

An important feature of the transmission electron microscope studies is that diffraction spots from b.c.c. structure alkali metal are not seen. Thus it is concluded that the colloids have an f.c.c. structure that very closely matches the f.c.c. metal ion lattice in the rocksalt structure (see also Figure 1). The f.c.c. structure is presumably favoured because it involves less interfacial energy and introduces less strain energy into the lattice. It can be shown that rather small strains are involved if the colloids are f.c.c. structure, whereas the b.c.c. structure, with larger atomic volume, would introduce substantial strain energy to

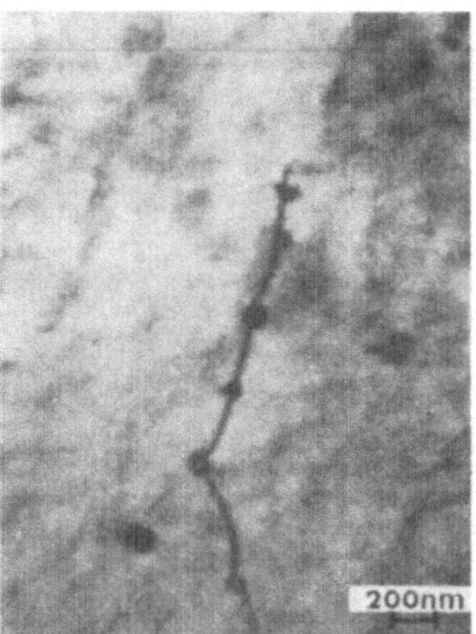

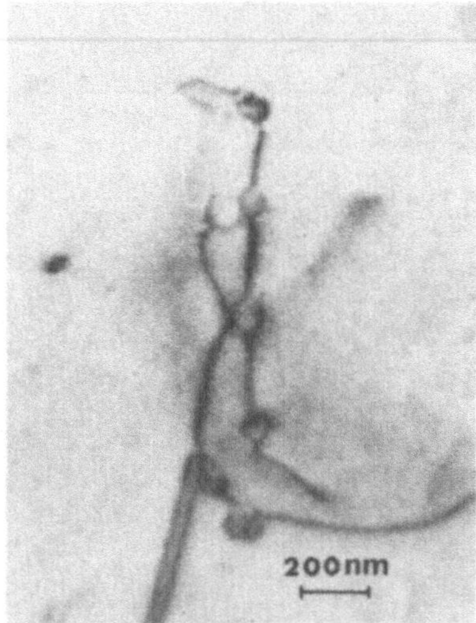

Fig. 4. Direct transmission electron microscope images of potassium colloids in additively coloured KCl. Note that the colloids form at and near dislocations (after Chassagne et al, ref. 15).

accommodate the colloid in the available volume of crystal (12,15). In a different system, Li, Na and K colloids formed by ion implantation in MgO, Treilleux and Chassagne (16) have analysed epitaxial relationships between the colloid and the matrix and shown how interfacial and strain energies influence coherency with the MgO lattice. It is clear from these studies that the structure of colloids is sometimes different from that of the bulk metals. This has also been observed for Ni colloids formed by thermochemical treatment of Ni-doped MgO (17).

Li colloids formed by neutron irradiation of LiF to doses $> 5 \times 10^{18}$ n cm^{-2} have also been found by X-ray diffraction to have f.c.c. structure (18), but they convert to b.c.c. structure on heating. At lower neutron doses (100) platelets of Li were also found, and were the subject of some interesting studies by magnetic resonance (19). The melting of the f.c.c. and b.c.c. Li colloids has also been studied with differential thermal analysis (DTA). Some results of Lambert et al (18) are shown in Figure 5. Peak B, corresponding to a melting temperature of nearly 190°C, is attributed to f.c.c. colloids since it disappears after heating, when the colloid structure is transformed to b.c.c. Figure 5(b) shows a series of DTA curves for b.c.c. colloids after the crystal has been heated to ever higher temperatures. It can be seen that after heating at 720°C the shape of the peak resembles that for bulk Li metal, Figure 5(c), whereas the DTA curves following earlier treatments at lower temperatures are broader. This is interpreted as an effect of particle size: smaller colloids melt at a lower temperature (20).

Optical spectroscopy is one of the most widely used methods of studying colloids in transparent matrices. The optical extinction spectrum of a spherical metal colloid can be calculated using the theory originally developed by Mie (21) and modified to take into account small particle effects (such as scattering of electrons at surfaces) by later authors (e.g. Doyle and Agarwal (22), Kreibig and Fragstein (23), Smithard and Tran (24); see Hughes and Jain (2), Papavassiliou (25), Van de Hulst (26) and Kreibig (27) for detailed discussions of the Mie theory). As an example Figure 6 shows calculations for Na colloids in NaCl for various colloid sizes by Smithard and Tran (24). The effects of increasing size are typical for colloids of nearly-free electron metals. The peak is broad for very small particles of radius < 5 nm because surface scattering reduces the mean free path of conduction electrons. As the particle size increases the band first narrows, and then broadens again and shifts to longer wavelengths as the radius increases to $\gtrsim 20$ nm and thus higher order multipole terms in the Mie theory are required.

Several interesting studies of conduction electron spin resonance (CESR, see Yafet (28)) have been made on metal colloids. The Li platelets studied by Taupin (19) showed features associated with

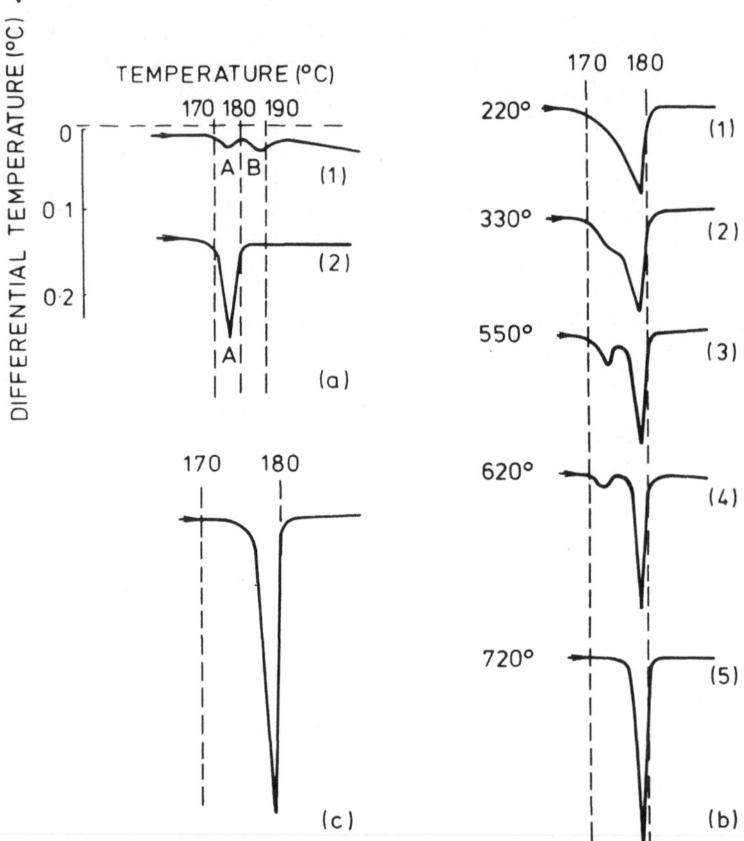

Fig. 5. DTA results for LiF irradiated to a dose of 6×10^{18} n cm^{-2}.
(a) Curve 1 is the signal observed during the first heating after irradiation. Curve 2 is the second run after recooling. Before curve 2 the sample had therefore been annealed to about 200°C, sufficient to convert f.c.c. Li to the b.c.c. form.
(b) The effect of successive heat treatments on the DTA results. The peak corresponds to the melting of b.c.c. Li precipitates and is the same as peak A in (a).
(c) DTA signal from a sample of bulk Li metal. (after Lambert et al ref. 18).

the quantized nature of the electron energy levels when the particles are very small (dimensions $\lesssim$ 5 nm). Kawabata (29) predicted that when the separation of electron levels δ_e exceeded the Zeeman energy $\hbar\omega_L = g\beta H$, the CESR linewidth would vary as R^2 where R is the particle radius. This contrasts with the effects of surface scattering of electrons which gives a contribution to the linewidth varying as R^{-1} (see Hughes and Jain (2)). The dependence of CESR

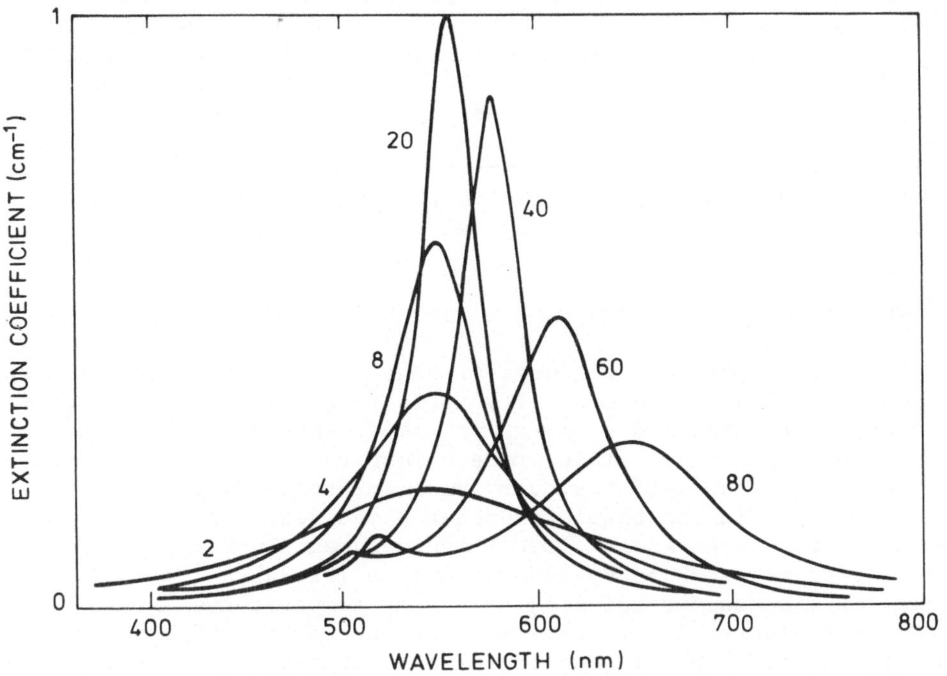

Fig. 6. Extinction spectra for a 10^{-7} volume fraction of sodium colloids in NaCl as calculated by Smithard and Tran, ref. 24. The numbers by the curves are colloid diameters in nm.

linewidth on R is thus not simple. The line is usually motionally narrowed because the characteristic time for electron scattering, τ_R, is much smaller than the reciprocal of the Larmor frequency ω_L. Under these conditions the spin-lattice and spin-spin relaxation times T_1 and T_2 are about equal and the linewidth is given by

$$\Delta\omega = T_2^{-1} \simeq (\Delta g)^2/\tau_R \tag{2}$$

where Δg, the g-shift, depends on the spin-orbit coupling in the metal, being larger for metals of high atomic number. For a small particle (but not so small that quantization effects are important) τ_R will be limited by surface and phonon scattering of electrons, so that, to a reasonable approximation

$$\Delta\omega \propto \tau_R^{-1} \propto \frac{a}{R} + bF(T) \tag{3}$$

where a and b are constants and F(T) would follow a Gruneisen law like the resistivity ($F(T) \propto T^5$ for $T \ll \theta_D$ and $\propto T$ for $T \gg \theta_D$).

The observation of CESR with a linewidth whose magnitude and temperature dependence fits with eqn (3) is often useful confirmation of the existence of metal colloids. Figure 7 shows the CESR linewidth observed by Vitol et al (30) in additively coloured NaCl. Notice that the linewidth increases discontinuously at the melting point of Na. It is also larger than the linewidth observed in bulk metal, consistent with a contribution from surface scattering. It is not known whether the Na colloids studied by Vitol et al were f.c.c. or b.c.c. by the time they melted.

4. NUCLEATION OF COLLOIDS FROM F CENTRES

If at a temperature T a crystal contains F centres at a concentration $c_F > c_F^e$ given by eqn. (1), it is in a supersaturated condition and a fraction $(c_F - c_F^e)/c_F$ of the F centres should cluster to form metal colloids. This process must take place by the formation of stable 'nuclei' i.e. very small clusters, which then grow by the condensation on them of further F centres. It is well known that at temperatures around 300K F centres can cluster to form well-known aggregates such as F_2 and F_3 centres (31,32). The obvious nucleation process for colloids at higher temperatures (typically $\gtrsim$ 200°C see Figures 2 and 3) is thus the nucleation of F aggregate centres that turn into larger colloidal clusters. If this process were to occur homogeneously in the bulk crystal, we can make the following very simple estimate of the density of stable clusters that grow into colloids, adapting the methods of Greenwood et al (33) and Jain (34).

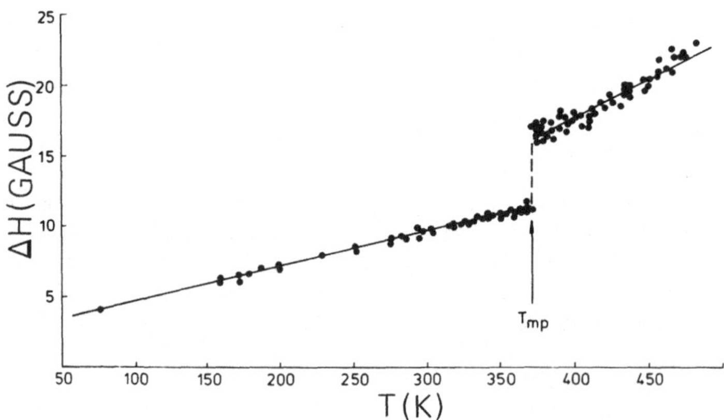

Fig. 7. CESR line width for sodium colloids in NaCl as a function of temperature (after Vitol et al, ref. 30).

Let C_c be the density of clusters at some stage in the process, R their effective radius and D the diffusion coefficient of F centres in the crystal lattice. The number of F centres arriving at existing clusters per unit volume per second will then be given by

$$\dot{c}_F \text{ (to clusters)} = 4\pi DRC_c c_F \tag{4}$$

We now calculate the rate of formation of new nuclei, assumed to form if two F centres meet to form an F_2 centre. Let Γ be the jump rate of an F centre and Z be the number of other sites sampled for each jump. The probability that one particular F centre will meet another per second is then given by $Z\Gamma(c_F/N)$ where N is the number of anion sites per unit volume. The rate of formation of new nuclei per unit volume per second is therefore

$$\dot{c}_F \text{ (to new nuclei)} = Z\Gamma\ c_F^2/N = 6ZDc_F^2/(Ns^2) \tag{5}$$

since $D = s^2\Gamma/6$ with s the jump distance. The growth of existing clusters will therefore dominate the formation of new nuclei only when

$$4\pi DRC_c c_F > 6ZDc_F^2/(Ns^2) \tag{6}$$

$$\text{i.e. } C_c > \frac{6Z}{4\pi}\, c_F\, (\frac{s}{R}) \text{ since } N \simeq s^{-3} \tag{7}$$

Since to form colloids c_F must be $\simeq 10^{17}-10^{18}\ cm^{-3}$ (see Figure 3) it follows that eqn. (7) predicts a colloid density C_c of $\gtrsim 10^{16}-10^{17}\ cm^{-3}$. This is more than a thousand times larger than the densities found experimentally (35). Also, the experimental evidence favours the inhomogeneous nucleation of colloids at, for example, dislocations, rather than homogeneous nucleation throughout the crystal, and no significant number of F_2 or F_3 centres is found during colloid growth at high temperatures. It thus appears that the simple assumption that cluster formation progresses homogeneously through stable dimers, trimers etc. does not apply simply to these systems. Rather, it looks as if stable nucleation is favoured at dislocations (and perhaps other defect sites such as impurities, e.g. see Durand et al (36)), followed by cluster growth at these sites. Only the development of a proper rate theory approach to nucleation, together with information on the stability of intermediate clusters, will be able to deal properly with the nucleation problem and indicate the conditions under which homogeneous or inhomogeneous nucleation will dominate. This is also important in the production of colloids by irradiation (34).

5. GROWTH OF COLLOIDS

Once nucleated, colloids will grow by the addition of further F centres. The F centres must diffuse to the colloids and then be transformed into extra metal atoms at the colloid-matrix interface. This is a classical precipitation process as treated theoretically by Zener (37), Wert and Zener (38,39), Frank (40), Ham (41,42) and others. Reviews may be found in Bullough and Newman (43), Christian (44) and Jain and Hughes (45). These theories consider the growth of a fixed number of precipitate particles separated from each other by a distance λ. Three limiting cases can be defined: (1) the rate controlling process is a reaction at the interface; (2) the rate controlling process is a bulk diffusion of the solute (F centres in our case); (3) the rate controlling process is diffusion of solute to and along dislocations or grain boundaries. In each case the equation for the rate of change of particle radius can be written in terms of the concentrations of the solute species (denoted by C's) as

$$C_p \frac{dR}{dt} = \frac{A}{R^{m-1}} (C_a - C_R); \quad (m = 1,2,3) \tag{8}$$

Here C_a is the solute concentration in the crystal, C_R is the solute concentration in equilibrium with the particle, C_p is the concentration in the precipitated phase and A is a rate constant (see Jain and Hughes (45); for case 2 A is simply the bulk diffusion coefficient D). From eqn. (8) the rate of change of the fraction W of original solute precipitated onto a network of precipitates separated by the distance λ can be worked out as:

$$\tau_m \frac{dW}{dt} = (1-W)\, W^{(1-m/3)} \tag{9}$$

$$\text{with } \tau_m = \lambda^3/(4\pi\, A\, R_F^{\,3-m}). \tag{10}$$

Note that, since R_F is the precipitate radius at the end of the process (i.e. when C_a has dropped to C_R), the rate constants τ_m depend on both the density of initial nuclei (through λ) and on the total amount of solute available for precipitation (through R_F).

The full solutions of eqn. (9) may be found in Jain and Hughes (45) and Hughes and Jain (2). They are plotted in Figure 8 as curves 1, 2 and 3 for m = 1, 2 and 3. In the early stage of precipitation ($W \ll 1$) the curves can be described by

$$1-W = \exp\{-P(t/\tau_m)^{3/m}\} \tag{11}$$

where P is a constant. In the later stages all three curves approximate to exponential decays (as does case 3 over the whole of the range of W) of the form

$$1-W \simeq \exp(-t/\tau_m). \tag{12}$$

Note that if we express 1-W in the often used form of $1-W = \exp(-\alpha t^n)$, eqns. (11) and (12) show that the theories all give $n \geqslant 1$.

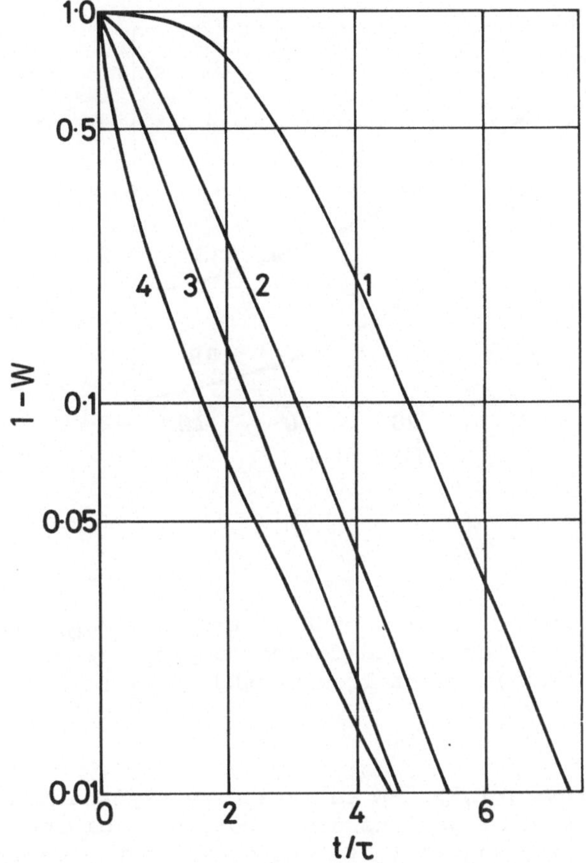

Fig. 8. Theoretical values of the unprecipitated fraction of solute 1-W as a function of time t for cases 1, 2 and 3 as discussed in the text. Curve 4 is constructed as the sum of several exponential curves and approximates very closely to $\exp(-\alpha t^n)$ with $n = {}^2/3$.

Durand et al (36) have studied precipitation of F centres into colloids at 200°C in KCl, including in deformed crystals. Their results, some of which are shown in Figure 9, require values of n in the range 0.6-0.9 and show that the precipitation rate is faster in deformed crystals. These features are closest to case 3 for precipitation controlled by dislocations, but even then should have n = 1. Values of $n < 1$ have frequently been observed for precipitation in metallic alloys with high dislocation densities (e.g. C in deformed Fe) and have been the subject of much debate

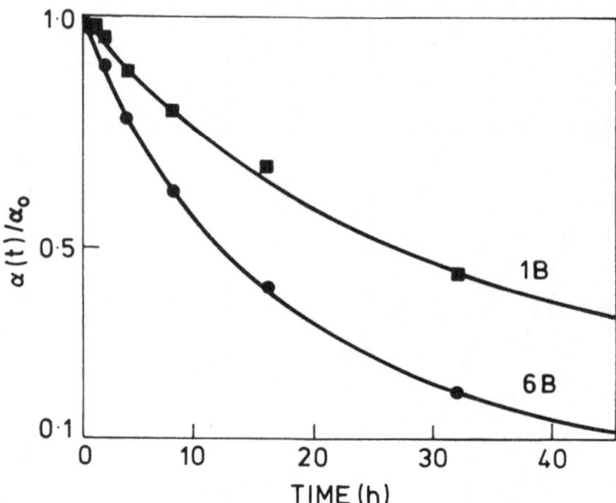

Fig. 9. Unprecipitated fraction of F centres $\alpha(t)/\alpha_0$ versus annealing time at 200°C for an undeformed sample 1B and for a deformed sample 6B (3.2%), taken from Durand et al ref. 36.

(see Bullough and Newman (43). A satisfactory explanation can be given in terms of an inhomogeneous distribution of dislocations and precipitates. Jain and Hughes (45) have shown that a variation in the inter-precipitate distance λ or the local dislocation density can lead to a situation in which different regions of the crystal are effectively precipitating at different rates. A superposition of exponential precipitation curves can then be obtained which fit an equation with $n < 1$. An example is plotted as curve 4 on Figure 8. There are very few other experimental measurements of precipitation curves for colloid formation in alkali halides, but some early work by Penley with Witte (46) on KCl also gives some curves with $n < 1$. It thus appears that studies of the precipitation stage of colloid growth also are consistent with inhomogeneous nucleation and the important part played by dislocations.

A distribution of particle sizes during precipitation will also lead in general to a superposition of precipitation curves with different rates, since eqn. (8) shows that the rate of growth of a particle depends on its radius. This appears in eqn. (9) as the term $W^{(1-m/3)}$. Note, however, that in case 3 ($m = 3$) this explicit dependence on radius vanishes, and that in all cases an inspection of eqn. (8) shows that any particle size distribution expressed in terms of particle radius tends to narrow as precipitation proceeds.

6. PARTICLE COARSENING

During the precipitation stage that we have just discussed it is assumed that all particles are growing as solute comes out of solution. This will continue until C_a falls to near C_R. However, we must recognize that the solute concentration in equilibrium with a small particle of radius R, C_R, is larger than it is for a plane surface C_∞, because of a contribution to the free energy from the interfacial energy. C_R will be given by

$$C_R = C_\infty \exp(\alpha/R) \tag{13}$$

where $\alpha = 2\Omega\sigma/kT$. Ω is the atomic volume of the solute and σ the interfacial energy of the precipitate. Since eqn. (13) shows that C_R is larger for small particles, there will come a time when C_a falls below the value of C_R appropriate for particles that are somewhat smaller than average, whilst larger particles are still growing and causing C_a to continue to fall. The smallest particles will then start to dissolve and there will be a flow of solute from small to large particles as well as a continued overall decrease in C_a. This process, known as Ostwald ripening, gives rise to a coarsening of the precipitate distribution since large particles grow as smaller ones disappear. The effect is vitally important in engineering alloys that rely on a distribution of fine precipitates for their high temperature strength (e.g. nickel-based alloys that rely on γ' precipitates). Once the main precipitation stage discussed in Section 4 is over, a ripening or coarsening stage follows in which the average particle size increases and the particle density falls as small particles disappear. Figure 10 shows some results for additively coloured KCl (35) that rather clearly illustrate the slow increase in mean colloid radius during the ripening stage (stage II on the figure), following a more rapid growth during precipitation (stage I).

The theory of Ostwald ripening is rather complicated and was developed by Lifshitz and Slezov (47-49) and Wagner (50). Reviews are given by Greenwood (51), Kahlweit (52,53) and, specifically for ionic crystals and glasses, by Jain and Hughes (54). The rate of shrinkage or growth of an individual particle is given by eqn. (8) into which eqn. (13) is substituted after linearizing it to the form

$$C_R \simeq C_\infty (1 + \alpha/R) \tag{14}$$

A critical radius R_c is defined such that $C_a = C_{R_c}$; thus R_c is the radius of a particle that is instantaneously neither shrinking nor growing. Eqn. (8) then becomes

$$C_p \frac{dR}{dT} = \frac{AC_\infty\alpha}{R^{m-1}} \left(\frac{1}{R_c} - \frac{1}{R}\right). \tag{15}$$

The particle size distribution function f(R,t), which the theory aims to calculate, must satisfy the continuity equation

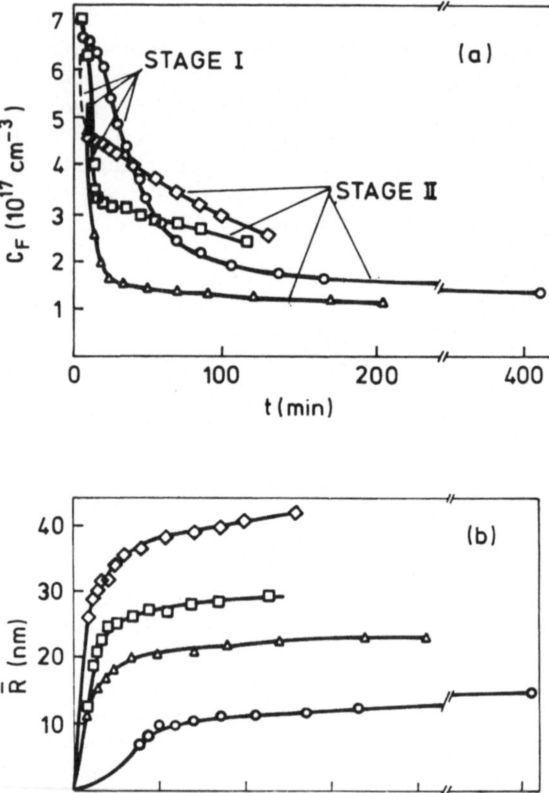

Fig. 10. Variation of the F centre concentration (a) and growth of mean colloid radius (b) in KCl initially containing 7 x 10^{17} F centres cm^{-3}. The points refer to different temperatures as follows:- circles 250°C, triangles 305°C, squares 348°C, diamonds 392°C (after Calleja and Agullo-Lopez, ref. 35).

$$\frac{\partial f}{\partial t} + \frac{\partial (f.\dot{R})}{\partial R} = 0 \; ; \quad \dot{R} = \frac{dR}{dt} \tag{16}$$

Lifshitz and Slezov describe a prescription for calculating f(R,t) from eqns. (16) and (15) and it is possible to show that there is a quasi-steady-state condition such that the size distribution function expressed as $F(R/R_c)$ is independent of time, and R_c evolves like $t^{1/(m+1)}$ for the three rate-limiting processes discussed earlier in Section 4. $F(R/R_c)$ for the three cases is plotted in Figure 11: the mathematical expressions may be found in Jain and Hughes (54).

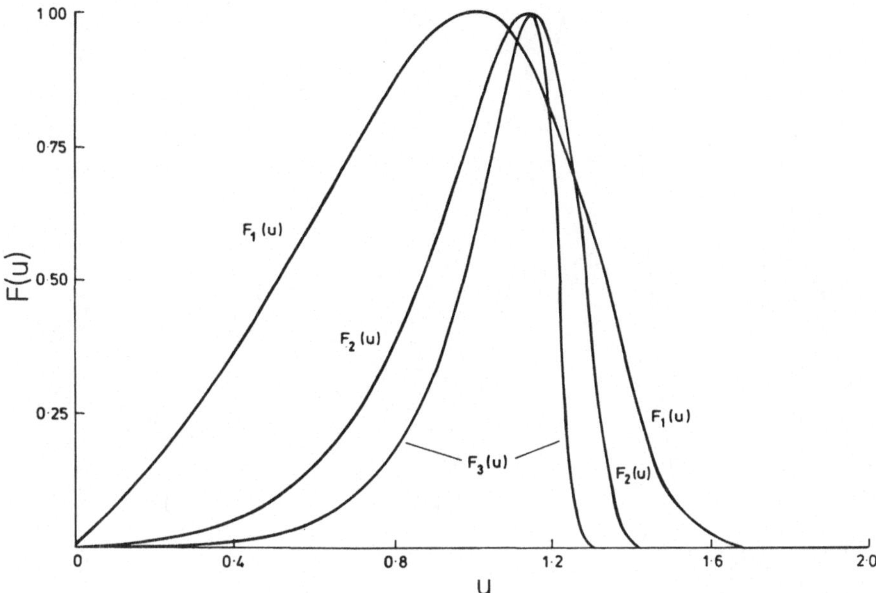

Fig. 11. Theoretical size distribution F(u) for Ostwald ripening for the three cases 1, 2 and 3 as described in the text. The maximum values of the three functions occur at u = 1.00, u = 1.135 and u = 1.142 for cases 1, 2 and 3 respectively.

The mechanism of ripening that we have just described relies on the transport of solute from small particles to large particles, and is the process usually assumed to work in solids. However, there is another mechanism, originally considered by Smoluchowski, (55,56), that is more common in fluid suspensions and involves the coalescence of particles in binary collisions or interactions. For this process the particles themselves must be capable of motion, or there must be a mechanism of mass exchange between special pairs of (perhaps close) particles. The theory is described by Baroody (57) and Geguzin and Krivoglaz (58) and leads to a size distribution approximating to the function

$$F_S(R) \simeq \exp\{-(\ln R/R_m)^2/2\xi^2\} \qquad (17)$$

This is shown in Figure 12, where it is clear that it is oppositely skewed from the Ostwald ripening size distribution functions in Figure 11.

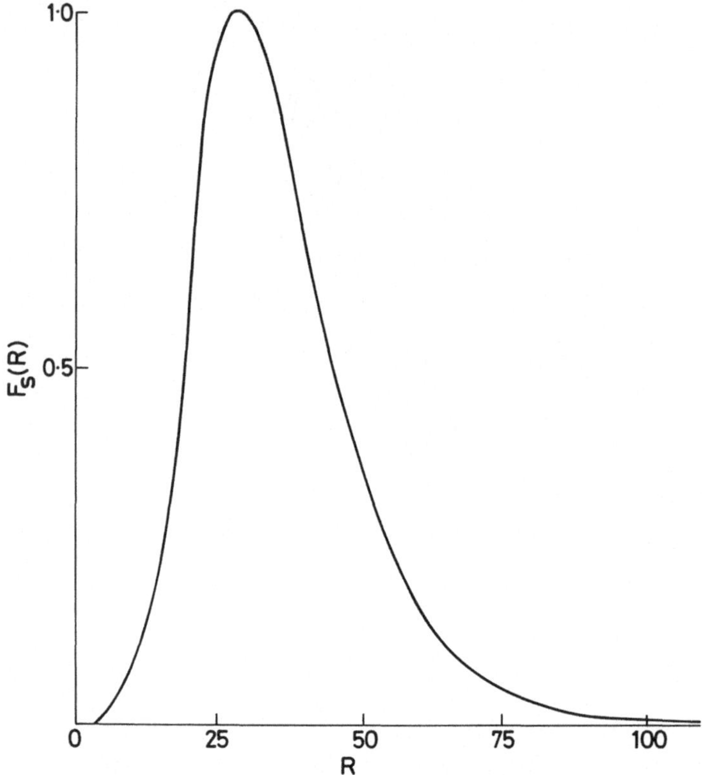

Fig. 12. Size distribution $F_S(R)$ for particles ripening by binary coalescence.

Experimental results on the ripening of colloids generally display the features expected from the theories just described, although quantitative agreement is often lacking. Calleja and Agullo-Lopez (35) plotted their data shown in Figure 10 as $\bar{R}^3$ vs. t, as for case 2, and found reasonable agreement at the highest temperature of 665K. At the lower temperatures their data probably fit $\bar{R}^4 \propto t$ (case 3) rather better (see Hughes and Jain (2)). Figure 13 shows the results of Smithard and Tran (24) for the ripening of Na colloids in NaCl. At each temperature the mean radius fits a power law with time, but it can be seen from the added reference lines that the slope seems to change from $t^{1/4}$ to $t^{1/3}$ and then to $t^{1/2}$ as the temperature increases. This might be interpreted as a change from a dislocation controlled process as the temperature increases.

There are only a few cases where size distributions have been measured by electron microscopy. Data for silver colloids in KCl obtained by Jain and Arora (59) can be fitted satisfactorily by the

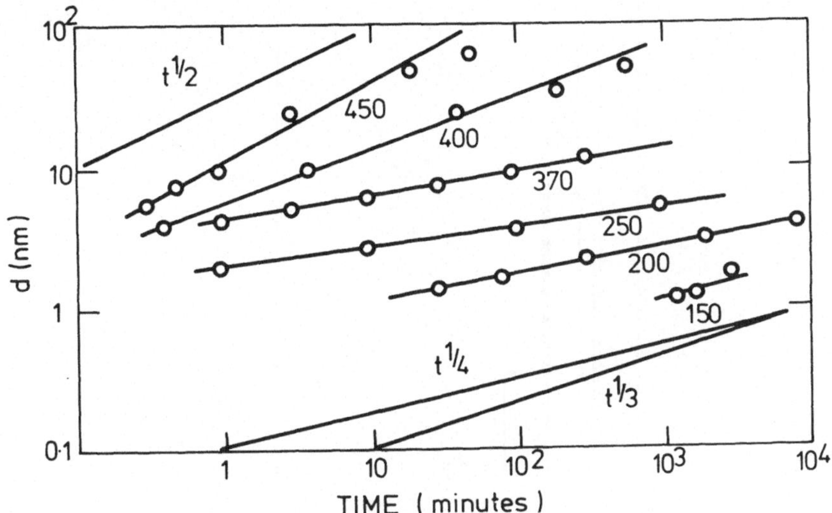

Fig. 13. Variation of the mean colloid diameter as a function of time for Ostwald ripening of sodium colloids in NaCl (after Smithard and Tran, ref. 24). The lines indicate slopes of curves corresponding to different powers of time for the three theoretical cases.

distribution function for case 3 shown in Figure 11 (see Hughes and Jain (60)). However, the data on K colloids in KCl presented by Chassagne et al (14) are more complicated. Figure 14 shows some of their results; it is clear that the observed size distributions are more complicated than those predicted by ripening theory and there seems to be a tendency for (i) a skewness more like that in Figure 12 (ii) groups of (large) particles that fall outside the main distribtuion and (iii) the absence of a smooth change in the size distribution with ageing time. Chassagne et al (15) and Jain and Hughes (54) have discussed these features (that are also found in other cluster or precipitate systems, e.g. Kreibig (51), Shvarts eg al (62), Treilleux et al (63)) and concluded again that they might be understood in terms of the inhomogeneous distribution of nucleation and growth sites. This agrees with evidence from electron microscopy, and therefore seems to be a common feature of the behaviour of the systems that we have discussed in the last two sections. It shows that although many features of colloid formation and growth can be understood in general terms, they do not provide easy routes to quantitative information about precipitation and ripening processes since it is often difficult to be sure that the correct rate limiting steps have been identified.

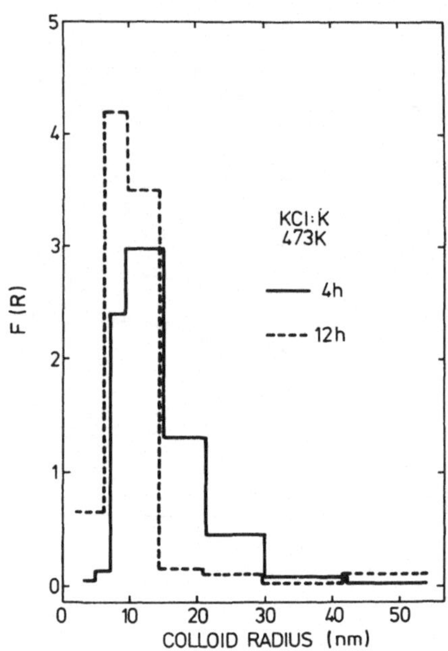

Fig. 14. Size distribution of colloids in additively coloured KCl as measured by Chassagne et al, ref. 14, using transmission electron microscopy. Two histograms are shown: one after annealing for four hours at 473K and the other after annealing for 12 hours at the same temperature.

7. FORMATION OF CLUSTERS BY IRRADIATION

One of the ways in which metal clusters or colloids can be produced in ionic crystals is by implantation with the appropriate metal ions (e.g. Davenas et al (64), Treilleux and Chassagne (16), Matzke et al (65), Turos et al (66)). Usually some heat treatment is required to convert the implanted ions, distributed within the implantation depth usually of less than 1 μm, into clusters that grow to become identifiable metal particles. This can be a useful way to make special systems not easily obtained by other methods, e.g. a high concentration of K or Pt in MgO, and to study the coherency relationships of such systems with the matrix. In the ion implantation process radiation damage may play a role, but is not usually the central purpose of the investigation.

Metal colloids are, however, one of the possible products of radiation damage in ionic crystals and glasses, one obvious example being in the photographic process in silver halides. In alkali halides radiolysis can produce atomic displacements on the anion sublattice through a mechanism that involves non-radiative decay of exciton states (see Williams (67) and Townsend and Agullo-Lopez (68) for recent reviews). At low temperatures the products are F centres and interstitial halogen atoms that form H centres (the H centre is essentially an X_2^- molecular ion occupying a halogen ion (X^-) site). The H centre is mobile at quite low temperatures (69) so that irradiation at room temperature produces interstitial clusters. Electron microscopy of these clusters led Hobbs et al (70) to conclude that they result from the aggregation of H centres and have the structure of a perfect dislocation loop surrounded by halogen molecules occupying anion-cation vacancy pairs. This two-part model for interstitial clusters can explain many features of radiation damage behaviour in alkali halides near room temperature (see for example Hobbs et al (70), Hobbs (12), Catlow et al (71) and Hughes (72)).

Colloids can be formed from F centres in crystals irradiated to fairly high doses ($\gtrsim$ 50 Mrad of ionizing radiation) near room temperature by heating them to temperatures (e.g. $\sim$ 150°C in NaCl) where the F centres are mobile and cluster (73). The colloidal centres thus produced take part in the recombination process by which the defect-free crystal is recovered by annealing (72). Irradiation at temperatures above room temperature can produce colloids directly, and moreover under such conditions much larger defect concentrations can be reached than are possible at lower temperatures. The colloids provide stable sinks for F centres so that damage can accumulate to a high level (several percent of the lattice converted to colloids and interstitial defect structures). This process has been studied particularly in NaCl, originally because it was a convenient system to study (5) and more recently because of the projected use of salt deposits to store radioactive waste (6,7,74-77). Figure 15 shows some growth curves of F centres and colloids for irradiation at several temperatures and illustrates

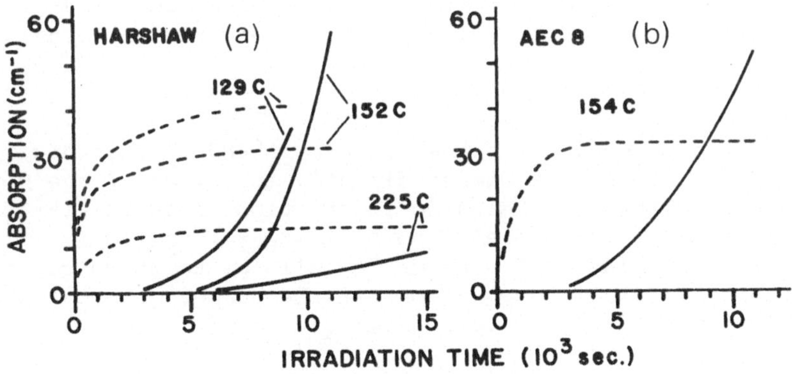

Fig. 15. Growth curves for the F centres (dashed lines) and colloids (full lines) during irradiation of synthetic NaCl (a) and natural rocksalt (b). The curves were measured at a dose rate of 120 Mrad h^{-1} by Swyler et al, ref. 74.

how the F centre concentration saturates as the colloid growth, after an 'incubation' period, starts. The efficiency of colloid growth in NaCl is highest at about 150°C. The details depend rather critically on the source, purity and state of deformation of the crystals and on the irradiation conditions (7,78).

The qualitative features of colloid growth in irradiated NaCl are reminiscent of void growth in neutron irradiated metals (79). Jain and Lidiard (80) have adapted the theory of void growth to treat colloid growth in irradiated alkali halides. They write the rate of change of the concentrations of F and H centres during irradiation in the form[1].

$$\frac{dc_F}{dt} = K' - K_1 c_F - K_2 c_F c_H \tag{18}$$

$$\frac{dc_H}{dt} = K - K_3 c_H - K_2 c_F c_H. \tag{19}$$

Here K is the rate of production of F-H pairs by the irradiation, K_2 is the rate of their recombination in the lattice (controlled mainly the rapid diffusion rate of the H centres) and K_1 and K_3 are the rate constants for disappearance of F and H centres at interstitial dislocation loops (of dislocation line density ρ_d) and colloids (of concentration C_c). They are given by

$$K_1 = D_F (4\pi R_c C_c + z_F \rho_d) \tag{20}$$

$$K_3 = D_H (4\pi R_c C_c + z_H \rho_d) \tag{21}$$

where D_F and D_H are the diffusion coefficents for F and H centres respectively. The forward rate of production of F centres K' includes a term for the evaporation of F centres from colloids:

$$K' = K + 4\pi R_c C_c D_F c_F^e \tag{22}$$

where c_F^e is given by eqn. (1). The factors z_F and z_H that describe trapping at dislocations are biased slightly in favour of the H centres because of the larger elastic interaction of interstitials with the dislocation strain field, so that $(z_H - z_F)/z_F \simeq 0.1$. This bias provides a net flux of F centres to colloids to drive colloid growth through the following equation for the number of metal atoms per unit volume in the form of colloids, c_A:

1. Note that Jain and Lidiard work in terms of mole fractions of F and H centres rather than concentrations as used here.

$$\frac{dc_A}{dt} = 4\pi\ R_c\ C_c\ \{D_F(c_F - c_F^e) - D_H\ c_H\}. \tag{23}$$

Jain and Lidiard solve these simultaneous differential equations in approximate analytical form and also numerically. They predict that the F centre concentration grows and then saturates according to

$$c_F = c_F(\text{sat})\ \{1 - \exp\ (-2K_1 t)\}^{\frac{1}{2}} \tag{24}$$

and that the number of atoms in colloids varies like t^n where $1 < n < 2$. They were able to fit some of the experimental data by choosing appropriate values for the parameters such as D_F; an example of a fit is shown in Figure 16. Lidiard (81) has also applied the theory to the stored energy data of Jenks and Bopp (6).

Levy et al (7) have now done many more experiments than were available when the theory was originally published, and have compared their findings with the predictions. They note some areas where discrepancies seem to remain. Firstly, Levy et al find that colloid growth follows a higher power of t in some cases than the theory predicts. This almost certainly reflects the way the theory ignores the colloid nucleation stage and assumes a constant colloid density. Secondly the required activation energy for the F centre diffusion coefficient D_F was taken as 0.8 eV by Jain and Lidiard, but Levy et al find, from $c_F(\text{sat})$ which is proportional to $D_F^{-\frac{1}{2}}$, that their data require rather lower values over the temperature

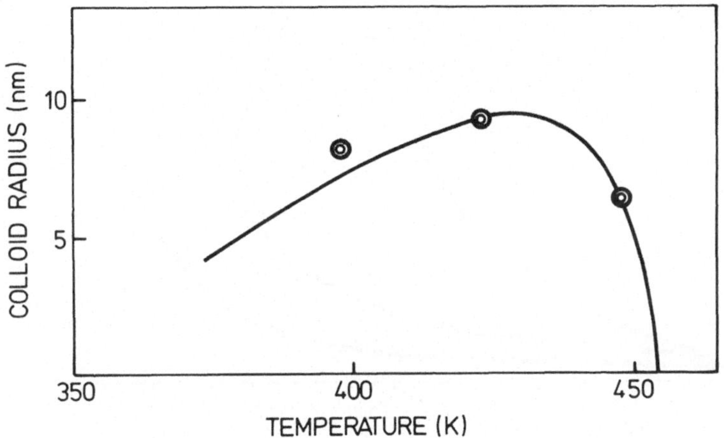

Fig. 16. The full line shows the temperature dependence of the colloid radius in sodium chloride after a total dose of 6 displacements per anion as calculated by the theory of Jain and Lidiard (ref. 80) for a dose rate of 10^{-4} dpa s^{-1}. The circles are experimental points at a total measured radiation dose of 15 Grad.

range where colloid growth is most efficient. Some of the quantitative differences between the original theory and recent experiments may be reduced by modifying the choice of the values used for some of the parameters (78). The theory and experiment do agree that the colloid induction period and growth depend on dislocation density and hence on plastic strain, as shown in Figure 17.

Quite apart from the direct relevance to radioactive waste repositories in salt, the work on NaCl provides a useful testing ground for the concepts involved in the formation of colloids (and other clusters) during irradiation of ionic crystals. Much less work has been carried out in other systems. Colloids can be formed by high temperature ionizing irradiation of alkaline earth fluorides (5,82) and in ambient temperature irradiation of alkali azides (83). In oxides radiation damage occurs predominantly by atomic collisions rather than by ionizing events, so that atoms on both sublattices will be displaced. There has been some doubt over whether metal colloids form in oxides such as Al_2O_3 and MgO irradiated to high doses with fast neutrons or electrons (84,85). Recently, Shikama and Pells (86) have irradiated single crystal foils of Al_2O_3 in a high voltage electron microscope to 1 MeV electron fluences of $> 10^{23}$ cm^{-2} at temperatures of 900-1130K, and have found evidence

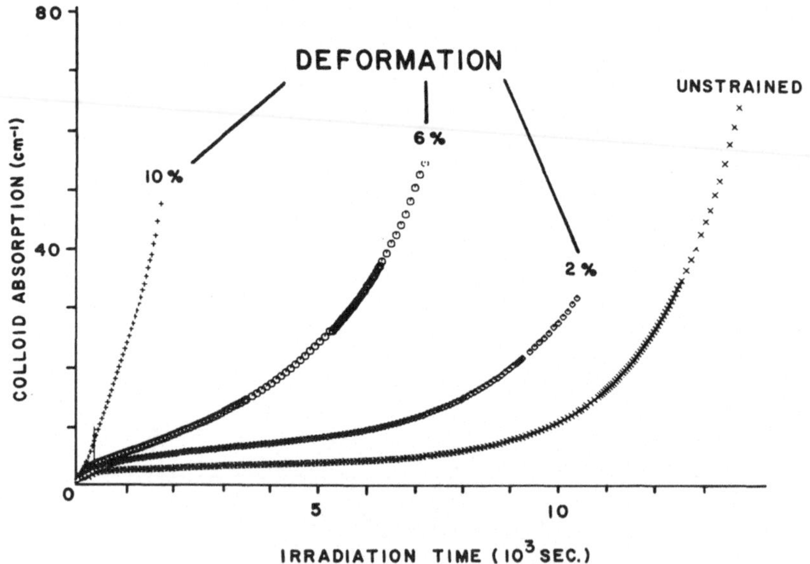

Fig. 17. Growth of the colloid absorption during irradiation of plastically deformed synthetic NaCl, measured at a dose rate of about 120 $Mradh^{-1}$ by Swyler et al, ref. 74.

for the production of Al colloids from the appearance of electron diffraction patterns from f.c.c. Al. The colloids have dimensions of a few tens of nm and the diffraction spots disappear near the melting temperature of Al. The mechanism of formation of the Al colloids is not certain; Shikama and Pells suggest that 1 MeV electron irradiation creates many more Al displacements than oxygen displacements, because the displacement energy of Al is much lower than for oxygen (85), and that the excess aluminium then precipitates at irradiation-induced voids. If this is the case neutron irradiation may not produce colloids since the number of displacements on the two sublattices will be more equal when large collision cascades are involved.

8. SUMMARY

In this article cluster formation in bulk materials has been discussed using as examples the formation of metallic colloids in ionic crystals. We have seen how the broad features of the precipitation and growth of such clusters can be related to theories of these processes that generally have rather wider applicability to the growth of second phases in solids. The theories provide a good framework for interpreting experiments, but care must be taken before quantitative information is extracted. This is particularly the case for the nucleation stages of cluster formation, where simple theories of homogeneous behaviour do not seem to reproduce the experimental situation, which may be strongly influenced by the formation of clusters at special sites in the solid such as dislocations.

These points also apply to the formation of metallic colloids by irradiation of ionic crystals. Most work has been done on NaCl where it is clear that the rate of formation of sodium colloids depends critically on the microstructure of the crystal as well as on the irradiation conditions. In the light of these results studies of the formation of clusters in other materials such as oxides must take note of possible sample-to-sample variations and the need to work with well-characterized specimens.

REFERENCES

1. Kelly, A., and Nicholson, R.B. Prog. Mat. Sci. 10 (1963) 151.
2. Hughes, A.E. and Jain, S.C. Adv. Phys. 28 (1979) 717.
3. Schulman, J.H. and Compton, W.D. Color Centers in Solids (Oxford: Pergamon Press, 1962).
4. Markham, J.J. F Centers in Alkali Halides Suppl. 8 to Solid State Physics ed. F. Seitz and D. Turnbull (New York: Academic Press, 1966).
5. Hobbs, L.W. and Hughes, A.E. A.E.R.E. Harwell Report R8092 (1975).
6. Jenks, G.H. and Bopp, C.D. Oak Ridge National Laboratory Report 5058 (1977).
7. Levy, P.W., Loman, J.M., Swyler, K.J. and Klaffky, R.W. in Advances in the Science and Technology of the Management of High Level Waste ed. P.L. Hofmann (Brookhaven National Laboratory Report 29909, 1981, to be published).
8. Clinard, F.W. in Critical Materials Problems in Energy Production ed. C. Stein (New York: Academic Press 1976) p.141.
9. Hobbs, L.W. J. Amer. Ceram. Soc. 62 (1979) 267.
10. Scott, A.B., Smith, W.A. and Thompson, M.A. J. Phys. Chem., 57 (1953) 757.
11. Amelinckx, S. Phil. Mag. 1, (1956) 269.
12. Hobbs, L.W. in Surface and Defect Properties of Solids, ed. M.W. Roberts and J.M. Thomas (London: The Chemical Society 1975) Vol. 4, p.152.
13. Hobbs, L.W., Hughes, A.E., and Chassagne, G. Nature 252 (1974) 383.
14. Chassagne, G., Durand, D., Serughetti, J., and Hobbs, L.W. Phys. Stat. Sol. (a) 40 (1977) 629.
15. Chassagne, G., Durand, D., Serughetti, J., and Hobbs, L.W. Phys. Stat. Sol. (a) 41 (1977) 183.
16. Treilleux, M. and Chassagne, G. J. de Physique 41 (Suppl.) (1980) C6-391.
17. Narayan, J., Chen, Y. and Moon, R.M. Phys. Rev. Letters 46 (1981) 1491.
18. Lambert, M., Mazieres, C. and Guinier, A. J. Phys. Chem. Solids, 18 (1961) 129.
19. Taupin, C. J. Phys. Chem. Solids 28 (1967) 41.
20. Couchman, P.R., and Jesser, W.A. Nature 269 (1977) 481.
21. Mie, G. Ann. Physik 25 (1908) 377; trans.: Royal Aircraft Establishment, Farnborough, Library Translation, No. 1873 (1976).
22. Doyle, W.T. and Agarwal, A. J. Opt. Soc. Amer. 55 (1965) 305.
23. Kreibig, U., and Fragstein, C.V. Z. Physik, 224 (1969) 307.
24. Smithard, M.A. and Tran, M.Q. Helv. Phys. Acta, 46 (1974) 869.
25. Papavassiliou, G.C. Prog. Solid State Chem. 12 (1979) 185.
26. Van de Hulst, H.C. Light Scattering by Small Particles (New York: John Wiley and Sons 1957), Chapter 14.

27. Kreibig, U. in Impact of Cluster Physics in Materials Science and Technology, NATO Advanced Study Institute (The Hague: Martinus Nijhoff Publishers, 1983).
28. Yafet, Y. in Solid State Physics ed. F. Seitz and D. Turnbull (New York: Academic Press 1963) Vol. 14, p.l.
29. Kawabata, A. J. Phys. Soc. Japan, 29 (1970) 902.
30. Vitol, A. Ya., Kharakhashyam, E.G., Cherkasov, F.G., and Shvarts, K.K. Fiz. Tver. Tela, 13 (1971) 2133; trans.: Sov. Phys. Solid State 13 (1971) 1787.
31. Fowler, W.B. (ed.) Physics of Color Centers (New York: Academic Press, 1968).
32. Von der Osten, W. in Defects and their Structure in Non-Metallic Solids ed. B. Henderson and A.E. Hughes, (New York: Plenum Press 1976) NATO Advanced Study Institute Series Vol. B19, p.237.
33. Greenwood, G.W., Foreman, A.J.E. and Rimmer, D.E. J. Nucl. Mat. 1 (1959) 305.
34. Jain, U. Ph.D. Thesis, Indian Institute of Technology, Delhi; A.E.R.E. Harwell Report TP 693 (1977).
35. Calleja, J.M. and Agullo-Lopez, F. Phys. Stat. Sol. (a) 25 (1974) 473.
36. Durand, D., Chassagne, G., and Serughetti, J. Phys. Stat. Sol. (a) 12 (1972) 389.
37. Zener, C. J. Appl. Phys., 20 (1949) 950.
38. Wert, C., and Zener, C. Phys. Rev., 76 (1949) 1169.
39. Wert. C., and Zener, C. J. Appl. Phys. 21 (1950) 5.
40. Frank, F.C. Proc. Roy. Soc., A201 (1950) 586.
41. Ham, F.S. J. Phys. Chem. Solids 6 (1958) 335.
42. Ham. F.S. J. Appl. Phys. 30 (1959) 915.
43. Bullough, R. and Newman, R.C. Rep. Prog. Phys. 33 (1970) 101.
44. Christian, J.W. The Theory of Transformations in Metals and Alloys (Oxford: Pergamon Press 1975).
45. Jain, S.C. and Hughes, A.E. Proc. Roy. Soc. A360 (1978) 47.
46. Penley, J.C. and Witte, R.S. J. Appl. Phys. 33 (1962) 2875.
47. Lifshitz, I.M. and Slezov, V.V. J. Exp. Theor. Phys. (U.S.S.R.) 35 (1958) 479; trans.: Sov. Phys. J.E.T.P. 35 (1959) 331.
48. Lifshitz, I.M. and Slezov, V.V. Fiz. Tver. Tela 1 (1959) 1401; trans.: Sov. Phys. Solid State 1 (1959) 1285.
49. Lifshitz, I.M. and Slezov, V.V. J. Phys. Chem. Solids 19 (1961) 35.
50. Wagner, C. Z. Electrochem., 65 (1961) 581.
51. Greenwood, G.W. in The Mechanism of Phase Transformations in Crystalline Solids, Inst. of Metals Monograph No. 33 (1969) p.105.
52. Kahlweit, M. in Physical Chemistry, ed. H. Eyring, D. Henderson and W. Jost, (New York: Academic Press, 1970) Vol. 10, Chapter 11.
53. Kahlweit, M. Adv. Colloid and Interface Science 5 (1975) 1.
54. Jain. S.C. and Hughes, A.E. J. Mat. Sci. 13 (1978) 1611.
55. Smoluchowski, M. Physik Z., 17 (1916) 557 and 585.

56. Smoluchowski, M. Z. Physik Chem. 92 (1917) 129.
57. Baroody, E.M. J. Appl. Phys. 38 (1967) 4893.
58. Geguzin, Y.E. and Krivoglaz, M.A. Migration of Macroscopic Inclusions in Solids (New York and London: Consultants Bureau, 1973).
59. Jain. S.C. and Arora, N.D. J. Phys. Chem. Solids 35 (1974) 1231.
60. Hughes, A.E. and Jain, S.C. Phys. Letters 58A (1976) 61.
61. Kreibig, U. J. Phys. F: Metal Phys. 4 (1974) 999.
62. Shvarts, K.K., Ekmanis, Y.A., Udod, V.V., Lyushina, A.F., Tiliks, Y.E. and Kan, R.A. Fiz. Tver. Tela, 12 (1970) 879; trans.: Sov. Phys. Solid State 12 (1970) 679.
63. Treilleux, M., Thevenard, P., Chassagne, G. and Hobbs, L.W. Phys. Stat. Sol. (a) 48 (1978) 425.
64. Davenas, J., Perez, A. and Dupuy, C.H.S. J. de Physique 37 (Suppl.) (1976) C7-531.
65. Matzke, Hj., Turos, A., Rabette, P. and Meyer, O. J. Phys. C: Solid State Physics 14 (1981) 3333.
66. Turos, A., Matzke, Hj. and Rabette, P. Phys. Stat. Sol. (a) 64 (1981) 565.
67. Williams, R.T. Semiconductors and Insulators, 3 (1978) 251.
68. Townsend, P.D. and Agullo-Lopez, F. J. de Physique 41 (Suppl.) (1980) C6-279.
69. Sonder, E. and Sibley, W.A. in Point Defects in Solids ed. J.H. Crawford and L.M. Slifkin (New York: Plenum Press, 1972) Vol. 1, p.201.
70. Hobbs, L.W., Hughes, A.E. and Pooley, D. Proc. Roy. Soc., A332 (1973) 167.
71. Catlow, C.R.A., Diller, K.M. and Hobbs, L.W. Phil. Mag. A 42 (1980) 123.
72. Hughes, A.E. Comments on Solid State Physics, 8 (1978) 83.
73. Pappu, S.V. and McCarthy, K.A. J. Phys. Chem. Solids 32 (1971) 1287.
74. Swyler, K.J., Klaffky, R.W. and Levy, P.W. in Scientific Basis for Nuclear Waste Management, ed. G.J. McCarthy (New York: Plenum Press, 1979) Vol. 1, p.349.
75. Swyler, K.J., Klaffky, R.W. and Levy, P.W. in Scientific Basis for Nuclear Waste Management, ed. C.J.M. Northrup Jr. (New York: Plenum Press, 1980) Vol. 2, p.553.
76. Loman, J.M., Levy. P.W. and Swyler, K.J. in Scientific Basis for Nuclear Waste Management Vol. 4, to be published (1982).
77. Arnold, G.W., in Scientific Basis for Nuclear Waste Management ed. J.G. Moore (New York: Plenum Press, 1981) Vol. 3, p.435.
78. Hughes, A.E. in Proceedings of the Conference on Lattice Defects in Ionic Crystals, Dublin 1982, to be published in Radiation Effects.
79. Bullough, R. and Nelson, R.S. Phys. in Technol. 5 (1974) 29.
80. Jain, U. and Lidiard, A.B. Phil. Mag. 35 (1977) 245.
81. Lidiard, A.B. Phil. Mag. A, 39 (1979) 647.

82. Alcala, R. and Orera, V.M. J. de Physique 37 (Suppl.) (1976) C7-520.
83. Wiegand, D.A. Phys. Rev., B10 (1974) 1241.
84. Bunch, J.M., Hoffman, J.G. and Zeltmann, A.H. J. Nucl. Mat. 73 (1978) 65.
85. Pells, G.P. and Phillips, D.C. J. Nucl. Mat. 80 (1979) 207 and 215.
86. Shikama, T. and Pells, G.P. Phil. Mag. (to be published, 1983).

OPTICAL PROPERTIES OF SMALL PARTICLES IN INSULATING MATRICES

U. Kreibig

FB 11 - Physik - der Universität des Saarlandes
D 6600 Saarbrücken / Germany

1. INTRODUCTION

"Physics of small particles" covers, both, molecular physics and condensed matter physics; it is the transition region between the two. In a somewhat idealized kind of growth process of a small particle, we may add atoms, one by one, to a nucleus atom, starting, thus, with a molecular aggregate of few atoms and ending with a condensed matter particle of, say, several millions of atoms. Observing physical properties of such aggregates as changing during growth we are witnesses of the formation of the solid state.
It may help to a better understanding of the confusing variety of phenomenological effects which have been observed on particles with different sizes, to file them into different size regions. A tentative such scheme is shown in table 1, with four distinct size intervals. I should mention, that, when establishing them, I mainly had metallic particles in mind. (By the way, this allows to give definitions for the obscure terms "cluster" and "particle".)
In several of the presented lectures, various optical properties of region I - aggregates have been reviewed. Throughout the following synopsis I shall restrict myself to the size regions II to IV, i.e. to the solid state approach.
Optical properties, in general, reflect, in a quite sensitive and distinct manner, the atomic arrangement of particles and their electron energy spectra and excitation states. Up to now, however, they can by no means be measured from a single particle small enough to be interesting in view of the formation of the solid state. Instead, samples of many particles - typically more than 10^{10} - are required to obtain measurable optical effects.
Some of the numerous methods of producing such particle systems are compiled in the classification scheme of table 2.

Table 1: Classification of "Small particles"	Size region I "Molecular Clusters"	Size region II "Solid state Clusters"	Size region III "Micro-crystals"	Size region IV "Bulky particles"
Number of atoms per aggregate	$N \lesssim 10^1$	$10^2 \lesssim N \lesssim 10^3$	$10^3 \lesssim N \lesssim 10^5$	$N > 10^5$
Number of surface atoms N_s compared to number of atoms of the inner volume (below surface layer) N_v	"surface" and "inner volume" not separable	surface-to-volume-ratio: $\frac{N_s}{N_v} \sim 1$	surface-to-volume-ratio: $\frac{N_s}{N_v} < 1$	surface-to-volume-ratio: $\frac{N_s}{N_v} << 1$
Atomic arrangements and electron energies	atomic arrangements and energy spectra vary strongly with n; discrete electron energy spectra	structure of the "volume" may deviate from bulk crystal structure; size dependencies of electronic surface and volume properties	bulk-like volume structure; size effects of volume material properties; surface properties similar to those of plane surfaces	bulk-like material properties of the volume; surface only causing polarization effects
Description	molecular theories	solid state theories with size- and surface effects	solid state theories with size effects	solid state theories

It is a common feature of small particles that they are in a kind of unstable state: though their binding energy lowers the total energy per atom compared to separated single atoms, the latter is higher than in the appropriate macroscopic bulk. In size regions III and IV, the excess energy is mainly the surface energy, which, in a system of many particles, gives a driving force, both for

- coagulation (clustering of m separated particles to a m-particle-cluster, fig. 1b) and for
- coalescence (m separated particles → m* larger particles with m*<m; fig. 1c).

In both cases, the excess energy is lowered. Thus the lifetime of separated particles is limited in a many particle system, if no care is taken to avoid these processes. This can be done, e.g., by

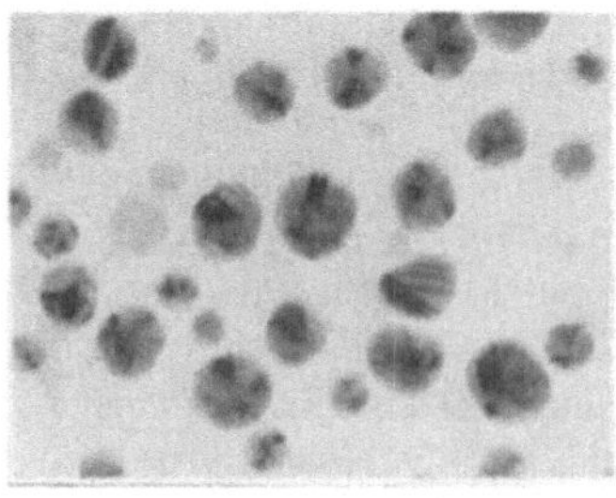

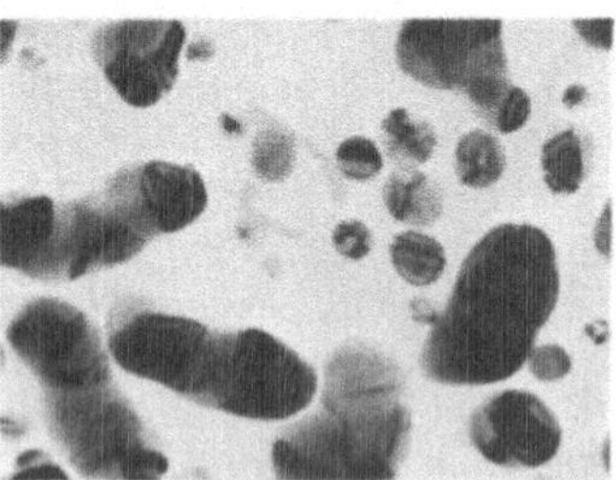

a) Separated particles b) Coagulation (Clustering) c) Coalescence

Fig.1 Ag particles ($2R \approx 2o$ nm)

Table 2.

condensation of atoms				breaking up of bulky material	
supersaturation in vapour		supersaturation in liquids	supersaturation in solid matrices	mechanically	arc-dispersion
atomic beam	evaporation on substrate				
(1) thermic (2) jet	(1) in vacuum (2) in inert gas ("blacks") (3) co-evaporation (matrix-isolation)	(1) chemical reduction with nuclei (2) without nucl. (3) solvent extraction (spraying techn.)	(1) diffusion in the matrix (2) diffusion into the matrix (3) phase separation	(1) grinding (2) forcing into porous matrix (3) dispersion by ultrasound	

a) suppression of the (usually, thermal) particle motion and particle collisions by embedding the particles in matrices of high viscosity or by fixing them on a support ("island films"), or
b) by forming electric double layers around the particles (ions or hydrate groups) which maintain sufficient particle separations by Coulomb repulsion ("stable" aqueous colloidal solutions).

Alternatively, measurements may be done in a quickly moving particle stream instead of a stable particle system ("steady flow colloids", "supersonic cluster jets"). Since the latter methods are at the very beginning for larger particles of size regions II to IV, the methods a) and b) still are the mostly used, today. Main disadvantage common to all: the samples consist of the system of particles plus surrounding (gaseous, liquid or solid) medium/media, and, both, physical and chemical interactions may occur at the various kinds of interfaces.
It is mainly the strength of this interaction (i.e. its influence on the observed physical quantity), which makes a special particle system production method more or less well suited for a given particle material. For the special case of optical experiments the very matrix/support is most suited, the influence of which can be described simply by the dielectric function ε_m of the bulk matrix/support material. (It is an open question, in principal, to which extent the use of ε_m becomes inadequate when particle sizes are decreased below, say, 10^2 atoms per particle.)

2 LECTURE I: OPTICAL MATERIAL PROPERTIES OF SMALL PARTICLES

Fig.2a gives examples for the abrupt changes of the discrete molecular electronic states of metal aggregates of size region I when the aggregates grow. There are several experiments, which show (fig. 2b and c) this discrete structure to be reflected in the optical excitation spectra only for extremely small numbers of atoms $(N<10)$.
For larger particles of size regions II to IV, the linear response of a single particle onto the incident electromagnetic field can, following the solid state approach, be described by introducing
a) a dielectric function of the particle MATERIAL $\varepsilon(\omega) = \varepsilon_1 + i\cdot\varepsilon_2$, which, in general, depends on particle size and shape,
b) a dielectric function of the surrounding medium $\varepsilon_m(\omega) = \varepsilon_{m1} + i\cdot\varepsilon_{m2}$ and
c) the inner electric and magnetic fields E_i and H_i in the particle which, also, depend on particle size and shape.
In the following, mainly a) and c) will be considered, separately, to describe optical properties of a SINGLE particle. In the 1st Lecture ε will be discussed for different size regions. Then E_i will be treated in various steps of approximation during the 2nd Lecture. At last we will see in Lecture 3, that the linear response of systems of MANY particles needs further considerations, conditioned by interaction effects among the particles. We will express the optical properties of the whole many particle system in terms of EFFECTIVE material functions of the whole system $\bar{\varepsilon}$ and $\bar{\mu}$, which depend, both, on single particle properties and on properties of the particle system.
It is not my purpose, to give an extensive compilation of theoretical and experimental work, done in this field - there are several recent excellent reviews (4) -, nor am I able to cite even all of the most interesting contributions. Instead, I want to present a selfcontained introduction to the correlations between numerous optical effects which are operative in small particle systems.

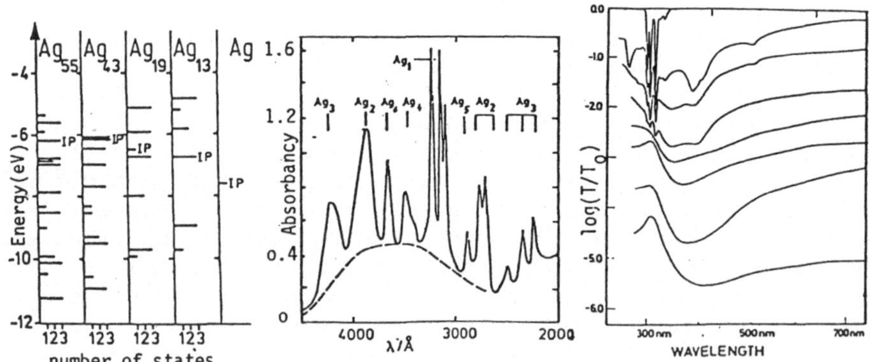

Fig.2 Silver clusters . Left: Energy levels of various clusters (after Baetzold(1))
Center:Experimental absorbancy of Ag clusters in solid Ar (after (2)).
Right:Experimental log(Transmission) of Ag clusters in solidAr.Mean cluster size increasing from top to bottom (after Welker(3)).

2.1 Material Functions of the Bulk Like Matter (Size Regime IV of Table 1).

The relative permeability $\mu(\omega)$ equals unity in the optical region for all particle and matrix materials, considered. (See, however, $\bar{\mu}$ of eq. (46)!). So we are left with the dielectric function $\varepsilon(\omega)$. In the frame of Maxwells theory it is a phenomenologically defined quantity and, so, we need additional microscopic models to predict frequency spectra (and, what is important here, dependencies on particle geometry).

Following the simple atomic oscillator model, the polarization P of (isotropic) matter as due to an electric field E is obtained by summing up the atomic polarizabilities α_j:

$$P/E = \sum_j n_j \cdot \alpha_j \quad (n_j\text{: numbers of atoms j per volume}) \tag{1}$$

$$\text{with} \quad \alpha_j = A_j \cdot (\omega_{rj}^2 - \omega^2 + i\omega\gamma_j)^{-1}$$

where ω_{rj} and γ_j are the resonance frequencies and relaxation frequencies of the j-th kind of atomic oscillator. The remaining problem is: which field E has to be chosen?

(a) In the case of DIELECTRICS, E is obtained at a given atom from the macroscopic Maxwellian field ($P = \varepsilon_o(\varepsilon - 1)\, E_{Max}$) by

$$E = E_{local} = E_{Max} + P/(3\varepsilon_o) + \Delta \quad (\varepsilon_o\text{: electric field constant}), \tag{2}$$

where the term $P/(3\varepsilon_o)$ includes the contributions of all atoms far away from the given atom ("Lorentz-sphere"-contribution). Δ is due to the atoms within this sphere and equals zero if they are arranged in a cubic lattice or purely statistical. Combining eqs. (1) and (2) with $\Delta = 0$, the famous Lorenz-Lorentz-formula is obtained

$$\frac{\varepsilon - 1}{\varepsilon + 2} = \frac{1}{3\varepsilon_o} \sum_j n_j \alpha_j \quad \text{or} \quad \varepsilon = \frac{1 + 2\frac{1}{3\varepsilon_o}\sum_j n_j \alpha_j}{1 - \frac{1}{3\varepsilon_o}\sum_j n_j \alpha_j}, \tag{3}$$

which means the Clausius-Mosotti relation for the case of optical frequencies.

The term A_j in eq.(1) has not yet been defined: it is $A_j = e^2/m_j \cdot f_j$ (f_j: oscillator strength; e, m: electron charge and mass) for electronic excitations and $A_j = \alpha(\omega{=}o) \cdot \omega_{rj}^2$ for ionic materials.

(b) For the case of METALS, it is usually concluded from conduction electron shielding that $E_{loc} = E_{Max}$ and, hence,

$$\varepsilon - 1 = \sum_j n_j \alpha_j \tag{4}$$

(b1) Specializing eq.(1) to CONDUCTION ELECTRONS, $A_j = e^2/m_{eff}$ and, since energy is purely kinetic, $\omega_{rj} = 0$. Thus, we arrive at the Drude-Sommerfeld formula for the susceptibility, χ^{DS}:

$$\varepsilon - 1 = \chi^{DS} = -\omega_p^2 \cdot (\omega^2 + i\omega\gamma)^{-1} \tag{5}$$

with $\omega_p = [n_e \cdot e^2 / \varepsilon_o m_{eff})]^{1/2}$ the plasma frequency and $\gamma = 1/\tau = v_{Fermi}/l$ the relaxation frequency (n_e: electron density; m_{eff}: effective mass; τ: relaxation time; v_{Fermi}: Fermi-velocity; l: electron mean free path).

(b2) INTERBAND TRANSITIONS give additive contributions to the susceptibility:

$$\varepsilon = 1 + \chi^{DS} + \chi^{interb} \tag{6}$$

If mainly direct transitions are important, the imaginary part, χ_2^{interb}, is roughly

$$\chi_2^{interb}(\omega) \approx \text{const} \cdot |M_{if}|^2 \cdot J(\omega) \tag{7}$$

with M_{if} the transition matrix element between initial ("i") and final ("f") state, which here is assumed to be independent of ω in the interesting spectral region, and $J(\omega)$ the joint density of states.

(b3) Under special experimental conditions, collective ELEMENTARY EXCITATIONS are excited by electromagnetic waves, as plasmon- or

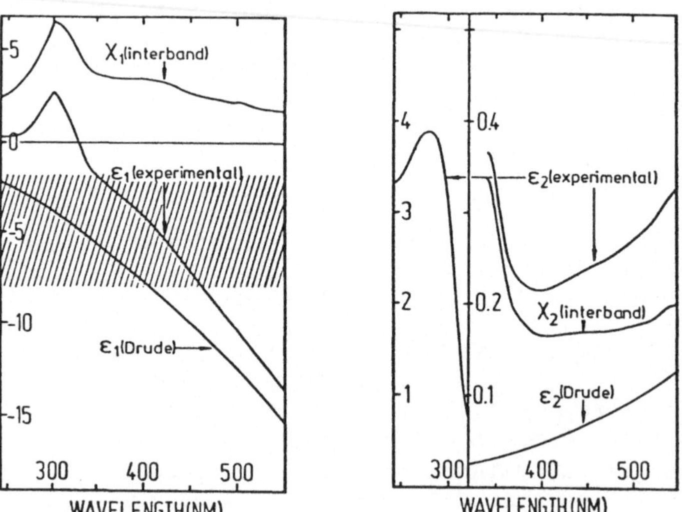

Fig.3 Dielectric function ε of bulk Ag in the visible, decomposed into interband and Drude-Sommerfeld contributions (eq.(6)). Left: $\varepsilon_1(\lambda)$; Right: $\varepsilon_2(\lambda)$. Small particle resonances occur in the hatched region of ε_1 (see eq.(33)).

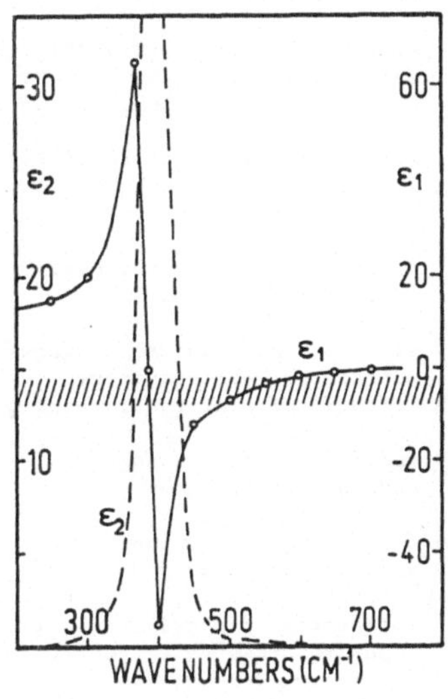

Fig.4 Dielectric function $\varepsilon = \varepsilon_1 + i\varepsilon_2$ of bulk MgO in the infrared. (After (18)) Small particle resonances occur in the hatched region of ε_1 (see eq.(33).

phonon-polaritons. We will see later that such kinds of excitations are of special importance for small particles.
Figs. 3 and 4 show, as examples, the ε-spectra for a metallic and an ionic material, respectively.
Summarizing, solid state parameters are listed up, informations about which may be drawn from optical experiments in general and, in special also from experiments on small particles:

- eigenfrequencies of electronic and ionic, single and collective excitations
- relaxation parameters γ_j
- conduction electron parameters n, m_{eff}, γ
- band structure details $|M_{if}|^2$, $J(\omega)$.

2.2 Material Functions of Small Particles of Size Regions III and II

All parameters, listed above, develop size dependencies in particles of sufficiently small size. Hence, particle size acts as an additional parameter for investigation and for interpretation of condensed matter properties. In Fig. 5 various effects causing such dependencies are sketched. (This figure is not meant, by far, to give a complete compilation of effects which are expected or already measured.) The largest characteristic length for electronic properties of metals is due to the conduction electrons; in terms of the Drude-Sommerfeld model it is their mean free path l, amounting e.g. to some 10 nm in noble metals at room temperature. So, when particle sizes are reduced, the conduction electron contributions to ε will be the first to show size dependencies.

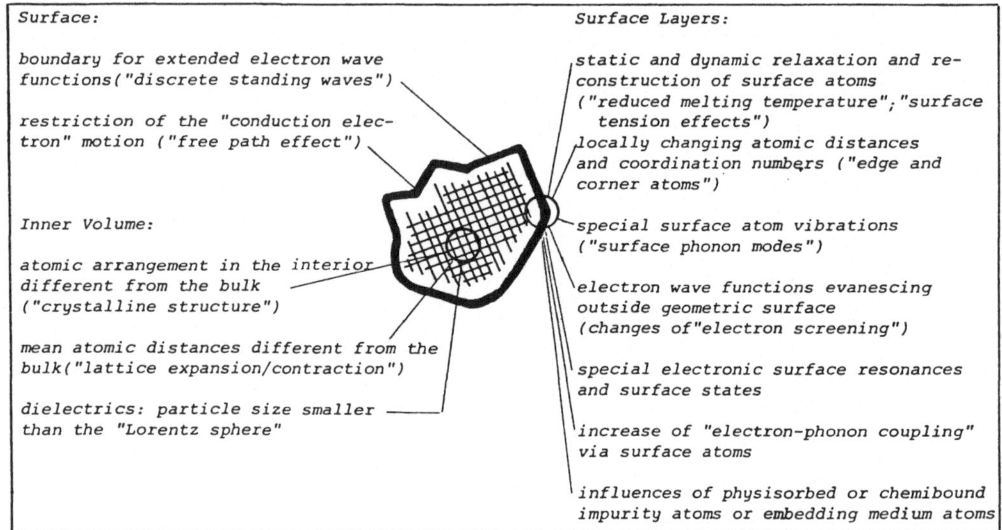

Fig. 5: Some size and surface effects in small particles which may influence their optical properties. (The terms in quotation marks (") describe the particle properties in the "solid state approximation".)

2.2.1 Such dependencies are described most easily in terms of the Drude-Sommerfeld model. They are due, then, to increasing amounts of electron scattering processes at the particle surfaces, augmenting the conduction electron relaxation. Assuming

- quasi continuous conduction bands near Fermi energy E_F and
- spherical particle shapes with radius R,

we find for the contributions l_s of this surface scattering to the mean free path

$$l_s = A \cdot R \tag{8}$$

with A = 1 (isotropic probability of electron directions after a scattering event), or A = 4/3 (diffuse scattering according to a Lambert cosine law).
Assuming, further, Mathiessens rule to hold, we obtain the resulting total relaxation frequency

$$\gamma(R) = v_{Fermi} \cdot (l_{bulk}^{-1} + l_s^{-1}) \tag{9}$$

Both, ε_1 and ε_2 of eq.(5) include γ and, hence, both depend on R via eq.(9). In the high frequency region ($\omega \gg \gamma$), however, mainly ε_2 is influenced:

$$\varepsilon(\omega,R) \approx \varepsilon^{bulk}(\omega) + i \cdot \frac{\omega_p^2}{\omega^3} \cdot A \frac{v_{Fermi}}{R} \tag{10}$$

This dependency ε(R) usually is called FREE PATH EFFECT (e.g.(5)). Without going into details, it should be pointed out, that

- a more thorough derivation yields influences to ε_1^{DS}, too, which are more important at low frequencies, and that
- the other Drude-Sommerfeld quantities n, m_{eff} and γ_{bulk} may also change in sufficiently small particles, thus inducing further size dependencies of ε^{DS}, and that
- optical experiments to determine ε(ω,R) are simplified by the fact, that there is no anomalous skin effect if the particles are sufficiently small.

Complications arise, if particles consist of several crystallites instead of being monocrystalline, (as e.g. the famous multiple-twin particles (14)), since, then the grain boundary scattering in the inner of the particles develop size dependencies, too (6).
Fig. 6 shows for the example of small Ag particles, both, experimental spectra of ε(λ,R) and computed spectra as resulting from the free path effect. There is nearly quantitative correspondence for the case of isotropic surface scattering. Marked changes of n and m_{eff}, on the other hand, are not consistent with these results. (Analogous experiments with multiple-twin particles might, in addition, give detailed information about grain boundary scattering of conduction electrons.)

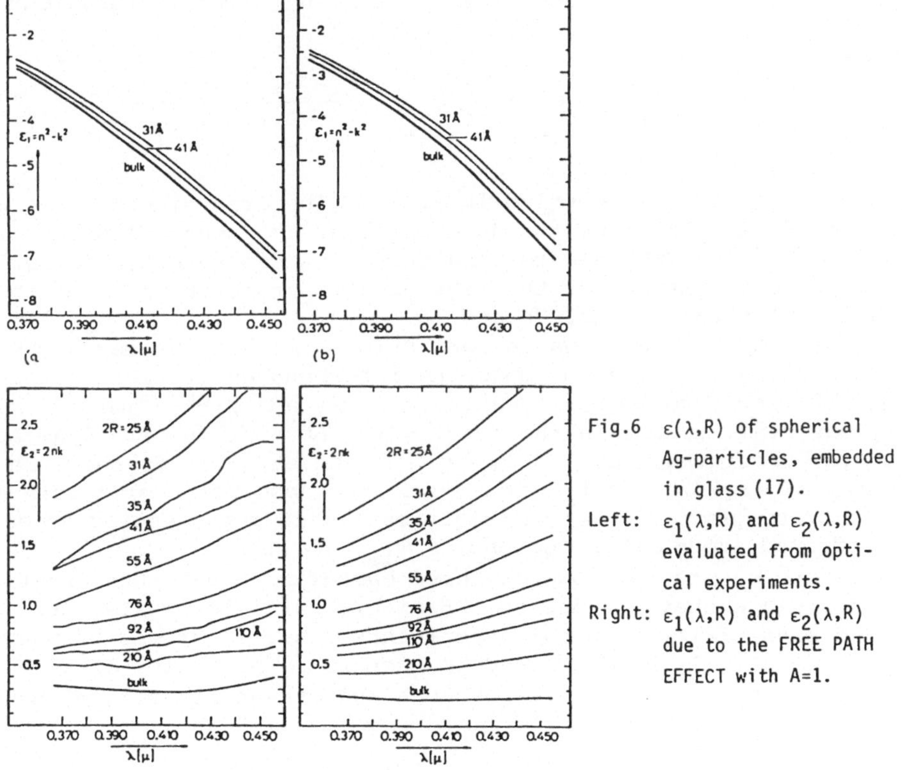

Fig.6 ε(λ,R) of spherical Ag-particles, embedded in glass (17).
Left: $\varepsilon_1(\lambda,R)$ and $\varepsilon_2(\lambda,R)$ evaluated from optical experiments.
Right: $\varepsilon_1(\lambda,R)$ and $\varepsilon_2(\lambda,R)$ due to the FREE PATH EFFECT with A=1.

2.2.2 Assuming the metal electron wave functions to be extended over the whole particle and assuming, further, their particle surface reflections to follow well defined conditions, the wave mechanical eigenstates are given by standing waves in the particles and, hence, the electron energy levels are discrete. The spacings between such levels depend, both, on the geometric arrangement of the atoms and on particle size. Due to lifetime limitations, these levels are broadened, and, thus , two cases may be distinguished:

- Quantization case: Level width < level distances (→discrete energy spectra)
- Drude-Sommerfeld case: Level width > level distances near E_F (→ quasicontinuous energy spectra).

The most simple model to calculate optical properties in the quantization case is to disregard the geometric atomic arrangement in the particles, further, to disregard localized states and to replace the particle by an empty potential box with infinite wall height which is filled with free electrons. Calculations of eigenfunctions and energy eigenvalues have been performed for cubic and for spherical box shapes (7,8).

The resulting dielectric function of cubic particles (7a) resembles formally eqs. (1) and (4) with $\omega_{rj} = \omega_{if}$ (i:initial; f:final state) and $A_j = s_{if}(F_i - F_f)$ (s_{if}:oscillator strength ; F:Fermi functions):

$$\varepsilon(\omega) = 1 + \chi^{interb}(\omega) + \frac{\omega_p^2}{N} \sum_{i,f} \frac{s_{if}(F_i - F_f)}{\omega_{if}^2 - \omega^2 - i\,\omega\gamma_{if}} \qquad (11)$$

Numerical results are compiled in fig. 7. The free electron energy spectrum of fig. 7a shows the degeneracy of the levels instead of the density of states as used for the bulk. Its discreteness is reflected in the oscillations of the free electron contributions of eq. (11), which are plotted in figs. 7b and 7c.

Going one step further towards realistic materials, we may assume the size/surface influences onto the interband excitations, included in eq. (11), too, to be negligeable, and may add $\chi^{interb}(\omega)$ of the bulky material, to obtain the total $\varepsilon(\omega)$. Absorption of a 4 nm Ag particle, resulting, thus, from eq. (11) and eq. (35), is shown in the spectrum of fig. 7d. It is compared to the absorption of a thin Ag film without quantum effects. (Additional size dependencies appear in the infrared around the frequencies ω_{if} (7b).)

Such particle spectra, however, have esoteric rather than practical meaning, since they are due to ONE single particle, while, due to limited sensitivity, typical experiments require systems of many particles. As will be discussed in Lecture 3, particles of such systems differ in their sizes, shapes and orientations, and, hence,

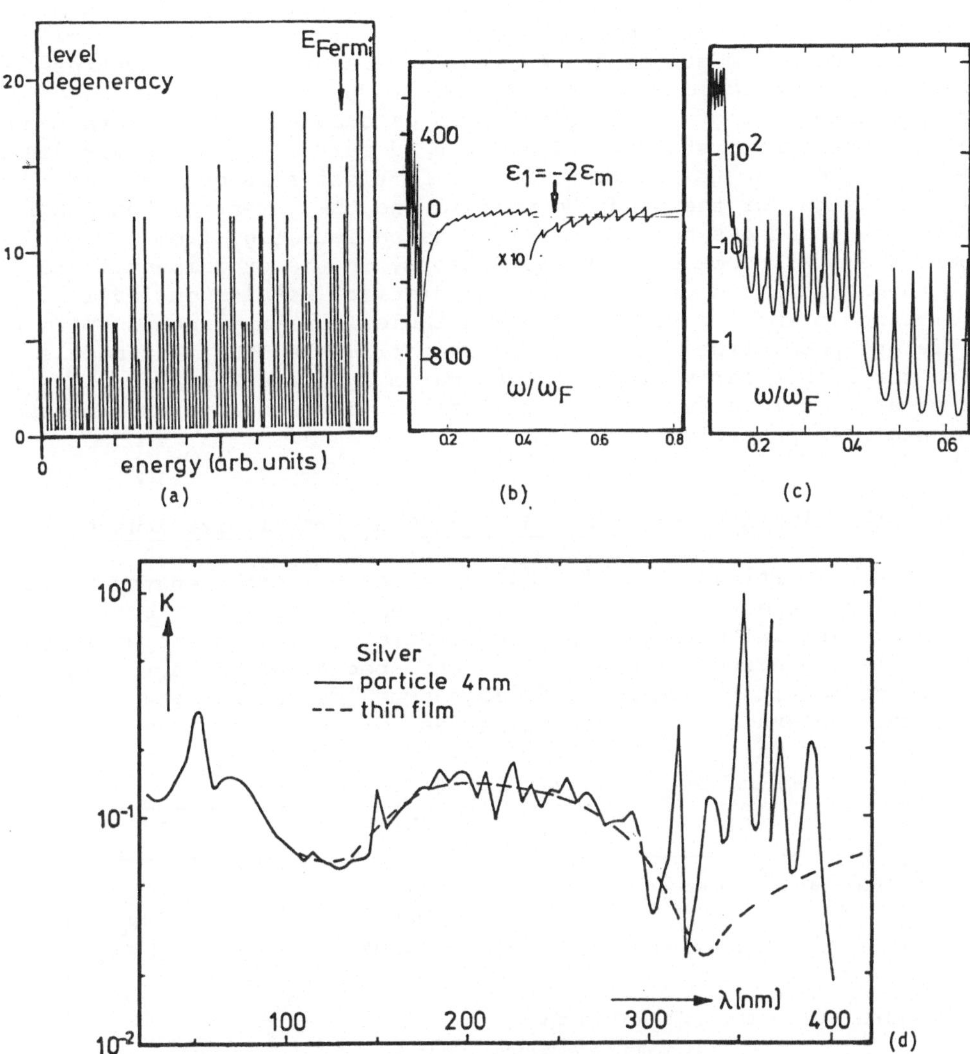

Fig.7 Cubic potential box model

(a) Energy level spectrum of 10^3 electrons (occupied states)
(b) Real part of ε(ω) of eq.(11) for 4096 free electrons (χ^{interb}=o)(after(7a))
(c) Imaginary part (after(7a))
(d) Absorption spectrum of Ag-particles (size 4nm), computed by inserting ε(ω) of eq.(11)with bulk χ^{interb} into eq.(35).
For comparison: absorption 4πk/λ of a thin Ag-film (k:bulk absorption coefficient)

also in their energy level combs. The marked oscillations, therefore, are smeared out, totally, in the resulting optical response, and for the mean ε_2 a monotonous function is obtained, which equals eq.(10) of the Drude-Sommerfeld model quantitatively with A = 1. Though basing on contradictory assumptions, the two models, thus, give identical results. Other values of A (all of the order of unity) were obtained by other calculations. They are compiled in table 3. Yet, the differences between these values being of the order of accuracy of experimentally evaluated ε_2, no distinct conclusions on energy level structures in small particles could - up to now - be drawn therefrom.
Size dependencies of the ε_1 spectra,as calculated from different models,differ more distinctly. E.g., for Ag particles in the visible, changes of ε_1 due to a diminuation of particle size even are of opposite sign for the Drude-Sommerfeld and the potential box model. Since these changes are very small - as obvious,e.g.,from fig. 6 -, there are still some controversies, both, on the experimental and on the theoretical side, and, again, no distinct results on level splitting in small particles are available from experiments up to now. Different predictions have been derived from the different models also for higher derivatives of ε to various parameters. E.g., the temperature derivative of $\varepsilon_2(\omega,R)$ has been investigated experimentally for small Ag particles (5b).

2.3 Material Functions of Solid-State-Clusters of Size Region II

Diminishing particle sizes further , other critical lengths beside those of the electrons, which are labelled "conduction electrons" in the bulk material,become important compared to the particle dimensions (fig. 5). There are surface layers with special atomic arrangements due to geometric relaxation and reconstruction, or surface layers of chemical compounds as oxides, sulphides, or due to surface enrichment

Table 3: Size effect constant A of eq.(1o)

MODEL	AUTHORS	A
Potential box, cubic	(7)	o.79
" " "	(8b)	$o.75\cdot g_c(\omega)$
" " ,spherical	(8a,8b)	$o.61\cdot g_s(\omega)$
" " "	(7)	1
free path effect (spherical particle)		
" " " isotropic scattering	(5b)	1
" " " diffuse scattering	(5b)	o.75
" " " multiple twin particles	(6)	$1 + B\cdot\Phi$
experiments, nearly spherical Ag particles	(5)	~ 1
" " " Au "	(5c)	~ 1
" " " Ag "	(16)	~ o.5

($g_c(\omega)$ and $g_s(\omega)$ are monotone functions of ω of the order of 1)

of impurities (segregation). Thicknesses of the former layers are of the order of lattice parameters, in the case of metals, because of shielding effects, and this is also the order of diffuseness of the electron wave functions evanescing outside the particle. In ionic material, shielding might be determined by the Lorentz sphere radius, which usually is assumed to be large compared to lattice parameters. Local polarizabilities may in all these cases depend markedly on the distance from the geometric surface, and, so, the definition of a uniform dielectric function for the particle material becomes roughly approximative. Its meaning is restricted in the sense, that <u>if</u> there were a particle with uniform polarizabilities, i.e. with uniform $\varepsilon(\omega,R)$, then, this particle would have the same overall linear re - sponse upon the incident field as the particle has, which really is observed in the experiment. Here we are near to the boarder, where solid state methods fail.

Smallest critical lengths compared to particle diameters are the atomic distances. In this size region size effects will also influence more or less localized electron states, and, consequently, size effects are observable in what is called the interband transition excitations in the bulky matter.

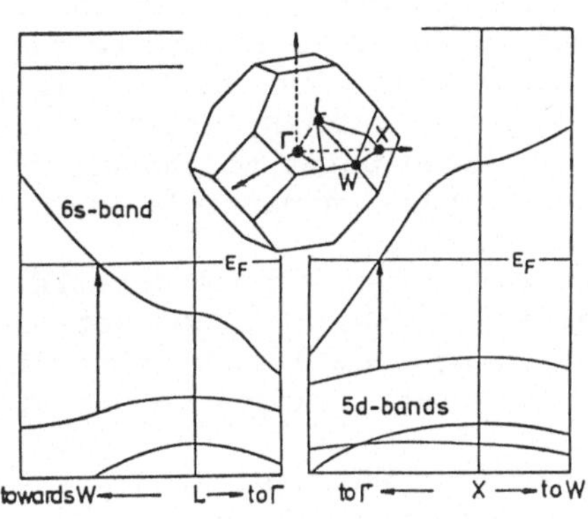

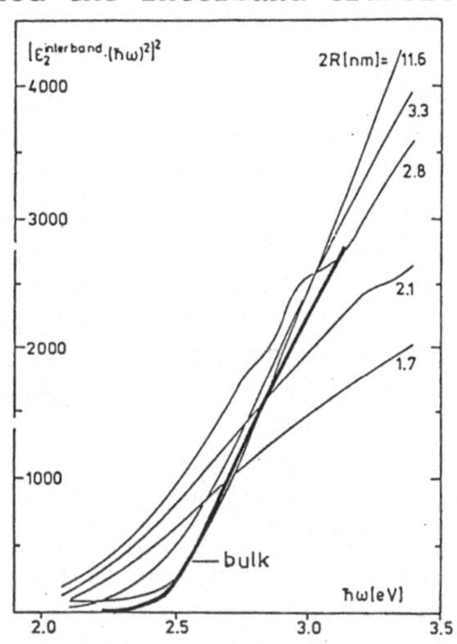

Fig.8 Size dependence of the low frequency interband transition edge of Au particles.
Left:Details of the band structure (after(19)) showing edge excitations.
Right: Experimental results showing the interband contributions to ε_2 (9).

As an example, the spectra in fig. 8 show changes of the low energy interband transition edge of small Au particles as due to a reduction of particle size (9). Both, the shift of the onset energy and the change of the slope of the edge are similar to those observed from amorphous Au films (1o). One might think of lifetime reductions of the excited state similar to the free path effect mentioned earlier, yet, the most surprising feature is the abrupt onset of a strong size dependency (fig. 9a) at a critical particle size of about 5×10^2 atoms/particle (9). (This step would be even more abrupt, if it were not smeared out by distribution of particle sizes in the many particle samples.)

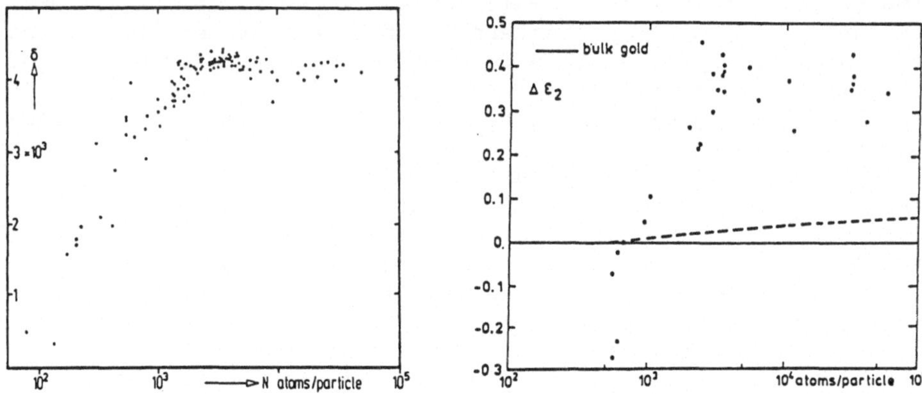

Fig.9 Size effects of optical properties of small Au particles.
Left: Slope of the optical interband transition edge (see fig.8)(9).
Right: Change of $\varepsilon_2(\omega,R)$ at $\hbar\omega$=2.5eV due to cooling from 3oo to 1.5K (11).
Dashed line: Contribution of "conduction"electrons (5b).

A critical size of same magnitude also was observed (11) in the temperature derivative of ε_2, which, in the case of Au is mainly due to temperature changes of the interband transition contributions (fig.9b).
There are several other experiments (e.g. XPS, EXAFS, optical), performed on various metals (12a,b; 13a,b) which point to changes of details of the bulklike electron bandstructure in particle size regions of the same order of magnitude.
Since interband transition excitations reflect much more details of the band structure and, hence, geometrical structure, than conduction electron excitations, the occurence of changes at a such critical particle size may be explained as due to abrupt changes in the atomic structure of the particles.
When going down the size scale, we know, from which structure these changes start: it is the crystalline structure of the bulk (plus surface layer). But we do not yet know, where this structural transition leads to. It might be a different crystalline structure,as observed in electron energy loss experiments on NaCl supported Au particles (15), or it may be a molecular cluster structure of size region I. We, then, observe the birth of the solid state band structure at that critical size of, say, 5 x 10^2 atoms, and this appears reasonable in view of the fact that, in such particles, there are as much atoms forming the surface as are in the interior(N_s/N_v~1 in table 1). For metallic particles, this is the lower limit of the solid state approach to their optical properties, i.e. the limit between size regions I and II in table 1.
Yet, there is the dilemma that clusters below this limit (say, of several 100 atoms) are still too large to be treated by molecular methods which take into account each atom in the particle, individually. Facing this fact, we may, therefore, try to extend the solid

state approximation further towards smaller particles by introducing "size effects", until molecular theories, applicable to such clusters, will be on the market.

---

3 LECTURE II: OPTICAL EXTINCTION AND DISPERSION OF ONE PARTICLE

The optical properties of a small particle result from the polarization of the particle as a whole, which on its part, is due to, both, material properties and the magnitude of the inner field. Having discussed the former in Lecture I in view of their particle size dependencies, we will now deal with the latter. Again, it is not my purpose to present a review of literature, which can be found in excellent books and articles (e.g.20,21,22) but to outline some of the physics.

3.1 Quasi-static Case (Ellipsoidal Particles)

It depends on the ratio of particle size to wavelength of the incident electromagnetic wave (which throughout the following is assumed to be a plane wave), whether the instantaneous phase of the wave is constant over the dimension of the particle (quasi-static or long-wavelength-case) or not (general case) (fig. 10). The quasi-static case is the most simple one; it leads to a homogeneous polarization of the particle, if

. an ellipsoidal shape is assumed,
. Maxwellian boundary conditions are fulfilled
. the dielectric function ε is homogeneous throughout the particle, or can be replaced by an average ε as discussed in chapter 2.3.

Then, the inner field is obtained as follows:

1) Incident field: $$\begin{matrix} E \\ H \end{matrix} = \begin{matrix} E_o \\ H_o \end{matrix} \cdot \exp(i(\omega t - kx)) \qquad (12)$$

2) Depolarization field: $E_1 = -N \cdot P/\varepsilon_o$ N: depolarization factor $P = \varepsilon_o(\varepsilon/\varepsilon_m - 1) \cdot E_i$ (13)

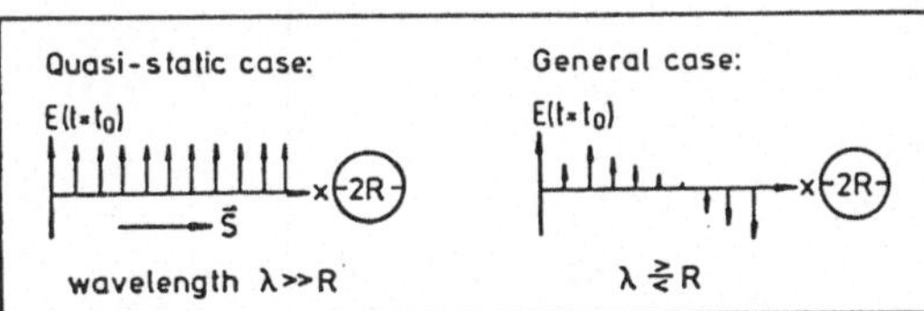

Fig.1o Excitation of a small particle (radius R) by electromagnetic waves (wavelength λ, Poynting vector $\vec{S}$).

3) Inner field: $E_i = E + E_1 = E \cdot \varepsilon_m/[\varepsilon_m + N(\varepsilon-\varepsilon_m)]$ (14)

4) Material functions:
General case: absorbing particle material $\varepsilon(\omega) = \varepsilon_1(\omega) + i\varepsilon_2(\omega)$ and absorbing embedding medium $\varepsilon_m(\omega) = \varepsilon_{m1}(\omega) + i\varepsilon_{m2}(\omega)$
permeability: $\mu \equiv 1$
Special cases: (a) $\varepsilon_2 \neq 0$; $\varepsilon_{m2} = 0$: absorbing particle in dielectric matrix (e.g. metal particles in a rare gase matrix)
(b) $\varepsilon_2 = 0$; $\varepsilon_{m2} \neq 0$: dielectric bubble in absorbing matrix

ad (a): Absorbing particle in a dielectric matrix:
For a given frequency ω we obtain the inner field:

$$E_i/E = \frac{\varepsilon_{m1}[\varepsilon_{m1}+N(\varepsilon_1-\varepsilon_{m1})] + i[N\varepsilon_{m1}\varepsilon_2]}{[\varepsilon_{m1}+N(\varepsilon_1-\varepsilon_{m1})]^2 + [N\varepsilon_2]^2} \tag{15}$$

While it holds always that $\varepsilon_2(\omega) \geq 0$, $\varepsilon_1(\omega)$ can be $\gtrless 0$ (see figs.3 and 4). If $\varepsilon_1(\omega) < 0$, E_i of eq. (15) can develop a resonance behaviour, the denominator reaching a minimum at a special frequency ω_r. If ε_2 depends weakly on ω near these frequencies, the resonance condition is [left] $\stackrel{!}{=} 0$, and this is equivalent to

$$\varepsilon_1(\omega_r) = -\varepsilon_{m1}(1-N)/N. \tag{16}$$

Then Re $\{E_i/E\} = 0$ and Im$\{E_i/E\}$ takes its maximum with the value $\varepsilon_{m1}/(N \cdot \varepsilon_2)$, i.e. the absorbed field energy is at its maximum. This resonance lies in the visible for many metals and in the IR for dielectric particles. It can (for a given particle shape) be shifted by choosing embedding media with different ε_m. In figs. 3 and 4 the resonance regions due to $1 \leq \varepsilon_{m1} \leq 4$ are shown off by hatching, for

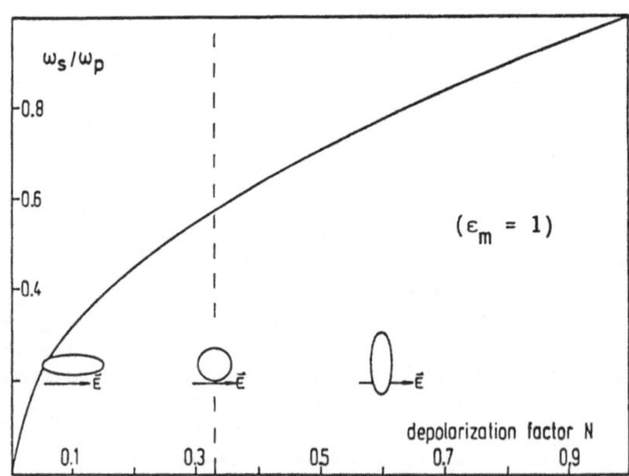

Fig.11 Resonance frequencies ω_s of small ellipsoidal free-electron particles according to eq.(16). N: depolarization factor; ω_p: Drude frequency

the case of spherical particles (N = 1/3).
Furthermore, fig. 4 shows that there are TWO resonances according to eq. (16), the lower one, however, being damped out by high ε_2. The other shifts towards higher wavenumbers for smaller ε_m, and, since then ε_2 decreases, its damping can be reduced for orders of magnitude by an apt choice of the embedding medium.
The frequency width of the resonance is determined, both, by ε_2 and by the steepness of the $\varepsilon_1(\omega)$ - spectrum around ω_r. One might even think of a combination of particle and matrix materials which fulfills eq. (16) over an extended frequency range and has, therefore, a resonance absorption band covering all that range. On the other hand, a large slope of $d\varepsilon_1/d\omega$ near ω_r gives a narrow absorption band, as is the case for Ag particles. Only for pure Drude-Sommerfeld particles, the absorption halfwidth is determined by ε_2 (i.e. by γ of eq. (5)) alone. The resonance frequency ω_r is shown for such particles in fig. 11 as a function of the depolarization factor N, i.e. the shape of the particle.
Absorption spectra for various realistic materials are compiled in figs. 12, 13, 16, 17, 2o, 21, 22 and 24. They demonstrate that such resonances occur as well in metallic as in ionic particles.

ad (b): Dielectric bubble in an absorbing matrix:
For a given frequency ω we obtain:

$$E_i/E = \frac{[(\varepsilon_{m1}^2+\varepsilon_{m2}^2)(1-N)+N\varepsilon_{m1}\varepsilon_1] + i[N\varepsilon_{m2}(\varepsilon_{m1}+\varepsilon_1)]}{[\varepsilon_{m1}+N(\varepsilon_1-\varepsilon_{m1})]^2 + [\varepsilon_{m2}(1-N)]^2} \tag{17}$$

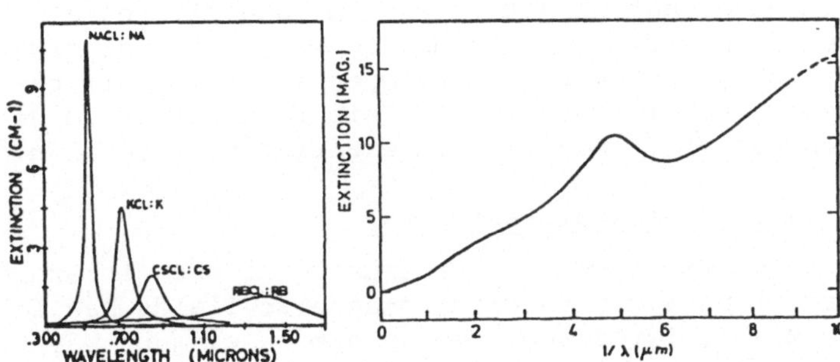

Fig.12 Extinction spectra of small particles. Left: computed for Na,K,Cs and Rb (after(34)). Right: measured extinction of interstellar grains(after (35)).

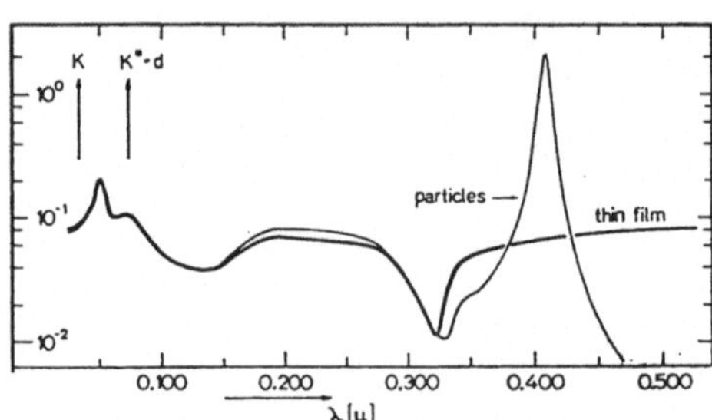

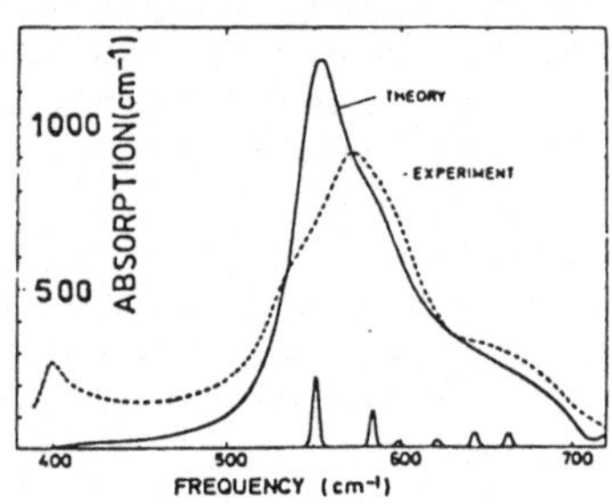

Fig.13 Extinction spectra of small particles.
Left: Ag particles, compared to thin film(17): strong differences in the conduction electron region (λ>o.32μm), small differences in the interband absorption region (λ<o.32 μm).
Right: Cubic MgO particles. Theory(32a), experiment(32b). The small peaks show the positions of the 6 Fuchs modes (chapter3.5.2).

With analogous assumptions as in *(a)* we find again the resonance condition

$$\varepsilon_1(\omega_r) = -\varepsilon_{m1} \cdot (1-N)/N \quad (18)$$

Then $\mathrm{Re}\{E_i/E\} = o$, and $\mathrm{Im}\,\{E_i/E\}$ takes its maximum with the value $N(\varepsilon_{m1} + \varepsilon_1^i)/[(1-N) \cdot \varepsilon_{m2}]$.

It needs considerably greater effort to evaluate eq.(14) for a system of absorbing particle and absorbing matrix. The resonance is then given by eq.(16) if ε_2 and ε_{m2} are small or depend weakly on ω near ω_r. It should be noted that in both cases *(a)* and *(b)* there is NO explicite dependency on particle size (since we assumed size/wavelength << 1). Implicitely, however, the size enters via ε, if particle sizes are sufficiently small (Lecture I). This fact allows to determine size dependencies of ε from optical experiments with high precision.

3.2 General Case (Spherical Particles)

The general case (wavelength $\lambda \gtrless$ radius R) of the linear response of ONE spherical particle on electromagnetic fields has first been developed by Mie (23), reformulated in a more elegant way by Debye(24) and extended by Gans & Happel(25). All these derivations being quite voluminous, I shall give an abridged sketch in order to show the assumptions, made, and the line of the proceeding.
Roughly spoken, the particle acts, both, as a receiving spherical antenna (absorption) and a reemitting or transmitting antenna (scattering). The linear reaction then is described by the extinction (i.e. absorption plus scattering), which is computed by a multipole expansion of the various fields (incident, penetrating and scattered/

diffracted field) apt to the spherical particle shape.

3.2.1 Sketch of the mathematics:

1) Spherical coordinates: r, θ, ϕ. Particle radius: R.
   Plane incident wave.
2) Solution of Maxwells equations with proper boundary conditions.

2a) Special assumption: divergency free curl fields (div. $E \equiv 0$; div $H \equiv 0$). (19)

2b) Boundary conditions ("int" stands for: interior of the particle):

For r = R: $E_\theta^{incident} = E_\theta^{int}$; $E_\phi^{incident} = E_\phi^{int}$ (2o)

2c) Two types of solutions for the fields:
"electric partial waves" ($H_r \equiv 0$)
"magnetic partial waves" ($E_r \equiv 0$) (21)

2d) Computation of the field components: To separate the variables, (scalar) Debye potentials Π_E, Π_M are introduced (E, M: electric, magnetic partial waves), which are solutions of the Helmholtz-equation:

$$\Delta\, \Pi_{E,M} + k^2\, \Pi_{E,M} = 0; \quad k^2 = \varepsilon \cdot \omega^2/c^2 \tag{22}$$

The latter equation is solved by using the following ansatz to obtain separate differential equations for the variables :

$$\Pi = F_1(r) \cdot F_2(\theta) \cdot F_3(\phi) \tag{23}$$

The results are listed in the table 4:

Variable	Differential equation	Solution functions
r	Bessel	cylinder-functions (Bessel + Neumann); Index l.
θ	spherical harmonics	Legendre polynomials; Indices l, m.
ϕ	harmonic oscillation	cos / sin - functions; Index m.

2e) Transformation of the Π's back to the fields which are divided into 3 x 2 groups:

. incident waves; electric and magnetic partial waves
. waves in the particle; " " " "
. waves behind the particle; " " " "

2f) General solution given, then, by $\sum_l \sum_m$ (partial waves).

3) Computation of intensities I from the field amplitudes.

3.2.2 The resulting intensities:

They are given in terms of the "Mie coefficients"a_l and p_l (see below):

1) Total extinction (absorption + scattering losses of the plane incident wave with intensity I_o) for ONE particle:

$$dI/I_o = \frac{\lambda^2}{2\pi\varepsilon_m} \cdot \mathrm{Im}\left\{ \sum_{l=1}^{\infty} (-1)^{l-1} \cdot (a_l - p_l) \right\} \qquad (24)$$

2) Scattering losses of the plane incident wave (i.e. due to waves behind the particles with directions other than of the incident wave) for ONE particle:

$$dI/I_o = \frac{\lambda^2}{2\pi\varepsilon_m} \cdot \sum_{l=1}^{\infty} (|a_l|^2 + |p_l|^2)/(2l+1) \qquad (25)$$

3) Consumptive absorption in the particle:

$$dI/I_o = 1) - 2) \qquad (26)$$

4) Assuming a system of Z particles per volume unit which shall NOT influence each other, the overall intensity losses are the Z-fold of the losses of one particle. Introducing, then, the volume concentration of the particle material in unit sample volume,

$$C = Z \cdot (4\pi/3)\, R^3 \,, \qquad (27)$$

we obtain
- the extinction constant $E = -C \cdot dI/I_o$ from eq.(24) (28)
- the scattering constant $S = -C \cdot dI/I_o$ from eq.(25) (29)
- the absorption constant $K = -C \cdot dI/I_o$ from eq.(26) (3o)

The "Mie coefficients" are for the electric partial waves

$$a_l = (-1)^l \frac{l+1}{l} \frac{(2\pi)^{2l+1} \cdot (R/\lambda)^{2l+1}}{1^2 \cdot 3^2 \cdots (2l-1)^2} \cdot u_l \cdot \frac{\varepsilon(\lambda) - v_l \cdot \varepsilon_m}{\varepsilon(\lambda) + ((l+1)/l) \cdot \varepsilon_m \cdot w_l} \qquad (31)$$

and for the magnetic partial waves

$$p_l = \text{- - - - - - - - - - - - - - - -} \cdot u_l \cdot \frac{\mu(\lambda) - v_l \cdot \mu_m}{\mu(\lambda) + ((l+1)/l) \cdot \mu_m \cdot w_l} \qquad (32)$$

with $\mu = \mu_m = 1$ in the optical region,

where u_l, v_l and w_l are complex valued power series in (R/λ) and ε, all beginning with 1.

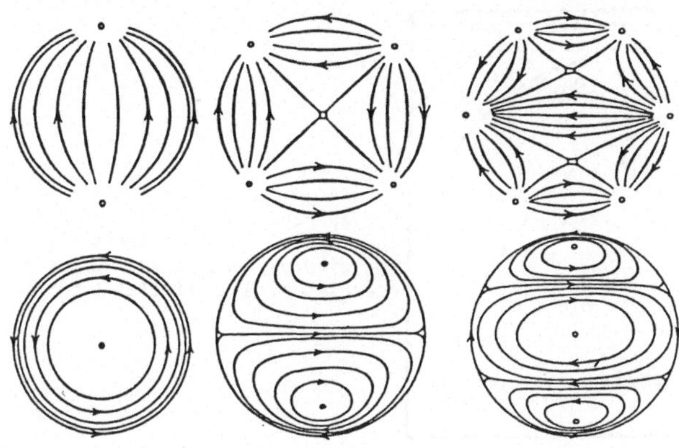

Fig.14 Electric (upper series) and magnetic fields (lower series) of the 1.,2.and 3. electric partial wave(l=1,2,3),i.e. the dipole,quadrupole and octupole mode (23).
The same pictures are obtained for the magnetic partial waves: then the upper series is due to the magnetic and the lower series to the electric fields.

3.2.3 Discussion of the Results:

1) *Partial waves:*
The sum index l gives the order of spherical harmonics (see table 4). l describes, thus, the order of spherical multipole excitations in the particle:
l = 1: dipolar fields; l = 2: quadrupolar fields;
l = 3: octupolar fields ...,etc.
Far away from the particle ("far-field"), the waves are identical to waves coming from equivalent point multipoles(25).(The electric dipole moment is $\propto a_1$, the magnetic one $\propto p_1$, etc.)
It should be pointed out, that the magnetic multipoles are not due to $\mu \neq 1$ (see Lecture I), but to electrical eddy currents in the particle.
Fig. 14 shows several such partial waves. Formulating more precisely, it shows the field lines on the surface of a arbitrarily chosen sphere with the particle in the center and the sphere diameter large compared to 2R. The dots in the electrical partial wave pictures indicate where the field lines leave the given surface towards the inner or outer space.
2) *Resonances:*
The losses eq.(28,29 and 3o) have resonance character for proper values of ε and ε_m due to the right hand term in a_l of eq.(31). This becomes more obvious if we specialize, first, to small particles ($R \ll \lambda$), where we obtain $u_l = v_l = w_l = 1$. Then, this right hand term in the Mie coefficients a_l reaches a maximum at frequencies $\omega_{r,l}$ which are determined by the condition

$$\varepsilon_1(\omega_{r,l}) = -\varepsilon_m(l+1)/l \quad , \; l=1,2,3.... \tag{33}$$

(if ε_2 depends weakly on λ near $\omega_{r,l}$). This is, evidently, a generalization of eq.(16) of the quasi static case, or, vice versa, eq.(16) of the quasi static approximation is the special case l = 1 of eq.(33) i.e. the case of the electric dipole mode with uniform polarization of the whole particle, if the particle shape is spherical.

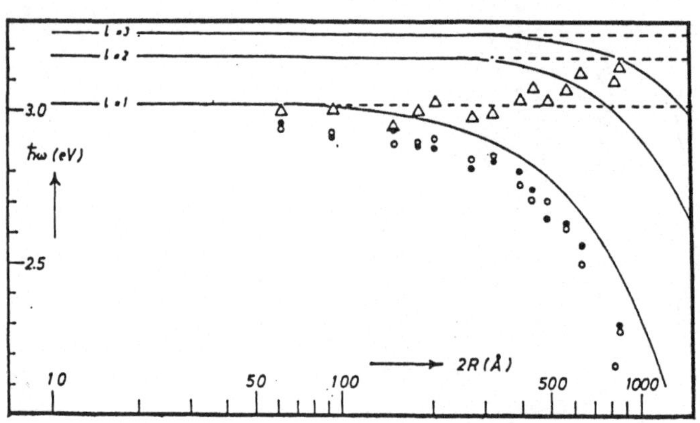

Fig.15 Particle size dependence of the dipole plasmon polariton (dots and circles) and of the HEEL-plasmon energies of Ag particles (36).

Thus, the comments concerning special behavior of ε and ε_m, made in the discussion of eq.(16), hold mutatis mutandis for eq.(33), too. If we drop the assumption $R \ll \lambda$, then $u_l = v_l = w_l \neq 1$, in general, and resonances occur at

$$\varepsilon_1(\omega_{r,l}) = -\varepsilon_m \cdot ((l+1)/l) \cdot \mathrm{Re}\{w_l\} \quad ,l=1,2,3.... \tag{34}$$

Since $w_l = w_l(R)$, the particle size influences the resonance frequencies in case of larger particles. This is shown in fig.15 for the dipole excitations (l=1) in Ag-particles of various sizes.
The magnetic multipoles exhibit no such resonance character, since the appropriate material functions μ and μ_m equal unity in the optical frequency region. Hence, the denominator of the right hand term of p_l in eq.(32) only includes weak frequency dependence via w_l.
3) *Contributions of the various multipoles:*
For dipole mode absorption of extremely small particles ($l = 1$; $u_1 = v_1 = w_1 = 1$) we obtain the well known formula

$$K = \frac{18\pi C\varepsilon_{m1}^{3/2}}{\lambda} \cdot \frac{\varepsilon_2}{(\varepsilon_1 + 2\cdot\varepsilon_{m1})^2 + \varepsilon_2^2} \tag{35}$$

which, often, has been misused for larger particles. In fact, it holds, e.g. for Ag particles in the resonance region, only if $2R \lesssim 10$ nm.
If particle sizes are larger, then,the dipole absorption is changed and scattering and higher multipole absorption contribute to the extinction From eqs.(31,32) follows that a_l and p_l are proportional to $(R/\lambda)^{(2l+1)}$. Introducing the volume concentration C of eq.(27), we find a remaining explicite size dependence proportional to $(R/\lambda)^{2l-2}$.
So, for the dipole mode ($l = 1$) this explicite dependence vanishes, leaving merely dependences via u_1, v_1 and w_1 (for larger particles) and via ε(for extremely small particles), respectively. For all higher multipole modes the according contributions are small if $R \ll \lambda$, and they are the smaller, the higher l. For the case of Ag particles,this

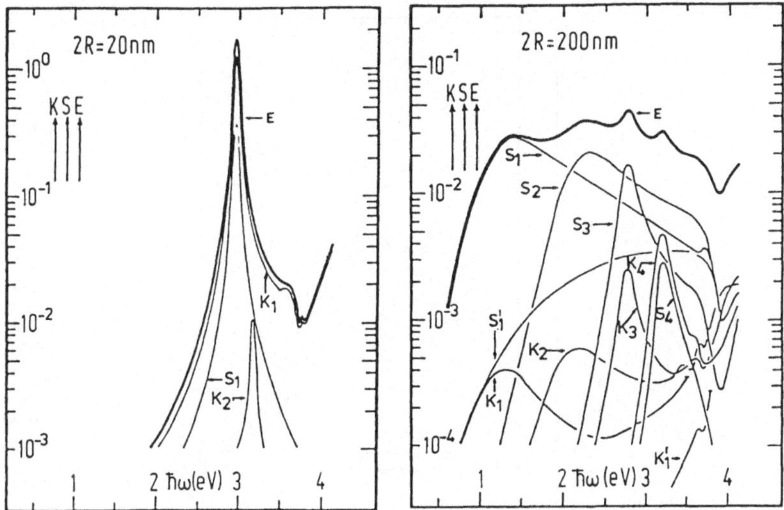

Fig.16 Extinction spectra due to Mie theory of 2R=2onm and of 2oonm Ag particles and their deconvolution into absorption (K) and scattering (S) contributions of the various electric (K_l,S_l) and magnetic (K_l',S_l') partial waves. (l=1,2,3,4)

is shown in fig. 16.
This fig. demonstrates that the extinction spectrum E(ħω) of 20 nm particles mainly is governed by the dipole absorption K_1, while for 200 nm particles contributions of the absorption constants K_l and the scattering constants S_l up to $l = 4$ are important. Comparing K_1 of both pictures, the size dependence entering via u_l, v_l and w_l is obvious: the resonance frequencies differ for about 1.5 eV. Also, it is an interesting point, that the peak positions of K- and S-resonances, coinciding for the small particles, differ distinctly for the larger particles. Hence, exact peak positions are to be determined from eqs.(24) to (26) rather than from eq.(34).
Fig. 17 gives the size dependencies of the peak heights max$\{K_l\}$ and max$\{S_l\}$ of the various contributions, again for our model particles of Ag.
Anew , we find that only the absorption of the dipole mode is left in extremely small particles and that it becomes independent of particle size. (Remember that this only holds, if ε is assumed not to exhibit an own size dependency, as discussed in Lecture I.) This resonance is easily visible, e.g., in noble metal particles, because of the uniquely high oscillator strength: Ag particles are brightly yellow, Au and Cu particles develop deep ruby colors.
The other multipole contributions to the extinction (S_1, K_2, S_2, K_3, S_3...) increase with increasing particle size and then decrease again, their maximum amounts at same time decreasing with increasing l. So, we obtain for large particles broad absorption regions due to very complex compositions of multipole contributions, instead of the sharp

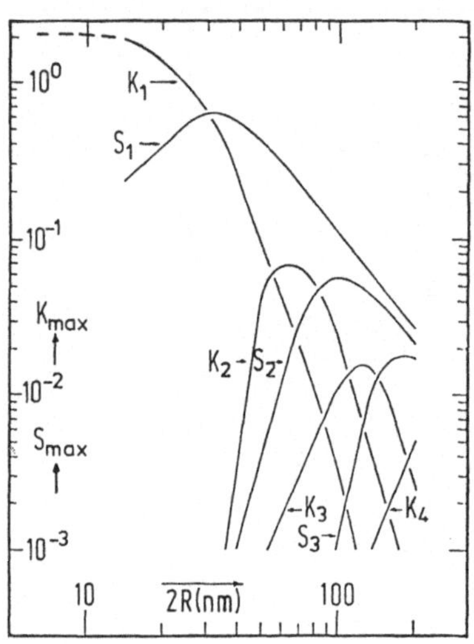

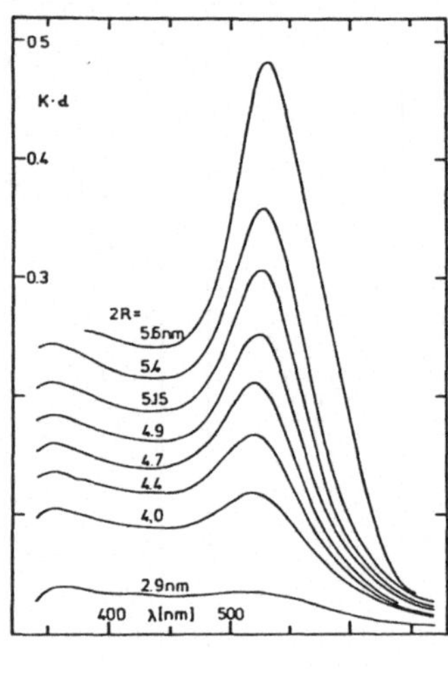

Fig.17 Left:Maxima of the K_1 and S_1 spectra (see Fig.16) versus particle size (36) for Ag particles.

Right:Experimental absorption spectra of Au particles which show size dependencies due to $\varepsilon(R)$.(9)

and very high absorption band of small particles. We may easily imagin how these regions in fig.16 further extend towards the infrared with increasing size, finally forming the metallic absorption of bulky Ag.

4) Scattered Light:

The scattering constant S, discussed in the preceding section,is obtain by integration over all scattering angles. The angular scattering characteristic itself is obtained by skipping this integration. To obtain some idea what we are going to expect, we may remind to the fact, that (in the far field), the particles act as point multipole antennae (chapter 3.2). Thence, it is not surprising that in case of the dipole mode we obtain a cosine-squared characteristic. Strong size dependencies occur, if the particles are large enough to allow of retardation effects (u_1, v_1, $w_1 \neq 1$) and higher multipole excitations. A very extensive comparison of experimental and theoretical results on Au particles was performed by the Fragstein group (26). Some results are compiled in fig. 18 which shows an increase of forward scattering in larger particles and a strong dependence on the polarization state of the incident light.

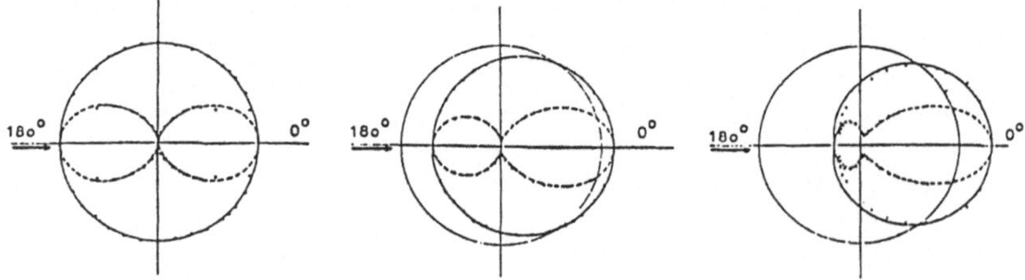

Fig.18 Scattering characteristics of Au particles of 2R= 35,12o and 16o nm. Comparison of experiment and Mie theory for both polarization directions (26).

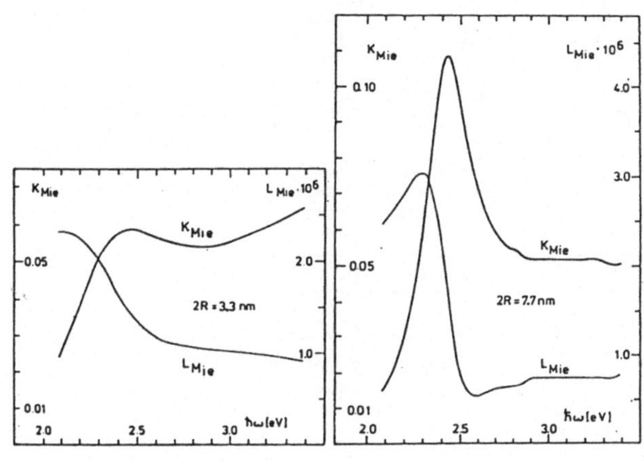

Fig.19 Measured absorption spectra (K) of spherical Au particles (2R=3.3nm and 7.7nm) and dispersion spectra (L), computed therefrom by Kramers-Kronig analysis(17).

5) Dispersion

Less well known than the Mie formulae for particle extinction are the according formulae concerning the dispersion of a particle system, which were derived by Gans & Happel (25) by introducing equivalent point multipoles for the particles. If a system of Z particles per volume unit in the embedding matrix acts as a quasi homogeneous material (- this problem will be treated in the third Lecture -), then a refractive index $n_s(\omega)$ of this material may be defined, and Gans & Happel found

$$n_s(\lambda,R,\varepsilon,\varepsilon_m,C) = \sqrt{\varepsilon_m}\cdot\left[1 + \frac{3}{4}\,\frac{3\,C\,\lambda^3}{(2\pi\sqrt{\varepsilon_m}R)^3}\cdot \mathrm{Re}\,\{a_1 - a_2 - p_1\}\right] \qquad (36)$$

In systems of low C, n_s differs very little from $\sqrt{\varepsilon_m}$ and it is very difficult to measure it. This is the reason, why experimental investigations are concentrated on the extinction, and only few exist for the refractive index (27). Spectra of the extinction and of the relative refractive index difference $L = (n_s - \sqrt{\varepsilon_m})/n_s$ of Au particles are shown in fig. 19 .

6) Determination of $\varepsilon(\omega, R)$ from Optical Experiments

To determine ε_1 and ε_2 of some material, in general two independent optical experiments are required. In the case of small particles, these may be the determination of $E(\omega, R)$ and $L(\omega, R)$. For different types of samples (bulky material, thin films, small particles) evaluation

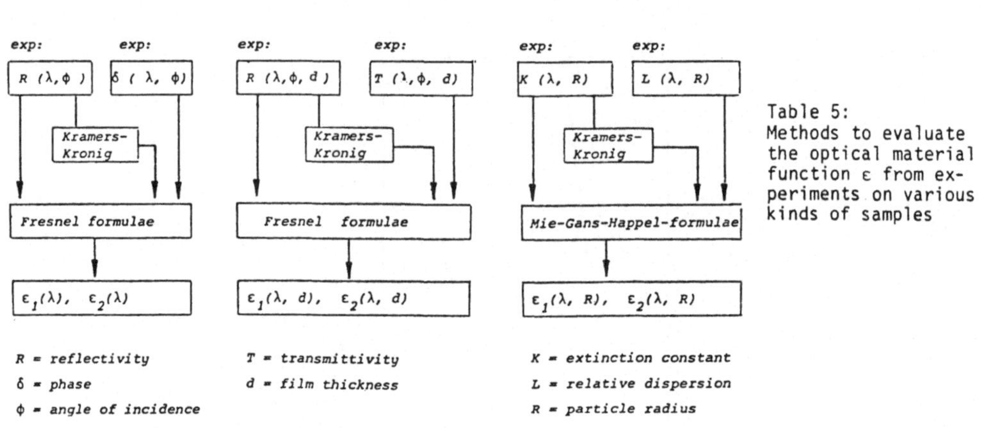

Table 5: Methods to evaluate the optical material function ε from experiments on various kinds of samples

procedures are listed in table 5. Alternatively, in all cases, one experiment may be replaced by a Kramers-Kronig-analysis of the other. (The L-spectra of fig. 19 were obtained in this way (17).
As stated before (chapter 3.2.3, no.3), the absorption spectra of extremely small particles depend on particle size ONLY via a size dependence of ε, which, thus, can be determined from optical experiments very accurately. $\varepsilon(\omega, R)$-spectra found by this way, were already presented in fig. 6.
Other methods to determine $\varepsilon_1(R)$ and $\varepsilon_2(R)$ for the special frequency $\omega=\omega_r$ have also been used to test models for size dependencies, as discussed in Lecture I (e.g.(5)).

3.3 Physics behind the Resonances

Maxwells theory contains ε and μ only as phenomenological quantities, leaving out of account the physical effects which give rise to these material functions. And so does the theory of Mie/Debye. Resonances occur at special values of the material constants ε, ε_m, irrespective of the material properties causing them in a special sample. We saw in figs. 3 and 4 that there are such values as well in ionic materials as in metals, and, hence, in both materials we observe the Mie resonances. Yet, there are quite different physical effects behind. This will become clearer when we regard the kinds of optical excitations which govern the ε-spectra near the resonance frequencies.

1) Dielectrics: The resonance conditions of eq.(34)are fulfilled in the FIR, where ε is caused by lattice excitations. At the resonances, we deal with multipole excitations of the particles as a whole They are, thus, due to elementary lattice excitations coupled to the electromagnetic waves inside and outside the particle, i.e. to an infinite series of phonon-polariton-modes, the symmetries of which are fit to the shape of the particle.

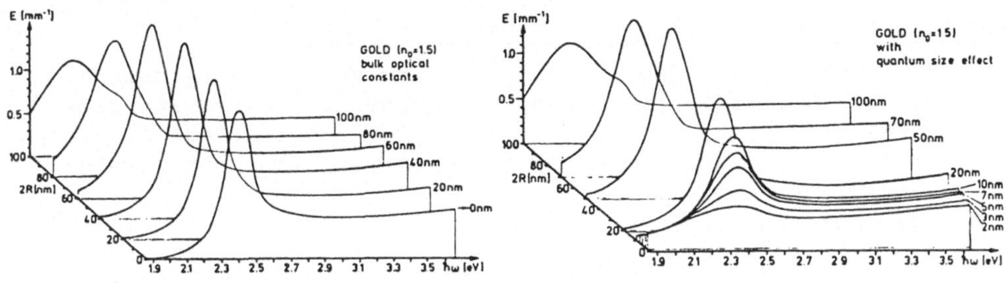

Fig.2o Extinction spectra of spherical Au particles, computed from eq.(24).
Left: with bulk $\varepsilon(\omega)$; Right: with $\varepsilon(\omega,R)$ of eq.(9)

2) In the case of metals, the resonance conditions eq.(34) are fulfilled in the region of conduction electron excitations, which are more (e.g. Au,Cu) or less (e.g. Na,Al,Ag) superimposed by interband transition excitations (see Lecture I). Now, special surface plasmon-polaritons are causing the resonances. There exists an infinite series of modes, labelled by l, which are due to the various multipole symmetries. The plasmon polaritons are more or less hybridizised with interband excitations. The influence of the latter e.g. effects the resonances in Ag particles to shift from the free electron value of ~ 5 eV downwards to about 3 eV. This shift becomes plausible from fig. 3a: By adding the susceptibility χ^{interb} to $\varepsilon_1^{Drude-Sommerfeld}$, the spectral region of ε_1-values within the hatched area is shifted towards longer wavelengths.

The size dependencies of ε_1 and ε_2, discussed in Lecture I, result in a peak shift and an increase of plasma resonance damping, respectively (fig. 2O).

It should be mentioned that the spherical plasmon modes were also observed by HEEL - (high energy electron loss -) experiments. In fig. 15, such results of Ag particles are compared with optical extinction resonances of the same particles. Clearly, the resonance energies differ for larger particles, the plasmon energies remaining constant, while the plasmon-polariton energies are lowered due to coupling to the electromagnetic fields.

The strengths of these elementary excitations are extraordinarily high, compared to samples with other geometries (thin films etc.). For the special case of Drude-Sommerfeld particles, it was shown (17) that the whole oscillator strength of the conduction electrons which, in bulky material is spread between ω=o and $\omega=\omega_p$, is concentrated to a narrow region around ω_r, i.e. no share is left beside the plasmon excitation for extra IR conduction electron absorption. So, metallic particles are transparent in the FIR (in contrast e.g. to thin films!). It is one of the actual open questions, both, from contradictory experimental and theoretical arguments, to which extent this IR-transparency is reduced in extremely small particles.

3.4 Volume Plasmons in Spherical Metal Particles

The plasmon modes discussed in the preceding section, are selected by the condition (eq.(19)), that only divergency free curl fields were regarded. That is, volume plasmons,being longitudinal electron density modulations, which are due to curl free divergency fields, curl E ≡ O div E ≠ O, were excluded.
To include them, the Mie theory has to be changed as follows(28,29):

1) Electrical fields: E = E(source-free) + E(curl-free) (37)

2) Introduction of an additional potential Γ:
$$E(\text{curl-free}) = -\,\text{grad}\Gamma \qquad (38)$$
Γ follows a wave-equation $\Delta\Gamma + k^2_{long}\cdot\Gamma = 0$, in analogy to eq.(22).

3) Γ is added to the Debye potentials $\Pi_{E,M}$.

4) An additional boundary condition is introduced, e.g.:
$$j^{\,normal}_{\,total} \text{ continuous at the surface (Sauter-condition)} \qquad (39)$$
(j: current density).

Then, the calculation follows the Mie procedure as sketched before.The results differ from Mie by a size dependent peak shift and slight oscillations in the extinction spectra above the volume plasma frequency ω_p, as shown in fig. 21. Mainly due to experimental difficulties in this frequency region, there is no experimental confirmation of the optically excited volume plasmons, yet.

3.5 Optical Properties of Particles of Various Shapes

3.5.1 Ellipsoidal Particles:

The single particle was treated already in the quasi-static approximation in chapter 3.1. A derivation was given by Gans (3o) of the dipole mode (l = 1 only) neglecting retardation effects ($u_1 = v_1 = w_1 = 1$), for the more realistic case of a system of many particles with arbitrary orientations. Also after orientation averaging the absorption maximum frequencies depend on the depolarization factor N; yet, e.g. for Au particles, now the resulting shifts are independent of polarization state of the incident light and are always towards lower frequencies compared to the sphere case (fig.22a). Results of

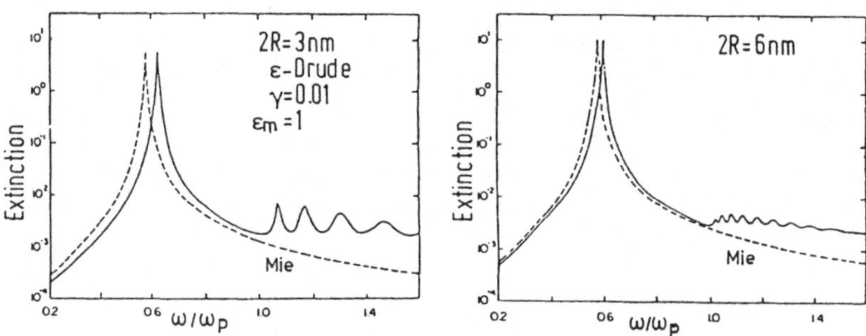

Fig.21 Extension of Mie's theory by including longitudinal plasmons (29).

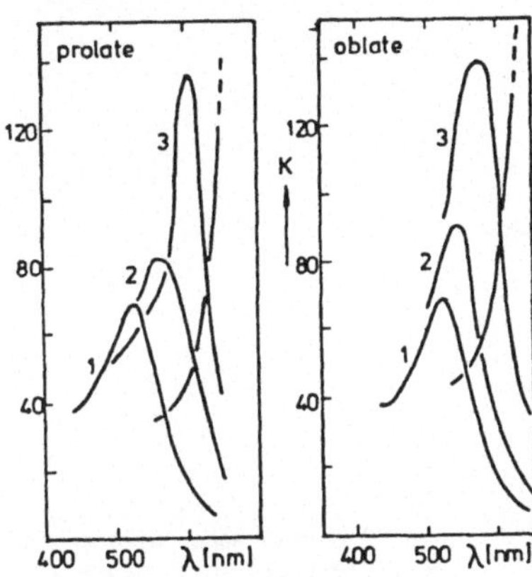

Fig.22a Extinction spectra of ellipsoidal Au particles after orientation averaging (excentricity increasing from 1 to 3)(3o).

an excellent experiment (31) with oblate ellipsoids of uniform orientation are shown in fig.22b .

3.5.2 Cubic Particles:

This shape is more commonly observed in ionic than in metallic particles. The main difficulty is that, in contrast to all the ellipsoidal shapes, even in the quasi-static case the polarization is not uniform in the particles, but deformed by contributions of edges and corners.

Fuchs (32a) computed 6 different discrete resonances (depending on particle orientation relative to the field direction) for values of ε_1 in the interval $-3.68\cdot\varepsilon_m \leq \varepsilon_1(\omega_j) \leq +0.42\cdot\varepsilon_m$, j=1..6 , (4o) instead of the sphere condition $\varepsilon_1(\omega_r) = -2\,\varepsilon_m$. These modes are sketched in fig. 23. They are superimposed and smeared out under experimental conditions of random orientation as shown in fig.13 (32b).

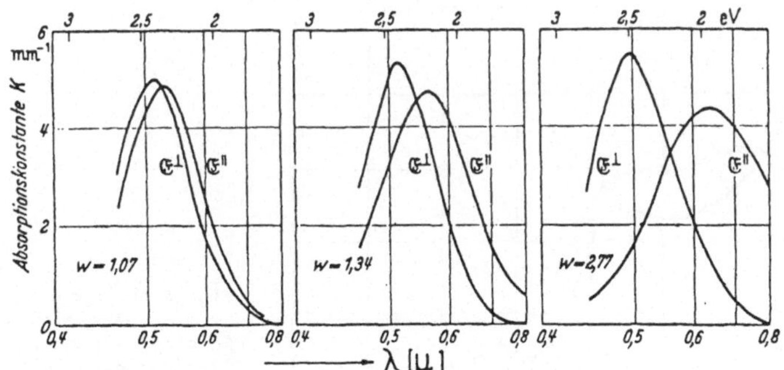

Fig.22b Experimental absorption spectra of ellipsoidal Ag particles with excentricities increasing from left to right (31).

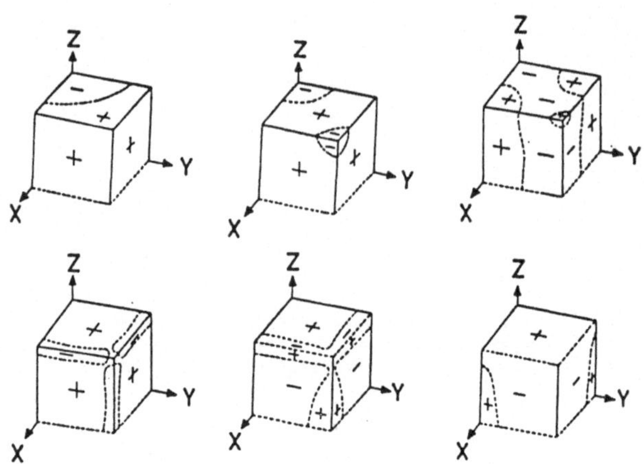

Fig.23 Normal modes of a cubic particle (32a) which cause six different absorption peaks (see fig.13).

3.5.3 Spherical Particles with Shells:

They have been treated for the quasi-static case ($l = 1$, $u_1 = v_1 = w_1 = 1$) (33a,b).The main difference to Mie is that integrations over the radial coordinate are now to be divided into two steps and boundary conditions have to be formulated also for the intrinsic boundary between nucleus and shell.
The particle polarizability can be analytically computed for ellipsoids; it is written down for spherical particles of two absorbing materials and absorbing matrix on the next page. Experimental results are shown in fig. 24 for Au particles with Ag shells.

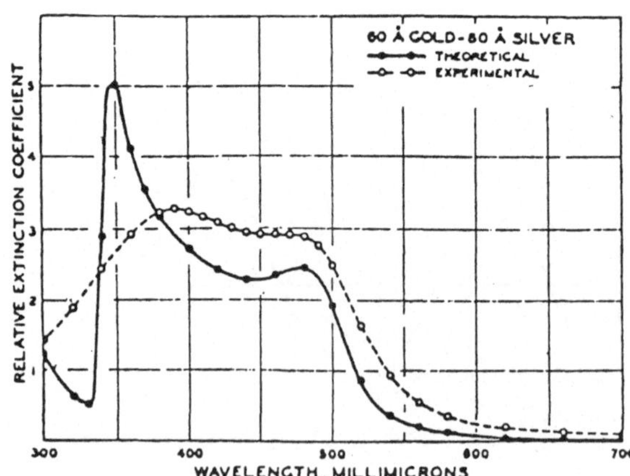

Fig.24 Absorption spectra of 6 nm Au particles with 1 nm Ag shell around (33b). Circles: experimental; Dots: theoretical.

3.6 Summary of Optical Single Particle Polarizabilities

The optical properties of single particles with various shapes will be required in the following Lecture. So, the polarizabilities as discussed above are compiled for the general case of

- absorbing particle material $\varepsilon = \varepsilon_1 + i\cdot\varepsilon_2$
- absorbing matrix material $\varepsilon_m = \varepsilon_{m1} + i\cdot\varepsilon_{m2}$
- absorbing shell material $\varepsilon_s = \varepsilon_{s1} + i\cdot\varepsilon_{s2}$.

(1) Sphere ($2R \lesseqgtr \lambda$) :

$$\alpha_l^{electric} \propto u_l \cdot \frac{\varepsilon - v_l \cdot \varepsilon_m}{\varepsilon + ((l+1)/l)\cdot\varepsilon_m\cdot w_l} \cdot \left(\frac{R}{\lambda}\right)^{2l+1} ;\ l = 1,2,3.. \quad (40)$$

$$\alpha_l^{magnetic} \propto u_l \cdot \frac{\mu - v_l\cdot\mu_m}{\mu + (l+1)/l)\cdot\mu_m\cdot w_l} \cdot \left(\frac{R}{\lambda}\right)^{2l+1} \text{ with } \mu=\mu_m=1 \quad (41)$$

(2) Ellipsoids (size $\ll \lambda$):

$$\alpha_1^{electric} \propto \frac{\varepsilon - \varepsilon_m}{\varepsilon + (\varepsilon-\varepsilon_m)\cdot N_i} \cdot V_{particle} ;\ \sum_{i=1}^{3} N_i = 1 \quad (42)$$

(3) Sphere (radius R) with shell (thickness d) ($D=R+d\ll\lambda$) :

$$\alpha_1^{electric} \propto \frac{(\varepsilon_s-\varepsilon_m)(\varepsilon+2\varepsilon_s)+(R/D)^3(\varepsilon-\varepsilon_s)(\varepsilon_m+2\varepsilon_s)}{(\varepsilon_s+2\varepsilon_m)(\varepsilon+2\varepsilon_s)+(R/D)^3(\varepsilon-\varepsilon_s)(2\varepsilon_s-\varepsilon_m)} \cdot D^3 \quad (43)$$

(4) Cubes: no analytical expressions!

The optical extinction is proportional to $\text{Im}\{\alpha_l\}$, the numerical factors being given for the example (1) in eq. (28).

4 LECTURE III: OPTICAL PROPERTIES OF SYSTEMS OF MANY PARTICLES

As stated before, two effects are acting together to make the unique optical properties of small particles depend on their size. The size dependencies of the material properties, $\varepsilon(R)$, were treated in the 1st Lecture, while size dependencies of the particle polarization (i.e. the inner field) were dealt with in the 2nd, the latter occuring even if ε is assumed to remain unchanged by particle size.
Though several experiments on single larger particles were successful, up to now optical response of a single particle with $R \ll \lambda$ is too small to be determined experimentally. (Such particles are of special interest, since size effects of ε become important below, say, 10 nm particle size.)
Instead, usual optical experiments on extinction, scattering or refraction require samples of about 10^{10} to 10^{15} particles, the reactions of the single particles thus being summed up to measurable quantities.
If all particles of a such system have

- uniform sizes,
- uniform shapes (and, if nonspherical, uniform or purely statistic orientation),
- uniform homogeneous embedding matrix (down to a scale smaller than R),
- neglectible influences by neighboring particles,
- no multiple extinction and interference effects of scattered light,

than holds that the extinction of the system of n particles equals n times the extinction of the single particle.
It is unrealistic to hope that a particle system,produced by one of the methods listed in table 2, might accomplish all these requirements, and therefore, problems arise, both,

- for the description of system properties by using single particle properties and
- for the deduction of single particle properties (e.g. size effects) from measured system properties.

Different sample production methods, thus, may be valued by the amount of approximation to our list of requirements, which is quite moderate for all the methods, used today.(It may be expected, however, that mass spectrometric particle separation methods, which already are very successful for small molecular clusters of size region I, will soon give rise to marked progress.)
On the other hand, there is a positive aspect in the fact that particle system properties are determined, aside from ε, ε_m, R and C, by the above mentioned additional parameters: the latter can be varied in a wide range, and, so, we have materials at hand with variable optical properties. Granqvist will show in his Lectures how to optimize such materials in view of their application as selective solar absorbers.
As a consequence of the large number of variables, theoretical treatments are highly difficult, and that the more, the more precisely the variables are treated. Granqvist will present several advanced

models in his Lectures and also will give an overview of the literature. The most detailed theoretical investigations were performed by the Liege group (e.g.37); their recent results, however, are too complex, by far, to be presented here. Instead, I shall confine myself to several principal problems and treat them in a introductory way with rough models which are accurate enough in the present experimental situation. The main object of this Lecture III will be the system of many interacting particles. Again, I shall mainly present experimental results of our group, and again they were obtained with Ag and Au particles. Before, however, some other system parameters will shortly be discussed.

4.1 Size Distribution

The distribution function of particle sizes in a given system (i.e. its shape and width) depends markedly on

- the particle production method (table 2) and the
- sample treatment during the experiments (e.g. exposure to air, increase of temperature, chemical surface reactions, Ostwald ripening etc.)

Some distribution histograms obtained by TEM, are compiled in fig. 25. For the method of gas evaporation, the distribution function has been determined to follow the "log-normal distribution" (38):

$$n(R) \propto \frac{R}{\sigma} \exp\left[-\log\left(\frac{R}{\bar{R}}\right)^2 / 2\sigma^2\right] \tag{44}$$

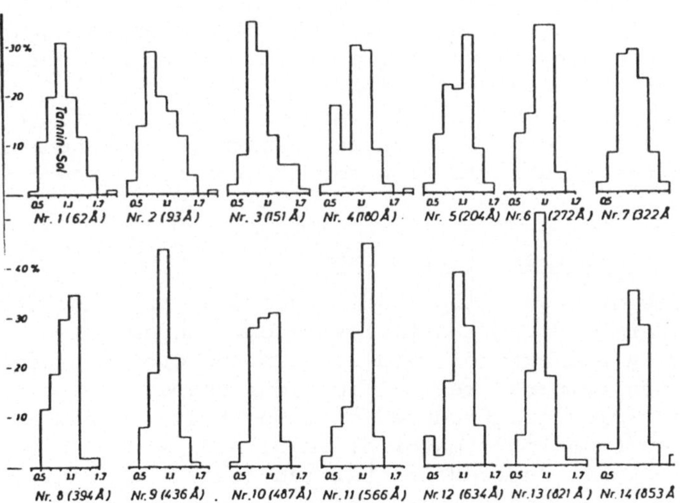

Fig.25 Particle size histograms of various samples of Ag particles produced in aqueous solution

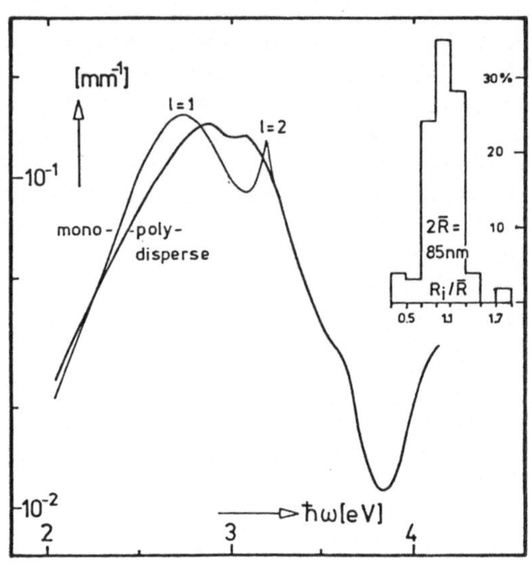

Fig.26 Computed extinction spectra of spherical Ag particles (2R=85 nm). a)monodisperse, b)with the experimental size distribution given in the insert.

As a general rule (with many exceptions, however) may be stated that heterogeneous nucleation processes give rise to sharper distributions than homogeneous nucleation. Perhaps, the size distribution problem may soon be eliminated by mass spectrometric particle separation methods.

As a consequence of size distribution for the optical extinction, size dependent spectral features of the single particle are smeared out. E.g., the discrete $\varepsilon(\omega, R)$ spectra of the potential box model (see fig. 7) should give rise to oscillating absorption spectra. Size distributions of realistic widths, however, produce monotonous spectra for the particle system. Another example is given in fig. 26. The Ag particles are large enough to show, both, dipole and quadrupole plasmon mode extinction, the separated peaks of which, however, are smeared out by the size distribution.

The mean extinction is obtained in case of dipole excitations by averaging over the R^3-distribution for absorption (since $K \propto C$ in eq. (3o)), and the R^6-distribution for scattering (since $S \propto R^6$ in eq.(29)).

4.2 Shape and Orientation Distribution

Spherical, cubic or even ellipsoidal shapes, as presumed for Lectures I and II, are rarely found. Particles resembling globules are preferably observed in the case of metals, when the growth process takes place at elevated temperatures. Ellipsoidal shapes can be obtained e.g. if the particles are produced on a substrate. Cubes are preferentially developed from ionic material (MgO, AgBr etc.) (fig. 27a). The formation of well defined crystal facettes, i.e. equilibrium shapes does not only depend on the particle material but also on the surrounding medium (gases, liquids etc.), which influences the relative growth rates of different crystal planes. Fig. 27b shows Au particles with widely differing shapes. (It is noteworthy that

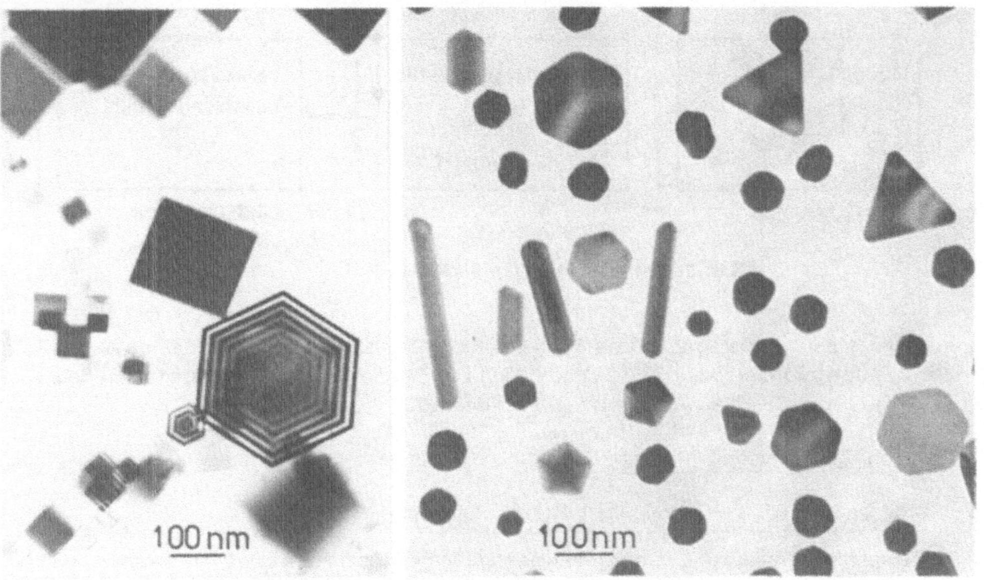

Fig.27 Electron micrographs of small particles.
Left: cubic MgO particles; Right: Au particles of various shapes

Ag particles produced under analogous conditions mainly had irregular, rounded shapes, similar to those found in the same Au particle systems for smaller sizes.)

Obviously, optical size effects, as computed for spherical symmetry in the preceding Lectures, are influenced by shape variations, too. As an example the discrete electron energy spectra of the potential box (fig. 7) vary markedly with shape(8b). Another example is the absorption peak shift of ellipsoidal particles in fig. 22. A system of ellipsoidal particles with various excentricities N will thus show up a broadened absorption band, which is shifted towards lower frequencies. (This broadening and shift may feign to be due to a size dependent $\varepsilon(\omega, R)$.) Beside a shape distribution, also variing orientations of particles with nonspherical shapes will cause selective spectral features to be smeared out, resulting in broad extinction bands (e.g. fig. 13).

4.3 Multiple Extinction Effects

If a sample is thick, the particle concentration is high and the particles are large enough for appreciable scattering intensities, then the scattered light may hit repeatedly other particles and, thus, higher order extinction processes occur, which are difficult to calculate (39).Such processes mainly cause the optical properties of metal blacks (multiple extinction) as well as of dispersion colors (multiple scattering). By using thin samples and/or low particle concentrations, these multiple extinction effects can be prevented,

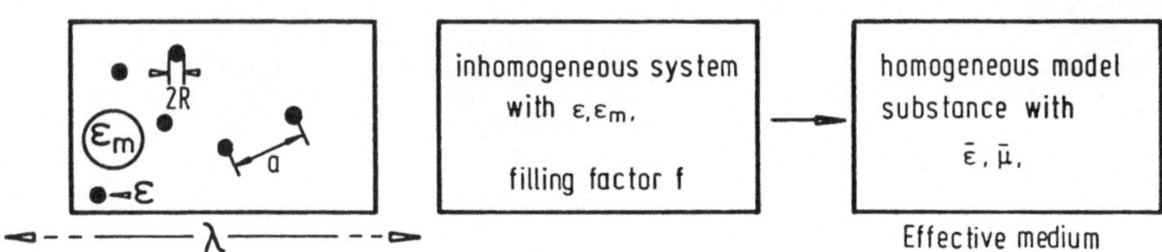

Fig.28 Definition of an Effective Medium

and so, we will disregard them in the following. Also, the case of coherent superposition of scattered light in systems of periodical particle arrangements will not be treated.

4.4 Interaction among Particles

4.4.1 Effective Media; Local Field Effects:

Let us assume a plane electromagnetic wave penetrating a sample with many particles and pick out a part of the sample, small compared to wavelength. Let us further assume that in this part are still many particles (long wavelength approximation). Then, in a first step, the influences of neighboring particles upon the optical excitation of a given particle are taken into account by regarding the polarization fields of the neighbors at that particle, which change its local field.

To simplify the description of this case, it is convenient to introduce an EFFECTIVE MEDIUM, which is defined as a homogeneous model substance with a (in general, complex) linear response $\bar{\varepsilon}$ and $\bar{\mu}$ to the external field identical with that of our real many particle system (fig. 28).

It is only admissible to define homogeneous, isotropic (i.e. scalar) "effective" material functions $\bar{\varepsilon}$ and $\bar{\mu}$ in the way shown in the following, if

- the number of particles n is large in the wavelength cube,
- the geometric particle arrangement is cubic or of statistical disorder,
- $2R \ll \lambda$; further holds that $a_j \ll \lambda$ (a_j: particle distances).

$\bar{\varepsilon}$ is defined by the electric displacement of the effective medium $D = \varepsilon_o \bar{\varepsilon} E$, and is connected to the real particle system by

$$D = \varepsilon_o\, \varepsilon_m\, E + P_E \text{ and } P_E = \varepsilon_o(\bar{\varepsilon} - \varepsilon_m) E_{Maxw} = \sum_i n_i \cdot \alpha_{Ei} \cdot E_{local} \qquad (45)$$

where $\alpha_{E,i}$ are the single particle polarizabilities as discussed in the preceding Lecture. Analogously holds for $\bar{\mu}$ that the magnetic induction $B = \mu_o\, \bar{\mu}\, H = \mu_o H + P_M$. (46)

The conditions which are given above, allow the Lorentz-sphere method to be applied, and, following the line of Lecture I, eqs.(2) and (3), we arrive at the Lorenz-Lorentz solution, which may be written in analogy to eq.(3) as

$$\bar{\varepsilon}/\varepsilon_m = \frac{1+2/(3\varepsilon_o\varepsilon_m)\sum_i n_i\,\alpha_{E,i}}{1-1/(3\varepsilon_o\varepsilon_m)\sum_i n_i\,\alpha_{E,i}} \quad . \tag{47}$$

The fundamental difference to eq.(3) is that now $\alpha_{E,i}$ are the polarizabilities of the WHOLE particles! These are well-known to us from their discussion in the preceding Lecture.

1) Spherical particles

For a many spherical particle system with embedding medium (ε_m) and volume concentration of the particles f("filling factor"), Gans and Happel (25) found:

$$\bar{\varepsilon} = \varepsilon_m \cdot \frac{1+2\cdot f\cdot\Lambda_E}{1-f\cdot\Lambda_E} \quad ; \quad \bar{\mu} = 1\cdot\frac{1+2\cdot f\cdot\Lambda_M}{1-f\cdot\Lambda_M} \tag{48}$$

with

$$\Lambda_E = \frac{1}{2}\,(\lambda/(2\pi R\sqrt{\varepsilon_m}))^3\cdot(a_1-a_2) \quad ; \quad \Lambda_M = \frac{1}{2}\,(\ldots)^3\cdot p_1 \tag{49}$$

where electric dipole and quadrupole (Mie coefficients a_1, a_2) and magnetic dipole (p_1) modes are taken into account. We see that these Λ are proportional to the single particle multipole polarizabilities given at the end of Lecture II.
The extinction constant E for the many particle system is computed from eq.(48) by using the Maxwell formula $\bar{\varepsilon}\,\bar{\mu} = (\bar{n} + i\,\bar{k})^2$ (5o)
and is given by

$$E = (4\pi/\lambda)\cdot\bar{k} \tag{51}$$

If $\bar{\mu} \equiv 1$, $f \ll 1$, then $\bar{n}\approx\sqrt{\varepsilon_m}$, and E is obtained more simply by

$$E \approx (2\pi/(\lambda\sqrt{\varepsilon_m})\cdot\bar{\varepsilon}_2 \tag{52}$$

Analogously, the dispersion of the many particle system, described by the Gans-Happel function L

$$L = (\bar{n} - \sqrt{\varepsilon_m})/\bar{n} \tag{53}$$

can be computed from eqs.(48) and (5o).

<u>Special cases:</u>

a) As shown by Gans & Happel, Mie's results of the extinction constant, eq.(28), and Gans-Happel's function for the refractive index (eq.(36), are obtained from eqs.(51) and (53), respectively for $f \ll 1$, i.e., then, the optical properties of the system of n particles equal n times the properties of the single particle.

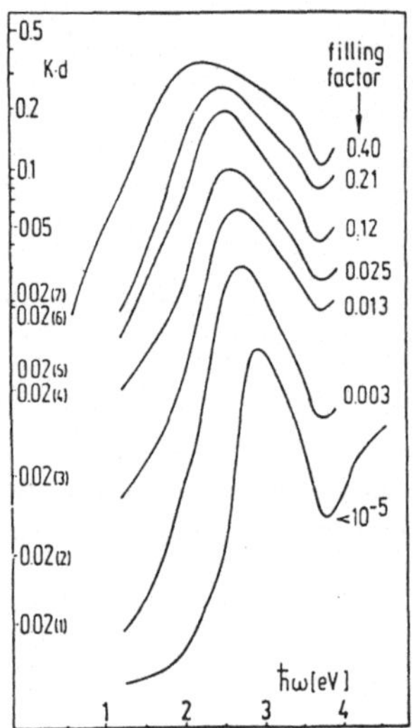

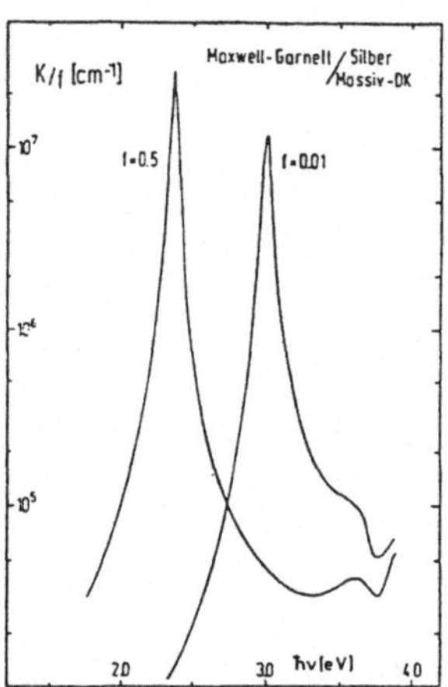

Fig.29 Left: Measured absorption spectra of a series of Ag particles (2R≈1onm) with varying filling factor. Right: Absorption spectra of Ag particles computed from Maxwell-Garnett,eq.(54),with low f(o.oo1) and high f(o.5).

b) Presuming extremely small particles (e.g. spherical particles with 2R < 15 nm), i.e. $a_2 = p_1 = 0;\ u_1 = v_1 = w_1 = 1$, we find the Maxwell Garnett formula

$$\bar{\varepsilon} = \varepsilon_m \frac{1+2\ f\ \eta}{1-\ f\ \eta}\ ;\quad \bar{\mu} \equiv 1 \qquad \text{with } \eta = \frac{\varepsilon - \varepsilon_m}{\varepsilon + 2\varepsilon_m} \tag{54}$$

2) Experimental Results of Effective Media

The experimental results, presented in the following, are due to our own group and are published only in part (4o a,b). They were performed with samples of (nearly) spherical Ag and Au particles prepared by the Zsigmondy method in aqueous colloidal solutions and then stabilized by gelatin . Drying these solutions, thin films were obtained with the particles embedded in a solid gelatin matrix. These kinds of samples were preferred for interaction experiments,since by changing amount of gelatin,the filling factor f could be varied as the ONLY parameter (i.e. with mean sizes, mean shapes, kind of embedding matrix etc. held constant). A further advantage of such samples is that they can be made thin enough to allow, both, optical experiments and direct transmission electron microscopical observation with one and the same sample.

The absorption spectra 2 to 7 of fig. 29a are due to 2R ≈ 10 nm Ag particles with filling factors variing between 3×10^{-3} and 0.40. (Curve 1 shows, for comparison, the liquid solution absorption ($f < 10^{-5}$).) Two effects are to be observed, when f increases:

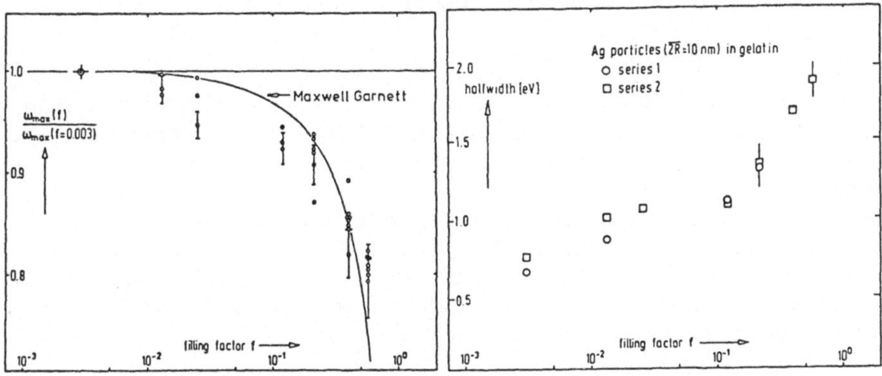

Fig.30 Absorption peak energies (left) and widths (right) of Ag particle systems with varying filling factor f.

. the peak position is shifted towards lower frequencies and
. the absorption band is broadened markedly.

The particles are small enough to compare with the simple Maxwell Garnett formula (eq.(54)), i.e. there is only consumptive absorption, due to the dipole mode,without retardation. Two spectra are shown in fig. 29b, which were computed using ε-values of bulk Ag. Measured peak shifts are compared again with the predictions of the Maxwell Garnett formula in fig. 30a. There is quantitative agreement. However, the large band widths and their marked increase, as compiled in fig. 30b disagree with theoretical predictions,since the halfwidths of fig. 29b are $\lesssim$ 0.1 eV. The large widths at low f's may be due to increased lattice defect density in the particles (6); the strong dependence on f, yet, remains unexplained in the frame of Maxwell Garnett theory. So we find

. Correspondence with Maxwell Garnett for peak positions,
. Discrepancy to Maxwell Garnett for band widths.

Several attempts have been made to confine the theoretical ansatz by skipping the condition $\Delta = 0$ in the Lorentz theory, (eq.(2)), i.e. including contributions to the local field from particles within the Lorentz sphere (41a,b) which are important if the particles are not arranged in cubic or purely statistic order. Indeed, better correspondence was thus obtained for the band widths by using special models for the particle arrangement, as shown in fig. 31a. Next step of improvement would be to introduce, instead, geometrical particle arrangements which are experimentally determined (e.g. fig. 31b). The electron micrographs of the samples show,how complex these arrangements are, even in those cases, where we tried to obtain arrangements as homogeneous as possible to be able to compare with Maxwell Garnett. Furthermore, we observed that there are spatial fluctuations of this arrangement and, even, of the filling factor. Both effects should be taken into account, since they give rise to additional (and f-dependent broadening of the measured absorption spectra.

3) Nonspherical Particles

We formulated the Gans-Happel equations (48) in terms of Λ which are proportional to the particle polarizabilities α, to show that they may easily be extended to ellipsoidal particles or shell particles by inserting the proper α's of chapter 3.6.
As well, absorbing embedding media, and other special cases may be treated analogously, by using the proper dielectric functions.
So, a wide variety of sample systems is described with eq.(48),as far as particles are sufficiently small and -as we will see in the next chapter- particle distances are not too small.

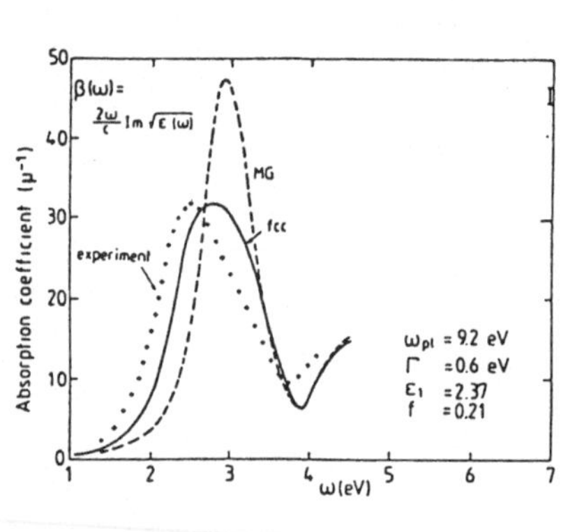

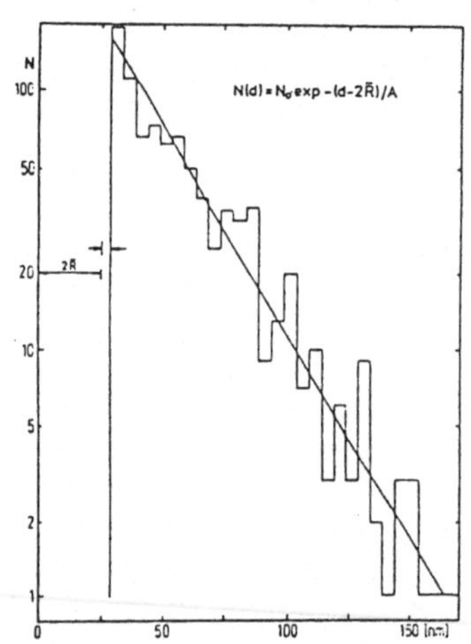

Fig.31 Left: Improvement of Maxwell Garnett theory (41b) for a fcc arrangement of the particles. Comparison with experimental results.
Right: Nearest neighbor distance distribution, measured from a mono-layer sample of Ag particles.

4.4.2 Direct Dipole Coupling of Neighboring Particles:

To summarize the preceding chapter 4.4.1 : the inner field acting upon one particle is changed by the polarization fields of neighboring particles, if their distance is not too high (i.e. the filling factor f is not too low) . The particle polarizabilities, both, of the re-garded particle and of its neighbors were assumed in our approximation to bethose of a SINGLE isolated particle as discussed in Lecture II. Consequently, we observe a shift of the dipole resonance down to lower frequencies and broadening if f is increased, however,there is still ONE resonance as assumed for the isolated single particle. For realistic geometrical particle arrangements this may be an only

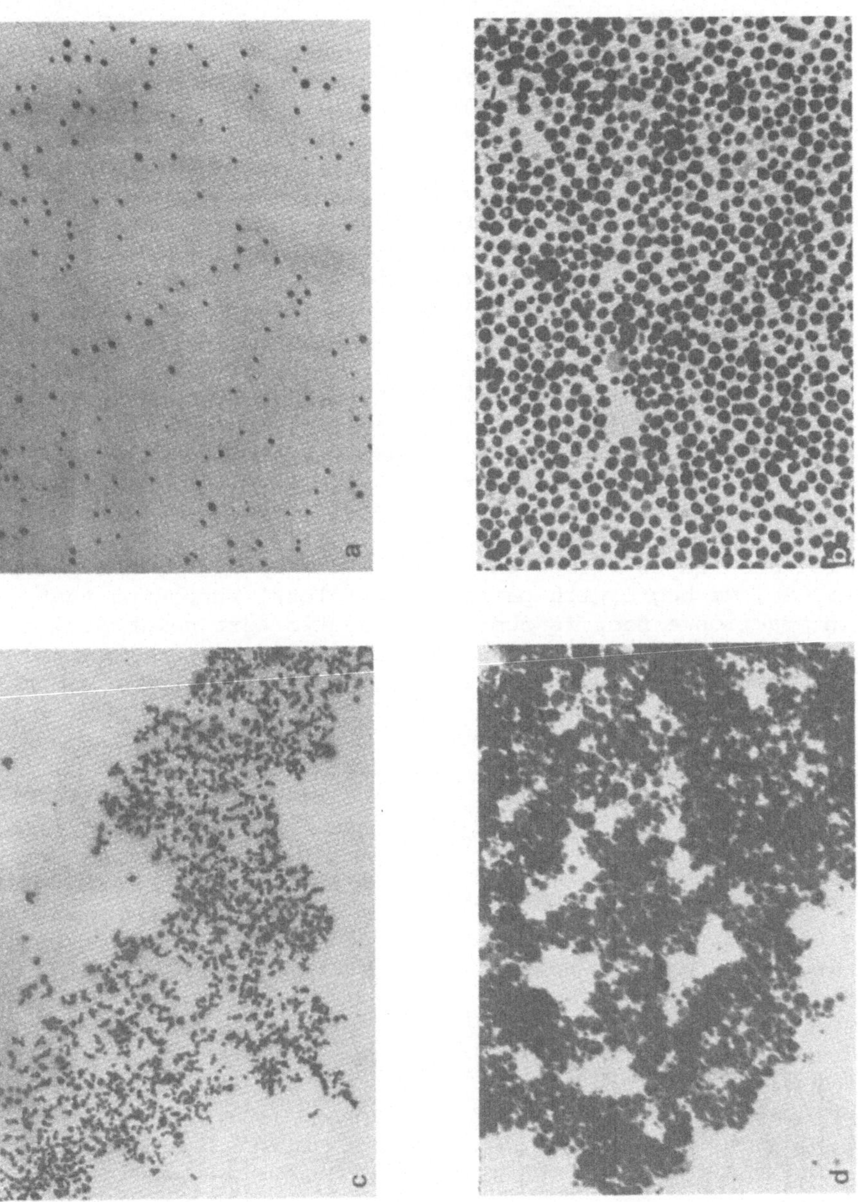

Fig.32 Electron micrographs of different Ag particle samples(4oa).
a)non-coagulated, low f, 2R≈8 nm; b)mainly non-coagulated,high f, 2R≈65nm;
c)partly coagulated, small clusters, 2R≈8 nm; d)large clusters, 2R≈15 nm.

crude approximation and, as the next step towards a more reliable description, it has to be taken into account that this single particle polarizability may be markedly influenced by interactions with neighboring particles, too.

Even reducing to the dipole case, the problem is very complex how to compute direct coupling between these dipoles in a many particle system of given particle arrangement. As we will see, the implications on the optical extinction are most graving, if clusters of particles in direct contact occur in the arrangement. Still more problems arise, if higher multipoles are taken into account. I confine myself in this context to refer to the Liege group papers(37).

The two lower TEM pictures of fig. 32 show complex and irregular particle clusters. As stated already in the Introduction, the formation of clusters (i.e. coagulation, fig. 1) is quite natural in many particle systems. There are countless examples, reaching from micrometeorids of loose aggregates of submicron particles (35) to emission products from automobile exhaust which are composed of chain-like aggregated small particles.

The diversity of cluster shapes impede quantitative computations.

So, in the following I shall discuss two simple models which have the advantage to be easily derived, and to give numerical results which may be compared to experimental ones.

The first step is to reduce the numbers of particles in a particle-cluster. So, we begin with pairs and triplets, supposing that the main interaction effect is due to one or two next neighbors. (Fig. 32 shows that samples can be produced with small, chain-like particle clusters where the interaction with the next one or next two neighbors dominate.)

1) Dipole Coupling of Particle Pairs

The following assumptions are made:

. spherical shape of the particles,
. $2R \ll \lambda$ (Thus only the dipole mode is excited without retardation.)
. Both particles have same size: $R_1 = R_2 = R$.

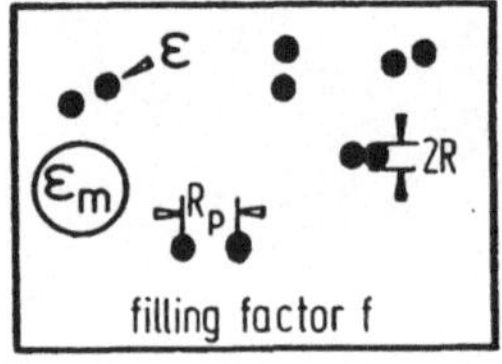

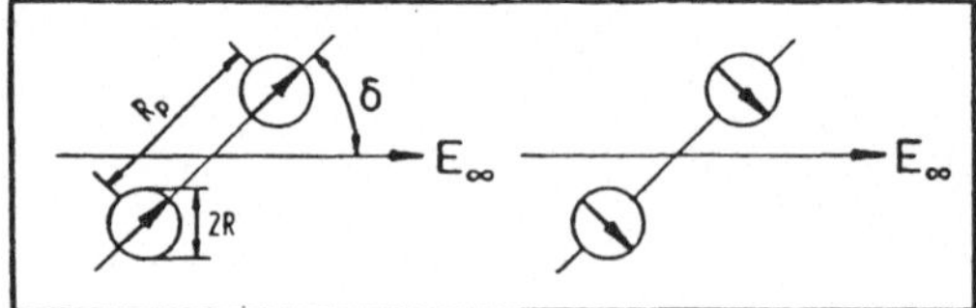

Fig.33 Particle pairs : optically relevant modes.

Then, the dipolar oscillation modes of the PAIR, which are optically relevant (i.e. which have large dipole moments) may be calculated for arbitrary interparticle distance R_p and arbitrary particle material in the following steps (42):

(1) Calculation of the dipolar fields of the $\frac{1st}{2nd}$ particle as induced in the $\frac{2nd}{1st}$ particle. The pair polarization, obtained, has TWO different resonance frequencies (instead of ONE dipole mode of the single particle) due to the two optical modes which are sketched in fig. 33.

(2) Orientation averaging (angle δ).

(3) Summation of the contributions of the two modes.

The resulting pair polarizability is then given by

$$\Lambda_{E,p} = \frac{\eta}{3} \left[\frac{1}{1-2\eta(R/R_p)^3} + \frac{2}{1+\eta(R/R_p)^3} \right] \text{ with } \eta = \frac{\varepsilon - \varepsilon_m}{\varepsilon + 2\varepsilon_m} \tag{55}$$

Dropping the assumption $R_1 = R_2$, Schönauer extended this computation to particle pairs of different sizes, $R_1 \neq R_2$, and obtained a similar expression for the pair polarizability (43).
Now, we assume our sample to consist of many such particle pairs (filling factor f). Then, following the procedure of chapter 4.4.1, we insert this $\Lambda_{E,p}$ into the Gans-Happel or the Maxwell Garnett formula and, thus, receive the resulting absorption spectrum $K(\omega, \varepsilon, \varepsilon_m, (R/Rp), f)$. We omit to present graphs of such spectra here since analogous ones for the triplet case will be shown and extensively discussed in the next chapter.

2) Dipole Coupling of Linear Particle Triplets (42)

The absorption of systems consisting of linear particle triplets with uniform particle radii R and identical distances R_p between particles 1 and 2 and particles 2 and 3, respectively, can be derived analogously to the pair case sketched above. As shows fig. 34, four different

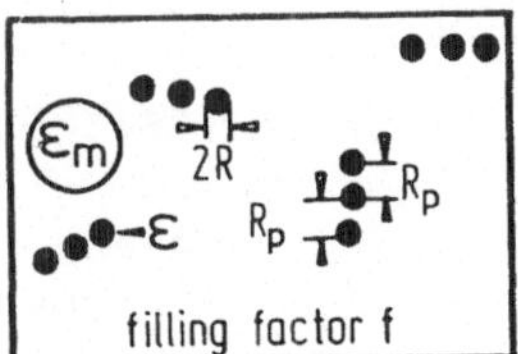

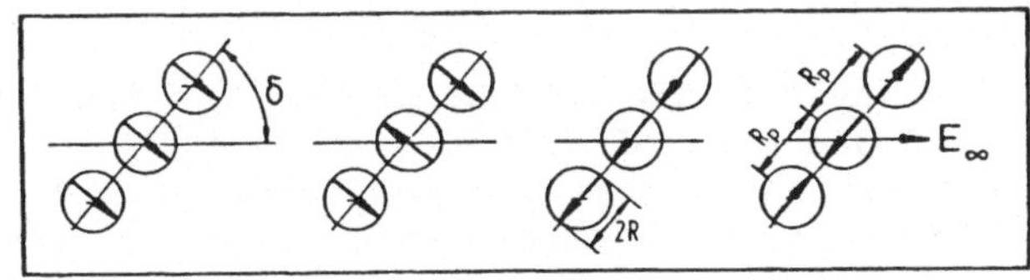

Fig.34 Linear particle triplets : optically relevant modes.

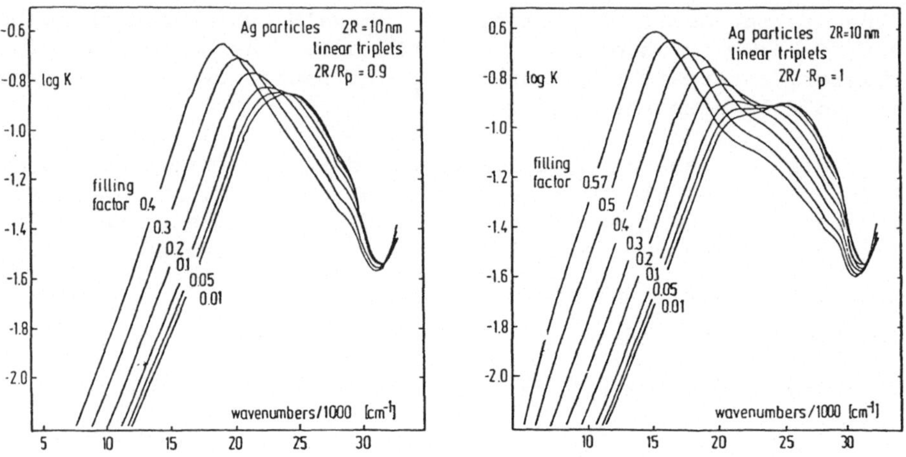

Fig.35 Absorption spectra of Effective media of Ag particle triplets (eq.(56) for two different particle distances R_p. Parameter: filling factor.

polarization modes are optically relevant and, assuming equal excitation probabilites, the expression for $\Lambda_{E,t}$ includes four terms:

$$\Lambda_{E,t} = \frac{16^2\eta}{3Z} \cdot [(1+2\eta(R/R_p)^3)[\frac{1}{(1-Z)(1+\lambda_{+2}\eta)} - \frac{1}{(1+Z)(1+\lambda_{-2}\eta)}] +$$

$$+ (2-2\eta(R/R_p)^3)[\frac{1}{(1-Z)(1+\lambda_{+1}\eta)} - \frac{1}{(1+Z)(1+\lambda_{-1}\eta)}]] \quad (56)$$

with $\eta = \frac{\varepsilon - \varepsilon_m}{\varepsilon + 2\varepsilon_m}$

$Z = \sqrt{257}$; $\lambda_{\pm\nu} = \frac{32}{1 \mp Z} \cdot F_\nu$; $F_\nu = \nu \cdot \eta \cdot (R/R_p)^3$; $\nu = 1,2$

The main contributions are due to the 1st and 3rd term of eq.(56), i.e. the 1st and 3rd mode of fig.34.
Again assuming a system with many such triplets, the resulting absorption spectra are obtained explicitely by inserting $\Lambda_{E,t}$ into the Gans-Happel formula and then computing $K(\omega, \varepsilon, \varepsilon_m, (R/R_p), f)$ from eqs.(5o) and (51).
Numerical results for Ag particles are presented in the spectra of fig. 35.

The main features are:

- Only if the particles are nearly touching (fig. 35b), the different modes cause a multiple peak structure of the absorption. The splitting of these peaks increases with decreasing particle distance and so does the overall width of the absorption.
- Increasing filling factors cause the peaks to shift to lower frequencies and cause the relative peak intensities to vary strongly (if the peaks are separated).

The peak positions, both, for varying R_p and f, are compiled in fig. 36. For large particle distances, these curves converge towards the Maxwell Garnett case with a single peak. (The computations were performed with reduced ε_2 compared to bulky Ag to improve separation of the two main peaks.)
It is because of the extremely increasing expenditure of computation, that we did not extend this method to larger linear chains. Besides, the condition of Gans-Happel or Maxwell Garnett that the cluster size be very small compared to wavelength, is less well fulfilled for such larger chains, even if the particles themselves are small. Then, higher multipole excitations of the particle clusters as a WHOLE and retardation effects become important (37).

3) Direct Dipole Coupling of Touching Spheres in Single Large Arrays

The theory of Clippe, Evrard, Lucas and Ausloos (44) allows to treat larger particle arrays in quasistatic approximation, if their symmetry is high (linear chains, tedrahedra etc.) and neighboring particles are touching directly (though without electric contact). Yet, instead of the explicite absorption spectra which were obtained in the preceding section, only the positions of their absorption peaks are obtained. The nice trick of this theory is to replace the particles with their dipole moment

$$e \cdot \vec{q} = \frac{\varepsilon(\omega) - \varepsilon_m}{\varepsilon(\omega) + 2\,\varepsilon_m} \cdot R^3 \cdot \vec{E}(\omega) \tag{57}$$

by a model oscillator with simple resonance behavior according to $1/(1 - \omega^2/\omega_r{}^2)$ and to re-introduce the special material properties by a function $F(\omega)$ without own resonance behavior:

$$e \cdot \vec{q} = \frac{F(\omega)}{1 - \omega^2/\omega_r^2} \cdot R^3 \cdot \vec{E}(\omega) \tag{57'}$$

(Note, that $\varepsilon(\omega)$ and $\varepsilon_m(\omega)$ in eq. (57) are, in general, not known as analytical functions but from point-to-point plots of experimental results!) Coupling, then, several such particles to form a cluster of particular symmetry, the eigenvalue problem of this cluster which gives the resonance frequencies of the coupled system, is solved,

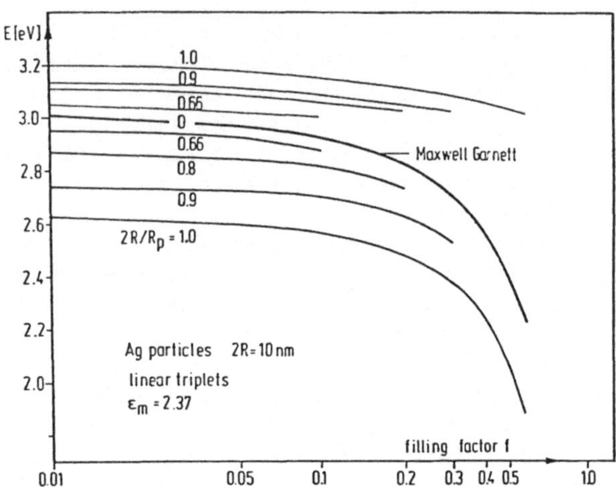

Fig.36 Ag particle triplet - Effective media: absorption peak positions. With increasing particle distance R_p, the positions of the main two peaks converge to the one of Maxwell Garnett (eq.(54)). (Clusters are only defined for $f < (4\pi/3)(R/R_p)^3$.)

by following the established methods for coupled harmonic oscillators. The Hamilton operator of the particle system is then given by:

$$H = \frac{1}{2}\sum_i \frac{e^2}{F(\omega)R^3\omega_r^2}\left(\left(\frac{d\vec{q}_i}{dt}\right)^2 + \omega_r^2\,\vec{q}_i^{\,2}\right) + \frac{e^2}{2}\sum_{i,j}\vec{q}_i\,T_{ij}\,\vec{q}_j\,) \tag{58}$$

with T_{ij} the dipole tensor with eigenvalues λ_μ.
The important feature is that these λ_μ are independent of particle material properties, only being determined by the cluster geometry. Only AFTER solving the eigenvalue problem , the material properties are incorporated via $F(\omega)$ again, to determine the appropriate eigenfrequencies ω_μ, which follow from

$$\omega_\mu = \omega_r\cdot(1 + \lambda_\mu\cdot F(\omega_\mu))^{1/2} \tag{59}$$

(It should be pointed out that the λ_μ are due to the undamped systems and, so, there are problems with materials of high plasma resonance damping.)
Table 6 gives some of the results, thus obtained, in terms of relative oscillator strengths and eigenvalues. Larger clusters, as there are linear chains, double strand chains, close packed planar arrangements, tetrahedra and 3-dimensional fcc lattices were treated analogously. Resonance peaks for small Au particles are shown in fig. 37. There are two main peaks (except for the double strand chain), the low energetic

	Oscillation mode	Multiplicity	Oscillator strength	Eigenvalue
(a)		1×	2	− 0.250
		2×	2	+ 0.125
(b)		2×	0.65 × 10⁻¹	− 0.338
		2×	2.93	+ 0.185
		1×	2.93	− 0.370
		1×	0.65 × 10⁻¹	+ 0.338
(c)		1×	3.83	− 0.436
			0.15	+ 0.176
		2×	0.16	− 0.088
		2×	3.83	+ 0.218

Table 6: Results of Clippe theory for small clusters: from left to right: oscillation mode; multiplicity; oscillator strength; eigenvalue λ_μ (44).

one varying strongly with cluster shape. In contrast, the high modes are all fixed at nearly the same position by the interband absorption edge. For the special case of Au particles, the low frequency modes divide into two groups:

. largely shifted peaks for all extended clusters, at about 1.8 eV, and
. slightly shifted peaks for small clusters at about 2.1 eV.

None of these regular clusters are observed in realistic particle systems. Instead, the clusters are usually arranged in a more or less statistically varying way. So, comparison to experiments remains

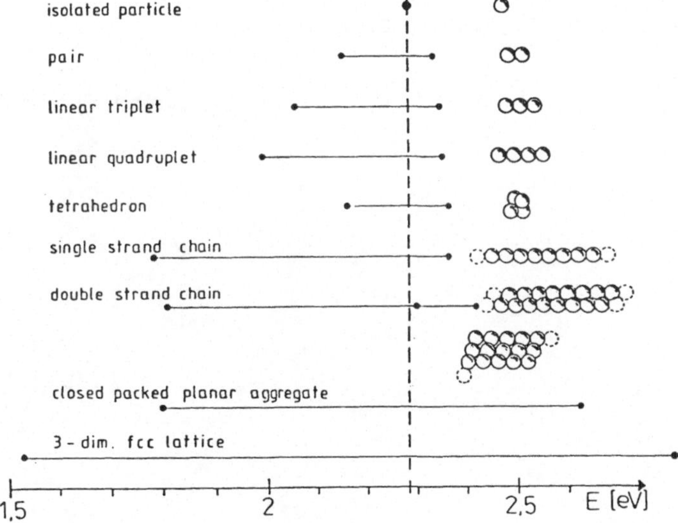

Fig.37 Absorption peak positions of various Au particle clusters, according to Clippe theory, eq.(59).

semiquantitative, and we can merely distinguish small clusters, long chain-like clusters, planar clusters and isometric large clusters. Each of these groups contains regular clusters as treated by the Clippe-theory, which we will regard as representative.
Comparisons will be made with experiments performed with samples of (nearly) spherical Au particles which were prepared analogously to the Ag particle samples, described in chapter 4.4.1. Again, the particles were embedded in a solid gelatin matrix with different filling factors. In addition, the amount of clustering during preparation of the samples could be influenced, and so we obtained samples with various filling factors AND with varying kinds of particle clusters (fig. 38). Such samples, hence, differed only in the geometric arrangement of the particles. They were analyzed by TEM and their optical extinction was measured.
The effects of increased clustering (i.e. formation of clusters with increased mean number of particles) on the absorption of one system of Au particles are shown in fig.39(left). Curve (a) belongs to non-interacting particles (no clusters, small filling factor, i.e. large particle distances), as described in Lecture II (fig. 17)), and curves (b) to (f) are due to stepwise increased clustering. The main features are that additional absorption structures arise which cause the overall spectra to broaden towards the infrared region. Several weak peaks point to the fact that these spectra are composed of various narrow absorption bands at different peak positions, which, according to the TEM analysis, are ascribed to clusters of different geometries. For comparison, the absorption spectrum $K = 4\pi\, k/\lambda$ of a thin plane Au film is added (curve (g)). Obviously, an increase of the mean cluster size causes the low frequency absorption wing to be shifted towards the IR, until, in case of the plane film, it extends to zero frequency, resulting in the usual Drude-Sommerfeld absorption.

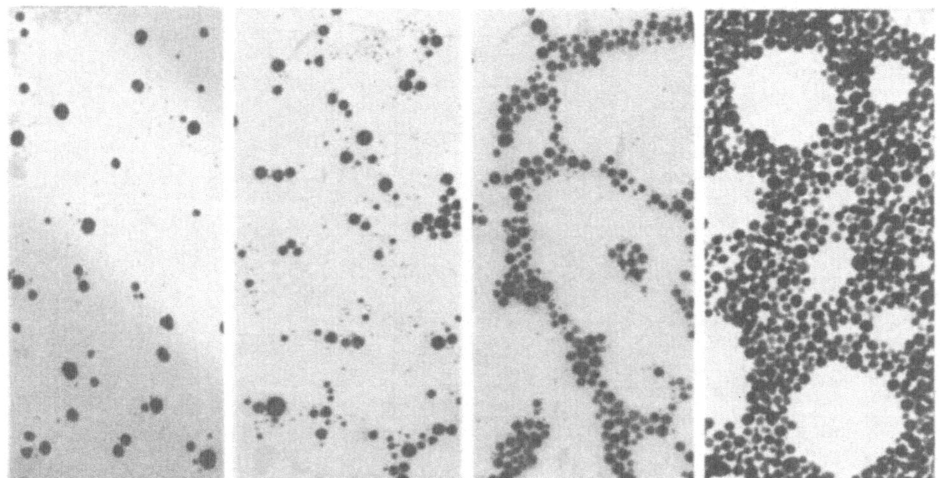

Fig.38 Electron micrographs of 10 nm Au particle samples with varying amount of clustering.

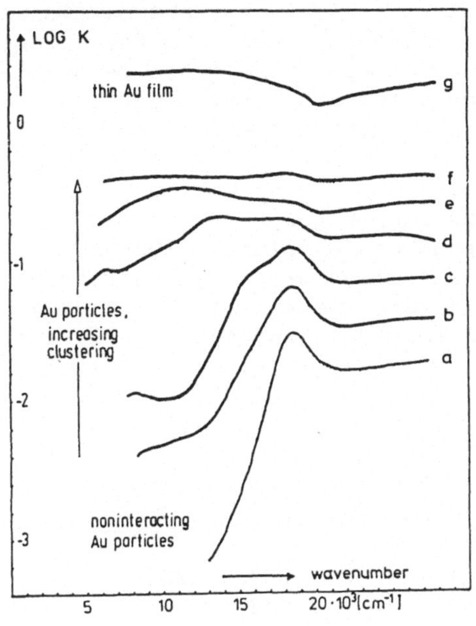

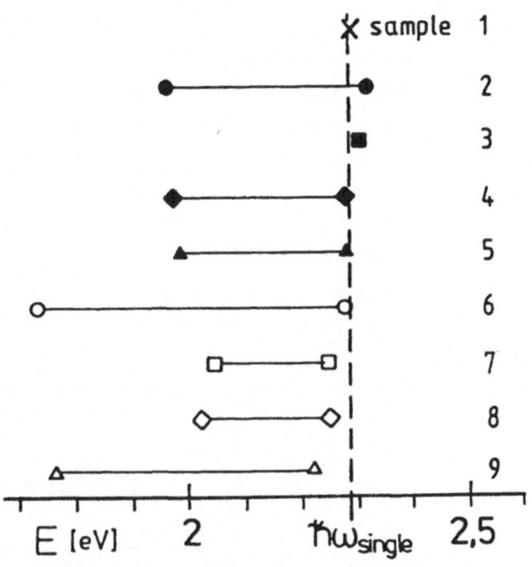

Fig.39 Measured absorption of Au particle systems (2R≈1o nm) with varying amount of clustering (43).
Left: Absorption spectra (a) to (f); Computed absorption of a thin Au film (g).
Right: Absorption peak positions of one sample series. $\hbar\omega_s$: single particle peak.

The experimentally found peak positions of fig.39(right) can be compared with the calculated ones of fig.37. Indeed, the two main frequency groups, mentioned above, which are due to extended clusters and to smaller clusters, respectively, are also found in the measured spectra. Analogous correspondence was published earlier for systems of Ag particles (4oa). It should be pointed out that the presented dipole coupling frequencies are not influenced by the (mean) particle concentration since the filling factor of the clusters was quite small in the investigated samples. A detailed discussion will be presented later(43).So we will finally state only that the observed multiple peak structures are a direct experimental confirmation of dipole coupling. Yet,more experimental work is needed in this field of inhomogeneous matt

4.5 Summary of Lecture III

. Only under very restrictive conditions holds that the extinction of a n-particle system equals n times the extinction of the single particle. Thus, deduction of single particle properties from experiments on many particle systems is to be done very cautiously because of size and shape averaging and of particle interaction effects. ("Cautiously" means that small particle experiments should always be accompanied by TEM characterization of the samples.)

. Interactions among particles via the inner field are described by the "effective medium" theory as far as quasistatic conditions are fulfilled. A large variety of samples (e.g. spheres, ellipsoids, bubbles in absorbing media, shell-particles etc.) can be treated this way by inserting appropriate single particle polarizabilities, as presented in Lecture II, into the Gans-Happel formulae.

. Interactions due to changes of the polarizabilities of the particles themselves by direct dipole coupling can be included into the effective medium theory by computing appropriate particle-cluster-polarizabilities. Explicite absorption spectra for arbitrary materials, however, have only been calculated for systems of densely packed particles pairs and linear triplets.

. The eigenfrequencies of single particle clusters with highly symmetric arrangement of touching particles can be computed for various cluster geometries (chain, double strand chain, planar cluster etc.) from the Clippe theory, if the resonance modes are only moderately damped. Corresponding mode frequencies were identified in measured extinction spectra of appropriately prepared Ag and Au particle samples

LITERATURE

(1) Baetzold, R.C. and R.E.Mack, J.Chem.Phys. 62(1975), 1573
(2) Schulze, W., Becker,H. and H.Abe, Proc.Int.Conf.Matrix Isolation Spectr. Berlin (1977)
(3) Welker, T. and T.P.Martin, J.Chem.Phys. 70 (1979), 5683
(4) a) Genzel, L. Festkörperprobleme XIV (1974), 183
b) Takasu, Y. and A.Bradshaw, Chem.Phys.Sol.and Thin Surf. 7 (1978), 59
c) "Growth and Properties of Metal Clusters" ed.by J. Bourdon, Elsevier (198o)
d) "Small Particles and Inorganic Clusters" ed.H.Gatos,Surf. Sci. 106 (1981)
e) Perenboom,J., Wyder P. and F.Meier, to be published
f) Baltes H. and E.Simanek "Physics of Microparticles" to appear (Springer)
(5) Kreibig U. and C.v.Fragstein, Z.Physik 224 (1969), 307
Kreibig U., J.Physics F. 4 (1974),999; J.Physique 38(1977), C2
(6) Kreibig U., Z.Physik B31 (1978), 39
(7) a) Genzel,L, Martin,T.P. and U.Kreibig, Z.Physik B21 (1975), 339
b) Genzel,L. and U.Kreibig , Z.Physik B37 (198o), 93
(8) a) Kawabata A. and R.Kubo, J.Phys.Soc.Japan 21(1966),1765
b) Ruppin R. and H.Yatom, Phys.Stat.Solidi(b) 74(1976),647
(9) Kreibig U.; in (4)c),
(10) Theye M.L. in "Opt.Prop.s of Solids" ed.B.O.Seraphin,North Holland (1976)
(11) Kreibig U., Sol. State Commun. 28 (1978), 767
(12) a) Hamilton I., Apai G., Lee S.and M. Mason; in (4)c)
b) Apai G., Lee S. and M. Mason, Sol.State Commun. 37 (1981),213
(13) a) Granqvist C. and O.Hunderi, J.Appl.Phys. 51 (1980),1751
b) Roulet H., Mariot I.,Dufour G. and C.Hague, J.Phys. F 10 (1980), 1025
(14) Gillet M., Surface Sci. 67 (1971), 139
(15) Klemke, I., Diplomarbeit, Hamburg(1978)
(16) Abe H., Schulze W. and B. Tesche,Chem.Phys. 47 (1980), 95
(17) Kreibig U., Z.Physik 234 (1970), 307
(18) Jasperse I., Kahan A., Plendl I. and S. Mitra, Phys.Rev. 146 (1966), 526

(19) Christensen N. and B. Seraphin, Phys.Rev. B4 (1972), 3321
(20) a) v.d.Hulst C., "Light Scattering by Small Particles"Wiley(1957)
b) Kerker M.,"The Scattering of Light" Academic (1969)
(21) a) Niklasson G., "Opt.Prop.s and Solar Selectivity of Inhomog. Metal-Insulator-Coatings", Göteborg(1982)
b) Yoshida S., Yamaguchi T. and A.Kinbara, J.O.S.A. 61 (1971),62
(22) a) Ruppin G. in "El.magn. Surf. Modes",ed. A.Boardman,Wiley(1981)
b) Lushnikov A.,Maksimenko V.and A.Simonov . " " " "
(23) Mie G., Ann. Physik 25 (1908),377
(24) Debye P., Ann. Physik 30 (1909),57
(25) Gans R. and H.Happel, Ann. Physik 29 (1909), 277
(26) Fragstein C.v.,Meingast I. and H.Hoch, Forsch.ber.Wiss.u.Verkehrs-ministerium Nordrhein-Westfalen 174, Köln (1955)
(27) Fragstein C.v., and F.Schoenes, Z.Physik 198 (1967), 477
(28) Clanget R., Optik 35 (1972), 180
(29) Ruppin R., Phys.Rev. B 11 (1975), 2871
(30) Gans R., Ann. Physik 37 (1912), 881
(31) Rohloff E., Z.Physik 132 (1952), 643
(32) a) Fuchs R., Phys.Rev. B 11 (1975), 1732
b) Genzel L. and T.P.Martin,Phys.Stat.Solidi(b) 51(1972),91
(33) a) Aden A. and M.Kerker, J.Appl.Phys.22 (1951), 1242
b) Morriss R. and L.Collins, J.Chem.Phys. 41 (1964), 3357
(34) Karlson A. and O.Beckmann, Sol.State Commun. 5 (1967), 795
(35) a) Mezger P., Phys.Blätter 31 (1975), 548
b) Huffman D., Adv. in Physics 26 (1977), 129
c) Nuth J.A., Dissertation, University of Maryland (1982)
(36) Kreibig U. and P. Zacharias, Z.Physik 231 (1970), 128
(37) Authors: Ausloos M., Clippe P., Evrard R., Gerardy J., Lucas A. Lit.: Phys.Rev.B18(1978),7176; B22(1980),4950; B26(1982) Surf.Sci 106(1981),319 etc.
(38) Granqvist C. and R.Buhrman, J.Appl.Phys. 47(1976), 2200, 2220
(39) a) v.d. Hulst H., "Multiple Scattering", Academic (1980)
b) Fragstein C.v., Optik 39 (1973), 58
(40) a) Kreibig U., Althoff A. and H. Pressman, Surface Sci. 106 (1981), 308
b) Schoenauer D, Genzel,L. and U.Kreibig, Verh.Deutsche Phys. Ges. 5 (1982), 841
(41) a) Genzel L., unpublished results
b) Persson B. and A. Liebsch, Sol.State Commun. 44 (1982),1637
(42) Genzel L. and U. Kreibig, to be published
(43) Schoenauer D., to be published
(44) Clippe ,P.,Evrard,L.and A.Lucas Phys.Rev.B14 (1976),1715
Corrections in: Ausloos,M,Clippe,P.and A.A.Lucas, Phys.Rev. B18 (1978), 7176.

INTRODUCTION TO PERCOLATION THEORY

R. BLANC

Département de Physique des Systèmes Désordonnés
Equipe de Recherche Associée au CNRS (ERA 1000)
Université de Provence-Centre de St Jérôme
13397 MARSEILLE CEDEX 13

SCHEME

As an experimentalist speaking about the theory of percolation, I shall adopt the point of view of an experimentalist dealing with percolation. So before giving a formal definition, it seems better to look at some examples.

1. TWO EXAMPLES

First, let us consider a well-stirred mixture of two powders : the first one, is an insulating powder, made of microballs of glass (like those used in the non skidding skin deposited on the deck of a sailing boat) and the second one is made of the same balls but silver coated. Let p the proportion of silvered spheres in the mixture.

$$p = \frac{\text{Volume of silvered spheres}}{\text{total volume of spheres (silvered or not)}}$$

We put this mixture in a cubic box, supplied with two electrodes (fig. 1) and we measure the d.c. conductivity of this mixture, for different values of p . Well, what are the results ? If σ_0 is the conductivity when all the spheres are conducting, we plot σ/σ_0 versus p. We can see on the figure 2 that :

1/ When p decreases from 1, we have, at first a linear variation of the conductivity. This linear variation is quite well interpreted by the effective medium theories (discussed by M.H. Cohen in this volume). But they predict that the conductivity must go to zero for a value of p equals to 1/3. As one can see

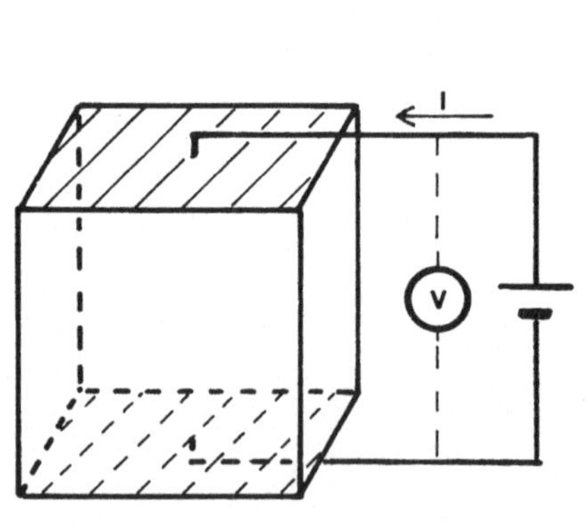

FIGURE 1-

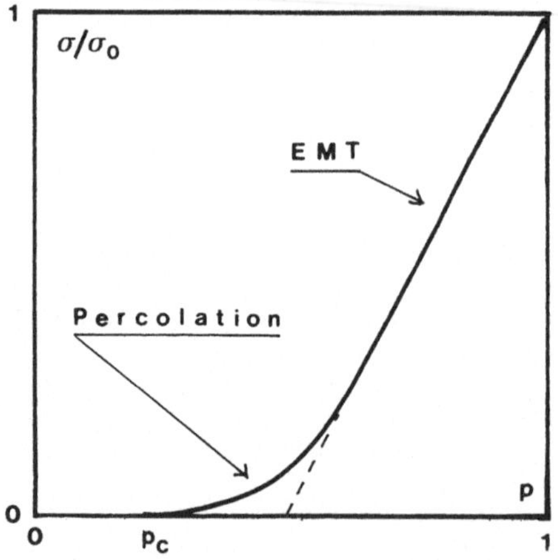

FIGURE 2- σ/σ_0 versus p. See e.g. references (10),(11),(12)

this is not the case and when p is of the order of 0.4 there is a difference between the actual and the predicted behavior.

2/ The conductivity stops for a value of p, p_c, equals to 0.27 (fig. 2) and for p close to p_c but larger than p_c there is a connecting path between electrodes but we no longer have a linear variation but rather $\sigma/\sigma_0 \sim (p-p_c)^t$ as one can see on the log-log plot (fig.3)

In our case ,that is for 3D experiments, exponent t is equal to 1.7. Some people made measurements in a 2D arrangement and they obtained a different exponent ($t \sim 1.1$).

Now let us imagine we have a great deal of time (and also a great amount of patience). If the spheres have the same radius, we should put them, one by one, following a regular, cristallographic lattice, a given cristallographic site being randomly occupied by a silvered sphere (with the probability p) or a sphere of glass (with the probability 1-p) (fig. 4).

But if,after our hard toil, we perform measurements of the conductivity, the results should be quite disappointing. The only one difference from our previous results is the critical value p_c which depends on the lattice.

But the behaviour of the conductivity is qualitatively the same as before that is to say that the conductivity behaves like $(p-p_c)^t$ with always the same valut of t . t does not depend on the lattice but only on the dimensionality D of space.

- in 2D we have $t \sim 1.1$
- in 3D we have $t \sim 1.7$
- in 6D and more (here there are computer experiments) we have t = 3.

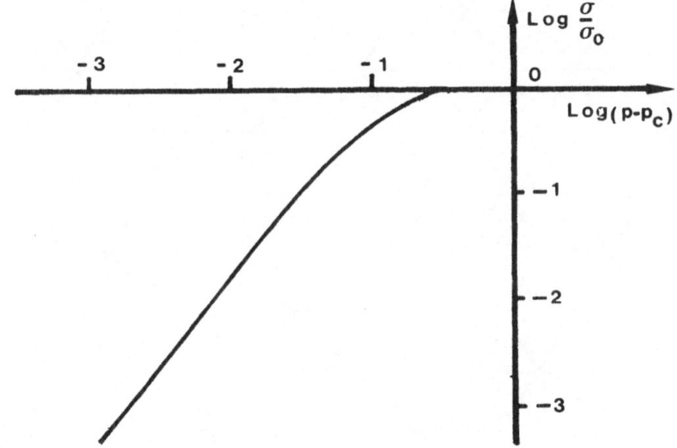

FIGURE 3 - Log-Log plot of σ/σ_0 versus p. The slope t of the straight line is $\sim$ 1.1 for 2-D case and $\sim$ 1.7 for 3.D. cases

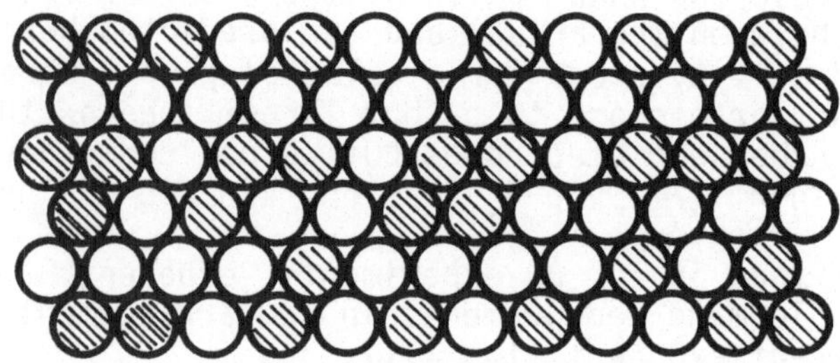

FIGURE 4- Sites of this regular lattice are randomly occupied by a silvered sphere with probability p or by a sphere of glass with probability 1-p

Now let us look at our second example. We take a triangular lattice, and using a computer, we can draw six branched stars on some sites of this lattice, these sites being chosen by a random process, so that each of them has the probability p to be occupied by a star. p goes from 0 to 1 and we look at the possibility of a connecting path. When p is very low (p equals 0.1) one can see that there is a great number of isolated stars. But there is also some groups of stars connected two by two by one of their arms. As you can well imagine, these groups are called "clusters", 2-clusters for doublets, 3-clusters for triplets and so on.(fig. 5)

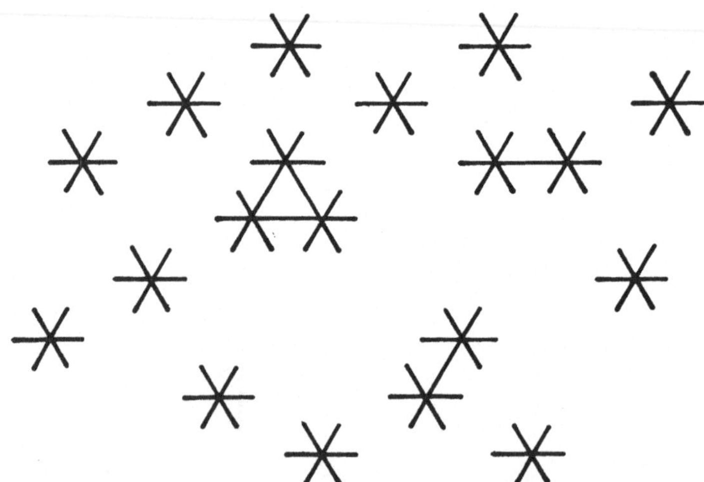

FIGURE 5- A part of a triangular lattice with twelve isolated sites (*), two doublets and one triplet

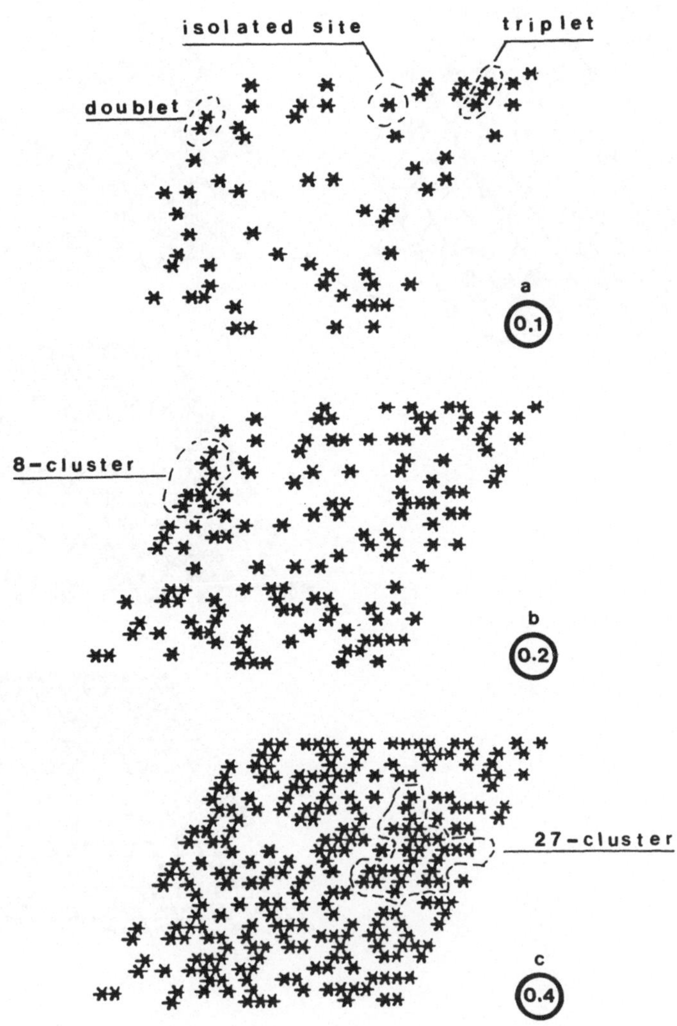

FIGURE 6 -a,b,c- The size of the largest cluster increases with p. See reference (13)

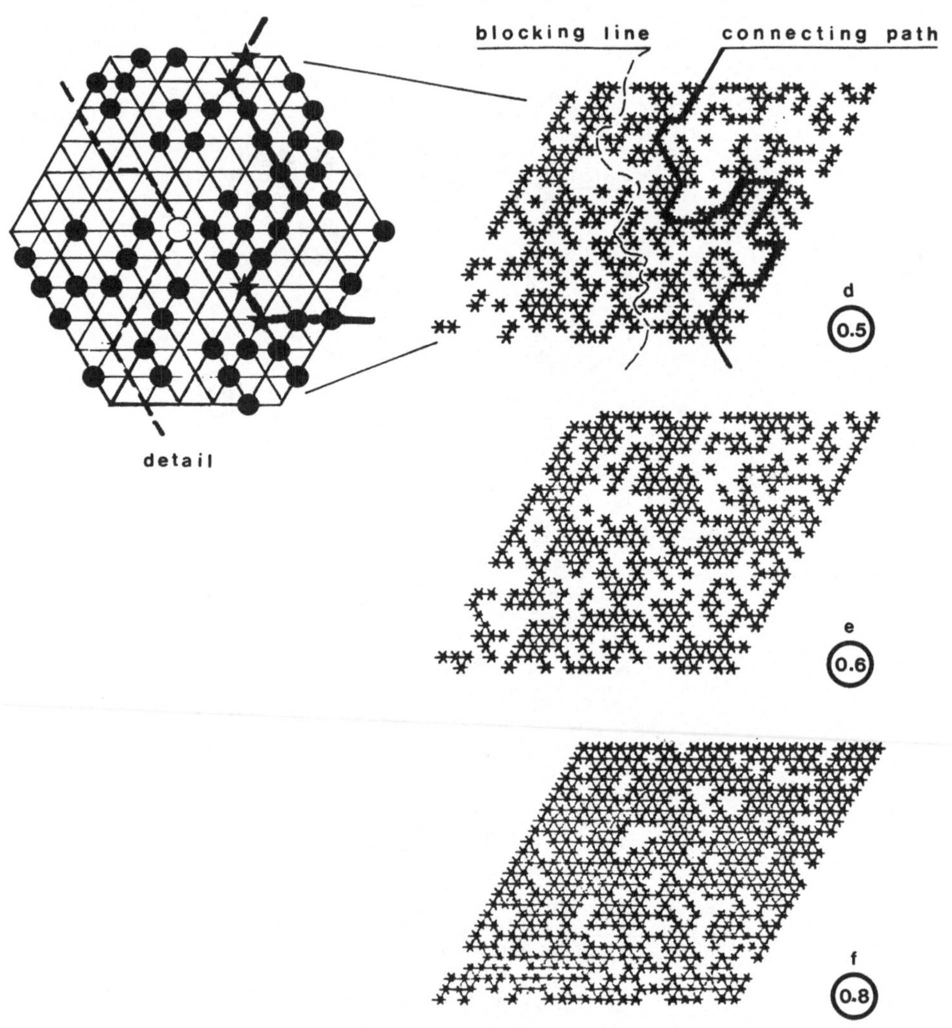

FIGURE 6 -d,e,f- Figure 6d shows one of the different connecting path (unbroken line). Details show the susceptibility of this percolation lattice to the addition of an active site ○ giving an "horizontal" connecting path and to the removal of an active site ★ breaking the "vertical" connecting path - See reference (13)

When p increases, one can see (fig.6) that the size of the largest clusters increases. We have a 8-cluster for p equals 0.2 (fig. 6b), a twenty seven-cluster for p equals 0.3 (fig .6c). But there is no cluster large enough to connect the top and the bottom or the left and the right of the lattice.

When p equals 1/2 (fig. 6d) there is almost such a cluster. "Almost", because one can go from top to bottom of this finite lattice following a connecting path but one cannot do the same thing from left to right because one can draw on this lattice this dotted line following non-occupied sites and, so, it is impossible to go from left to right following occupied sites only.

In fact, for this triangular lattice, $p_c=1/2$. So, in principle, it would be possible that we have a "vertical" and an "horizontal" connecting paths. Here we have only a vertical one : this is an effect of the finite size of this sample and later we will understand more about this size effect.

Along the connecting path there are some critical sites (see details in fig. 6d) and you can see that if we rub out one of these stars (labelled ★) there is no connecting path. Reversely if we add a star (labelled ○), there is an horizontal path. So, we have near p_c a great sensibility of the global connectivity of the sample with very weak modifications of our lattice. One can say that there is a very large <u>susceptibility</u>.

When p equals 0.6, we can go from left to right and from bottom to top without any difficulty.

If we extrapolate this finite sample towards an infinite lattice, we can see that when $p = p_c$ there is an infinite cluster and that we can go at the infinity in any direction following the sites of this infinite cluster.

Now let us suppose we have an infinite lattice and let $P_\infty(p)$ the fraction of occupied sites belonging to the infinite cluster. If we look at the variations of P_∞ with p we have someting like figure 7 :

- because for p below p_c, there is not an infinite cluster, we have $P_\infty = 0$.

- for p above p_c , P_∞ is different from zero and close to p_c, it behaves like $(p-p_c)^\beta$. β is an exponent, which is independent of the symmetry of the lattice (square lattice, triangular lattice ...) and depends only on the dimensionality of space.

So, P_∞ (p) plays the same role an as an order parameter in a thermodynamic phase transition. I remind you that for these phase

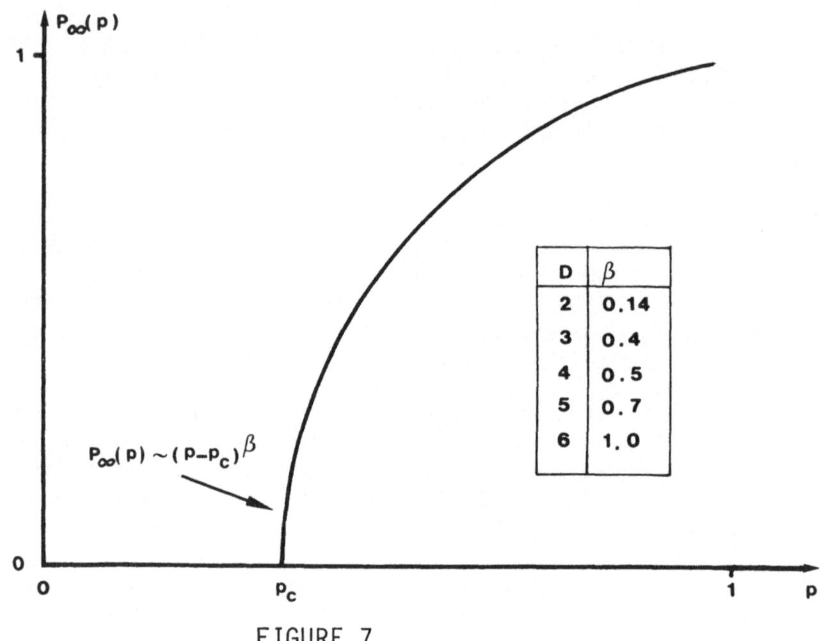

D	β
2	0.14
3	0.4
4	0.5
5	0.7
6	1.0

FIGURE 7

transitions, the order parameter is zero on one side of the critical point and different from zero on the other side. For instance, in the paramagnetic-ferromagnetic transition, the order parameter is the magnetization M of figure 8, which is zero above the Curie point and non-vanishing below. Near the Curie point, the magnetization behaves like $(T_c-T)^\beta$ with β in the range 0.3-0.5, and I remember that the mean field theories gave an exponent equals to 1/2.

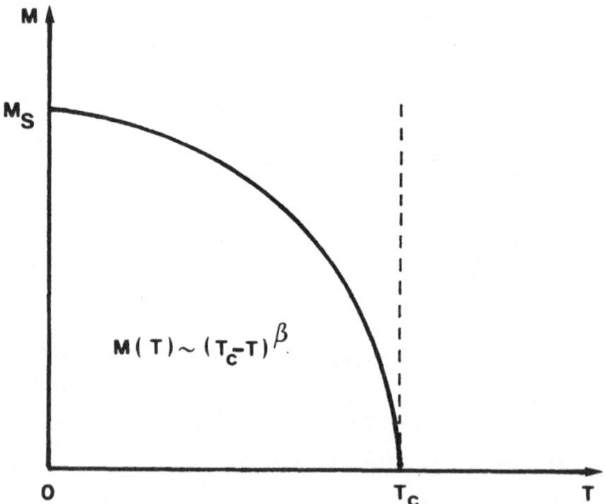

FIGURE 8- Magnetization M near the Curie Point, in the paramagnetic-ferromagnetic transition. Note the similarity with the behaviour of P_∞ (p) versus p (figure 7)

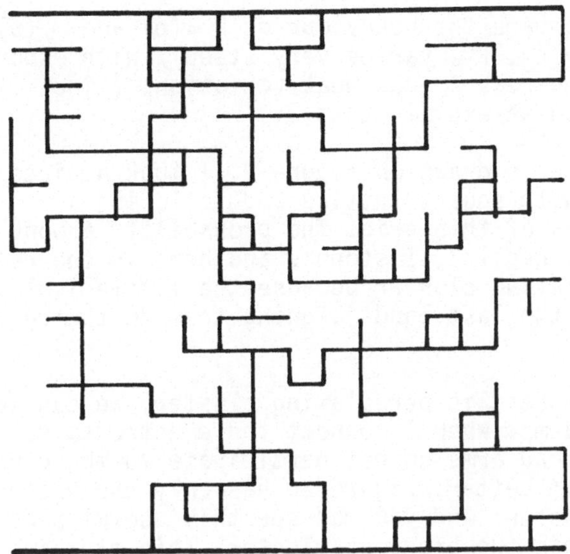

FIGURE 9- A percolating cluster

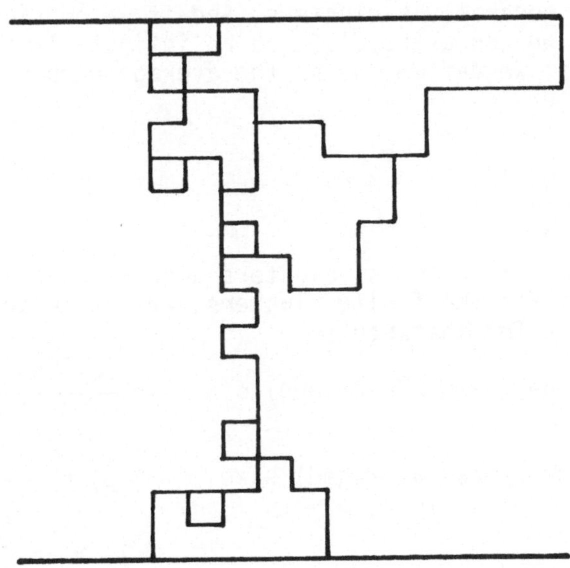

FIGURE 10- Same as figure 9 without dead arms

If we try to compare the behaviour of $P_\infty(p)$ and $\sigma(p)$, we have seen that, near p_c, P_∞ varies very steeply with exponent β (β lower than 1) whereas σ has weak variations (since t is greater than 1). Hcw can we explain that ?

An exercice is drawn on figure 9. I took a piece of grid, square shaped with square lattice and,following a random process, I cut some bonds of this grid. The probability a bond is cut was 1-p. When p reached 1/2, I stopped and here is the result. We have only the percolating cluster because the little isolated clusters went down when the last bond joigning them to the rest of the grid were cut.

If we look at this percolating cluster, we can see that there are many dead arms. When I connect the electrodes to a current generator the dead arms do not participate to the conduction. To demonstrate that better, figure 10 has only the useful part of the percolating cluster. And you can see this useful part is very much thinner than the previous cluster. This explains the different behaviours of P_∞ and the conductivity σ. You can see also on figure 10 that there is some loops in this cluster and also some critical bonds.

Now if we look again at the triangular lattice, we can make a detailed account, for each value of p, of the number of 1-cluster, 2-clusters, and generally s-clusters. And if we consider larger and larger samples, we can extrapolate to an infinite lattice. For this infinite lattice, we define n_s as the average number (per lattice site) of s-clusters.

$$n_s = \lim_{N \to \infty} \frac{N_s}{N}$$

where N_s is the number of s clusters and N is the total number of sites. For the finite clusters, we can define some average values, like for instance :

- the average (number averaged) size = $\dfrac{\sum_s s n_s}{\sum_s n_s}$

- the average (mass averaged) size = $S(p) = \dfrac{\sum_s s.s n_s}{\sum_s s n_s}$

- the average (z-averaged) size = $\dfrac{\sum_s s.s^2 n_s}{\sum_s s^2 n_s}$

and so on.

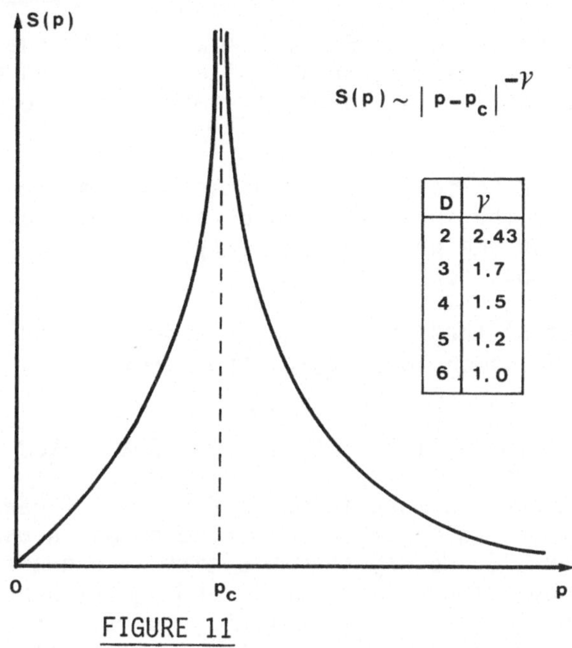

FIGURE 11

If we look at the variations of S(p) with p (fig.11), we see that close to p_c, this sum diverges following a power law $|p-p_c|^{-\gamma}$. S(p) behaves like the susceptibility of para-ferrotransition.Figure 11 gives the values of γ for the different dimensions of space. Here is a short digression on this subject.

In one dimension, all the clusters are linear chains, and the threshold p_c equals one. This case of 1-D percolation can be solved exactly but it is a somewhat pathologic case about which I shall say little.

At the other hand, we can see that in 6-dimensions and more, all the exponents are the same. An I remind you that for phase transitions, there is also an upper critical dimensionality which is 4, at which the mean field theories give the correct exponent, that is to say that the dimensionality of this space is sufficient for the fluctuations may be averaged. In the case of percolation, we have seen that there are, for instance, in the 2D dimensions, some critical sites or critical bonds. If we add a third dimension we can go beyond some of these critical sites. But in 3 dimensional cases, we have also some critical sites which can be surpassed in a fourth dimension, and so on. In 6 dimensions we have enough possibilities for make averaged all the fluctuations, and the mean field theory works well.

In summary, we have seen some essential characteristics of the percolation :

i- There is a threshold p_c. For p below p_c, all the clusters are finite but above p_c there is an infinite cluster and a global connectivity on the lattice. Moreover p_c depends on the geometry of the lattice.

ii- We defined statistical quantities as the fraction of occupied sites belonging to the infinite cluster or the mass averaged mean size S(p) and also one physical quantity $\sigma(p)$. These quantities follow, close to p_c, power laws.

iii- The exponent of these power laws does not depend on the lattice and depends on the dimensionality of the space.

2. FORMAL DEFINITION - SITE PERCOLATION - BOND PERCOLATION

Now, we return to a more formal definition of percolation. I recall that a graph is a set of sites and bonds. A part of a graph is drawn on figure 12. We say that two sites are nearest neighbours if there is a bond between them. So S_1 and S_3 are nearest neighbours, but S_1 and S_2 are not nearest neighbours. Moreover, between two sites, there is at the most one bond. So a bond like L'_{14} does not exist.

Now, give to each site one of two labels (occupied or empty) (active or inactive) or (up and down) or (black or white) etc... and assume that the probability for a site to be occupied is p_s and that, for the moment, the probability for a given site to be occupied or not does not depend on the state of the nearest neighbours.

In the same way each bond is conducting with the probability p_b or isolating with the complementary probability $1-p_b$, independently of the state of the others bonds. We are dealing with "random percolation".

In the most part of this paper I shall consider the two simplest cases :

i- site percolation (fig. 13). In this case, all the bonds are conducting, that is to say $p_b = 1$ and sites are "up" with probability p_s. For instance here, the sites S_1, S_3, S_5, S_7 are active (occupied) the others being inactive.

ii- bond percolation (fig.14). In this case, all the sites are active, p_s equals one, and the bonds are conducting or not with the probability p_b .

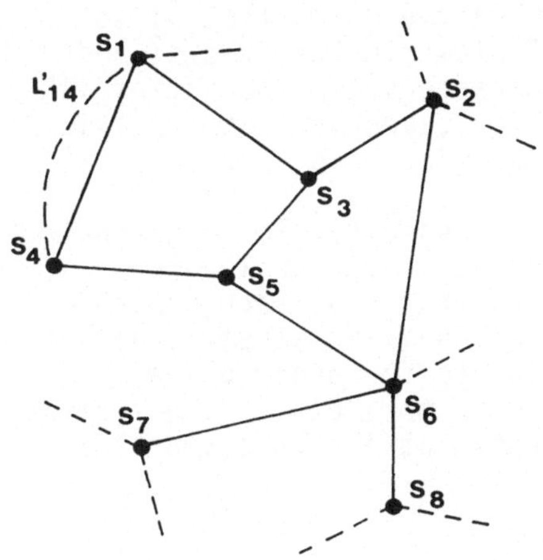

FIGURE 12

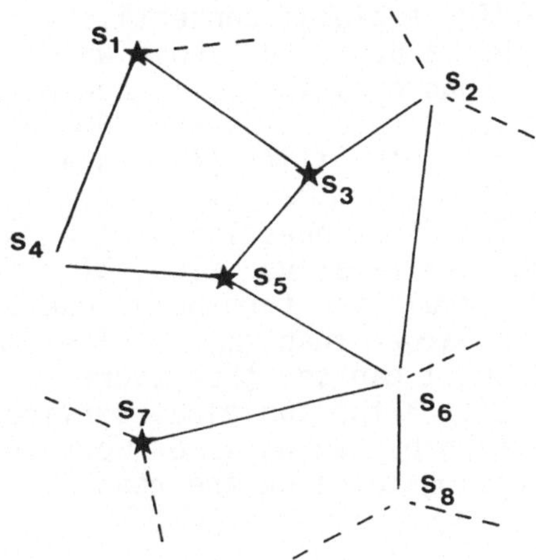

FIGURE 13- Same graph as in figure 12. All the bonds are conducting (p_b=1). Sites are conducting (★) with probability p_s

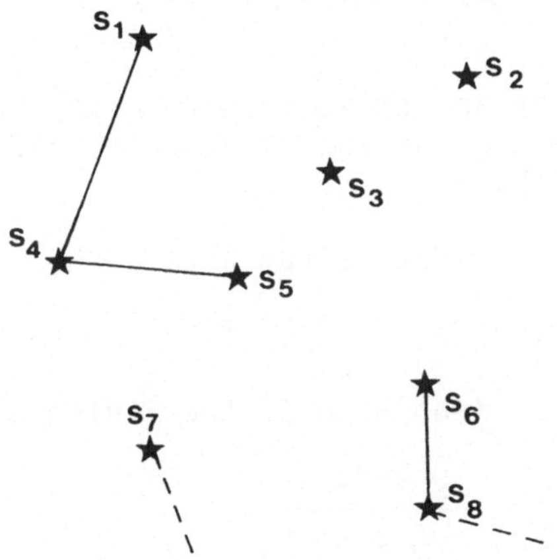

FIGURE 14- Same graph as in figure 12. All the sites are active (p_s=1). Bond are conducting(★—★) -or not (★ ★) with probability p_b (or 1-p_b)

In both cases, the percolation problem is the following : for a given graph, can we predict the value of the probability for which a global connectivity appears in the graph that is to say the value of p above which an infinite cluster exists? And the second question is : how do the characteristic quantities like the mean size of clusters, $P_\infty(p)$, $S(p)$,..., behave close to this critical probability (i.e. scale as a function of $p - p_c$)?

The first question we can solve is to specify, for a given graph, the relative values of site percolation threshold and bond percolation threshold. One can prove that, for each graph, p_{sc} is larger than p_{bc}. The demonstration is a mathematical one, but I can say that there is an evidence for every driver in a city : the municipal services can stop more easily the traffic in a town by having some "work in progress" at intersections that by driling holes on the roads.

3. DEFINITION AND BEHAVIOUR OF THE CHARACTERISTIC QUANTITIES

Now we define some characteristic quantities and their behaviours.

We shall argue about the site percolation. In the case of bond percolation, most of the definitions will be the same ; others will be in evidence altered. We have already seen some quantities.

1- The average number (by site) of clusters containing s sites : $n_s(p)$

2- The probability for an active site to belong to the infinite cluster which is the same as the fraction of active sites in the infinite cluster $P_\infty(p)$.

3- The total number of clusters (by site) : G(p)

$$G(p) = \sum_s n_s(p)$$

4- The (mass averaged) mean size of the finite clusters S(p)

$$S(p) = \frac{\sum_s s s n_s}{\sum_s s\, n_s}$$

and as we have the identity $1 = 1-p+pP + \sum_s sn_s$ which implies that either a site is non active with probability 1-p, or it belongs to the infinite clusters with the probability pP_∞(p for this site is active <u>and</u> belongs to the ∞-cluster) or it is active and belongs to one of the finite clusters , we can write :

$$S(p) = \frac{\sum_s s^2 n_s}{p(1-P_\infty(p))}$$

5- The correlation length. On a regular, periodical lattice, whose lattice constant is a, consider an arbitrary origin. Let any two sites identified by their radius vector $\vec{r}$ and $\vec{r}'$ (figure 15). Call $\omega(\vec{r},\vec{r}')$ a number which equals one if the two sites are active and belong to the same cluster and zero in the other cases.

For a given value of $\vec{\rho} = \vec{r}' - \vec{r}$ and of p, let us calculate the mean value of ω for all the sites and let $\pi(\vec{\rho},p)$ this mean value. The typical variations of $\pi(\vec{\rho},p)$ for $p < p_c$ show that π is going to zero when the distance ρ is going to infinity. And we can define ξ approximately as the distance for which one can consider π as negligible (cf. figure 16).

But when p is larger than p_c, π does not go to zero when ρ increases (figure 17) because an infinite cluster exists and as the probability a site is belonging to this infinite cluster is $P_\infty(p)$ we see that, when ρ goes to infinity, π tends to $P_\infty{}^2$. Here again, we can define ξ as the distance for which the difference between π and $P_\infty{}^2$ goes to zero.

When one looks at the linear size of clusters, one can see that ξ is a characteristic dimension of finite clusters. In fact, to characterize mean size of clusters, we have series like

$$\frac{\sum_s s.n_s}{\sum_s n_s} \quad \text{or} \quad \frac{\sum_s s^2.n_s}{\sum_s s.n_s} \quad \text{and so on.}$$

And we define a typical cluster size as the size of clusters which give the main contribution to these series. Excepting some unimportant numerical factors, this typical cluster size is assumed independant of the series used in its definition. And ξ is the linear dimension of those typical clusters. Close to p_c, ξ behaves like $|p-p_c|^{-\nu}$. ξ diverges because close to p_c, the size of the finite clusters becomes larger and larger, as we have seen in the example of the triangular lattice

Figure 18 gives some values of ν for different dimensions of space. All these values are approximate, except in 2D where the exact value is perhaps 4/3.[22]

The fact that, close to p_c, ξ becomes very large, is the key to the universality of the exponents that is to say that for a given space dimensionality, the exponents do not depend on the

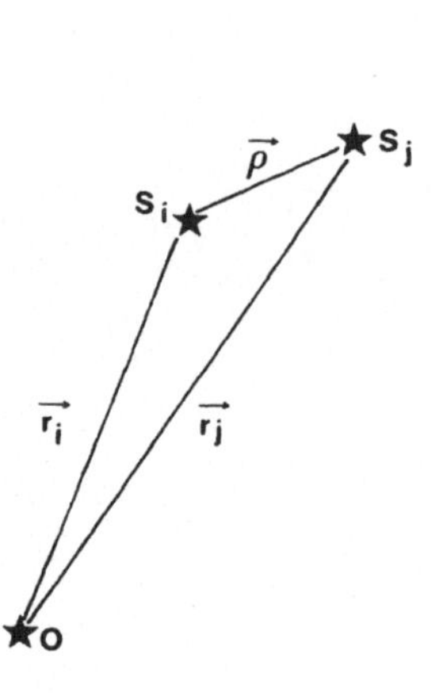

FIGURE 15

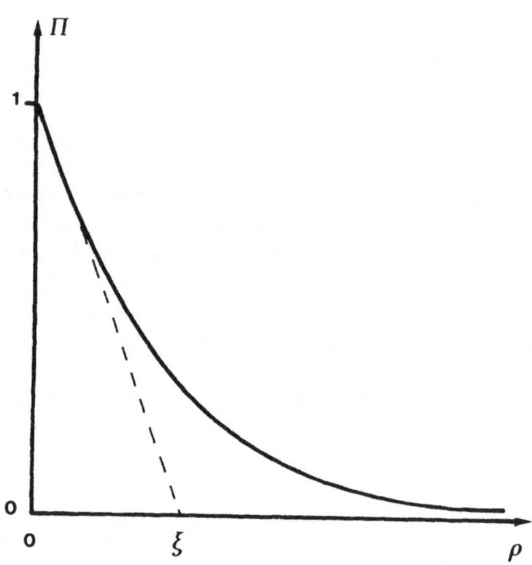

FIGURE 16- When $p<p_c$, ξ may be defined approximately as the distance for which $\pi(\vec{\rho},p)$ is negligible

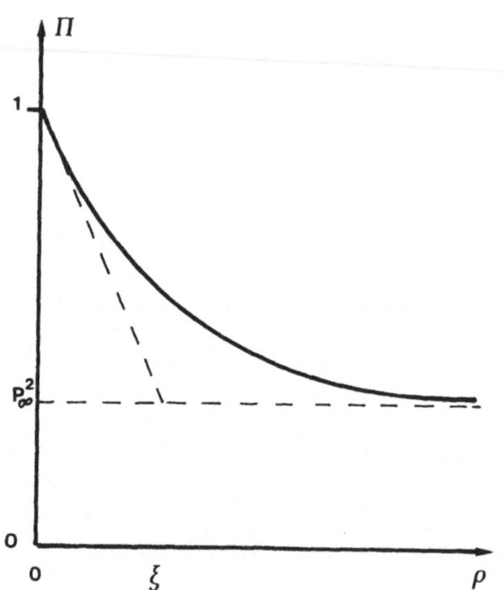

FIGURE 17- When $p<p_c$, ξ may be defined approximately as the distance for which $(\pi(\vec{\rho},p)-P_\infty^2(p))$ is negligible

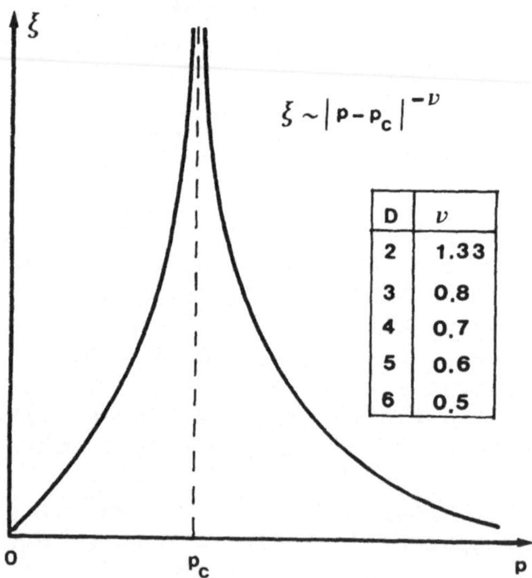

D	ν
2	1.33
3	0.8
4	0.7
5	0.6
6	0.5

FIGURE 18

geometry of the lattice. Near p_c, ξ is so large that we can ignore the details of the geometry of the lattice. This assertion is not mathematically proved but there is a general agreement on the validity of this law, which is also obeyed numerically whenever data are available.

So we can understand how it is possible to simulate on ordered lattices (for instance using a computer simulation or an analog simulation) some physical systems in which there are,at the same time,compositional disorder and geometrical disorder. I gave an example of these systems with the mixture of conducting and isolating spheres. When you put the mixture in a box, you have a random packing of spheres and in this random packing you have also randomness in the conductivity of the spheres themselves.

Recent experimental results (14, 15), both analog and computer experiment, showed that for continuous percolation, that is to say percolation without any lattice, the exponent ν of the correlation length is the same as for percolation on a lattice.

4. THERMODYNAMIC ANALOGY

We have seen that here is an analogy between some statistical quantities in the percolation problem, as for instance S(p),P(p), and corresponding physical quantities in magnet undergoing thermal phase transition : $P(p) \Leftrightarrow M(T)$ and $S(p) \Leftrightarrow \chi(T)$

and we can ask : Is there a real correspondance between percolation problem and phase transition or are these analogies between P and M or S and χ only fortuitous ? The answer is : there is really a correspondance between these two problems. And the mathematical demonstration was given by Kasteleyn and Fortuin in 1969 (16).

They used a general model : the Ashkin-Teller-Potts model. Consider a graph like that of figure 19, with n sites and N bonds. At each site S_i we put a moment μ_i which can take one of S different values

$$\mu_i = 1,2,\ldots,S$$

There is an interaction between nearest neighbours moments only and the interaction energy is

$$\begin{cases} -A \text{ if } \mu_i = \mu_j \\ 0 \text{ if } \mu_i \neq \mu_j \end{cases} \quad i, j \text{ nearest neighbours}$$

If S = 2, we recover the well-Known Ising Model.

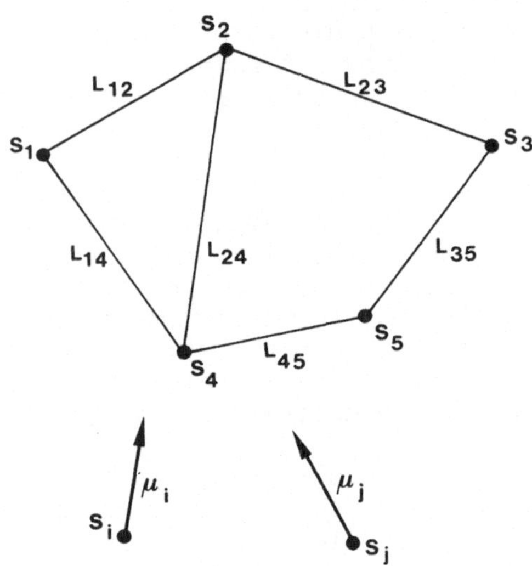

FIGURE 19- Finite graph with five sites and six bonds. On each site, the moment μ_i takes one of S different values. Each bond may be labelled 0 or 1

Furthermore, we can characterize each bond by zero or one. So on the graph of figure 19 we have :

- Magnetic configurations characterized by the values of the moments on each site. With n sites and S different values of the moment we have S^n different magnetic configurations

- Percolation configurations characterized by the values of the number labelling each bond. And as we have N bonds and two different values 0 or 1, we have 2^N percolation configurations.

And the theorem of Kasteleyn and Fortuin demonstrates how it is possible to go from the magnetic problem to the percolation problem(cf. figure 20).

As in the magnetic problem, one has magnetic energy, it is possible to calculate a partition function Z from which one can obtain, by differentiation, all the thermodynamic functions. And these functions depend on S.

But, if you want to go to the bond percolation problem, you have to make S = 1, because in the bond percolation problem, you have no choice on the state of a site. So, the bond percolation problem is the limiting case S = 1 of the Potts Model. Of course

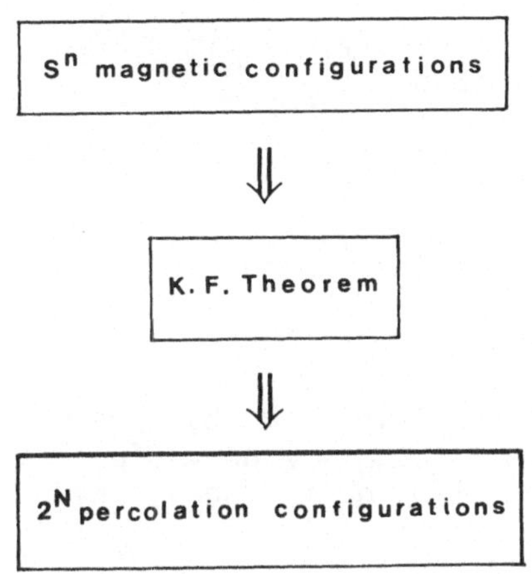

FIGURE 20

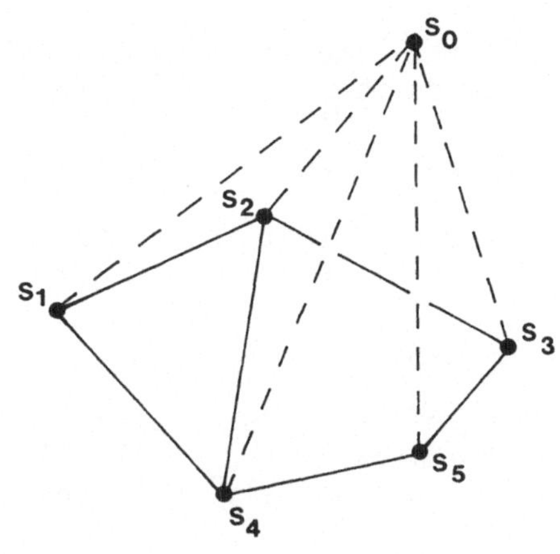

FIGURE 21- Same as figure 19 with with a ghost site (S_0) interacting with all the other moments

Z (S=1) = 1 so that one must differentiate before taking the limit S = 1 to obtain the statistical quantities P(p), S(p), etc.. There is a one-to-one correspondance between magnetic configurations in the limit S = 1 and percolation configurations.

More over, in the ferromagnetic-paramagnetic transition, it is useful to introduce an external field H. In the APT model, we can consider a ghost site, exterior to the lattice,which can interact with all the other moments and in the percolation problem we can consider a ghost site connected with probability p_0 to every other site of the lattice. By this means, one obtains the partition function Z' in the presence of an external magnetic field and calculates the thermodynamic functions.(fig.21)

When H = 0, the limit value $\lim_{s\to 1} \partial/(\partial S)$ (ln Z') is precisely $\sum_s n_s$. So there is a correspondance between the total number of clusters (by site) $G(p) = \sum_s n_s(p)$ and the free energy F $(-\beta F = \ln Z')$ in the Potts model with an external magnetic field.

When H $\neq$ 0, the magnetization M is defined in the magnetic problem as $\lim_{H\to 0} \partial F/\partial H$. Similarly, the limit value

$$\lim_{H\to 0} \partial/\partial H \left(\lim_{S\to 1} \partial F/\partial S\right) = 1 - \sum_s s.n_s = P_\infty(p) / p$$

establishes the correspondance between M and $P_\infty(p)$.

In the same way, the magnetic susceptibility $\chi = \lim_{H\to 0} \partial^2 F/\partial H^2$ corresponds to $\lim_{H\to 0} \partial^2/\partial H^2 (\lim_{S\to 1} \partial F/\partial S) = S(p)$.

We summarize these correspondances in table I.

Note also that the different quantities $G(p)$, $P_\infty(p)$, $S(p)$, etc... may be obtained (17) from the generating function

$$\Gamma(q,q_0) = \sum_s q_0^{\,s}\, n_s$$

where $q_0 = 1 - p_0$ is related to the external field H by $q_0 = e^{-\beta H}$ (when $H \to 0$, we have $q_0 \to 1$). Clearly $\lim \Gamma(q,q_0) = \sum_s n_s = G(p)$. Applying successively to Γ the operator $\lim_{q_0\to 1} q_0\,(\partial/\partial q_0)$, one obtains :

$$\lim_{q_0\to 1} q_0\,(\partial/\partial q_0)\Gamma = \sum_s s n_s = 1 - P_\infty(p)$$

$$\lim_{q_0\to 1} (q_0\,(\partial/\partial q_0))^2\Gamma = \sum_s s^2 n_s = S(p)(1-P_\infty(p)),\ldots.$$

Remark also that, in the bond percolation transition, there is no energy involved. This is evident from the definition : random percolation in a pure statistical problem. But with Potts model, in the limit $S \to 1$, the interaction energy is a constant equal to $-A$ for each pair of neighbouring sites.

We have seen that the Kasteleyn and Fortuin theorem allows us to make an analogy between phase transition and <u>bond</u> percolation problem. But what about the <u>site</u> percolation problem ?

Phase transition	Bond percolation
Free energy	$\sum_s n_s \sim (p-p_c)^{2-\alpha}$
Magnetization	$\sum_s s n_s \sim P_\infty(p) \sim (p-p_c)^{\beta}$
Susceptibility	$\sum_s s^2 n_s \sim S(p) \sim (p-p_c)^{-\gamma}$
Correlation length	Correlation length $\xi \sim (p-p_c)^{-\nu}$

TABLE I

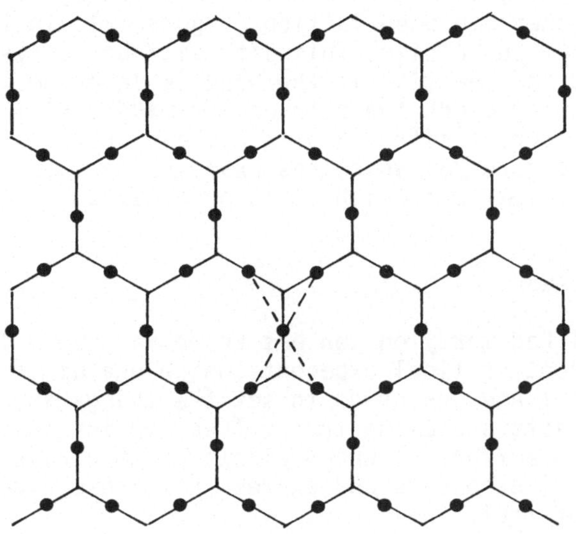

FIGURE 22- A site is put in the middle of each bond of this honeycomb bond lattice

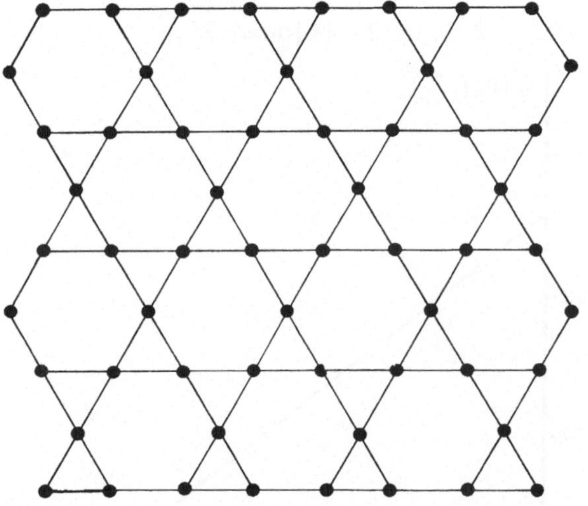

FIGURE 23-Kagomé site lattice obtained from the honeycomb bond lattice

Happily there is a very simple way to transform some bond percolation problem into site ones. For instance, consider an hexagonal or honeycomb bond lattice (figure 22). In the middle of each bond, I can put a site. This site will be active if the bond is active and inactive if the bond is inactive. As a bond has 4 neighbouring bonds the site on its middle will be connected to the 4 neighbouring sites. And, then, I can draw a site percolation problem (figure 23) which has exactly the same statistics as the bond problem from which it is originated.

5. SCALING LAWS

In fact I had mercy on you but there is a very large collection of different critical exponents which use up most of the greek alphabet ! And one needs to set the things in order. Happily, there exists a scaling theory (see for instance the review article of D. Stauffer(5))which allows us to derive some relations between critical exponents and express all these exponents in terms of two of them.

From experimental results (18), obtained by computer simulation, we know that at $p = p_c$, the cluster number $n_s(p_c)$ varies with s, following , for large s, the law

$$n_s(p_c) \sim s^{-\tau}$$

with $\tau = 2$ in 2D, ~ 2.1 in 3D (figure 24)

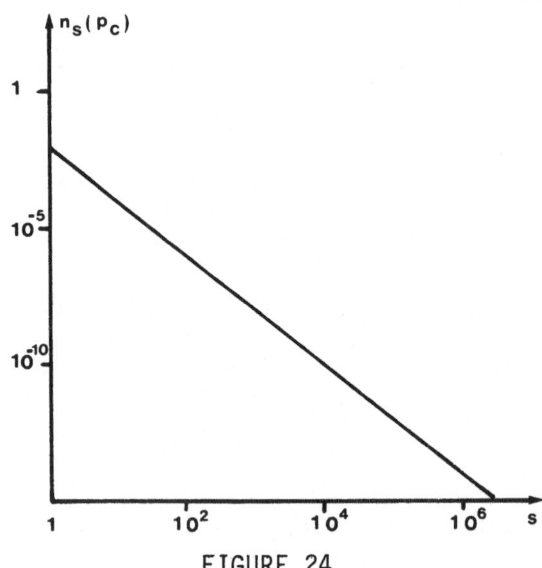

FIGURE 24

On the other hand, we assume that in the sums like $\sum_s s\, n_s$, $\sum_s s^2 n_s$ and so on, the main contribution near p_c is given by the clusters having the typical cluster radius ξ and the typical cluster size s_ξ with

$$s_\xi \sim |p - p_c|^{-1/\sigma}$$

The two exponents τ and σ may be used in order to express all the other exponents.

As s_ξ is the typical cluster size, we can express the size of a particular cluster by the ratio s/s_ξ. We can also express the cluster numbers using $n_s(p_c)$ as an unit:

$$n_s(p)/\, n_s(p_c).$$

The scaling hypothesis states that, as long as s is large and p close to p_c, the ratio $n_s(p)/n_s(p_c)$ is a function of the ratio s/s_ξ only.

As we can see on the figure 25, this is the case. When s increases, the shape of thecurve goes to a limit for large values of s. And you can see that f is a very nice function without any singularity (19).

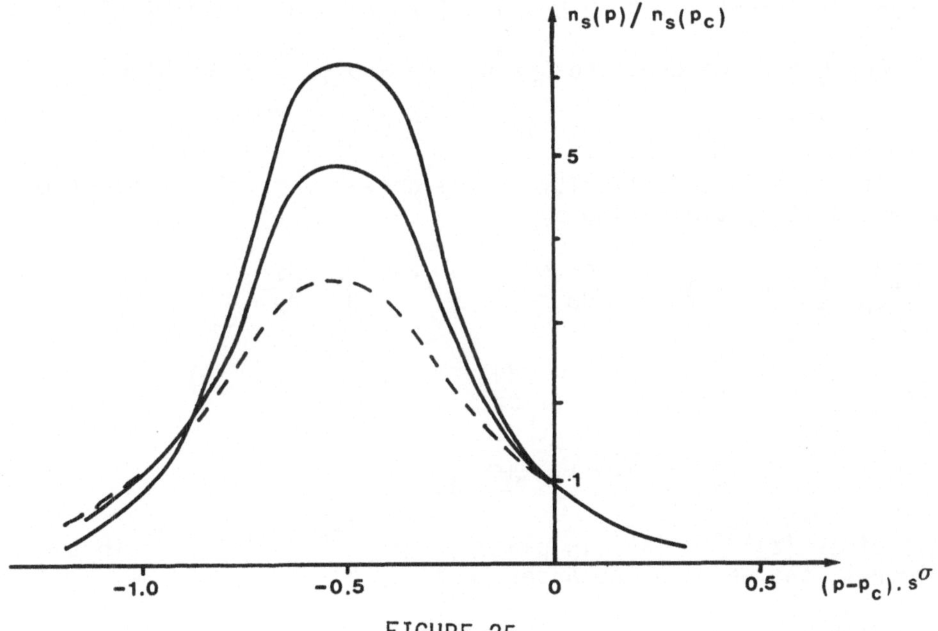

FIGURE 25

So we have :

$$\frac{n_s(p)}{n_s(p_c)} = f\left(\frac{s}{s_\xi}\right) = f\left(s\,|p - p_c|^{1/\sigma}\right)$$

$$= f\ \left|\{(p - p_c).s\}^{1/\sigma}\right|$$

$$= g(z) \text{ with } z = (p - p_c).s^{\sigma}$$

So we can write with $n_s(p_c) \sim s^{-\tau}$

$$n_s(p) \sim s^{-\tau}\ g(z)$$

There, if we calculate the sum $\sum_s s^k n_s$, we can write

$$\sum_s s^k n_s \sim \int_0^\infty s^k n_s(p)\, ds \sim \int_0^\infty s^{k-\tau}\ g(z)\, ds$$

As $z = (p - p_c).s^{\sigma}$, we have also by changing the variable

$$\sum_s s^k n_s \sim (1/\sigma)\,(p - p_c)^{(k-\tau+1)/\sigma} \int_0^{\pm\infty} z^{((k-\tau+1)/\sigma)-1}\ g(z)\, dz$$

As long as there is no difficulty in the convergence of this integral (that is to say whenever $((k-\tau+1)/\sigma) > 0$ or $k > (\tau - 1)$ we can write that $\sum_s s^k n^s$ behaves like $(p - p_c)^{(-k-\tau+1)/\sigma}$.

For $k = 2$, we know that $\sum_s s^2 n_s \sim (p - p_c)^{-\gamma}$. So we have

$$\gamma = (2-\tau+1)/\sigma = (3 - \tau)/\sigma$$

If there is a difficulty in the convergence, it is possible to bypass it by evaluating :

$$\frac{d}{dp} \sum_s s^k n_s = \sum_s s^k \frac{dn_s(p)}{dp} = \sum_s s^{k-\tau} \frac{dg}{dz}\frac{dz}{dp}$$

$$= \sum_s s^{k-\tau+\sigma} \frac{dg}{dz}$$

$$\sim \int_0^\infty s^{k-\tau+\sigma} \frac{df}{dz}\, ds$$

This gives $|z|^{(k-\tau+1)/\sigma}$ in place of $|z|^{((k-\tau+1)/\sigma)-1}$ in the integral and we gain one order in z.

With $k = 1$ we obtain $\beta = (\tau-2)/\sigma$
and with $k = 0$ we obtain $2-\alpha = (\tau-1)/\sigma$. (20)

How is the exponent ν related to the other exponents ?

The scaling hypothesis we have made is not sufficient for this purpose and we have to make another assumption.

If we look, above p_c, at a very large cluster and if we look at average properties in this cluster like, for instance, the density (i.e. the ratio between the number of sites belonging to the cluster in a volume element to the total number of sites in that volume) averaged over distances much smaller than the cluster radius, we cannot distinguish between the density of the finite cluster and the density of the infinite cluster.

If we take this postulate together with the assumption that the radius R of a cluster expressed in unit of ξ depends only on the ratio s/s_ξ, one can obtain (5) a new scaling law (hyper scaling)

$$2 - \alpha = D\nu$$

which is the Josephson's law. This law may be interpreted as follows : consider, near p_c, a volume ξ^D. As ξ is the characteristic length of clusters, the active sites in this volume are distributed among few clusters and the total number of clusters by site ($G(p) \sim (p-p_c)^{2-\alpha}$) varies like $\xi^{-D} \sim (p-p_c)^{D\nu}$, according to $2 - \alpha = D\nu$. This law is a quite good one for $D \leqslant 6$, as one can see on table 2.

6. EXACT RESULTS

6.1 Bethe Lattice

Exact results are known only for a few lattices. One of the simplest casesis the Bethe Lattice , which is a very special case of a lattice without loops. In this lattice each site has z and only z neighbours. For instance, in the case $z = 3$, we have the type of graph shown in figure 26. We are going to calculate for this lattice, the threshold p_c for the site percolation problem and the exponents.

D	1	2	3	4	5	6
$2-\alpha = \gamma + 2\beta$	1	$\sim$2.7	$\sim$2.5	$\sim$2.5	$\sim$2.6	3
$D\nu$	1	$\sim$2.7	$\sim$2.5	$\sim$2.8	$\sim$ 3	3

Table 2

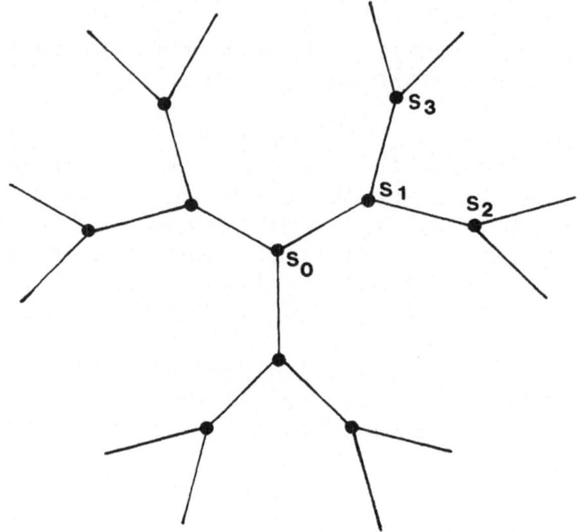

FIGURE 26- Bethe lattice with z = 3

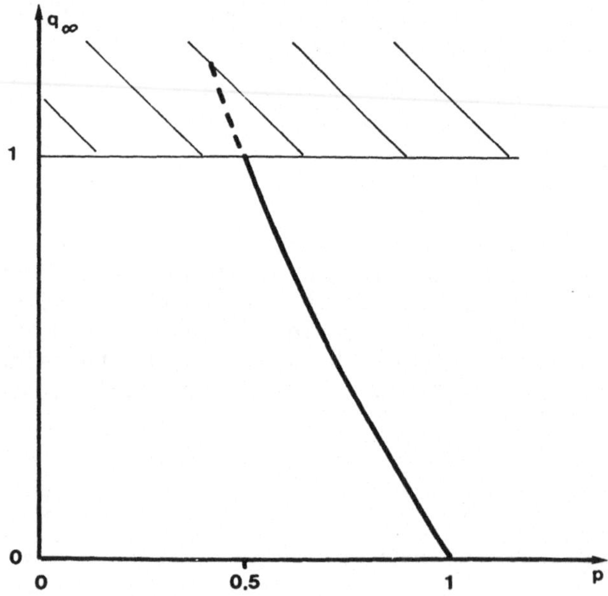

FIGURE 27-Threshold determination for Bethe lattice

Let us take any one site, S_0, and consider one of the branches originating from S_0. Let us call q_∞ the probability that the site S_0 is not connected to infinity along this branch. We have :

q_∞ = 1 - p (the first site (S_1) of this branch is inactive)

+

$pq_\infty q_\infty$ (the first site S_1 is active but none of the two branches originating from S_1 connect it to infinity).

So we have :

$$q_\infty = 1-p + pq_\infty^2 \Longrightarrow \begin{cases} q_\infty = 1 \\ q_\infty = (1-p)/p \end{cases}$$

and the threshold p_c is obtained when $(1-p)/p = 1$ (cf. figure 27) so that $p_c = 1/2$ (For $p < p_c$, $q_\infty = 1$ is the only physical solution). Now let us look at $P_\infty(p)$ which is the probability that the site S_0 is connected to infinity. Clearly, we have $P_\infty(p) = 1 - q_\infty^3$

If $p < 1/2$	$q_\infty = 1$	$P_\infty(p) = 0$
$p > 1/2$	$q_\infty = (1-p)/p$	$P_\infty(p) = 1-((1-p)/p)^3$

Near p_c, $P_\infty(p)$ behaves as follows: if we take $p = p_c + \varepsilon = 1/2 + \varepsilon$ we see that $(1-p)/p \sim 1-4\varepsilon$ and that $P_\infty(p) \sim 12\varepsilon \sim (p-p_c)$ so that the exponent β equals 1.

Now, let us investigate the behaviour of the mean size S and the exponent γ. For that, we have to compute the mean size. Let T, the mean size of the fraction of the cluster along any one of the 3 branches originating from S_0. We have then

S = 1 + 3T

(the site S_0) +(3 times the mean size along one branch)

Now $T = (1-p).0 + p(1+2T)$

The two terms corresponding to site S_1 empty and site S_1 occupied respectively. Thus $T = p(1+2T)$ which yields $T = p/(1-2p)$ and $S = 1 + (3p/(1-2p)) = (1+p)/(1-2p)$ and we recover $p_c = 1/2$ since $S \to \infty$ when $p \to 1/2$. Close to p_c, we put $\varepsilon = p_c - p$ and obtain

$$S = (3/2 - \varepsilon)/2\varepsilon \sim 3/4\varepsilon \sim \varepsilon^{-1} \Longrightarrow \gamma = 1$$

In the general case, for the Bethe lattice with z neighbours,

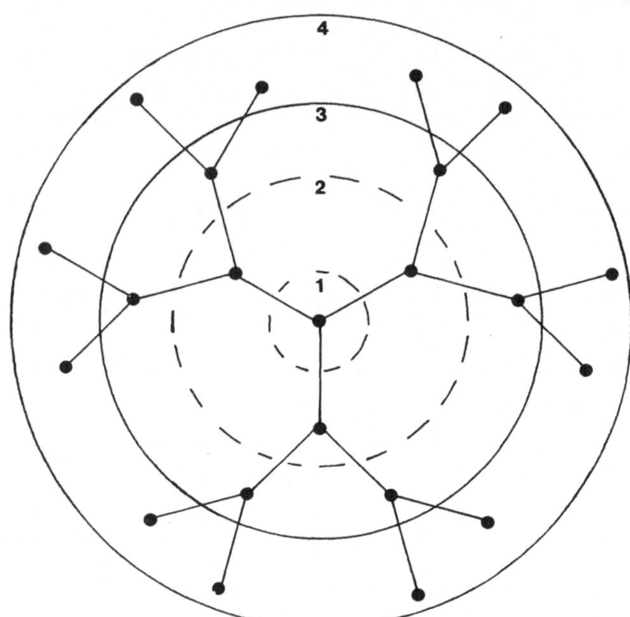

FIGURE 28- Surface sites (n^{th} generation : S_n sites) and volume sites (1st + 2nd+...+$(n-1)^{th}$ generation V_n sites). When $n \to \infty$, the ratio of $S_n/V_n \to 1/2$

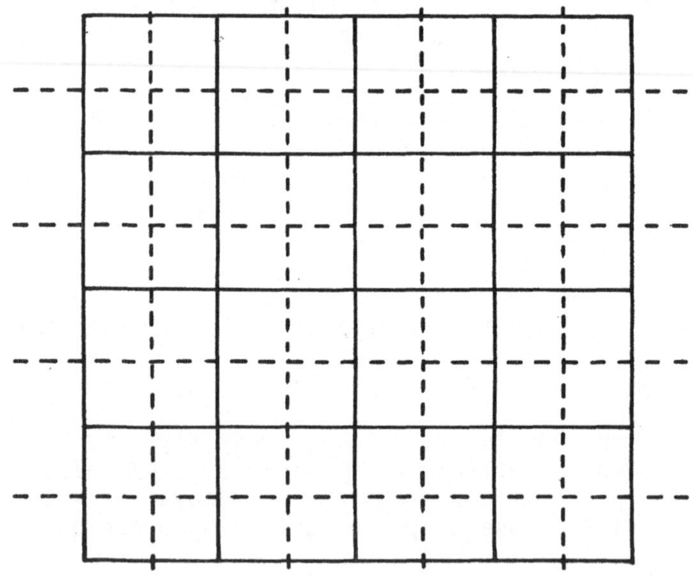

FIGURE 29- The dual lattice (---) of a square lattice(——) is a square lattice

the results are :

$$p_c = 1/(z-1) \qquad \beta = 1 \qquad \gamma = 1$$

The Bethe lattice is a very special lattice. It corresponds to a graph drawn in a space with an infinite number of dimensions. This can be easily seen if one looks (cf. figure 28) at the ratio of the number of sites on the surface over the number of sites in the volume, when the size of the lattice increases. This ratio S/V is going to 1/2 when the size L goes to infinity.

Now in a space at D dimensions, we have for compact figures $S \sim V^{(D-1)/D}$. If $S/V \rightarrow 1/2$, when $L \rightarrow \infty$, we have $(D-1)/D=1 \Longrightarrow D=\infty$

6.2 Other exact results

We have just seen the results on the Bethe Lattice. To my knowledge it is the only lattice for which exact results have been obtained for p_c and the exponents together. For 2-dimensional lattices, there are some other exact results known about the threshold and in some case one exact result about the exponent ν. For the square bond lattice, we have $p_c = 1/2$. In fact, some mathematicians still argue about some points of the demonstration, but, here, I shall give only an argument. If you draw, a square lattice, you can play a game with a partner (cf. figure 29). You can fill randomly the bonds of the lattice, trying to make a way going to infinity (the edge of the board). Your partner, in his turn, may cut the bonds at random, also attempting to join the edges of the board. In a square lattice, your lattice and your partner's one are the same. So you have exactly an even chance to win(tossing a coin to decide who starts). And the threshold is 1/2. Your partner's lattice,with bonds of his lattice cutting the bonds of your own lattice, is the dual lattice of your own. In any lattice, the threshold for the dual lattice is the complementary to one of the threshold of the direct lattice (they add up to 1). So, when the dual lattice and the direct one are identical, the threshold is 1/2.

The dual lattice of the honeycomb is a triangular lattice (figure 30). So if,as can be demonstrated, $p_c^b = 2 \sin \pi/18$ for the triangular lattice, one has $p_c^b = 1 - 2 \sin \pi/18$ for the honeycomb lattice.

We have seen earlier that one can transform a bond problem into a site problem. In that case, if p_c^b is the threshold for the bond problem, the threshold of the corresponding site problem is $p_c^s = p_c^b$. As an example, we have seen the transformation from a honeycomb to a kagomé lattice. Thus p_c^s (kagomé) = $1-2\sin \pi/18$.

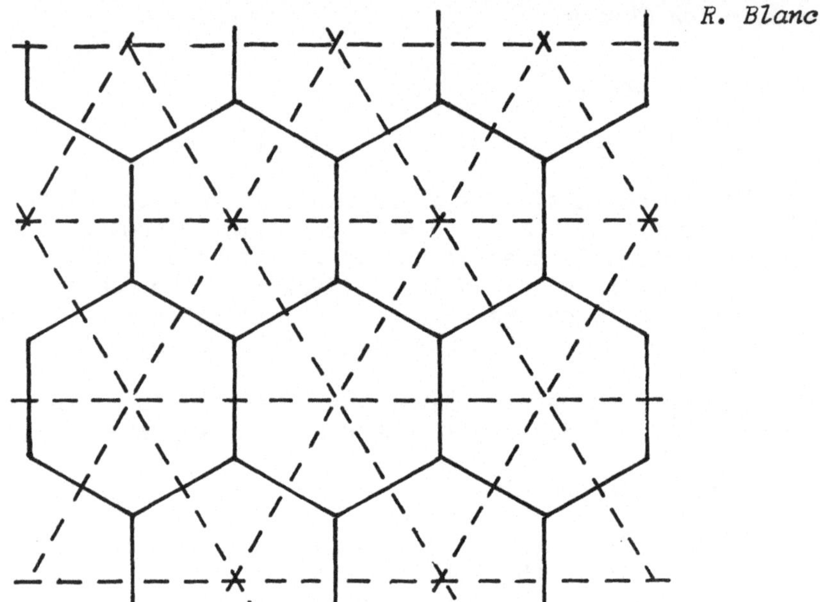

FIGURE 30- The dual lattice (---) of a honeycomb lattice (——) is a triangular lattice

7. APPROXIMATE RESULTS

7.1 Series expansion

The principle of this method is the following : for a given lattice, one computes directly the clusters numbers $n_s(p)$, for increasing values of their size s. Then one computes $\sum_s s^k n_s(p)$ which has the form of a series expansion and one looks at the convergence radius of this series. The convergence radius is equal to p_c, and the behaviour of this series near p_c gives the corresponding exponent.

For instance, let us take a triangular lattice. We can easily compute the values of the different n_s when s is small (figure 31) :

$n_1(p) = pq^6$ (as one has one occupied site with the probability p surrounded by 6 inoccupied sites with the probability q = 1-p)

$n_2(p) = 3p^2q^8$ (2 occupied sites p^2 and 8 inoccupied which give q^8 . And we have 3 different orientations for this dimer).

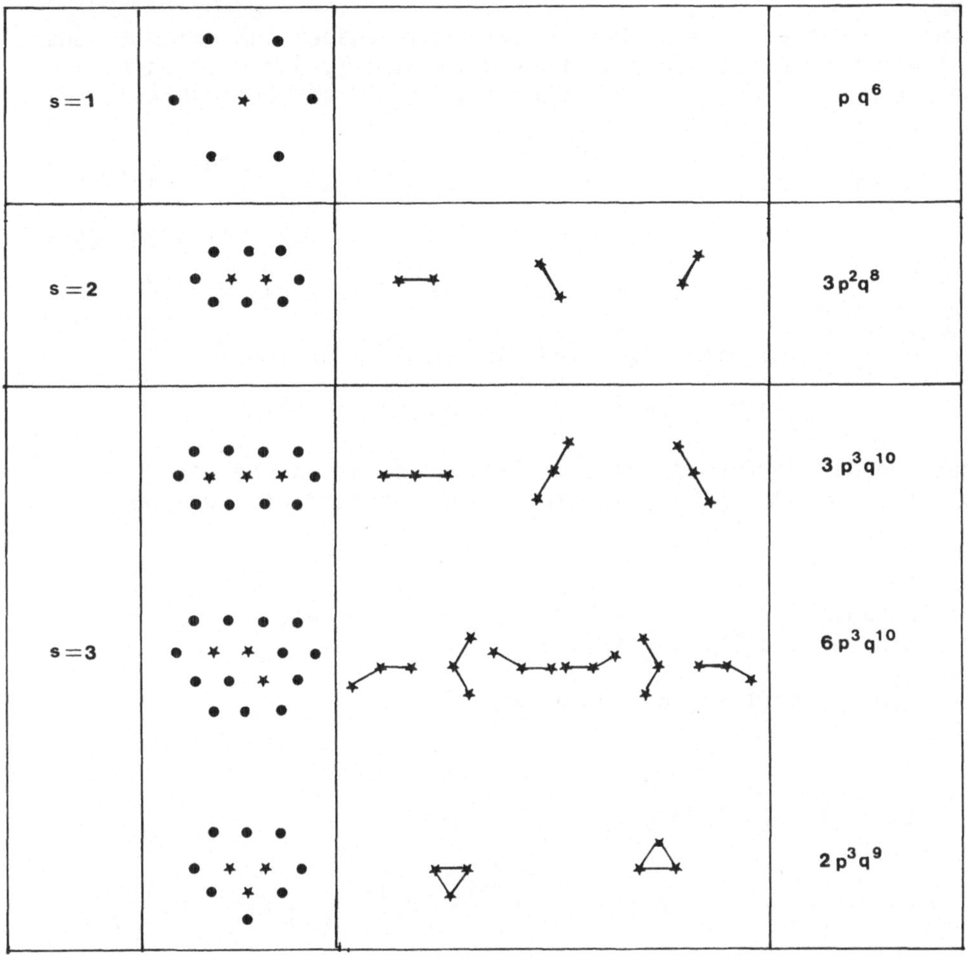

s = 1			$p\,q^6$
s = 2			$3\,p^2q^8$
			$3\,p^3q^{10}$
s = 3			$6\,p^3q^{10}$
			$2\,p^3q^9$

FIGURE 31

$n_3(p) = p^3(2q^9 + 9q^{10})$ (here we have three different shapes for a trimer :

- the 3 sites are on a row, thus $p^3\ q^{10}$ with 3 different orientations, giving $3p^3q^{10}$
- the 3 sites form an obtuse angle. So we have p^3q^9 and 6 different orientations giving an additional $6p^3\ q^9$.
- the 3 sites form a triangle giving $2p^3q^9$).

When s increases, the number of different shapes and orientations of the different clusters increases enormously and one has to use a computer. For instance when s = 14, in the triangular lattice, we have (21) :

$$n_{14}(p) = p^{14}(3q^{16} + 168q^{17} + 1\,524q^{18} + 10\,029q^{19} + 46119q^{20}$$
$$+ 185\,220q^{21} + 605\,766q^{22} + 1\,730\,943q^{23} + 4\,287\,699q^{24}$$
$$+ 9\,131\,949q^{25} + 16\,871\,550q^{26} + 26\,571\,525\,q^{27}$$
$$+ 35\,061\,399q^{28} + 3\,796\,541q^{29} + 32\,198\,928q^{30}$$
$$+ 19\,012\,074q^{31} + 5\,812\,482q^{32})$$

When one has performed the calculation as far as one can, one computes the sum $\sum_s s^k n_s(p)$ which takes the form of a polynomial

$$\sum_s s^k n_s(p) = \sum_i a_i p^i$$

For instance for k = 2, we have $S(p) \sim \sum_s s^2 n_s$ and as S(p) may be expanded as a Taylor serie :

$$S(p) = \sum_i p^i (1/i!)(d^i S/dp^i)_{p=0}$$

one has

$$a_i = (1/i!)(d^i S/dp^i)_{p=0}$$

$$(a_{i+1})/a_i = (i\,!\,/(i+1)!)\;(d^{i+1}S\,/dp^{i+1})_{p=0}\;/(d^i S/dp^i)_{p=0}$$

Then, if S, close to p_c, behaves like $(p-p_c)^{-\gamma}$ we get

$$(d^i s/dp^i)_{p=0} = (-1)^i \gamma\,(\gamma+1)\ldots(\gamma+i)\,p_c^{-\gamma-i}$$

$$(d^{i+1}S\,/dp^i)_{p=0} = (-1)^{i+1}\gamma(\gamma+1)\ldots(\gamma+i)p_c^{-\gamma-i-1}$$

$$(a_{i+1}/a_i) = -(1/(i+1))((\gamma+i)/p_c) = -(\gamma-1)/p_c\;(1/(i+1)) - 1/p_c$$

And so, if one plots this ratio versus 1/i (figure 32) one obtains a straight line, of slope $(\gamma-1)/p_c$ and the origin $1/p_c$. In fact, very often, there are large oscillations in the ratio a_{i+1}/a_i and one has to use more sophisticated techniques like Padé approximants.

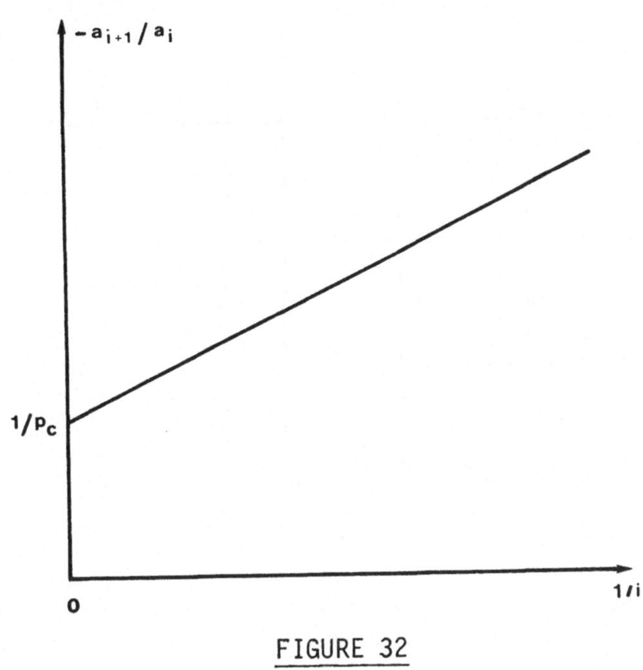

FIGURE 32

7.2 Renormalisation method

This method is an adaptation to the percolation problem of more general techniques used for instance, in phase transitions. The basic idea is that,near p_c, the correlation length ξ diverges. And then, all the finite lengths in the system, as for instance the radius of a cluster, become much smaller than $\xi$and the appearance of the lattice must be the same whatever the scale one uses. Thus, instead looking at each site of the lattice, we can average over regions much smaller than ξ (cf.figures 33 and 34) and build a new system for which each site takes the properties of one averaged region. But, as the new lattice constant is again much smaller than ξ , we can average again, and so on.

If we are able to go from one scale to another without modifications of the critical characteristics of the problem, this method allows us to calculate the threshold and the exponents.

Let us take one example of that transformation. This example deals with site problem on a triangular lattice (cf. figures 35 and 36).

From the initial lattice we can draw a super lattice by replacing three sites by a new one, a super site, which represents the average over the three original sites. We get again a triangular

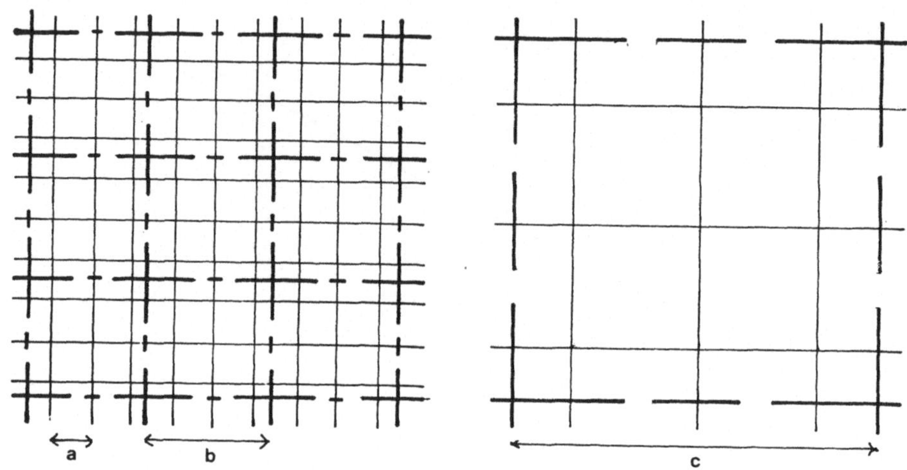

FIGURES 33-34 - Properties of initial square lattice (lattice constant $a<<\xi$) are averaged on regions of size $3a=b<<\xi$. The averages are used to obtain a new square lattice which in turn is averaged over regions of size $3b=c<<\xi$,and so on.

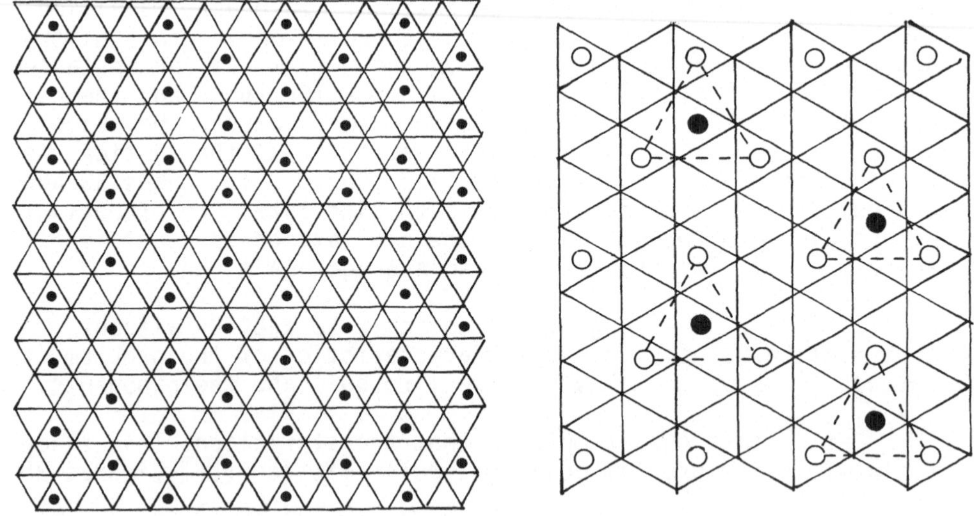

FIGURE 35-36 - Same as figures 33 and 34
initial sites : ✱ super sites : •
super-supersites: ○ (super)^{3}sites : ●

lattice. And we can draw a super-super lattice by replacing three super sites by a super-super site, which represents the average of the three super sites and so on.

Now, how can we take a suitable average over three sites? In our case, it is very easy to impose majority rules : one can say that a super-site will be active if 3 or 2 sites are active. So if p is the probability that a site is active, the probability that a super site is active is $p' = p^3 + 3p^2q$. Morover, the new lattice constant is $b = a\sqrt{3}$.

As we have $\xi = \alpha\, a\, (p-p_c)^{-\nu}$, where α is an unimportant pre-factor, we have also

$$\xi = \alpha b(p'-p_c)^{-\nu}$$

and then $((p-p_c)/(p'-p_c))^{-\nu} = b/a = \sqrt{3} \rightarrow \nu = \log\sqrt{3}/\log((p'-p_c)/(p-p_c))$
At the threshold $p'=p=p_c$ which gives

$$p^3 + 3p^2\,(1-p) = p \rightarrow p=0\,,\ p=1\,,\ p=1/2$$

And near the threshold $p = p_c - \varepsilon \rightarrow p'-1/2 = 3/2\,(p-1/2)$,
$\nu = \log\ 3/\log(3/2) = 1.355$ which is very close to the perhaps exact value $\nu = 4/3$. (22).

7.3 Computer simulations

The principle of computer simulations is the following. One works with a finite sized lattice, say a cubic lattice of edge L, having N^D sites. For each site, a random number generator gives a number x belonging to the interval $|0,1|$. For a given value of p, the site is active if x is less than or equal to p, and inactive when x is greater than p. I am overlooking the techniques allowing to save computer time or storage size. When the work is done, the computer gives the numbers of clusters n_s and the sums $\sum_s sn_s$, $\sum_s s^2 n_s$ ....

But, with finite-sized lattice, we have two sorts of errors :

- For a given value of p, there are some statistical fluctuations between different experiments. If, for instance, I consider the limiting case with one site only, I get sometimes 0 cluster and sometimes one and I have to make many experiments and take the average value. The importance of these fluctuations decreases as the number of sites increases, but goes up when one approaches the threshold, as we remember from the susceptibility which diverges near p_c.

- The second type of errors is altogether more tedious and more interesting. It is tedious because there is a fundamental limitation connected with the finite size of the sample. When p goes to p_c, the correlation length increases, and at some stage, it becomes larger than the sample and the boundaries of the sample take a more and more important role. These boundaries can cut some clusters, as shown, for instance, in figure 37 and this fact upsets the statistics in a fundamental way. But, by itself, it is a very interesting fact, because you can use it in order to compute the correlation length and the exponent ν . But if one is not interested in ξ and ν one can gloss over the size effect, using periodic limits, that is to say, using a lattice drawn on a torus. The first and the last rows are connected together, the left and the right too (figure 38).

But we are interested in the correlation length and the exponent and in their determination by use of size effects. For instance, let us perform computer simulation on a 2D square-lattice with n sites. For each value of p, one repeats many times the simulation and one counts the number of samples with a conductive path, or equivalently, with a cluster in contact with the four boundaries. Let $F_n(p)$ be the relative number of such samples If we look at the variations of $F_n(p)$ with n and p (23), we have the function plotted in figure 39 . For $n^2 = 400$, we have a smooth curve. As n increases, the variations of F become steeper and for $n^2 = 10^6$ we obtain nearly a step function.

Now, if quite arbitrarily we define the zone of the transition by the values p_1 and p_2, such as $F_n(p_1) = 0.1$ and $F_n(p_2) = 0.9$ and

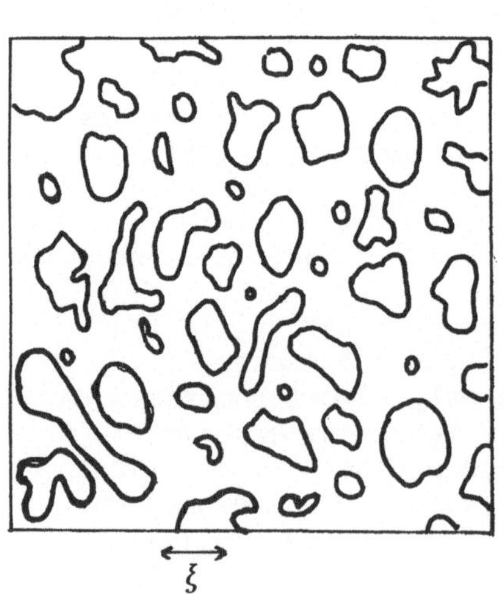

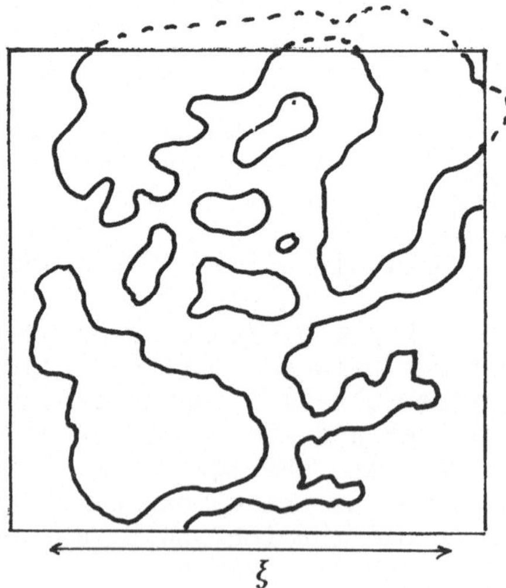

FIGURE 37

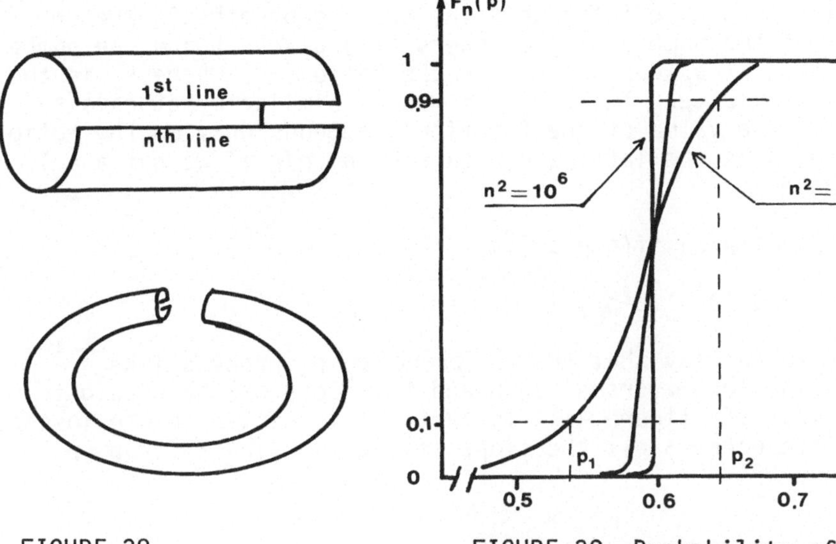

FIGURE 38

FIGURE 39- Probability of conduction of a sample versus p for different sizes.

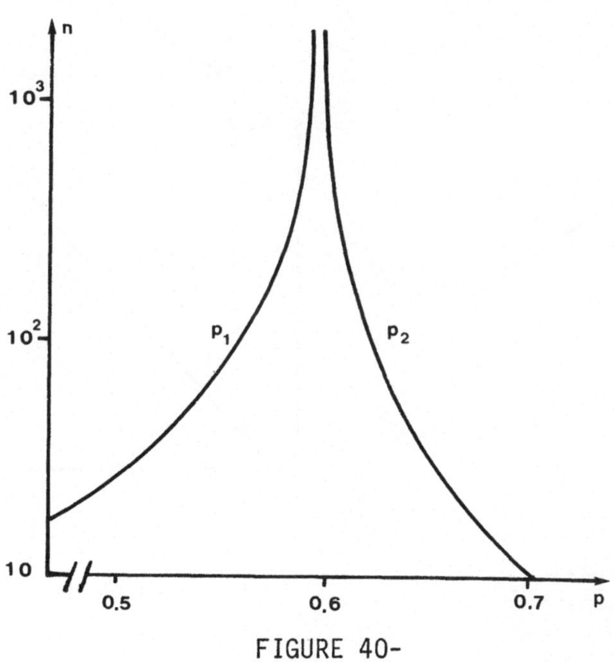

FIGURE 40-

if we look at the variations of p_1 and p_2 as function of n, we obtain the curves plotted in figure 40 (11)

Earlier, we have introduced the scaling hypothesis whereby the ratio of the number of s-clusters at a given p and the number of s-clusters at p_c depend only on the ratio s/s_ξ where s_ξ is the typical size cluster. There we can make a similar hypothesis and assume that the value of the function F depends only on the ratio na/ξ where a is the lattice constant or on n/ξ if we put a = 1. Thus, we have :

$$F(n,p) \sim f(n/\xi) \sim f(n(p-p_c)^{\nu})$$

$$\sim g\ |n^{1/\nu}(p-p_c)\ |$$

Then we can say that the difference p_2-p_1 behaves like $n^{-1/\nu}$ We could plot $\log \Delta p$ versus log n and the slope of the straight line is $-1/\nu$ (cf. figure 41). In fact, it is better to use, instead of the difference p_2-p_1, the slope of the function F for $p = p_c$ (cf. figure 42)

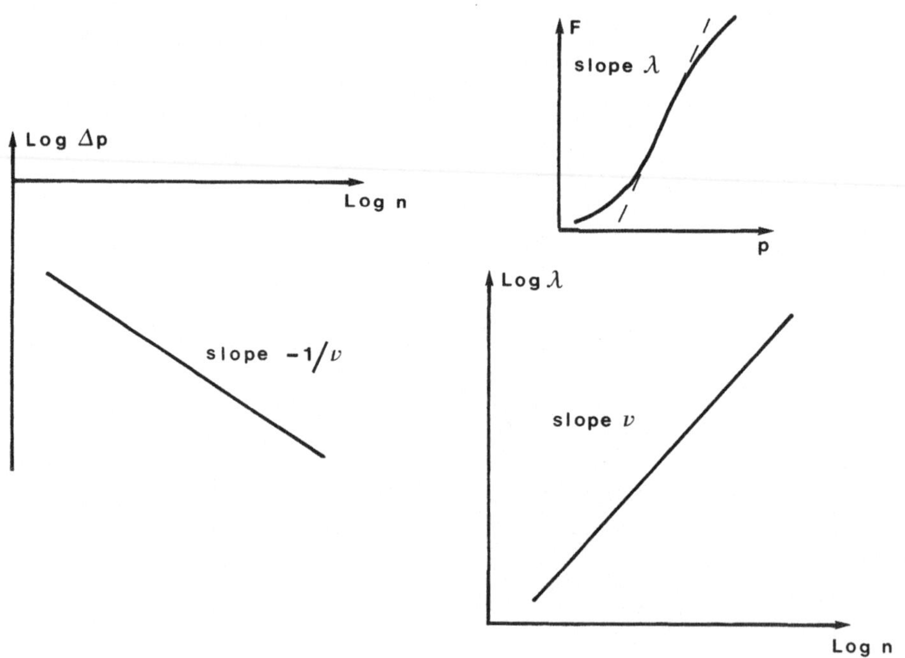

FIGURE 41 FIGURE 42

7.4 Percolation thresholds

On the one hand, the thresholds in a percolation problem are not very interesting because they have no universality character and depend on the symmetry of the lattice. But on the other hand, it is sometimes important for an experimentalist to know them, or, if there are not known to be able to estimate them. So, before leaving this part of my talk, I shall digress for a moment. I have listed on table 3 several thresholds for different bond problems, together with z the coordination number of the lattice.

We can see that the threshold value decreases for a given dimensionality when z increases. That is very easy to understand. When the coordination increases one has more possibilities to bypass a cut bond. For the same reason, for a given coordination number, for instance z = 4, the threshold is lower in 3 dimensions than in 2 dimensions. Moreover, if one computs the product of z and p_c for bond percolation, one gets nearby a constant value : 2 in 2D lattice and 1.5 in 3D. The product is not really a constant but this gives a useful relation.

Now if we look at the site percolation problem, we have another set of results (table 4).

We can again see that the larger the coordination number, z , is, the lower is the threshold. But if we try to calculate the product $z.p_c^s$, we do not obtain a constant but p_c varies regularly with z or 1/z (cf. figure 43).

On the other hand, if we take the filling factor f of the lattice, the product of f and p_c^s takes a nearly constant value

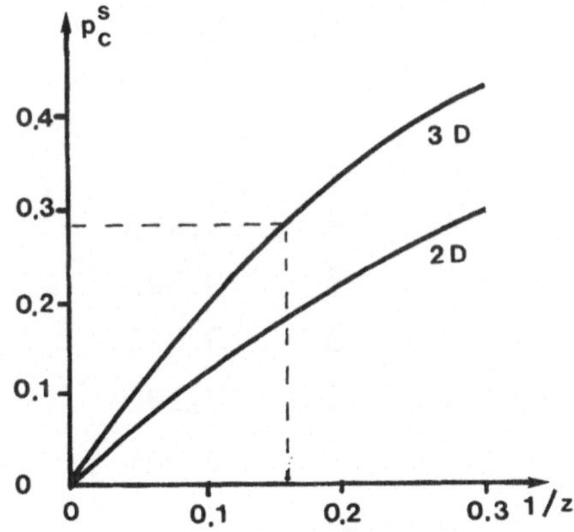

FIGURE 43- Site percolation thresholds versus the opposite of the coordination number z for 2-Dimensional and 3-Dimensional lattices.

	LATTICE		z	p_c^{bond}	$z.p_c^b$	
2D	Honeycomb		3	0.6527	1.96	2
	Kagomé		4	0.525	2.1	
	Square		4	0.500	2.0	
	Triangular		6	0.3473	2.08	
3D	Diamond		4	0.388	1.55	1.5
	Simple	cubic	6	0.247	1.48	
	Centered	cubic	8	0.178	1.42	
	Face centered	cubic	12	0.119	1.43	
	Hexagonal		12	0.124	1.49	

TABLE 3 - Bond percolation threshold

	LATTICE		z	p_c^{site}	f	
2D	Honeycomb		3	0.700	0.61	$f.p_c^s$ 0.44
	Kagomé		4	0.6527	0.68	
	Square		4	0.593	0.79	
	Triangular		6	0.5000	0.91	
3D	Diamond		4	0.425	0.34	$f.p_c^s$ 0.154
	Simple	cubic	6	0.307	0.52	
	Centered	cubic	8	0.243	0.68	
	Face centered	cubic	12	0.195	0.74	
	Hexagonal		12	0.204	0.74	

TABLE 4 - Site percolation threshold

($f.p_c^s \sim 0.44$ in 2D ; 0.154 in 3D). These relations may be very useful. For instance, if we ask about $\bar{z}$, the mean number of neighbours that a sphere has in a random packing, we can measure the threshold (with conductivity measurement for instance) and get the coordination number. (24)

With $p_c \sim 0.27$ we get $\bar{z} \sim 5.5$ assuming bond percolation, which is certainly a poor assumption in such a mixed bond site percolation problem on a random lattice, but nevertheless yields a useful estimate for $\bar{z}$. From the curve on figure 43, we obtain approximately $\bar{z} = 6$ when $p_c^s = 0.27$.

8. THE EXPONENT OF THE CONDUCTIVITY

So far we have examined statistical quantities like G(p), $P_\infty(p)$, S(p) or a geometrical one ξ . But in most physical cases, we are interested in other quantities as for instance the conductivity of a random network, the viscosity of a sol-gel phase transition, the elastic modulus of a gel, the dielectric constant of a mixture of powders, and so on. Some of these quantities exhibit a critical behaviour near a critical point in connection with the occurence of a global connectivity in the system. Is it possible to relate the critical exponents of these behaviours and the critical exponents we have seen earlier? It is up to now an open question.

The problem for which the work seems to me the most advanced is that of the conductivity of a random network.

8.1 Conductivity measurements

The measurements of the d.c. conductivity are very easy experiments, and many physicists have made this type of measurements. I mention here only the names of Last and Thouless(25) who performed the first experiment in 1971, making random holes in a graphited paper foil and the name of my colleagues, Clerc, Giraud and Rousseng, in Marseille,who performed many experiments on mixture of powders (6) Once , as a joke, they took a mixture of sugared almonds and silver coated sugared almonds and got a very good exponent ! (indicating that universality is a good law).

In 3D, the measurements are very easy and one can take a very large sample, but in 2D, it is not so easy. One has to make random holes as Last and Thouless, or cut the bonds of a grid as I did (cf. figures 9 and 10). So,some experiments have been performed on computer. But it is a quite complicated task since with a square lattice having a length of 200 lattice constants, and thus $2.(200)^2 = 80\ 000$ bonds of which at least 40 000 are left near p_c, you have a Kirchoff system with 40 000 equations to solve.

Moreover, we have finite size effects. And for a series of samples of finite size, one has large fluctuations on the values of the conductivity, together with the onset of conductivity below threshold when the size of the sample is very small. This onset of conductivity below threshold corresponds to the existence of a fraction of the samples being conducting.

In the spirit of the scaling laws, one looks at a scaling relation (26) (N is the number of sites on the sample)

$$\sigma \sim N^{-x} \quad f \, |(p-p_c).N^{1/\nu}|$$

with $f(0) \neq 0$ which gives at $p = p_c$ a variation of the conductivity following a law $\sigma \sim N^{-x}$ with $x > 0$. When $N \to \infty$ we know that $\sigma \sim (p-p_c)^t$. So when $z \to \infty$ $f(z) \sim z^t$. But as σ, in this case, does not depend on N, we must have $\sigma \sim N^{-x} (p-p_c)^t \, N^{t/\nu}$ independant of N. This gives $x = t/\nu$. So if we draw in a log-log plot σ $(p=p_c)$ versus log N, we get the exponent t/ν and t if ν is known.

8.2 3D Cross-over

Let us take a sample in 3D space, but with only 2 infinite sides, the third, the thikness of the sample, being d. If for different values of d, one performs conductivity measurements, one gets a threshold which depends on d (cf. figure 44).

On the other hand, when for a given value of d, one looks at the conductivity of the sample, as a function of p, one has the variation shown schematically in figure 45.

How can we understand these facts ? Concerning the threshold it is clear that, if we look at a 2D lattice, and if one has a non conducting sample because there is a missing bond, if we had a third dimension, we could bypass the missing bond by using some bonds in the third dimension (figure 46). So that, for a sample of finite thickness, the threshold is lower than for strictly two dimensional lattice.(27)

When we are as close to the threshold $p_c(d)$ for which the correlation length in the lattice is larger than the thickness of the sample, the surfaces of the sample play a very important role and the conductivity is strongly affected by them. In one sense, the lattice behaves like a 2 dimensional one and the exponent of the conductivity is very close to 1.1 (cf. figure 45). But when the concentration p increases, ξ decreases and there is a value of p for which ξ is of the order of the thickness of the sample. And when p increases further, ξ becomes lower than the thickness.

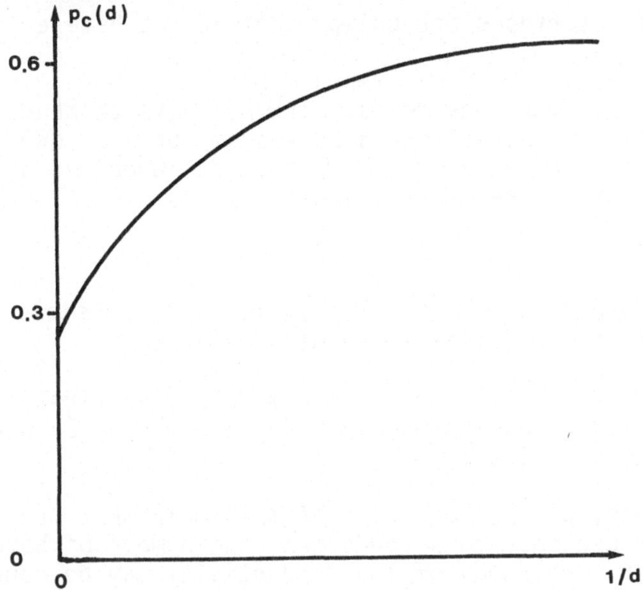

FIGURE 44- Threshold $p_c(d)$ for a finite (∞x ∞xd) sample

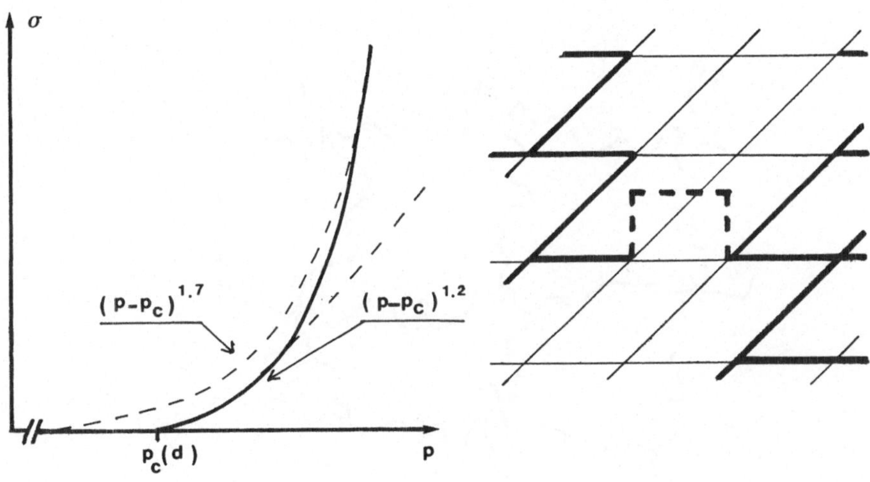

FIGURE 45

FIGURE 46

At that time, the surfaces are unimportant and we have a 3 dimensional behaviour.

So we see that when one deals with a finite thickness sample, the systems behaves like a 2D system, or a 3D one, depending on the relative values of ξ and d. When $\xi \sim d$ a crossover between the two dimensionalities takes place.

8.3 Models

Let us return to the question whether it is possible to relate the exponent of the conductivity to other exponents.

In order to do that, we need some models. The first one is the super-lattice model proposed independently by De Gennes (28) and Skal and Shklowskii (29).

We have seen, at the beginning of this talk that the conducting path use many less bonds than the number of bonds in the infinite cluster. The study of the conductivity may be done on a simplified infinite cluster, without dead arms. This lattice is called the super-lattice. And one assumes that this super-lattice can be approximated by a periodical one, with lattice constant ξ (cf. figure 47). But the length of the actual path between two sites is different from ξ , because this path is very twisty and this length behaves like $(p-p_c)^{-\zeta}$ where ζ is a new exponent called the tortuosity exponent.

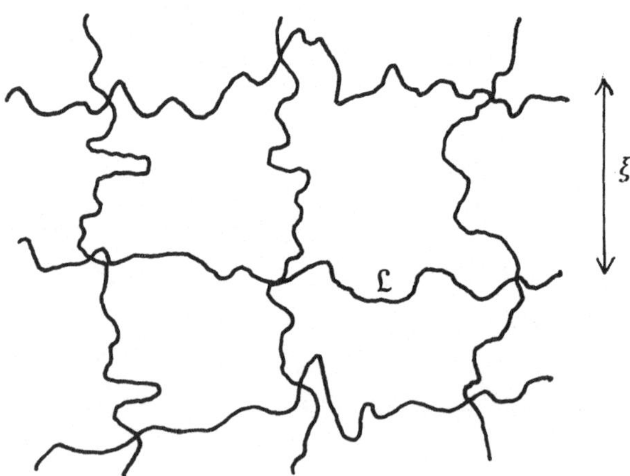

FIGURE 47- The super lattice constant is ξ .The actual length between two sites of the super lattice is $\mathcal{L}$

One builds a cubic lattice with N^3 elementary cubes, each of size ξ, the resistance of a side being $r = (\mathcal{L}/1)\, r_0$ (cf. figure 48) where 1 is the lattice constant of the initial lattice and r_0 the resistance of a bond in the initial lattice. If σ is the conductivity of the lattice, one has :

$$r/N \doteq (1/\sigma)\,(N\xi / N^2\xi^2)$$

so that $\sigma = (1/r)(1/\xi)$

and $\sigma = (1/\mathcal{L})(1/r_0)(1/\xi) = \sigma_0\,(1/\mathcal{L}\cdot\xi)$

More generally in a D dimensional space, we have

$$\sigma = \sigma_0\,(1/\mathcal{L})(1/\xi^{D-2})$$

Finally, one obtains :

$$\sigma \sim (p-p_c)^{\zeta}\ \ (p-p_c)^{\nu\,(D-2)}$$
$$\sim (p-p_c)^{\zeta + (D-2)\nu}$$

so that

$$t = \zeta + (D-2)\,\nu$$

Is the super lattice model a good one ?

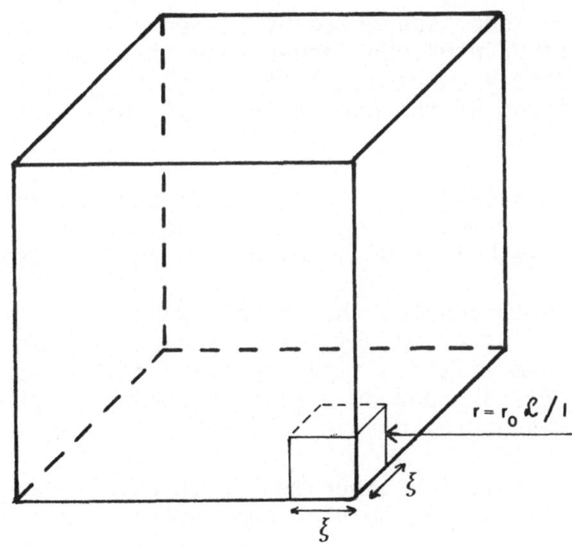

FIGURE 48

For a 6-dimensional space, for which the critical effects of percolation disappear, one expects a random walk between two sites of the super lattice so that $\zeta = 2\nu = 1$. Thus, $t = D\nu = 6.(1/2) = 3$ exactly. For D = 3, $\zeta = 1$ and $\nu \doteqdot 0.84$, we have t = 1.84 approximately as observed (1.72). For D = 2 and $\zeta = 1$ we have t = 1. But then $\zeta<\nu$ so that the length along the path between two sites would be shorter than the distance between these sites !!

So we see that the super-lattice model is a good model in 6 dimensions, a quite good one in 3 dimensions, and a very poor one in the two dimensional case.

If one analyses the reason why this model is better for high dimensionalities, one sees that when the dimensionality increases, the loops in the actual network play relatively a less and less important role, and the model of a network with one relevant wire becomes better and better.

The second model we can use is the fractal model. Kirkpatrick (30) and Aharoni (31) assume that the infinite cluster, without the dead arms, can be described in 2 dimensions as a self-similar system having a similar structure of loops at any scale, as long as this scale is lower than ξ. From a two dimensional square lattice one has a fractal dimension D = 1.58 and an exponant $t \sim 1$. But when D increases, the model becomes less and less good since the influence of the loops decreases.

The third model suggested by De Gennes (8) is called the "ant in the labyrinth". De Gennes suggested that a random walk in an incompletely connected lattice, might give some informations on the variations of the conductivity of this lattice, in the critical region.

One assumes that such an ant is dropped on an active site and then has a random walk going from one active site to another without the possibility of jumping over an inactive site.

Below the threshold, this random walk is limited by the size of the finite cluster on which the ant was dropped. Above the threshold, there exists a finite probability, which is $P_\infty(p)$, that the ant was dropped on a site belonging to the infinite lattice. And then it can go to infinity.

If one repeats the experiment with a very large number of ants with different starting points, one can take the average value of the distance from its starting point, which the ant has covered after N steps : $(\langle R_N^2 \rangle)^{1/2}$.

For p below p_c ($p = 0.5$), we obtain an experimental relation between the average value of $<R_N^2>$ and N (cf. figure 49). This relation is (32) :

$$<R_N^2> = A + B\, e^{-N/\Theta}$$

When N increases, one has a limit value $< R_\infty^2 > = A$ ($\sim$37) which is something like the average radius of gyration of the finite clusters. And it can be proved that $<R_\infty^2>$ behaves near p_c as $(p_c-p)^{-2\nu+\beta}$. (33)

For a value of p closer to p_c as e.g. $p = 0.55$ (cf. Figure 50) we get the same type of variations. But as we approach p_c, the average radius of gyration of the clusters increases ($A = 120$) and so does the number of steps needed before reaching the limiting value A.

When p is larger than p_c (cf. figure 51) it is now possible for the ant to go to infinity and we obtain the well known result of a random walk, $<R^2_\infty> \sim \mathcal{D}N$, where $\mathcal{D}$ is a diffusion constant. May I remark here as I mentionned earlier that above p_c and for a very large cluster we are unable, from isolated mean values, to say whether we are on the infinite cluster or not. One can see from the figures 50 and 51 that if the number of steps is not very large, the ant cannot determine whether it is or not on infinite cluster as it has to go near the boundary of the cluster to ascertain whether the cluster is finite or not.

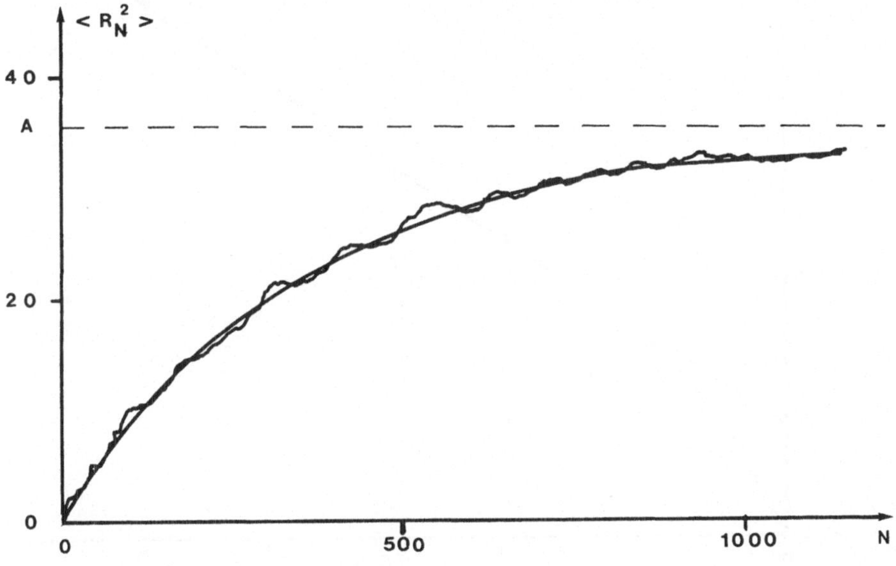

FIGURE 49- Random walk in a incompletely connected lattice $p\ (=0.5) < p_c (=0.593)$

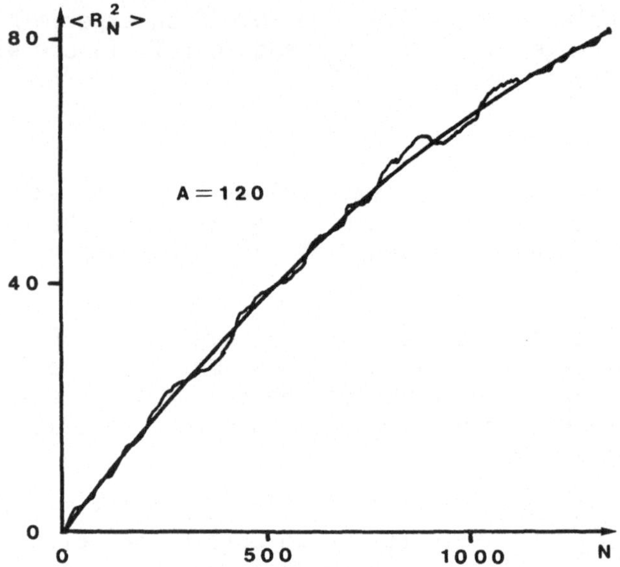

FIGURE 50- Random walk in a incompletely connected lattice $p(=0.55) < p_c(=0.593)$

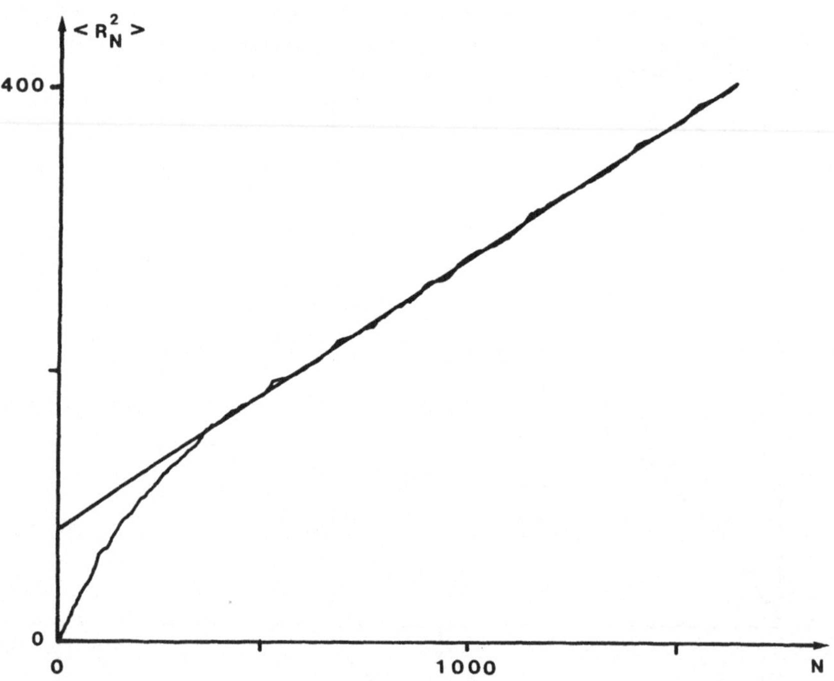

FIGURE 51- same as in figures 49 and 50 with $p = 0.64 > p_c$

Now let us derive the conductivity from these results. We have to relate the diffusion constant to the conductivity. This relation is simply

$$\mathcal{D} = \sigma$$

I do not give here the full demonstration (see, e.g.(8),(33),(6)), although it is very simple . I give only the beginning and the end.

Let $l_i(N)$ the probability that the ant is on site i after N steps. When the next step is performed, this probability can decrease , because the ant can jump away from site i : if s_{ij} is the probability of a jump from the site i to the site j , the decrease of l_i is $\sum_j s_{ij}\, l_i(N)$. But the probability l_i can also increase, because an ant at j after N steps may jump into site i ; the increase of l_i is $\sum_j s_{ij}\, l_j(N)$. So that the total change in probability is :

$$l_i(N+1) - l_i(N) = \sum_j s_{ij}(l_j(N) - l_i(N))$$

As the number N of steps is a linear function of the time we get

$$dl_i(t)/dt = \sum_j s_{ij}\ (l_i(t) - l_j(t))$$

which is an equation for the diffusion of the ant.

Now if we put

$$\int_0^\infty l_i(t)\ dt = V_i/I$$

one has

$$\int_0^\infty dl_i(t)/dt = \sum_j s_{ij}\ \left| \int_0^\infty l_i(t)\ dt - \int_0^\infty l_j(t)\ dt \right|$$

$$l_i(t=\ ) - l_i(t=\infty) = 1/I\ \sum_j s_{ij}(V_i - V_j).$$

so that, since the probability for the ant to be in a given site at the infinite time is zero and since the probability that for t = 0 the ant is on the site i is $\delta(i)$, if we label 0 the site where the ant is dropped we have

$$-\delta(i) = 1/I\ \sum_j s_{ij}(V_i - V_j)$$

which is a Kirchoff equation for the lattice when one injects the current at the site 0 and one retires it at the infinity. And for this lattice, the conductance of an individual bond is s_{ij}. So we can understand that $\mathcal{D} = \sigma$.

If we look at the different values of the diffusion constant obtained for different values of p , we can put that

$$\mathcal{D} \sim (p - p_c)^t \sim \sigma$$

Indeed, experiments (32) show that the exponent t obtained by that method is quite good, specially for 3 D case (cf. Figure 52).

Nevertheless, to my knowledge there exists as yet, no simple way to express, in this model, the exponent t as a function of the exponents of the percolation.

9. PERCOLATION AND MACROSCOPIC RANDOM MEDIA (MIAM)

Very often, one has physical situations in which a global connectivity appears in a system and together with this connectivity a critical behaviour appears. And in my talk, I have made very often reference to the conductivity of a random network. I shall say now a few words about some others macroscopic random media. In these media, the local properties may be different from a point to another but the word "macroscopic" indicates that the local properties are quite constant on an atomic scale.

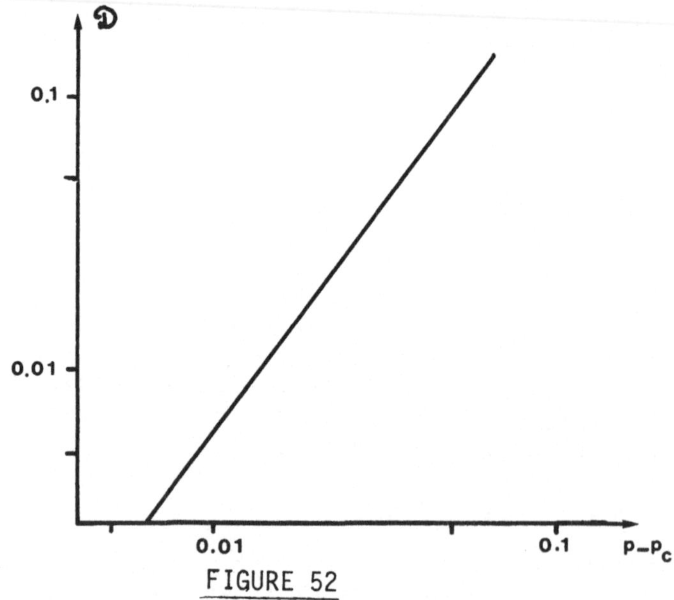

FIGURE 52

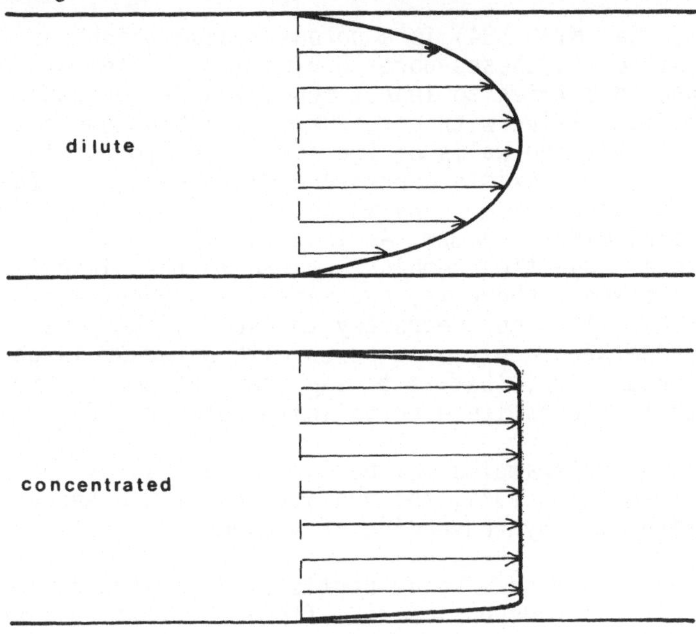

FIGURE 53

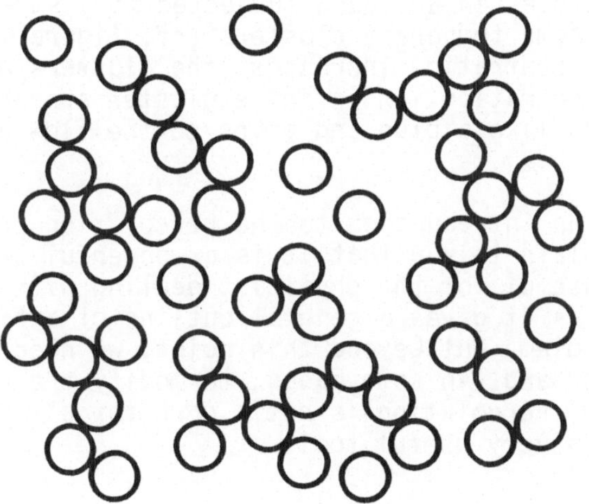

FIGURE 54

A typical MIAM (34) is a porous medium which has an infinite network of pores, theses pores having very different shapes and radii. When one tries to inject mercury under pressure in a porous medium filled with air, due to its high surface tension σ Hg does not wet the walls of the solid. In hydrostatic equilibrium conditions, there exists a pressure difference π at the curved boundary between the mercury which has penetrated a pore of radius r and air; with $\pi \sim \sigma/r$. As long as π is smaller than a critical value π_c injection remains limited to the vicinity of the injection plane . Above π_c, one can obtain penetration of the liquid (35). It is not necessary to use all the pores for penetration. One needs only use pores with a radius larger than a critical radius r_c, that is to say that the pores with radius larger than r_c constitute an infinite network (36).

Many other examples may be introduced as for instance, the sol gel transition (37), the propagation of a desease (38) or a forest-fire, the hydrodynamics of suspensions (39).

When one put some solid particles in a liquid one has a suspension. If this suspension is flowing through a pipe, the velocity profile has a shape depending on the concentration of the suspension. For weak concentration, the velocity profile follows a Poiseuille law, it has a parabolic profile. But when the concentrations of solid particles increases, the velocity profile becomes flat (40) (cf. figure 53).

How is it possible to understand this effect ? When one looks at solid particles in a liquid subjected to a shear, the particles may form some temporary clusters (cf. figure 54). When the volumic concentration increases, the clusters grow and, for a critical concentration, an infinite cluster appears and this cluster behaves like a plug and gives to the flow a flat profile (41).

This has been an introduction to the percolation problem. I tried here to explain to you that it is a wonderful problem which may be very useful for the physicist dealing with macroscopic random media as it gives a general outline of a first approach to the problem. But beyond this point, we need to specify some physics and, in some cases, to modify the pure percolation problem. Percolation is not a cure for all but is is nevertheless a very useful tool.

ACKNOWLEDGEMENTS

I have benefited from many discussions with J.P. Clerc, G. Giraud, E. Guyon and J. Roussenq.

N. Rivier made a critical reading of this lecture, which repeats some parts of a course on the "Physics of heteregenous media and disordered systems" given in Marseille by D. Stauffer (1980-1981) and myself (1981-1982).

REFERENCES (references 1 to 9 are review articles)

1. ESSAM J.W. in "Phase Transitions and Critical phenomena" vol 2,192 (1972)
2. SHANTE V.K.S. and KIRKPATRICK S. - Adv. Phys. 20, 325 (1971)
3. KIRKPATRICK S. - Rev. Mod. Phys. 45, 574 (1973)
4. ESSAM J.W. - Rep. Prog. Phys. 43, 833 (1980)
5. STAUFFER D. - Phys. Repts. 54, 1 (1979)
6. CLERC J.P., GIRAUD G., ROUSSENQ J., BLANC R., CARTON J.P. GUYON E., OTTAVI H. and STAUFFER D. - Annales de Physique (to appear)
7. Percolation structures and processes - Annals of the Israël Physical Society - to appear
8. De GENNES P.G. - La Recherche 72, 919 (1976)
9. HAMMERSLEY J.M. and WELSH D.J.A.- Contemp. Phys. 21, 593 (1980)
10. CLERC J.P., GIRAUD G. and ROUSSENQ J. - C.R. Acad. Sci. Paris B 281, 227 (1975)
11. ROUSSENQ J. , CLERC J.P., GIRAUD G., OTTAVI H. and GUYON E. J. Physique Lettres 37, L 99 (1976)
12. GUYON E., CLERC J.P., GIRAUD G. and ROUSSENQ J. - J. Physique 42, 1553 (1981)
13. ROUSSENQ J. - Thesis -Université de Provence - Marseille (1980)
14. VICSEK T. and KERTESZ J. - J. Phys. A 14, L 31 (1981)
15. GALIWSKI E.T. and STANLEY H.E. - J. Phys. A 14, L 291(1981)
16. KASTELEYN P.W. and FORTUIN C.M. - J. Phys. Soc. Jap. supp. 26, 11 (1969)
17. See for instance ref. 16 or CLERC J.P. Thesis - Université de Provence - Marseille (1980)
18. HOSHEN J., STAUFFER D., BISHOP G.H., HARRISSON R.J. and QUINN G.P.- J. Phys. A 12,1285(1979)
19. WOLFF W.S. and STAUFFER D. - Z. Physik B 29, 67 (1978)
20. So we recover with the exponents of percolation, Rushbrook's relation $2 - \alpha = \gamma + 2\beta$ which has been first established for phase transitions (see for instance H.E. STANLEY in "Introduction to phase transitions and critical phenomena" Clarendon Press - Oxford 1971).
21. SYKES M.F. and GLEN M.-J.Phys. A 9, 87 (1976)
22. BLOTE H., NIGHTINGALE M. and DERRIDA B. - J. Phys. A 14 L 45 (1981)

23. see ref. 13 or 11.
REYNOLDS J.P., STANLEY H.E. and KLEIN W.- Phys. Rev. B 21, 3, 1223 (1980)
24. GIRAUD G. - Thesis - Université de Provence - Marseille (1980)
25. LAST J.P. and THOULESS D.J. Phys. Rev. Lett. 27,1719(1971)
26. MITESCU C.D., ALLAIN C., GUYON E. and CLERC J.P. - J. Phys. A 15, 2523 (1982)
27. CLERC J.P., GIRAUD G., ALEXANDER S. and GUYON E. - Phys. Rev. B 22, 2489 (1980)
28. De GENNES P.G., J. Physique Lettres 37 L 1 (1976)
29. SKAL A.S. and SHKLOWSKII B.I. - Sov. Phys. Semicond. 8 1029 (1975)
30 KIRKPATRICK S. -"Les Houches 78" Ill condensed matter (North Holland ed.)(1979)
31. AHARONI A. - Proc. Conf. Int. Disordered systems and localisation - Rome (may 1981)
32. MITESCU C.D. and ROUSSENQ J.- C.R. Acad. Sci. Paris A 283 299 (1976)
MITESCU C.D., OTTAVI H. and ROUSSENQ J. - AIP Conf. Proc. 40, 377 (1979)
see also ref. 13
33. De GENNES P.G. - C.R. Acad. Sci. Paris 286 B, 131 (1978)
STEPHEN J. - Phys. Rev. B. 17, 4444 (1978)
STRALEY J.P. - J. Phys. C 13, 819 (1980)
J. Phys. C 13, 2991 (1980)
see also ref. 13
34. in french MIlieux Aléatoires Macroscopiques
35. LENORMAND R. and BORIES S. - C.R. Acad. Sci. Paris B 219 (1980)
LENORMAND R. - Thesis - Toulouse (1981)
PACSYRSKI J., GAUTIER C. and MARLE C. - Rev. I.F.P. 11 803 (1965)
36. De GENNES P.G. and GUYON E. - Jour. de Méca. 17, 403 (1978)
37. STAUFFER D., CONIGLIO A. and ADAM M.-Adv. Poly. Sci. 44,103 (1982)
38. MARCHANT J. and GABILLARD R. - C.R. Acad. Sci. Paris B 281, 261 (1975)
39 BOUILLOT J.L., CAMOIN C., BELZONS M., BLANC R. and GUYON E. Adv. In Coll. and Int. Sci. 17, 299 (1982)
CAMOIN C. , BELZONS M., BLANC R. and BOUILLOT J.L. submitted to Journal de Physique
BLANC R., BOUILLOT J.L., CAMOIN C. and BELZONS M. submitted to Rheologica Acta.
40. COX R.G. and MASON S.G. - Ann. Rev. Fl. Mech. 3 291 (1971)
41. De GENNES P.G. J. Physique 40, 783 (1979)
Phys. Chem. Hydr. III, 2, 31 (1981)

ELECTRONIC AND TRANSPORT PROPERTIES OF GRANULAR MATERIALS I: EFFECTIVE MEDIUM THEORIES OF TRANSPORT IN INHOMOGENEOUS MATERIALS

Morrel H. Cohen

Exxon Research and Engineering Company
Corporate Research Science Laboratories
P. O. Box 45, Linden, New Jersey 07036

Abstract

After a brief introduction to motivate the use of statistical methods for calculating the macroscopic properties of inhomogeneous materials, we derive the effective medium approximation (EMA) for random resistor networks. We then compare the results of the EMA with accurate numerical simulations, establishing the surprisingly large range of validity for the EMA. We next show that the EMA can be embedded within an exact multiple scattering formalism which leads naturally to extensions of the EMA which substantially improve its accuracy. We then discuss the EMA for continuous random media and examine the effects of spatial correlations on the results. Finally, we briefly review the results of the EMA for physical properties other than the electrical conductivity.

1. INTRODUCTION

The title of the Advanced Summer Institute of which this lecture is a part is "Impact of Cluster Physics in Materials Science and Technology," and the section to which this lecture belongs is "Granular Materials." Materials in which clusters occur are inhomogeneous, and, when clustering is so pronounced that the materials become granular, the inhomogeneities are large. We are therefore addressing questions which can only be answered within the larger context provided by the theory of inhomogeneous materials. My task here is to review the theory of the electronic and transport properties of inhomogeneous materials.

I shall select the electrical conductivity for detailed discussion, treating other properties briefly at the end of this lecture. The local current density $\vec{j}(\vec{r})$ is given by

$$\vec{j}(\vec{r}) = \int d^3r' \overleftrightarrow{\sigma} (\vec{r}, \vec{r}') \bullet \vec{E}(\vec{r}') , \quad (1)$$

where $\vec{E}(\vec{r}')$ is the local electrostatic field and $\overleftrightarrow{\sigma} (\vec{r}, \vec{r}')$ is the nonlocal conductivity tensor. For simplicity, we shall restrict outselves to macroscopically isotropic materials. We define the macroscopic current density $\vec{j}$ and electrostatic field $\vec{E}$ as volume averages over the system or ensemble averages over a set of equivalent inhomogeneous systems,

$$\overleftrightarrow{j} = \langle \vec{j}(r) \rangle, \ \vec{E} = \langle \vec{E}(\vec{r}) \rangle . \quad (2)$$

The macroscopic conductivity σ is then defined through

$$j_x = \sigma E_x, \quad (3)$$

or

$$\sigma = \frac{\langle \int d^3r' \Sigma_i \sigma_{xi} (\vec{r}, \vec{r}') \ E_i(\vec{r}') \rangle}{\langle \ E_x(\vec{r}) \rangle} \quad (4)$$

The stationary local current density obeys the continuity equation

$$\vec{\nabla} \bullet \vec{j}(\vec{r}) = 0, \quad (5)$$

and the local field, being irrotational, is the gradient of the electrostatic potential $\phi (\vec{r})$

$$\vec{E}(\vec{r}) = -\vec{\nabla} \phi(\vec{r}) \quad (6)$$

Thus (4) can in principle be evaluated from the solution of

$$\vec{\nabla} \bullet \int d^3r' \overleftrightarrow{\sigma} (\vec{r}, \vec{r}') \bullet \vec{\nabla} \phi(\vec{r}') = 0 \tag{7}$$

subject to suitable boundary conditions on $\phi(\vec{r})$ and $\vec{j}(\vec{r})$. However, we do not know $\overleftrightarrow{\sigma} (\vec{r}, \vec{r}')$ in detail. Even if we knew $\overleftrightarrow{\sigma} (\vec{r}, \vec{r}')$, we could not solve the partial differential equation (7). Even if we could solve it, the averaging would be nontrivial to carry out.

Thus, we are forced to make a series of simplifications. The first is that mean free paths are small compared to the physical scale of the inhomogeneities and that tunneling is unimportant. $\overleftrightarrow{\sigma} (\vec{r}, \vec{r}')$ then reduces to a local conductivity, which we assume for convenience to be isotropic,

$$\overleftrightarrow{\sigma} (\vec{r}, \vec{r}') = \sigma(\vec{r}) \overleftrightarrow{I} \delta(\vec{r} - \vec{r}') \tag{8}$$

Introducing (8) into (7) converts it from an integro-differential equation to a differential equation, but a direct attack is still impossible. Instead statistical treatments are used, of which the effective medium approximation (EMA)[1-4] is the simplest accurate one known. Discretization of the continuum on which $\vec{j}(\vec{r})$, $E(\vec{r})$ and $\sigma(\vec{r})$ are defined leads to a lattice or resistor network with random values of the resistance. The transition from a continuous to a resistor network can sometimes be done by an exact mapping, as discussed in the second lecture of this series, without the approximation of discretization. We shall deal directly with continua later in this lecture.

2. EFFECTIVE MEDIUM THEORY FOR RANDOM RESISTOR NETWORKS

A simple example of a random-resistor network is shown in Fig. 1. It is a simple square lattice of edge length a. The lattice points are nodes in the resistance network and are joined by conductances along each square edge i. The values of the conductances, g_i', are randomly distributed. The resistors on different bonds are uncorrelated,

$$P(g_i', g_j') = P(g_i') P(g'_j) \tag{9}$$

where P indicates a probability distribution.

Consider first the periodic case in which all conductances are equal and have the value g_m. Let us examine a situation in which a current I is introduced into the lattice through node a, and taken out of the lattice through node b, as shown in Fig. 2. Let us couple to the current flowing into a a current I flowing out at infinity and to that out of b one flowing in at infinity. The original current flow is then the superposition of these two independent current flows. If Z is the coordination number, then,

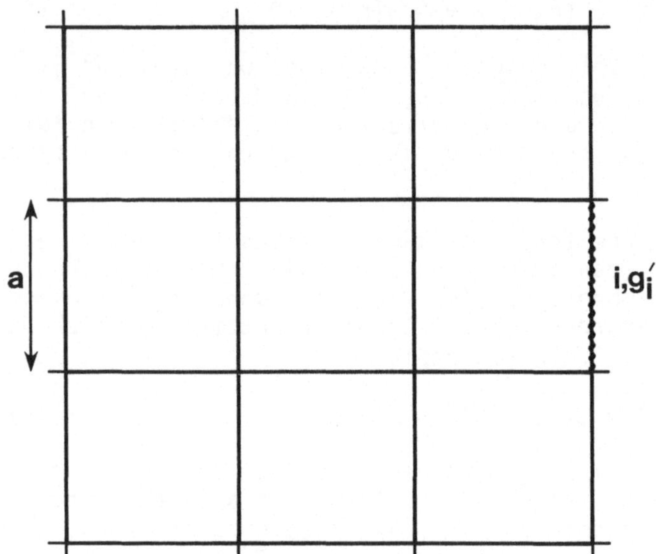

Figure 1. Simple square random resistor network. g_i' is the value of the conductance of the ith bond. The values of the g_i' are randomly distributed.

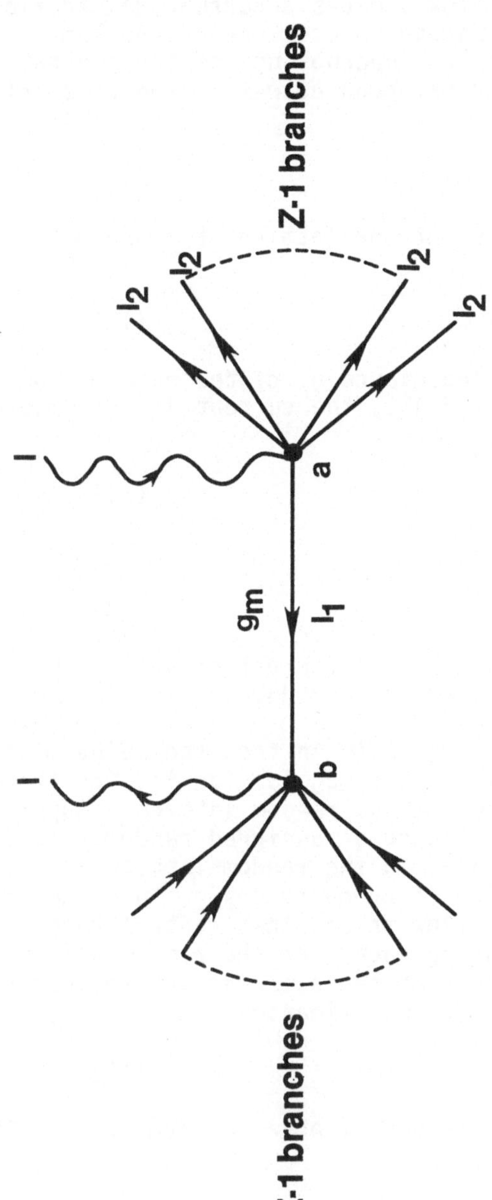

Figure 2. Current flowing into node a and out of node b, neighboring it, in a lattice with coordination number Z and all bonds occupied by conductance g_m.

by symmetry, the first current flow induces a current I/Z to flow out of each bond at a, and the second induces one of the same value to flow into each bond at b. Superposing the two current flows, the current I_1 in the central bond between a and b is, cf. Fig. 2,

$$I_1 = 2I/Z. \tag{10}$$

The current flowing into the rest of the lattice at a and out of it at b is, cf. Fig. 2,

$$(Z-1)\ I_2 = I - 2I/Z \tag{11}$$

Introducing the notion of the conductance g_r of the rest of the system (excluding that of the bond ab), the current in (11) must be given by

$$I - 2I/Z = \frac{g_r}{g_m + g_r}\ I \tag{12}$$

or

$$g_r = \left(\frac{Z}{2} - 1\right) g_m \tag{13}$$

Eq. (13) is important in our analysis of the effective medium approximation, which we now proceed to introduce.

Suppose now that the g_m of Fig. 2 is in fact the value of the bond conductance one has to assign to a regular lattice for it to have the same conductivity as the actual random lattice. Suppose further that a bond with a conductance g' assigned randomly to it with the same distribution $P(g')$ as in the random lattice is embedded into the regular lattice of conductances g_m in place of the bond ab. A current I' will flow across this central bond, generating a voltage V'. Averaging each over the random values of the conductance g' and requiring that the averages are related by g_m introduces the effective medium approximation (EMA) to g_m,

$$\langle I' \rangle = g_m \langle V' \rangle \tag{14}$$

An explicit formula for g_m is easily obtained from (14). The current I' is given by

$$I' = \frac{g'}{g' + g_r}\ I \tag{15}$$

and the voltage V' by

$$V' = \frac{1}{g' + g_r}\ I \tag{16}$$

Substituting (13) into (15) and (16) and the results into (14) give the EMA equation for the lattice.

$$\left\langle \frac{g' - g_m}{g' + (\frac{Z}{2} - 1)\, g_m} \right\rangle = 0 \tag{17}$$

Explicit forms of (17) are

$$\frac{1}{g_m} = \left\langle \frac{1}{g'} \right\rangle \tag{18}$$

for one dimension (1d) (Z = 2), which is exact;

$$\left\langle \frac{g' - g_m}{g' + g_m} \right\rangle = 0 \tag{19}$$

for the simple-square lattice in 2d (Z = 4), and

$$\left\langle \frac{g' - g_m}{g' + 2g_m} \right\rangle = 0 \tag{20}$$

for the simple cubic lattice in 3d (Z = 6). Given a value for g_m, one computes the value of the bulk conductivity of the lattice in 3d from

$$\sigma_m = na\, \overline{\cos^2 \theta}\, g_m \tag{21}$$

where $\underline{a}$ is the bond length, as before, n is the number density of bonds crossing a plane and $\overline{\cos^2 \theta}$ is the mean square value of the cosine of the angle between the bonds and the normal to the plane. For the simple cubic lattice, the result is

$$\sigma_m = g_m/a \tag{21'}$$

The conductivity can be calculated numerically to a high degree of accuracy for such random lattices. For example, Kirkpatrick(3) has obtained explicit results for the simple cubic lattice with the resistance $r' = (g')^{-1}$ a binary random variable,

$$\begin{aligned} r' &= r_1 \text{ with probability } p \\ &= r_2 \quad \text{"} \qquad \text{"} \qquad 1-p \\ x &= r_1/r_2 < 1 \end{aligned} \tag{22}$$

The corresponding values of the conductivities for the pure cases are

$$\sigma_1 = (r_1 a)^{-1}, \quad \sigma_2 = (r_2 a)^{-1}, \quad x = \sigma_2/\sigma_1 \qquad (23)$$

Normalizing the conductivity σ in the random case to σ_1, we see that we have a two parameter problem

$$y \equiv \sigma/\sigma_1 = f\ (p,\ x) \qquad (24)$$

which can be solved analytically in the EMA:

$$f\ (p,\ x) = \alpha + \sqrt{\alpha^2 + \tfrac{1}{2}x}$$

$$\alpha = \tfrac{1}{4}[\ 3p\ (1-x) + (2\ x-1)\ ] \qquad (25)$$

A comparison of (25) with the numerical results is shown in Fig. 3. Those results and broader experience with the EMA have shown that the EMA is accurate to within 3% for all x for $p \gtrsim 0.4$ and that it is similarly accurate for all p, $x > 0.03$.

It has been proved rigorously by e.g., G. Papanicolau,[5] that the EMA is exact in the asymptotic limits $p \to 1$ or $x \to 1$. Thus, a deep question is raised by the results reported in the previous paragraph. Why does the EMA work so well over such a large range of x and p if it is only asymptotically correct? Similar results are found for different geometries, for continuous systems, when there are correlations between the bonds, and so on. We shall answer this question in the next section with a multiple scattering formalism developed by Yonezawa, Webman, and myself.[6]

3. THE MULTIPLE SCATTERING FORMALISM

Consider a simple cubic lattice of random resistors. Let the x, y and z axes be chosen parallel to the cube axes. Label the nodes by the indices k or ℓ. The current flowing from k to ℓ when they are nearest neighbors is given by

$$i_{k\ell} = \sigma_{k\ell}\ (\phi_k - \phi_\ell)\ , \qquad (26)$$

where ϕ_k and ϕ_ℓ are the voltages at the nodes. We have set the lattice constant equal to unity, and make no distinction between conductance and conductivity. Thus $\sigma_{k\ell}$ in (26) is the conductance of the bond $k\ell$.

It is convenient in connection with specifying the boundary conditions to divide the lattice into two domains C' and C, as shown in Fig. 4. C' includes the entire lattice with bounding planes normal to the x axis B_1 at $x_k = 0$ and B_2 at $x_k = L$. The

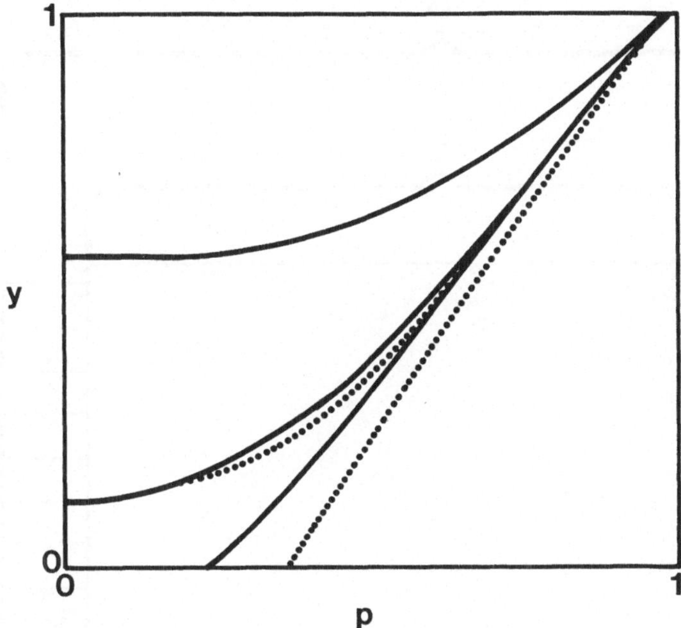

Figure 3. Comparison of EMT results for y (Eq. 24), dotted lines, with those of numerical simulations, solid lines, for x = 0, 0.1 and 0.5, respectively.

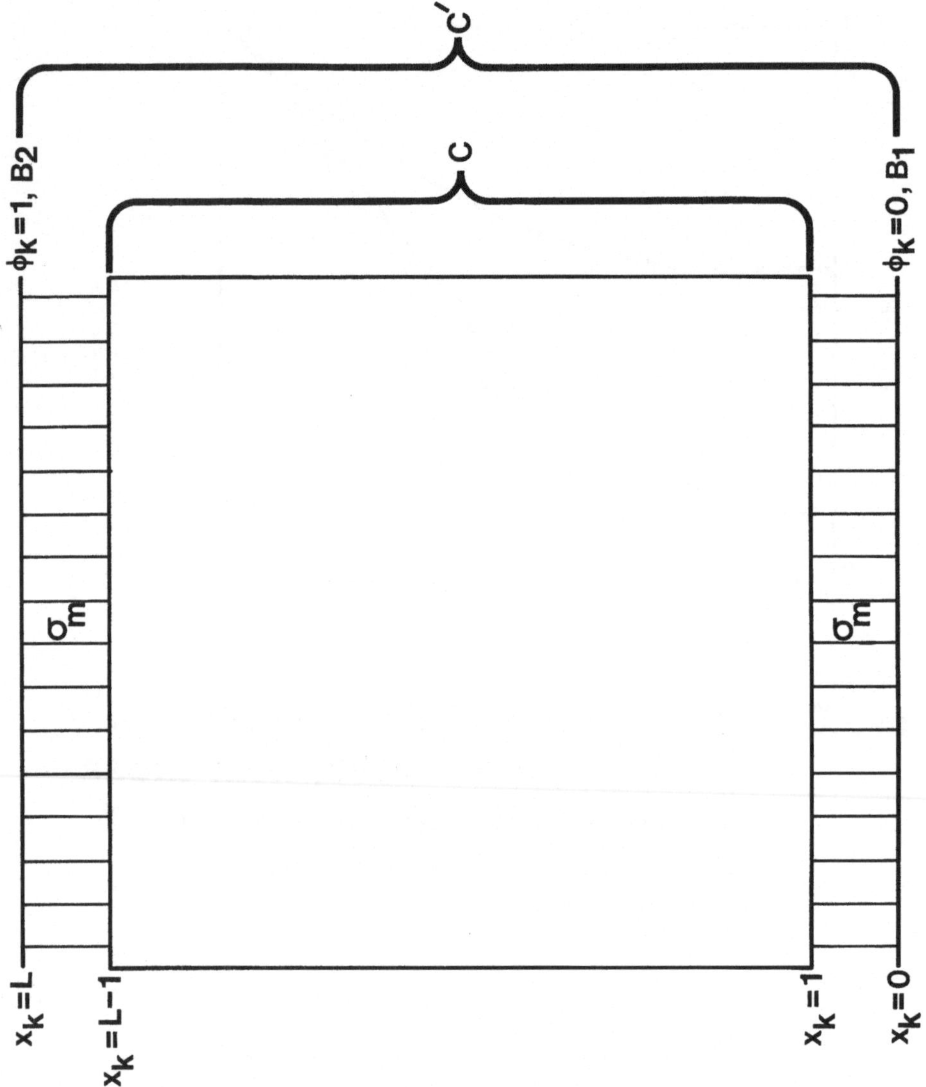

Figure 4. **The boundary conditions imposed on the simple cubic lattice. Definition of the domains C, C′, B_1 and B_2.**

electrostatic potential ϕ_k has the boundary values,

$$\phi_k = 1, \; k\varepsilon B_2 \tag{27}$$

$$\phi_k = 0, \; k\varepsilon B_1 \tag{28}$$

We use periodic boundary conditions on the corresponding bounding planes normal to the y and z axes which together with B_1 and B_2 define a cube with $(L + 1)^3$ sites. Sites in B_1 and B_2 are connected to their nearest neighbor sites in C, that part of C′ exclusive of B_1 and B_2, by constant conductances σ_M.

Kirchhoff's laws can now be written out explicitly for $k\varepsilon C$,

$$\sum_\ell A_{k\ell}\phi_\ell = \sigma_M \, \delta_{x_k, L-1}; \; k,\ell\varepsilon C \tag{29}$$

$$A_{k\ell} \equiv \delta_{k\ell} \sum_n \sigma_{kn} - \sigma_{k\ell}; \quad k,\ell\varepsilon C \quad n\varepsilon C' \tag{30}$$

In Eq.(30), the fact that $n\varepsilon C'$ eliminates the translational invariance of $A_{k\ell}$ and therefore the existence of a vanishing eigen-value, giving $A_{k\ell}$ an inverse. The formal solution of (29) is, therefore,

$$\phi_k = \sum_{X_\ell = L-1} G_{k\ell} \, \sigma_m \tag{31}$$

where G, the inverse of A in C, plays the role of a Green's function.

We now develop a formal expression for the macroscopic conductivity. The local field across the bond $k\ell$ is

$$E_{k\ell} = \phi_k - \phi_\ell \tag{32}$$

The local current through that bond is

$$i_{k\ell} = \sigma_{k\ell} \, E_{k\ell} \tag{33}$$

Only bonds which are parallel to the x direction contribute to the macroscopic current flowing in the x direction. Such bonds we indicate by underlining their indices:

$$i_{\underline{k\ell}} = \sigma_{\underline{k\ell}} \, E_{\underline{k\ell}} \text{ (current flow along X)} \tag{33'}$$

The macroscopic average of a quantity associated with a bond such as $i_{k\ell}$ or $E_{k\ell}$ is

$$< Q_{k\ell} > = \frac{1}{L^3} \sum_{k\ell} Q_{k\ell}, \quad k,\ell \varepsilon C' \tag{34}$$

In analogy with (4), the macroscopic conductivity σ^* is therefore defined by

$$< i_{\underline{k\ell}} > = \sigma^* < E_{\underline{k\ell}} > , \tag{35}$$

or

$$\sigma^* = \frac{\sum_{\underline{k\ell}} \sigma_{\underline{k\ell}} E_{\underline{k\ell}}}{\sum_{\underline{k\ell}} E_{\underline{k\ell}}} , \quad k\varepsilon C' \tag{36}$$

The sum on $E_{\underline{k\ell}}$ is easily evaluated, using its definition and the boundary conditions,

$$\sum_{\underline{k\ell}} E_{\underline{k\ell}} = \sum_{\underline{k\ell}} (\phi_k - \phi_\ell) = L^2 \tag{37}$$

Inserting (37) into (36) we obtain

$$\sigma^* = \frac{1}{L^2} \sum_{\underline{k\ell}} \sigma_{\underline{k\ell}} E_{\underline{k\ell}} + o \left(\frac{1}{L} \right), \quad k\varepsilon C \tag{38}$$

Using (31) in (38), we obtain the desired formal result

$$\sigma^* = \frac{1}{L^2} \sum_{\substack{k\varepsilon C \\ X_i = L-1}} \sigma_{\underline{k\ell}} < \underline{k\ell}|G|i > \sigma_m \tag{39}$$

where

$$|k\ell > = |k > - |\ell > \tag{40}$$

in the Dirac notation.

Suppose we had an effective medium in which

$$\sigma_{k\ell} = \sigma_M \quad \text{all } k, \ell \text{ in } C \tag{41}$$

The corresponding matrix $A_{k\ell}$ would be

$$\begin{aligned} < k|A_M|\ell > &= \sigma_M(Z\, \delta_{k\ell} - \Delta_{k\ell}) \\ \Delta_{k\ell} &= 1 \quad \text{if } k\ell \text{ are nn} \\ &= 0 \quad \text{otherwise} \end{aligned} \tag{42}$$

The corresponding local fields are

$$E_{k\ell}^{M} = \sum_{X_i=L-1} < \underline{k\ell} \mid G_M \mid i > \sigma_M, \; k\varepsilon C \tag{43}$$

Because of the uniformity of the resistor network,

$$\begin{aligned} E_{k\ell}^{M} &= 0, \; k\ell \neq \underline{k\ell} \\ &= \frac{1}{L}, \; k\ell = \underline{k\ell} \end{aligned} \tag{44}$$

which is intuitively obvious, but can be proven by explicit evaluation of $G_M = A_M^{-1}$.

Eqs. (43) and (44) provide a convenient way of combining multiple scattering and effective medium theory. They yield the identity

$$\sigma_M = \frac{1}{L^2} \sum_{\substack{k\varepsilon C \\ X_i=L-1}} \sigma_M < \underline{k\ell} \mid G \mid i > \sigma_M \tag{45}$$

Subtracting (45) from (39) gives

$$\sigma^* - \sigma_M = \frac{1}{L^2} \sum_{\substack{k\varepsilon C \\ X_i=L-1}} (\sigma_{\underline{k\ell}} - \sigma_M) < \underline{k\ell} \mid G \mid i > \sigma_M \tag{46}$$

We now express A in terms of A_M

$$\begin{aligned} A &= A_M + \delta A \\ \delta A &= \sum_{\langle k\ell \rangle} |k\ell > (\sigma_{k\ell} - \sigma_M) < k\ell| \end{aligned} \tag{47}$$

The sum in (47) is over distinct bonds only. Expanding G in powers of δA provides the basis for a multiple scattering expansion,

$$G = G_M + G_M(-\delta A)G_M + G_M(-\delta A)G_M(-\delta A)G_M + \cdots \tag{48}$$

Inserting (48), (47), (43), and (44) into (46) gives the desired result

$$\sigma-\sigma_M = \langle \sigma_{\underline{k\ell}}-\sigma_M) \rangle + \langle \sum_{\underline{mn}} (\sigma_{\underline{k\ell}}-\sigma_M)\Gamma(\underline{k\ell},\underline{mn})(\sigma_{\underline{mn}}-\sigma_M) \rangle$$

$$+ \langle \sum_{\substack{pq\\ \underline{mn}}} (\sigma_{\underline{k\ell}}-\sigma_M)\Gamma(\underline{k\ell},pq)(\sigma_{pq}-\sigma_M)\Gamma(pq,\underline{mn})(\sigma_{mn}-\sigma_M) \rangle$$

$$+ \text{--------} \tag{49}$$

The sums in (49) are over distinct bonds in C, and

$$\Gamma(k\ell,mn) = - [G^M_{km} - G^M_{\ell m} - G^M_{kn} + G^M_{\ell n}] \tag{50}$$

is the bond to bond propagator. Its elements are simply related to those of the lattice Green's function for the simple cubic lattice, $\sigma_M G^M$.

4. THE HIERARCHY OF EFFECTIVE MEDIUM APPROXIMATIONS

The notation in which Eq. (49) is written can be simplified by using a single index μ for the bond indices $k\ell$, retaining the underline to indicate a bond parallel to the x direction:

$$\begin{aligned}\sigma^*-\sigma_M &= \langle \sigma_{\underline{\mu}}-\sigma_M \rangle + \sum_{\nu} \langle (\sigma_{\underline{\mu}}-\sigma_M)\Gamma_{\underline{\mu\nu}} (\sigma_{\underline{\nu}}-\sigma_M) \rangle \\ &+ \sum_{\underline{\nu},\tau} \langle (\sigma_{\underline{\mu}}-\sigma_M)\ \Gamma_{\underline{\mu}\tau} (\sigma_{\tau}-\sigma_M)\Gamma_{\tau\underline{\nu}} (\sigma_{\underline{\nu}}-\sigma_{\underline{M}}) \rangle + \text{---}\end{aligned} \tag{49'}$$

σ_M will shortly be chosen so as to approximate σ^*. The multiple scattering series is thus an expansion in powers of the fluctuations of the bond conductances around the macroscopic conductivity. These may be large, so that the series (49') must be resummed before approximations can be introduced. It is convenient to sum all terms in (49') in which the index $\underline{\mu}$ appears repeatedly at the left without the intervention of a distinct bond, and then repeat the process with the next index, proceeding from left to right. This process replaces the fluctuation in the bond conductance $\sigma_{\mu}-\sigma_m$ by the corresponding t-matrix,

$$t_{\mu} = \frac{\sigma_{\mu}-\sigma_m}{1-(\sigma_{\mu}-\sigma_M)\Gamma_{\mu\mu}} , \tag{51}$$

and introduces the restriction that successive indices cannot be equal,

$$\sigma^* - \sigma_\mu = \langle t_\mu \rangle + \sum_{\nu \neq \mu} \langle t_\mu \Gamma_{\mu\nu} t_\nu \rangle + \sum_{\substack{\tau \neq \mu \\ \nu \neq \tau}} \langle t_\mu \Gamma_{\mu\tau} t_\tau \Gamma_{\tau\nu} t_\nu \rangle + \cdots \tag{52}$$

The advantage of (51) over (49') is that no matter how large $(\sigma_\mu - \sigma_M)$ becomes, $|t_\mu|$ is never larger than $\sigma_I Z/2$. The next step is to set σ_M, as yet unspecified, equal to the macroscopic conductivity σ^* on the left of (52),

$$\langle t_\mu \rangle + \sum_{\nu \neq \mu} \langle t_\mu \Gamma_{\mu\nu} t_\nu \rangle + \cdots = 0 \tag{53}$$

Thus far the treatment has been exact. Truncation of (53) or the summation of selected terms generates approximations of varying accuracy.

To recapture the EMA in general it is necessary to drop from (53) all terms higher than third order in the t_μ,

$$\langle t_\mu \rangle + \sum_{\nu \neq \mu} \langle t_\mu \rangle \Gamma_{\mu\nu} \langle t_\nu \rangle + \sum_{\substack{\tau \neq \mu \\ \nu \neq \tau}} \langle t_\mu \Gamma_{\mu\nu} \langle t_\tau \rangle \Gamma_{\tau\nu} t_\nu \rangle = 0 \tag{54}$$

Because of the restrictions on the sums in (54), it reduces to

$$\langle t_\mu \rangle = 0 \tag{55}$$

which can straightforwardly be shown to be the EMA by inserting the relation

$$\Gamma_{\mu\mu} = -\frac{2}{Z\sigma_m} \tag{56}$$

into (51). Thus, the EMA works as well as it does in 3d because the first nonvanishing term in (53) resulting from the imposition of (55) is fourth order in the t_μ in 3d and, because of an additional symmetry, fifth-order for the simple-square lattice in 2d. Moreover, one can calculate the values of these first nonvanishing terms; they turn out to be small within the domain within which the EMA works, but become significant near the percolation threshhold.

Eq. (55) is precisely analogous to the defining equation for the Coherent Potential Approximation (CPA) developed for microscopically disordered systems, alloys, etc.[7-9]

The EMA includes all single-bond terms in the multiple scattering expressions. In 3d, the first nonvanishing terms involve 2 bonds. This suggests reorganizing the multiple scattering theory into a hierarchy in which first all single-bond terms are set equal to zero, next all two-bond terms are added to these with the result set equal to zero, then the three-bond terms included, etc. This hierarchy is different from that of the cluster EMA, in which clusters of bonds are embedded in an effective medium. The present scheme is much easier to implement and can give analytical results in simple cases. Table I shows results obtained for the percolation threshhold p_c of the simple cubic lattice by analyzing the case where $\sigma_\mu = 1$ with probability p and o with probability 1-p. The first entry in the table is the result of numerical simulation.[10-12] The next entry is the simple EMA result. This is followed by the results obtained by including all pair terms in 4th order, in 5th order, and in all orders. The next two entries replace the simple average for all single bonds and all pairs of bonds, respectively, by cumulant averaging.[13,14] One sees a systematic improvement to p_c as the sophistication of the approximation increases which leads to an order of magnitude decrease in the error in p_c.

TABLE I

IMPROVEMENTS TO THE EMA RESULT FOR p_c, SIMPLE CUBIC LATTICE

Theory	p_c	Δp_c
Numerical	0.249	-
EMA	0.333	0.084
Pairs, 4th Order	0.316	0.067
Pairs, 5th Order	0.304	0.055
All Pairs	0.296	0.047
Cumulant EMA (Single Bond)	0.283	0.034
Cumulant EMA (All Pairs)	0.257	0.008

5. EMA FOR CONTINUOUS RANDOM SYSTEMS

One sets up the EMA for continuous random systems in a way analogous to that for the random resistor networks. First one imagines a uniform, continuous medium, the effective medium, with a conductivity σ_M. One then embeds within the medium a sampling volume of a shape appropriate to the microstructure of the actual random system. If the latter is granular with the grains nearly spherical in shape, one chooses a spherical sampling volume. If the grains are eccentric, one can choose an ellipsoid and average over shapes and orientations. Let us consider specifically the spherical case. One gives the conductivity within the sampling volume a random value σ' with the same probability $P(\sigma')$ with which it occurs in the actual material. One then solves the electrostatic problem for a sphere (or other shape) of σ' embedded in a continuum of σ_M. One obtains the current density j_x and field E_x within the sampling volume and requires that their averages over σ' satisfy

$$\langle j_x \rangle = \sigma_M \langle E_x \rangle , \tag{57}$$

where

$$\langle Q \rangle = \int P(\sigma') Q(\sigma') d\sigma' , \tag{58}$$

so that the sampling volume is, on average, the same as the effective medium. There results, for the sphere

$$\left\langle \frac{\sigma' - \sigma_M}{\sigma' + 2\sigma_M} \right\rangle = 0 \tag{59}$$

which is precisely what we found for the simple cubic lattice.

One can also do multiple scattering theory for the continuum(15,16) and set up hierarchies of improvements to the EMA. The arguments differ in detail from those used for the random resistor lattices, but the logical structure is identical and the results similar.

6. EFFECTS OF SPATIAL CORRELATION

For $\sigma_{k\ell}$ in a lattice or $\sigma(\vec{r})$ in a continuum a random variable with no short range order, there is always a percolation threshhold when there is a finite probability that $\sigma_{k\ell}$ or $\sigma(\vec{r})$ be zero. This result is unaffected by the introduction of short range order, that is, by a finite correlation length for the values of σ. Such is the case, for example, in a granular material which is a random mxture of conducting and nonconducting grains. However, there are problems in which spatial correlations eliminate the percolation threshold, and the question of how to develop an EMA for such materials arises.

Porous media can provide examples of the above. Consider sedimentary rocks and, in particular, clean sandstones. These are made up of sand grains with an interstitial pore space filled with an aqueous electrolyte of conductivity σ_w. The contacts between grains remain small as the rocks become compacted under the ground so that the pore space remains continuous. No percolation threshhold has been observed. The conductivity σ of the rocks is related empirically to the porosity ϕ by

$$\sigma = a\sigma_w\phi^m , \tag{60}$$

known as Archie's law,[17] where

$$a \cong 1, \; 10^{-4} < \phi < 0.3, \; 1.5 \lesssim m \lesssim 4.5 \tag{61}$$

Sen, Scala and Cohen[18] and Mendelson and Cohen[19] have derived (60) with a = 1 by the use of a self similar version of the EMA. One starts with water and reconstructs the rock by adding the grains in stages. One applies the EMA at each stage, embedding the newly added rock grains and the effective medium of the previous stage into the effective medium of the current stage. The result is Archie's law, Eq. (60), with a = 1 and with m given by

$$m = \frac{1}{3} \sum_{p=1}^{3} \langle (1-L_p)^{-1} \rangle \geqslant 1.5 \tag{62}$$

where L_p is a principal depolarizing factor for the ellipsoidal grains and the average is over grain shapes. The lower limit 1.5 occurs for spheres. The upper limit found empirically probably arises from limitations on the eccentricities which can survive compaction.

7. EMA FOR OTHER PHYSICAL PROPERTIES

Here we briefly review what is known about the treatment of other properties of random media by the EMA. We consider the general problem of linear response,

$$A = BC, \tag{63}$$

where C is a vector or tensor stimulus, A is the corresponding response, and B is the tensor response function. The material can be intrinsically isotropic, made anisotropic by the application of an additional external field as in the magnetoconductivity, or intrinsically anisotropic. The tensor B can be correspondingly isotropic or anisotropic. Its rank normally ranges from second through fourth, and it can be a tensor or a pseudotensor. There can be real or complex responses, the latter occurring when the stimulus is time dependent, i.e., of finite frequency. There can be more than one stimulus and response, and the responses can then be coupled. We shall take up simple examples of these various cases in the following.

First we consider the case of an isotropic, second-rank tensor response function which is complex, the dielectric function[20] $\varepsilon(\omega)$. The effective medium theory for $\varepsilon(\omega)$ is the same as that for σ despite the complex nature of $\varepsilon(\omega)$,

$$\left\langle \frac{\varepsilon' - \varepsilon_M}{\varepsilon' + 2\,\varepsilon_M} \right\rangle = 0 \tag{64}$$

Next we consider the case of the magnetoconductivity and, in particular, the Hall effect.[21] The response function is then a second-rank anisotropic tensor so that

$$\overleftrightarrow{\sigma}' = \sigma' \overleftrightarrow{I} + \overleftrightarrow{\sigma}_a' \tag{65}$$

In (65), σ' is the isotropic part and $\overleftrightarrow{\sigma}_a'$ the antisymmetric part of the conductivity tensor $\overleftrightarrow{\sigma}'$. $\overleftrightarrow{\sigma}_a'$ is linear in H, the magnetic field, and contains the Hall effect. The effective medium conductivity $\overleftrightarrow{\sigma}_M$ has a similar form. The EMA proceeds in precisely the same manner as before, with a spherical sampling volume, except that care must be taken with the anisotropy of $\overleftrightarrow{\sigma}'$ and $\overleftrightarrow{\sigma}_M$. The effective medium condition becomes

$$\left\langle \left(\sigma_M \overleftrightarrow{I} - \frac{1}{3}\overleftrightarrow{\sigma}_M + \frac{1}{3}\overleftrightarrow{\sigma}'\right)^{-1} \bullet \left(\overleftrightarrow{\sigma}' - \overleftrightarrow{\sigma}_M\right) \right\rangle = 0 \, . \tag{66}$$

to first order in H one obtains

$$\overleftrightarrow{\sigma}_{Ma} = \frac{\langle \sigma_{a'} / (\sigma' + 2\sigma_M)^2 \rangle}{\langle (\sigma' + 2\sigma_M)^2 \rangle} \tag{67}$$

with σ_M given by (59). From (67), one readily calculates the Hall constant.

The case of coupled isotropic second-rank tensors arises in connection with the analysis of the thermopower of a continuous random medium.[22] Here the heat and electrical fluxes are coupled, and the resulting EMA is complicated thereby. Both the electrical conductivity and the thermal conductivity are given by (59). If K' is the local thermal conductivity and K_M the effective medium value, then

$$\left\langle \frac{K' - K_M}{K' + 2K_M} \right\rangle = 0 \tag{68}$$

On the other hand, the thermopower (S' the local value and S_M the effective medium value) is given by the more complex relation

$$S_M = \frac{\sigma_M K_M \left\langle \dfrac{S'\sigma'}{(K' + 2K_M)\ (\sigma' + 2\sigma_M)} \right\rangle}{1-3 \left\langle \dfrac{K'\sigma'}{K' + 2K_M)\ (\sigma' + 2\sigma_M)} \right\rangle} \tag{69}$$

Finally, we note that there is no difficulty other than with the somewhat greater complexity in extending the results to higher-order tensors, for example, to the elastic constants which form a fourth-rank tensor. We do not give details here, but note that even in this more complex case, results can become quite simple. The bulk modulus B_M of a binary mixture of two constituents B_1 with volume fraction ϕ and B_2 with $1-\phi$ becomes, when the two shear moduli are equal,[23]

$$\frac{1}{B_M} = \frac{\phi}{B_1} + \frac{1-\phi}{B_2} \,. \tag{70}$$

References

1. D. A. G. Bruggeman, Ann. Phys. Lpz <u>24</u>, 636-79 (1935).

2. R. Landauer, J. Appl. Phys. <u>23</u>, 779-84 (1952).

3. S. Kirkpatrick: (a) Phys. Rev. Lett. <u>27</u>, 1722-5 (1971); (b) Rev. Mod. Phys. <u>45</u>, 574-88 (1973).

4. J. A. Krumhansl in "Amorphous Magnetism," ed. H. O. Hooper and A. M. deGaaf (Plenum, New York, 1973), p. 15.

5. W. E. Kohler and G. C. Papanicolau, "Multiple Scattering and Waves in Random Media," ed. P. L. Chow, W. E. Kohler, G. C. Papanicolau, North Holland, Amsterdam, 1981, p. 199.

6. F. Yonezawa, I. Webman and M. H. Cohen in "Anderson Localization," in Springer Series in Solid State Physics, 1982, ed. Y. Nagaoka and H. Fukuyama, to be submitted to Phys. Rev. B.

7. P. Soven, Phys. Rev. 156, 809 (1967); D. W. Taylor, Phys. Rev. 156, 1017 (1967).

8. R. J. Elliott, J. A. Krumhansl and P. L. Leath, Revs. Mod. Phys. 46, 465 (1974).

9. F. Yonezawa and K. Morigaki: Progr. Theor. Phys. Suppl. 53, 1 (1973).

10. S. R. Broadbent and J. M. Hammersley, Proc. Comb. Phil. Soc. 53, 629-41 (1957).

11. J. W. Essam in "Phase Transitions and Critical Phenomena," Vol. II, eds. C. Domb and M. S. Green (Academic Press, New York), pp. 197-270 (1972).

12. V. K. S. Shante and S. Kirkpatrick, Adv. Phys. 20, 325-571 (1971).

13. F. Yonezawa and T. Matsubara, Prog. Theor. Phys. 35, 357, 759 (1966).

14. F. Yonezawa, Prog. Theor. Phys. 40, 734 (1968).

15. E. Kroner, J. Mech. Phys. Solids 15, 319-29 (1967).

16. M. Hori and F. Yonezawa: J. Math. Phys. 15, 2, 1177-85 (1974); J. Math. Phys. 16, 352-64 and 365-77 (1975); J. Phys. C: Solid State Phys. 10, 229-248 (1977).

17. G. E. Archie: Trans. AIME 46, 54 (1942).

18. P. N. Sen, C. Scala and M. H. Cohen: Geophysics 46, 781-795 (1981).

19. K. S. Mendelson and M. H. Cohen: Geophysics 47, 257-263 (1982).

20. I. Webman, J. Jortner and M. H. Cohen, Phys. Rev. B 12, 5712 (1977).

21. M. H. Cohen and J. Jortner, Phys. Rev. Lett. 30, 696 (1973); I. Webman, J. Jortner and M. H. Cohen, Phys. Rev. B 15, 1936 (1977).

22. I. Webman, J. Jortner and M. H. Cohen, Phys. Rev. B 16. 2959 (1977).

23. P. Sheng and A. Callegari, unpublished.

ELECTRONIC AND TRANSPORT PROPERTIES OF GRANULAR MATERIALS II: SPECIFIC INHOMOGENEOUS MATERIALS

Morrel H. Cohen

Exxon Research and Engineering Company
Corporate Research Science Laboratories
P. O. Box 45, Linden, New Jersey 07036

Abstract

Materials can be inhomogeneous on macroscopic or microscopic scales or both. We describe briefly the ways in which materials can be macroscopically inhomogeneous. We then describe fluctuations in microscopically inhomogeneous materials in sufficient detail to clarify the criteria for dealing with them as though they were macroscopic. The concept of local electronic properties in microscopically inhomogeneous materials emerges naturally, and the methods reviewed in the first paper of this series can then be brought to bear on the discussion of their transport and electronic properties. We use these ideas to develop the theory of metal-nonmetal transitions in microscopically inhomogeneous materials. We review three examples of such transitions in expanded liquid Hg; in liquid Te, Se and Se Te alloys; and in metal-ammonia solutions. Finally, we return to macroscopically inhomogeneous materials, discussing porous media as an interesting and representative example.

1. INTRODUCTION

In the first paper of this series of three (I), we addressed the problems of accurately calculating the transport coefficients of macroscopically inhomogeneous materials treated as continua or as discrete networks approximating or representing the continua. In the present paper, we review the nature of inhomogeneities both on the macroscopic and the microscopic level. We show that under certain conditions the transport and electronic properties become local on the scale of the microscopic inhomogeneities. Macroscopic and microscopically inhomogeneous materials can then be treated in a unified way. We review from that point of view metal-nonmetal transitions in microscopically inhomogeneous materials in general, and in particular in expanded liquid mercury, in the liquid chalcogenides Se, Te and Se Te alloys, and in metal-ammonia solutions. Returning to macroscopically inhomogeneous materials, we discuss porous media both as typical examples and with emphasis on their special features.

2. MACROSCOPICALLY INHOMOGENEOUS MATERIALS

Macroscopically inhomogeneous materials are locally homogeneous with regard to their microscopic constitution and consequently their macroscopic properties over macroscopic distance scales. The boundary between the macroscopic and microscopic distance scales is ill defined and depends on the property under consideration. If one arbitrarily chooses 100Å as the boundary, one encounters situations in which normally microscopic lengths, e.g. coherence lengths or mean free paths, become longer. One then has to deal with constitutive relations which are nonlocal on a macroscopic scale. That, while an acceptable complication, is one not dealt with in the present paper, where we confine our attention to local relations. In the present paper, we take the point of view that fluctuations on the scale of tens of Å or less are microscopic, and grains of size 1000 Å or more are macroscopic.

Macroscopic inhomogeneities can occur in atomic structure. Grains of different orientation can be separated by grain or twin boundaries. Grains can be displaced relative to one another by a stacking fault. The composition can vary so that one has a mixture with two or more constituents dispersed on a macroscopic scale. In an ordered material, the order parameter can change its orientation or magnitude, so that the material is broken up into macroscopic domains.

In general, one is interested in understanding the effect of such inhomogeneities on material properties such as the linear response A to an external stimulus C as expressed by the linear-response function B,

$$A = BC \tag{1}$$

The quantities A, B, C are functions of position $\vec{r}$ within the material. We are interested in the macroscopic averages of $A(\vec{r})$ and $C(\vec{r})$ and in particular in the macroscopic response function B^* relating them,

$$\langle A(\vec{r})\rangle = B^* \langle C(\vec{r})\rangle \tag{2}$$

We know something of $B(\vec{r})$ from our knowledge of the structure, composition and order of the material. The basic problem, then, is to find B^* from $B(\vec{r})$. This problem is discussed in the first paper, and especially in §§ 5, 6 and 7. It is not generally realized that large classes of microscopic inhomogeneities can be treated as effectively macroscopic so that those methods become available for the microscopic problems as well. We review the nature of microscopic inhomogeneities in the next section and then, in the following section, give the arguments which reduce the microscopic to the macroscopic case.

3. MICROSCOPICALLY INHOMOGENEOUS MATERIALS[1]

In a microscopically inhomogeneous material, properties such as the density, the valence (i.e., the number of valence electrons), the structure (e.g., the coordination), or the composition can fluctuate on a microscopic distance scale. Some of these properties themselves require a finite distance scale for their definition, and we suppose that the fluctuations occur on a longer though still microscopic scale.

Let us represent a fluctuating property by X. X is position dependent in a quenched solid and also time dependent in a liquid. We shall suppose that the fluctuation time scale is long relative to electronic time scales so that $X = X(\vec{r})$ in all cases.

A snapshot of $X(\vec{r})$ in thermodynamic equilibrium, Fig.1, would display two characteristic lengths. The value of $X(\vec{r})$ remains roughly constant over distances of about 2b, where b is the Debye short correlation length. b would be the mean cluster radius in clustering problems. $X(\vec{r})$ changes to a different value over a distance of 2ξ, where ξ is the Ornstein-Zernike fluctuation decay length. 2ξ would be the thickness of the cluster surface in clustering problems. These two lengths are clearly displayed in the correlation function, Fig. 2,

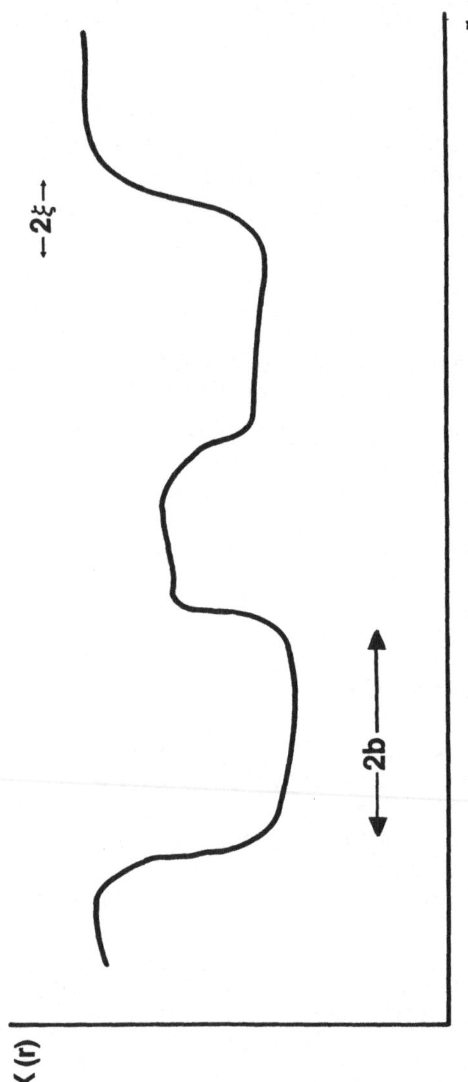

Fig. 1 The position dependence of a micro-scopically fluctuating quantity, displaying the two characteristic lengths b and ξ.

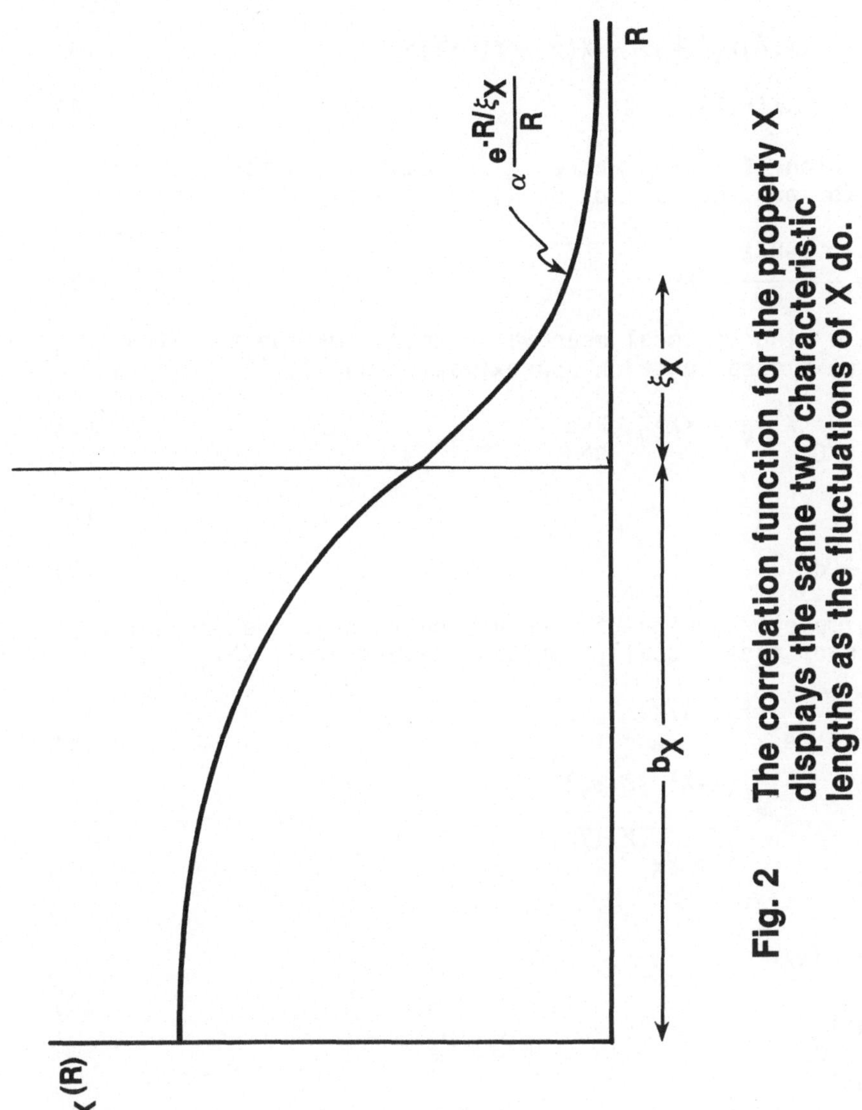

Fig. 2 The correlation function for the property X displays the same two characteristic lengths as the fluctuations of X do.

$$C_X(\vec{R}) = \langle X(\vec{r})X(\vec{r}+\vec{R})\rangle - \langle X(\vec{r})\rangle\langle X(\vec{r}+\vec{R})\rangle \tag{3}$$

$$\langle X(\vec{r})\rangle = \langle X(\vec{r}+\vec{R})\rangle = \bar{X} \tag{4}$$

The inner region of $C_X(\vec{R})$, where it is roughly constant, corresponds to $R<b$, and the tail of $C_X(\vec{R})$ has the asymptotic form

$$C_X(\vec{R}) \sim \frac{e^{-R/\xi}}{R}, \quad R\to\infty \tag{5}$$

We now make a kind of local mean value approximation to $X(\vec{r})$ equivalent to a step function approximation to $C_X(\vec{R})$, Fig. 3,

$$\begin{aligned} C_X(\vec{R}) &= \langle X^2\rangle_V - \langle X\rangle_V^2, \quad R<d \\ &= 0 \qquad\qquad\qquad\quad , \quad R>d \end{aligned} \tag{6}$$

$$d_X = b_X + \xi_X \tag{7}$$

$$V = \frac{4\pi}{3} d^3 \tag{8}$$

Using the simplest version of fluctuation theory, the probability distribution of these locally constant values of X, $\langle X\rangle_V$, is

$$\begin{aligned} P(X) &= e^{-F_V(X)/kT} \\ &= \frac{e^{-(X-\bar{X})^2/2(\sigma_X)^2}}{(2\pi(\sigma_X)^2)^{1/2}}, \end{aligned} \tag{9}$$

where

$$\begin{aligned} \sigma_X^2 &= kT\chi/V \\ \chi &= \frac{\partial^2 F_V}{\partial Y^2} \end{aligned} \tag{10}$$

F_V being the fluctuation free energy for the volume V and Y the thermodynamic variable conjugate to X.

Near a critical point, the behavior of the fluctuations becomes particularly interesting because both ξ and χ grow rapidly as the critical point is approached,

$$\begin{aligned} \xi &\propto \varepsilon^{-\nu} \\ \chi &\propto \varepsilon^{-\gamma} \\ \varepsilon &\equiv |(T-T_c)/T_c| \end{aligned} \tag{11}$$

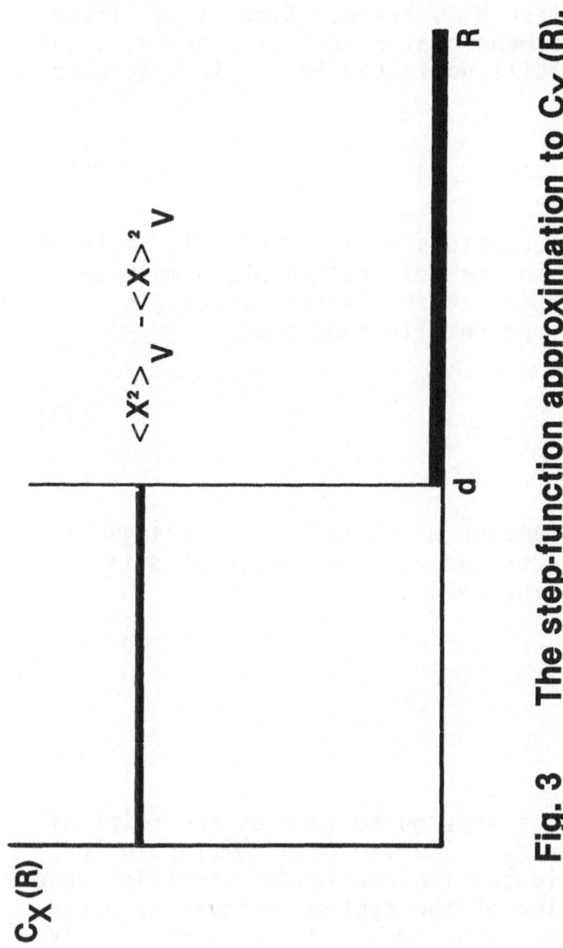

Fig. 3 The step-function approximation to C_X (R).

There are two cases. In the first b is large. Since γ is larger than ν, e.g. 1 vs. ½ in three dimensional mean-field theory, χ can become quite large while d is still dominated by b. In that case, we have

$$\sigma_X^2 = kT\ \chi/\tfrac{4\pi}{3}\ b_X^3 \qquad (12)$$
$$\propto \varepsilon^{-\gamma}$$

and, correspondingly, large fluctuations in X. When b is small or as one approaches still closer to the critical point, d must be dominated by ξ_X because b, relating to short-range order, is slowly varying in the critical region. In that case,

$$\sigma_X^2 = kT\ \chi/\tfrac{4\pi}{3}\ \xi_X^3 \qquad (13)$$
$$\alpha\ \varepsilon^{3\nu-\gamma}$$

Below T_c, there is an order parameter which relates to changes in the mean value of X and which grows as ε_β. The exponent β is related to ν and γ via the scaling relation

$$\gamma + 2\beta = 3\nu \qquad (14)$$

thus, we have

$$\sigma_X^2 \propto \varepsilon^{2\beta} \to 0 \text{ at } T_c, \qquad (15)$$

The mean fluctuation amplitudes always go to zero at the critical point because of averaging over ξ_X. The two cases differ in that when b_X is large enough, the fluctuation amplitudes first increase greatly, because of the softening of the system, before the divergence of ξ_X takes over and reduces the amplitudes to zero. It is important to realize that the fluctuation amplitudes need not have a large effect on physical properties near a critical point.

4. LOCAL ELECTRONIC PROPERTIES[1]

We now introduce the notion of local electronic structure and local response functions. We treat the electronic motion in the independent quasiparticle approximation in which the quasiparticle wave-function ψ is determined from an effective single-particle Hamiltonian H,

$$H\psi = E\psi \qquad (16)$$

$$H = \frac{p^2}{2m} + V \qquad (17)$$

The potential V is derived from the self-energy of the one-electron Green's function including both electron-electron interaction and electron-atom interaction with atomic recoil, but the imaginary part of the self-energy (lifetime effects) is ignored as is its energy dependence (nonlocality in time). This is the simplest realistic interpretation of the independent-electron picture normally used to simplify such discussions as we give here.

The wave function ψ can be written as

$$\psi = Ae^{i\phi} \tag{18}$$

Both the amplitude A and the phase ϕ are random functions of position because of the fluctuations of X, and ϕ is additionally randomized by structural and compositional variation leaving X unchanged. Each will have a correlation function $C_A(R)$, $C_\phi(R)$ with a range d_A, d_ϕ in the step function approximation. Detailed examination of the consequences for ψ of the variation of V with X shows that

$$d_A \cong d_X \tag{19}$$

in the cases of interest to us. On the other hand,

$$C_\phi(R) \sim \frac{f(R)e^{-R/\xi_\phi}}{R} \tag{20}$$

where f(R) can be smooth or oscillatory and $C_\phi(R)$ reaches its asymptotic form for

$$R > b_\phi \cong \xi_\phi \tag{21}$$

Here ξ_ϕ, the phase coherence length, is approximately equal to the mean free path when both are greater than atomic separations. However, while A is sensitive primarily to variations in X, ϕ is sensitive to all irregularities in the structure. Thus, in a system in which fluctuations in X are unimportant, ξ_ϕ can still be smaller than interatomic separations in a strong scattering regime.

Let us define the Green's function corresponding to (16),

$$G(\vec{r},\vec{r}\,';E^-) = \lim_{\delta \to 0^+} \Sigma \frac{\psi_j(\vec{r})\ \psi_j^*(\vec{r}\,')}{E - E_j - i\delta} \tag{22}$$

where

$$H\ \psi_j = E_j\ \psi_j \tag{23}$$

Because of the special properties of quasiparticles, the total density of states for single-particle excitations is given by

$$N(E) = \frac{1}{\pi} \int d^3r \ \mathrm{Im}\ G(r,r;\ E^-) \tag{24}$$

at the temperatures of interest to us. We can use (24.) as a basis for a formal definition of a local density of states,

$$N(E) = \int d^3r \ n(\vec{r},E) \tag{25}$$

$$n(\vec{r},E) = \frac{1}{\pi} \mathrm{Im}\ G(\vec{r},\vec{r};E^-) \tag{26}$$

$n(\vec{r},E)$ depends only on the amplitudes $|A_j|^2$ and the energies E_j, and therefore fluctuates with correlation length d_χ. On the other hand $G(\vec{r},\vec{r}';\ E^-)$ depends on $(\vec{r}-\vec{r}')$ through the phases as well. It therefore can fluctuate more rapidly, decaying towards zero over the phase coherence length ξ_ϕ. If we restrict ourselves to cases in which

$$\xi_\phi \ll d_\chi, \tag{27}$$

the $n(\vec{r},E)$ accurately portrays the local electronic structure since phase coherence decays over distances small compared to those on which the local electronic structure changes.

The kind of information about local electronic structure contained in $n(\vec{r},E)$ is shown in Fig. 4. It is possible that $n(\vec{r},E)$ has a low minimum at some E_0. When the Fermi energy E_F is well away from E_0, the system is metallic. When it is near E_0, the system is nonmetallic.

In addition to establishing the existence of local electronic structure when (27) holds, it is possible to show that the linear response to stimuli is local as well. The local current density $\vec{j}(\vec{r})$ has, in general, a nonlocal relation to the local electrostatic field $\vec{E}(\vec{r})$ when an electronic current is caused to flow in an inhomogeneous material,

$$\vec{j}(\vec{r}) = \int \overleftrightarrow{\sigma}(\vec{r},\vec{r}') \cdot \vec{E}(\vec{r}')\ d^3r' \tag{28}$$

Introduce

$$\overleftrightarrow{\sigma}'\left(\frac{\vec{r}+\vec{r}'}{2},\vec{r}-\vec{r}'\right) \equiv \overleftrightarrow{\sigma}(\vec{r},\vec{r}') \tag{29}$$

$$\vec{\rho} \equiv \vec{r}-\vec{r}' \tag{30}$$

Inserting (30) and (29) into (28) permits rewriting it in the form

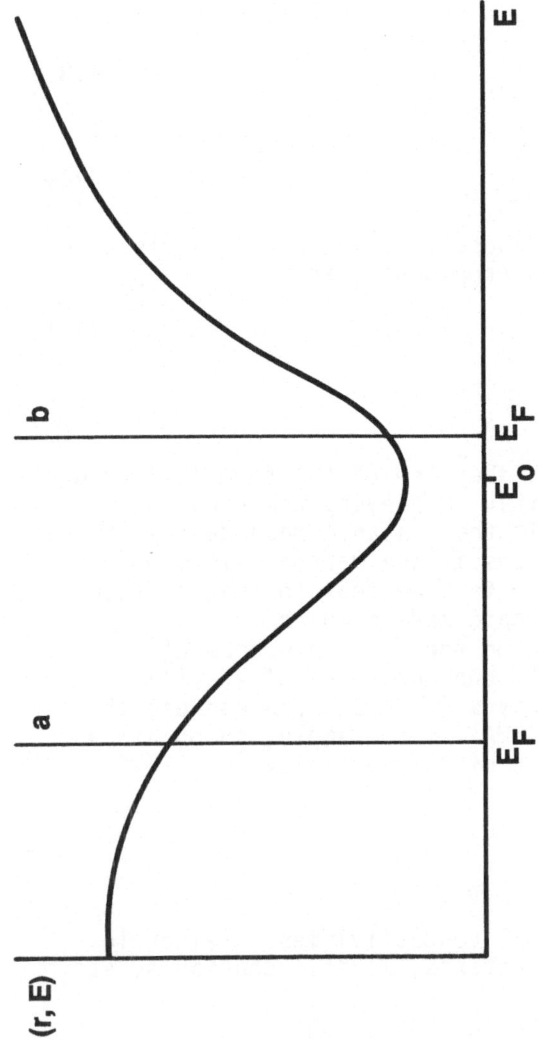

Fig. 4 Local density of states as a function of energy. a.) Metallic case: the density of states is high at E_F, in the metallic range. b.) Nonmetallic case: the density of states is low at E_F, in the nonmetallic range.

$$\vec{j}(\vec{r}) = \int \overleftrightarrow{\sigma}' \left(\vec{r}-\frac{\vec{\rho}}{2}, \vec{\rho}\right) \cdot \vec{E}\,(\vec{r}+\vec{\rho})d^3\rho \tag{31}$$

The Kubo relations for $\overleftrightarrow{\sigma}\,(\vec{r},\vec{r}\,')$ show that the range of σ' in ρ is ξ_ϕ while the variation of σ' with $\frac{1}{2}(\vec{r}+\vec{r}\,')$ is on the scale of d_X. The formal solution of $\vec{\nabla}\cdot\vec{j}(\vec{r})=0$ then shows that both $\vec{j}(\vec{r})$ and $\vec{E}(\vec{r})$ vary on the scale of d_X. Thus, eq. (28) simplifies to

$$\vec{j} = \overleftrightarrow{\sigma}(\vec{r})\cdot\vec{E}(\vec{r}), \tag{28'}$$

where

$$\overleftrightarrow{\sigma}(\vec{r}) = \int \overleftrightarrow{\sigma}(\vec{r},\vec{r}\,')\, d^3r' \tag{32}$$

The cases of interest to us are those in which over the distance d_X the material is on average isotropic so that

$$\overleftrightarrow{\sigma}(\vec{r}) = \sigma(\vec{r})\,\overleftrightarrow{I} \tag{32'}$$

and

$$\vec{j}(\vec{r}) = \sigma(\vec{r})\,\vec{E}(\vec{r}) \tag{28''}$$

Eqs. (28') and (28") are local. They are of the form (1.) occurring for macroscopic inhomogeneity. Moreover, the step function approximation for $C_X(R)$ parallels the distance dependences of the corresponding correlation functions in the macroscopic problems. Thus the problem of determining σ is identical to that of determining σ in a continuous macroscopic medium which is randomly heterogeneous. From information on the $\vec{r}$ dependence of X, we can obtain information on the $\vec{r}$ dependence of $\sigma' = \sigma(\vec{r})$ through $\sigma' = \sigma'(X)$. Then P(X) gives $P(\sigma')$ and one can use the effective medium approximation (EMA), for example, to obtain an approximation, σ_M, to the macroscopic conductivity:

$$\int P(\sigma')\,\frac{\sigma'-\sigma_M}{\sigma'+2\sigma_M}\,d\sigma' = 0 \tag{33}$$

We can calculate similarly thermal conductivities, dielectric constants, thermopowers, Hall constants, elastic constants, etc.

5. THE INHOMOGENEOUS METAL-NONMETAL TRANSITION(1)

One of the most dramatic effects of macroscopic inhomogeneity is the percolation transition in the conductivity which occurs when the volume fraction of conductor in a binary random mixture of conductor and insulator increases above the percolation threshold. Analogous phenomena can occur on the microscopic scale.

Suppose that we have, as described in the last section, a fluctuating variable X which causes fluctuations in the local conductivity σ. Suppose further that there is a critical value of the conductivity σ^* at which, for example, the temperature dependence of σ changes from metallic to nonmetallic in character or some other characteristically metallic behavior ceases. σ^* should be close to Mott's minimum metallic conductivity. When $\sigma(r) > \sigma^*$, the material is locally metallic. When $\sigma<\sigma^*$, the material is locally nonmetallic. We thus have a heterogeneous mixture of metallic and nonmetallic regions. The metallic volume fraction is

$$C = \int_{\sigma^*}^{\infty} P(\sigma)d\sigma \tag{34}$$

Two examples of how σ^* might relate to $P(\sigma)$ are shown in Fig. 5, a unimodal case and a bimodal case. The unimodal case might occur above a critical point. The bimodal case might occur below a critical point in association with a first-order phase transition in which the surface tension was still too low because of proximity to the critical point to force macroscopic phase separation.

We now give examples of specific systems in which inhomogeneous metal-nonmetal transitions occur on the microscopic scale as C varies from small values towards unity.

A. EXPANDED LIQUID Hg[2]

The conductivity σ of liquid Hg falls smoothly by a factor of 30 as the density decreases from 9.3g/cc to 8.0 g/cc. In addition, there are anomalies in the Hall constant, Hall mobility and the Knight shift. The behavior of all quantities is consistent with a continuous metal-nonmetal transition from metallic to semiconducting behavior via an inhomogeneous regime. In this picture, the microscopic inhomogeneities in the electronic properties are associated with fluctuations in the atomic number density ρ. Detailed fits to the data can be obtained with a Gaussian distribution of ρ and a value of d_ρ of about 15Å. The more recent data, when combined with Warren's Knight Shift data, make the transition substantially sharper.

The success of the proposed inhomogeneous metal-nonmetal transition in providing a fit to the data raises deep questions. What is the origin of large density fluctuations so far from the critical point? Why should these density fluctuations occur on a 15 Å scale?

B. LIQUID Te, Se AND Te-Se ALLOYS[3]

Liquid Te, Se and Te-Se alloys all show the same pattern, which I shall describe only for Te. There is a smooth five-fold

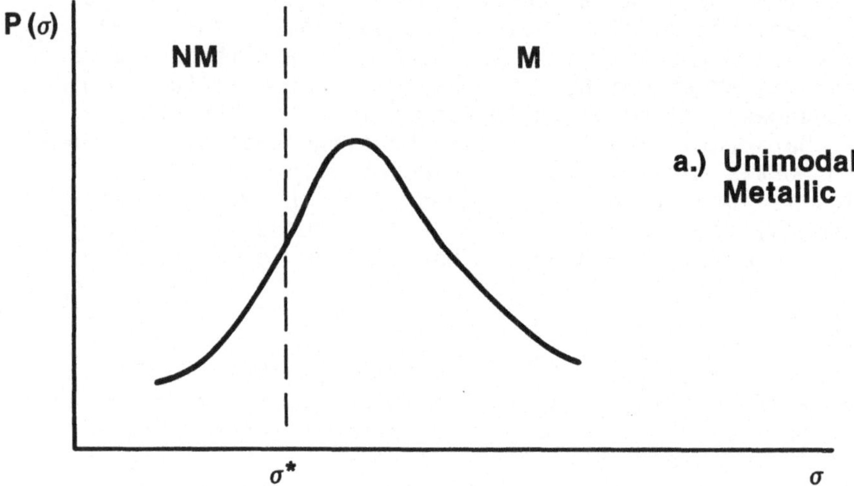

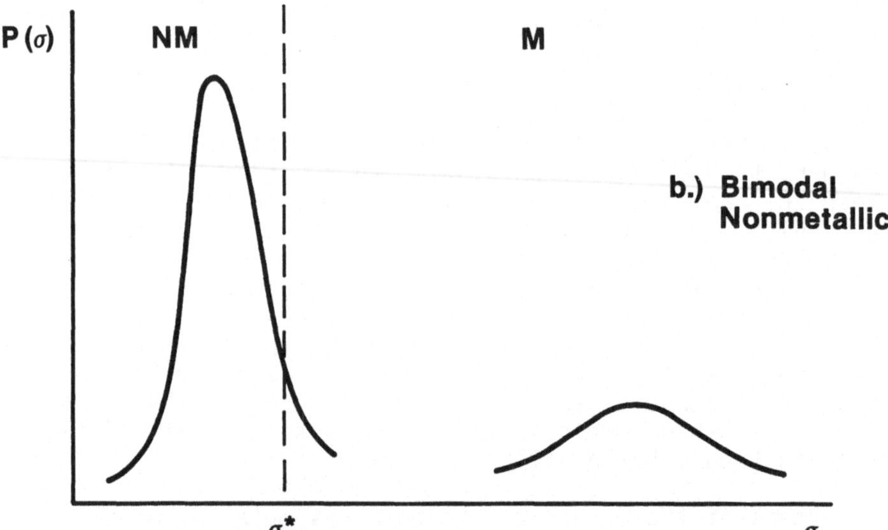

Fig. 5 Examples of possible relationships between P (σ) and σ^*.

drop in the conductivity around 1000 K. The conductivity, the Hall constant, the Hall mobility, the Knight shift, the nuclear relaxation rate, X-ray and neutron scattering are all quantitatively consistent with an inhomogeneous metal-nonmetal transition. In this case, the inhomogeneities are associated with bonding fluctuations. The Te are two-fold coordinated and semiconducting at lower temperatures and three-fold coordinated and metallic at higher temperatures. Detailed analysis of the data yields a correlation length d in excess of two atomic separations. Once again, a deep question is raised. How and why does the bonding change come about?

C. METAL-AMMONIA SOLUTIONS[4]

In Li and Na solutions in liquid NH_3 at about 10K above the consolute temperature, the conductivity drops about three orders of magnitude as the concentration decreases from about 9 to 3 mole percent metal. The conductivity, Hall effect, Hall mobility, thermal conductivity, optical dielectric constant, magnetic susceptibility, Knight shift, and velocity of sound are all quantitatively consistent with an inhomogeneous metal-nonmetal transition. In this case the microscopic inhomogeneities are concentration fluctuations, with a d of about 30 Å for the Na solutions. Yet again a deep question arises. Are critical fluctuations near the consolute point sufficient? A positive answer has been proposed to that question, but there is some indication that a bimodal distribution of concentration fluctuations is required to understand the concentration dependence of the thermopower.

The questions raised about expanded liquid Hg, the liquid chalcogenides, and the metal ammonia solutions all relate to the physical origins of fluctuations, of clustering. We address these questions in the third paper of this series and return now to macroscopically heterogeneous materials.

6. POROUS MEDIA

Sedimentary rocks, i.e., sandstones, limestones, shales, clays, etc. provide interesting and typical examples of porous media. To be explicit, we consider a clean sandstone which consists of grains of sand, fused at small contacts with a continuous interstitial pore space. Let G be the space occupied by the grains and P be the pore space. P and G are complementary spaces. Let ϕ be the porosity, that is, the fraction of the total volume of P+G occupied by P. G is obviously connected because of the intergranular contacts. However, the processes by which the rocks are formed and transformed are such that the contact areas remain small on the scale of the grain size even for very low porosities ϕ. As a consequence, P remains connected down to values of ϕ as low as 10^{-4} to 10^{-5} (vid. ref. 5).

When the rocks are in the earth, the pore space can be occupied with water, gas, oil or any mixture of these. The water is in fact an electrolyte because of dissolved minerals. In attempting to characterize the rocks both in the laboratory and in situ in the earth, a very broad range of measurements is carried out. The properties of interest include elastic properties (e.g., longitudinal and transverse sound velocities),[6] electrical properties (e.g. d.c. conductivity and complex dielectric constant over a range of frequencies),[5,7] nuclear magnetic resonance (e.g. proton relaxation spectra),[8] and hydrodynamic properties (e.g. permeability).[9] One can measure the properties and attempt to infer from them the random geometry and topology of the rocks. To implement that program, one needs to understand in general terms the geometry and topology and how it relates to the properties of the rocks and then to measure the geometry and topology of samples, compute the properties, and compare with measured values to establish the existence of an adequate understanding. This process is proceeding, but it is still in its early stages. Accordingly, we present here some of the conceptual framework one needs to understand the geometry and topology with a few examples of how it helps to understand properties.[10,11]

Skeletization is one of the most useful of the topological concepts. It is a process by which P or G can be mapped onto a deformation retract of itself. A deformation retract is a lower dimensional entity, e.g., a graph or a surface, which has the same topology as the original space. Consider first the grain space. G. Lin and I have given a precise way to identify grains, grain centers, contacts between grains and contact centers.[10,11] One can construct a deformation retract of G by taking as vertices the grain centers and drawing edges from vertex to vertex through the contact centers. The deformation retract is thus a three dimensional network in which the vertices are in one to one correspondence with the grains and the edges with the contacts. This construct can be used to make an exact mapping of the mechanical properties of the dry rock into the dynamics of a molecular glass having as its structure a continuous random network equivalent to the above deformation retract of G. This mapping is useful in turn for pushing our understanding of the mechanical properties of such porous media from wavelengths large compared to the grain size, where it is beginning to become adequate, to wavelengths comparable to and smaller than the grain size, where it is nonexistent.

With regard to the pore space, Lin and I have shown how to divide both P and G into cells each containing one grain.[10,11]

The deformation retract of P is the network made up of the vertices and edges of these cells and is the dual of the deformation retract (DR) of G, the faces of the DR of G corresponding to the edges of that of P and the cells of the DR of P corresponding to the vertices of the DR of G. In the DR of P, the vertices correspond to the pores, and the edges correspond to channels between pores. This construction can be used to map the problem of proton relaxation in the water in P and at the P-G interface onto a set of coupled relaxation equations defined on a network equivalent to the DR of P. This leads immediately to contact with localization theory, the study of the motion of electrons in random lattices, a much more highly developed subject, and to a clear understanding of how the proton relaxation spectrum relates to the geometry of P. One can also make an approximate mapping of the electrical conductivity problem into a resistance network on the DR of P and an approximate mapping of the problem of flow through P into a hydrodynamic resistance network on the DR of P.

Another important problem is why the self-similar EMA described in I works so well. The justification given in the Sen, Scala and Cohen[5] paper was that each grain is well clothed by the effective medium at each stage of reconstruction of the rock. The way the argument was carried out required self-similarity in both the grain space and the pore space. However, their theoretical result from the self-similar EMA for a sphere pack,

$$\sigma = \sigma_w \phi^m, \; m = 3/2 \tag{35}$$

where σ_w is the water conductivity, fitted their own experimental data for a monosized sphere pack very well. A dense random packing of spheres of unique radius in no way gives a self-similar G. On the other hand, Thompson and Katz[12] have shown the P-G interface of a sandstone containing clay to be a fractal and therefore self-similar over four orders of magnitude of feature size. They found

$$P(\ell) \quad 1/\ell^s, \; s \cong 2/3 \tag{36}$$

for the probability distribution $P(\ell)$ of feature sizes ℓ. They argued from this that P was a fractal. However, they did not find fractal behavior in a monosized sphere pack. One concludes that the fractal behavior observed was associated with the arrangement of small clay particles in the sandstone, and that Archie's law can be observed when P is not a fractal.

References

1. M. H. Cohen, J. Jortner, and I. Webman, in Electrical Transport and Optical Properties of Inhomogeneous Media," eds. J. C. Garland and D. B. Tanner, American Institute of Physics, New York (1978).

2. M. H. Cohen and J. Jortner, Phys. Rev. A 10, 978 (1974); Phys. Rev. B 15, 1227 (1977).

3. M. H. Cohen and J. Jortner, Phys. Rev. B 13, 5255 (1976).

4. M. H. Cohen and J. Jortner, J. Phys. Chem. 79, 2900 (1975); J. Jortner and M. H. Cohen, Phys. Rev. B 13, 1548 (1976); M. H. Cohen, I. Webman, and J. Jortner, J. Chem. Phys. 64, 2013 (1976). Cf. also I. Webman, J. Jortner, and M. H. Cohen, Phys. Rev. B 16, 2959 (1977).

5. P. N. Sen, C. Scala, and M. H. Cohen, Geophys. 46,781 (1981).

6. P. Sheng and A. Callegari, to be published.

7. K. S. Mendelson and M. H. Cohen, Geophys. 47, 257 (1982).

8. M. H. Cohen and K. S. Mendelson, J. Appl. Phys. 53, 1127 (1982).

9. J. Koplik and T. J. Lasseter, to be published.

10. M. H. Cohen and C. Lin, Proceedings, Conference on the Macroscopic Properties of Disordered Media, eds. R. Burridge, S. Childress, and G. Papanicolau, p. 74 (1981).

11. C. Lin and M. H. Cohen, J. Appl. Phys. 53, 6, 4152 (1982).

12. A. Thompson and A. Katz, to be published.

ELECTRONIC AND TRANSPORT PROPERTIES OF GRANULAR MATERIALS III. THE ORIGINS OF CLUSTERING

Morrel H. Cohen

Exxon Research and Engineering Company
Corporate Research Science Laboratories
P. O. Box 45, Linden, New Jersey 07036

Abstract

In the present paper, the third and last of this series, we deal with the origins of granularity, that is, of clustering. The physical or chemical origins of macroscopic inhomogeneities are usually obvious so we confine our attention to elucidating the driving forces behind clustering on a microscopic scale. We consider in turn the origins of the inhomogeneities responsible for the continuous metal-nonmetal transitions discussed in the second paper of this series (II): density fluctuations in expanded liquid Hg, concentration fluctuations in metal-ammonia solutions, and valence fluctuations in Te, Se and TeSe alloys. In each of these cases, the clustering is electronically driven with energy considerations dominating over entropy considerations. In our final example of clustering, density fluctuations in supercooled liquids and glasses, however, the driving force is entropic.

1. INTRODUCTION

The overall theme of this volume is the impact of cluster physics on materials science and technology, and the focus of this particular section is on granular materials. Our discussion with the exception of brief treatments of the properties (in I), topology (in II), and geometry (in II) of porous rocks, has not appeared to concern itself explicitly with clusters or with granular materials. In I we dealt with the properties of inhomogeneous materials in general, and in II we primarily discussed microscopically inhomogeneous materials. However, most commonly, both clusters and grains are homogeneous over their extent, and we have focused our attention on microscopic inhomogeneities for which the same is true and within which there are a well defined local electronic structure and local response functions. In fact, the microscopic homogeneities which we have discussed and of which we have given examples are indeed clusters and may be regarded as small grains. Thus, the theme of the present paper, the physical origins of microscopic inhomogeneities is of relevance to the physical origins of clustering in general and of granularity on a microscopic scale. The particular cases we shall deal with, density fluctuations in expanded liquid Hg, concentration fluctuations in metal-ammonia solutions, and valence fluctuations in Te, Se and TeSe alloys are those already introduced in II as examples of inhomogeneous metal-nonmetal transitions. They provide us with examples of electronically driven clustering. For balance and contrast, we discuss also density fluctuations in supercooled liquids and glasses, which are entropically driven, the entropy arising from atomic motions and the electrons playing no immediate role.

All of the above examples are drawn from thermal equilibrium situations with the partial exception of the last. Clustering and the formation of granular materials frequently occur far from thermodynamic equilibrium, so the examples chosen here do not have universal relevance. Nevertheless, they are illuminating, and understanding of equilibrium situations must normally precede understanding of nonequilibrium phenomena.

2. DENSITY FLUCTUATIONS IN EXPANDED LIQUID Hg(1)

We reported in II that the anomalies observed in the electronic properties of expanded liquid Hg could be understood quantitatively in terms of a continuous metal-nonmetal transition associated with density fluctuations. The dependence on mean density of the metallic volume fraction C introduced in II was taken from the fit of the EMA to the density dependence of the conductivity, assuming that at a local density of 8.0g/cc the material would be in transition from locally metallic to locally nonmetallic behavior. This density dependence of C(E) was then interpreted in terms of the probability distribution of local densities P(ρ) using Eq. (31) of II and the Gaussian from Eq. (9) of II for P(ρ). Inserting the observed compressibility into Eq. (10) of II then led to a value of 15 Å for d_ρ throughout the entire density range of the transition. In that density range ξ_ρ is much smaller than 15 Å, so that d_ρ has to be dominated by b_ρ. Thus, Cohen and Jortner in positing the existence of an inhomogeneous metal-nonmetal transition in Hg in fact, apart from the approximations made in carrying out their calculations, introduced only one new element, a short-range correlation length $b_\rho \cong d_\rho$ of ~ 15 Å. No explanation of its origin was proffered.

Several recent experimental developments have illuminated the situation, and a basis for a more soundly based model is beginning to emerge. First, Schmutzler[2] has found a bend in the coexistence curve at 9.0g/cc, just where the proposed transition begins. Next, Endo has found a linear dependence of pressure on temperature in the isochores of the liquid. The temperature coefficient of the pressure also changes rapidly at 9.0g/cc. Even more striking Hubbard and Ross[3] have carried out a very precise study of the vapor pressure curve and find a sharp kink at a density of 9.0g/cc. The vapor pressure data was then used by Schumtzler to improve the dependence of density on pressure and temperature. Warren reinterpreted his Knight shift data on the basis of the new density data and found that the transition in the Knight shift from a metallic shift to a nonmetallic chemical shift occurred over a smaller range of densities. Thus, as the data have been improving, there seems to be increasing evidence of a sharp change in the thermodynamic as well as the electronic properties across a narrow strip in the liquid sector of the phase plane.

Still more dramatic are recent discoveries of Hefner and Hensel[4] following from their measurement of the density and frequency dependence of the dielectric constant of Hg vapor. Their results for ε^1 versus density are shown schematically in Fig. 1 for a temperature of 1800K and a photon energy of 0.6eV. As the density increases, ε^1 increases above the value expected from the Claussius-Mosotti relation at first slowly and then abruptly with a subsequent slower falloff. As the frequency is decreased the peak sharpens. Hefner and Hensel have derived the frequency dependence of the optical conductivity, shown schematically in Fig 2 from a Kramers-Kronig analysis of their ε^1 data. At the density of the peak, the d.c. conductivity remains small: $\sigma_{dc} = 10^{-3}\Omega^{-1}\text{cm}^{-1}$ at 1800K. There is an absorption

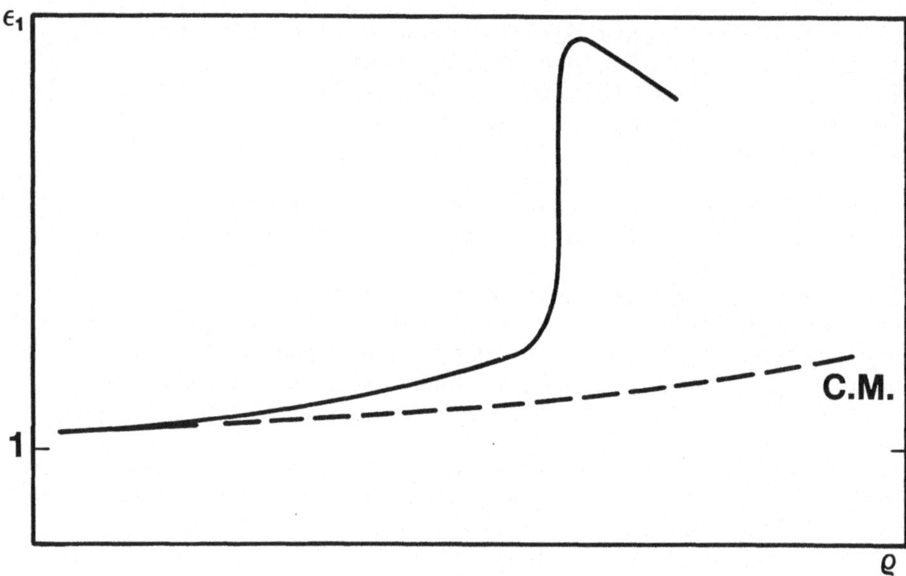

Figure 1. The density-dependence of the real part of the dielectric constant of Hg vapor shown schematically for T = 1800 K and photon energy 0.6 eV.

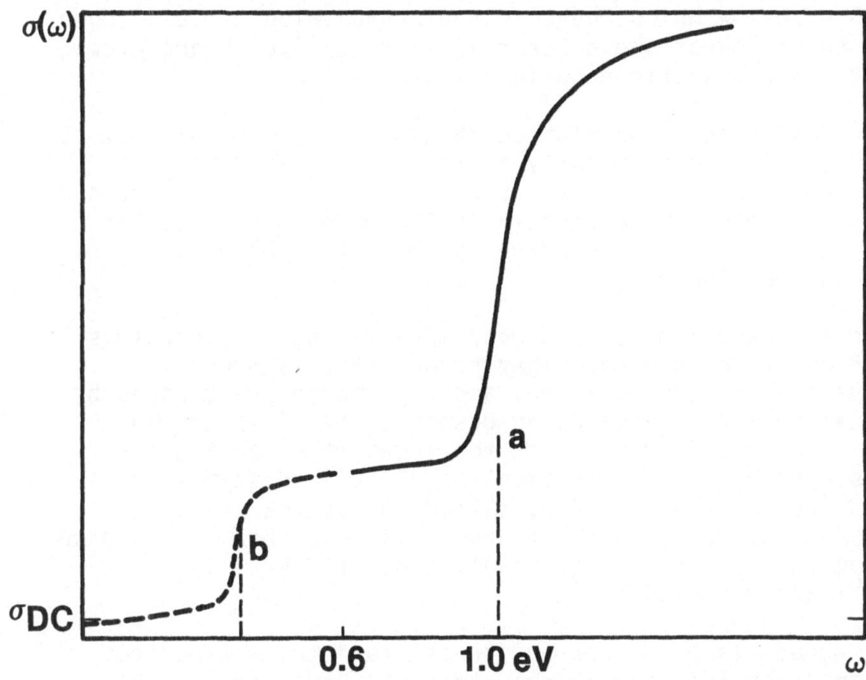

Figure 2. The frequency dependence of the optical conductivity $\sigma(\omega)$ found at T = 1800 K in Hg vapor by Hefner and Hensel from a Kramers-Kronig inversion of their $\epsilon_1(\omega)$ data. The density is that of the peak in Figure 1. The dc conductivity is very small, $10^{-3}\ \Omega^{-1}\ \mathrm{cm}^{-1}$. The ϵ_1 data only extend down to 0.6 eV. We have indicated two absorption edges: a.) one found by Hefner and Hensel at 1.0 eV and b.) one that we propose to intervene between their lowest frequency (0.6 eV) and dc.

edge at 1.0 eV below which there is a plateau value of $\varepsilon(\omega)$ down to 0.6 eV, their lowest photon energy. Between 0.6 eV and d.c., there has to be a dramatic drop in $\sigma(\omega)$.

Hefner and Hensel have plotted the locus of the peak in ε^1 in the ρT plane, shown schematically in Fig. 3, and find that it intersects the coexistence curve at the kink found by Hubbard and Ross. Thus, the dielectric anomaly in the vapor occurs at the same point on the coexistence curve as the metal-nonmetal transition in the liquid.

Hefner and Hensel propose a most interesting and ingenious interpretation of their data. They propose that there are electrons present in the Hg vapor, and that these are trapped by induced local density increases as described by Lifshitz and Gredescul. They then propose that the locus of ε^1 maxima corresponds to the phase transition from localized states to extended states for the electron, called the plasma state by Lifshitz and Gredescul(5). Their proposals have been followed up by extensions of the theory of Lifshitz and Gredescul by Hernandez(6) and Popielawski(7).

These proposals by Hefner and Hensel have two weaknesses. First a transition into the plasma state will not explain the remarkable frequency dependence of $\sigma(\omega)$: a very low d.c. value, a transition (b in Fig. 2) to a much higher plateau value below 0.6 eV, and a second transition (a in Fig. 2) at 1.0 eV. Instead, one would expect Drude-Lorenz behavior above the transition locus, and a negative value of ε^1 at low frequencies. Second, the phase transition found by Lifshitz and Gredescul is a spurious consequence of the approximations they used in treating the coupling of an electron to density fluctuations in an ideal gas. Turkevich(8) has shown that the problem considered by Lifshitz and Gredescul can be mapped onto Toyozawa's(9) deformation potential polaron problem. Toyozawa's results, using the scaling arguments of Emin and Holstein(10), then permit one to show that transition from localized to extended behavior is smooth.

Turkevich and I have proposed an alternative but closely related explanation of Hefner and Hensel's results. First, we also propose that the locus of ε^1 peaks corresponds to a phase transition. Second,the phase transition consists of a spontaneous ionization of a fraction of the Hg atoms so that electrons and holes are formed. Third, the electrons are stabilized by polaron formation to describe which the more precise theory of Turkevich is to be preferred. Fourth, the holes are stabilized by the formation of an $(Hg)_n^+$ ion with a density increase around it, as well. Fifth, transition a of Fig. 2 at 1.0 eV corresponds to the

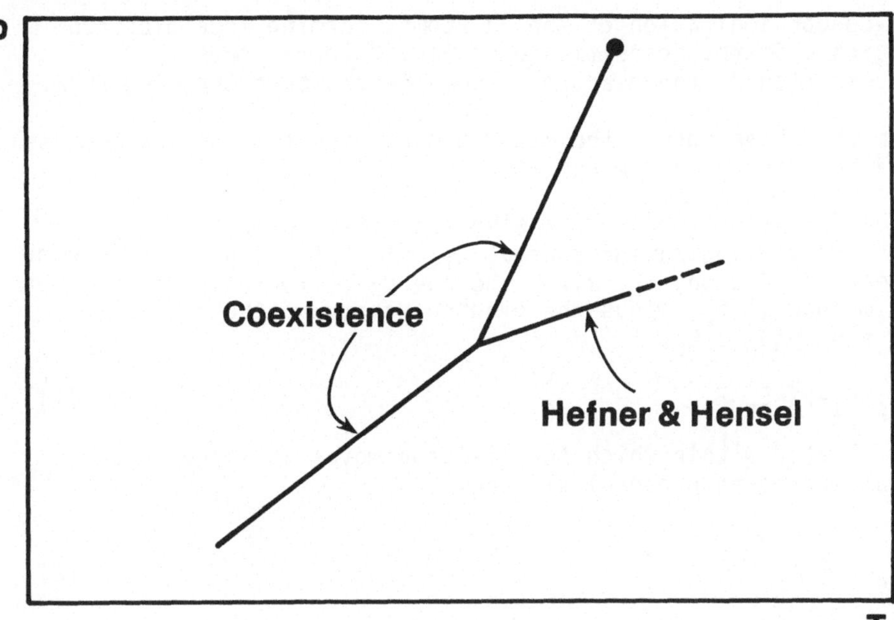

Figure 3. The p-T phase plane of Hg with the locus of the peak in ϵ_1 found by Hefner and Hensel drawn partly as a solid line and partly dotted to indicate that its extent is unknown.

Franck-Condon ionization of the Hg atoms, forming free electron-hole pairs. Sixth, transition b of Fig. 2 corresponds to photoionization of the electrons from the clusters trapping them.

In the above model, the d.c. conductivity remains low because all charged species are massive.

The formation energy of a single electron-hole pair, E_p, consists of the ionization energy, E_I, minus the binding energy of the electron polaron, E_e, minus the energy of formation of the $(Hg)_n^+$ molecule, E_n, minus the binding energy of the cluster around the $(Hg)_n^+$, E_h.

$$E_p=E_I-E_e-E_n-E_h \tag{1}$$

If the cluster within which the electron moves is large enough, then, according to proposal six above,

$$E_e \simeq \frac{1}{2} E_a \tag{2}$$

The value of E depends on n, and the value of n to be used is that which minimizes E_p. For the present, we seek only to see whether the proposals above are in fact consistent with the optical data. Accordingly, we assume that

$$E_b \simeq E_h \ . \tag{3}$$

We have already proposed that

$$E_I = E_b. \tag{4}$$

We can expect that

$$E_n > E_2, \tag{5}$$

and the Hartnee-Fock value[11] for E_2 is 0.67 eV. Spectroscopic evidence[12] suggests that E_2 <0.9 eV. Using E_2 = 0.67 eV for E_n in (1) leaves us only with E_a to be estimated. E_a has the bounds

$$kT = 0.15 \text{ eV} < E_a < 0.6 \text{ eV} \ , \tag{6}$$

and a reasonable estimate for E_a is the average of these bounds, or 0.4 eV. Using the above estimates for E_I, E_e, E_h, and E_n leads to a value of -0.1 eV for E_p. Thus the vapor would be unstable against spontaneous ionization and condensation of electron-hole pairs would occur. In other words, our model leads to an internally consistent interpretation of $\sigma(\omega)$.

Hefner and Hensel have pointed out that their interpretation of the peak in ε^1 implies the existence of substantial density fluctuations in the vapor at the same point on the coexistence curve as the metal-nonmetal transition occurs in the liquid. They argue that this strengthens the case for the Cohen-Jortner picture of an inhomogeneous metal-nonmetal transition in the liquid, suggesting that the density fluctuations are present in the vapor and persist along a path around the critical point from vapor to liquid.

Turkevich and I have taken our model somewhat further but have as yet no formal results for the liquid. We can, however, make the following observations. Within our model, the phase diagram of Hg stands as shown in Fig. 4. The vapor is an atomic insulator at low densities and temperatures. Across the transition line discovered by Hefner and Hensel, it changes over into an electrolyte, the charged species being ionic in general character. As the critical region is approached from the vapor side, near the coexistence curve, the compressibility increases and with it E_e and E_h and the sizes of the electron and hole clusters. The density increase may decrease E_I as well as increase E_e and E_h. The number density of electron-hole pairs thus must increase to the point where the electrons and holes all start overlapping. Dielectric screening of the electrostrictive forces occurs. The hole-hole overlap implies that valence bonding of the Hg atoms takes place. Deionization can occur either gradually, or via a phase transition near the critical point. Continuing around the critical point leads into a liquid semiconducting state within which there are hybridized s-p valence bands which are predominantly s-like, with a gap to the predominantly p-like conduction bands.

What does all this have to do with the inhomogeneous metal-nonmetal transition? Turkevich and I have speculated that there may be an approximate hidden symmetry in the problem. Just as the coexistence curve is symmetric in the lattice-gas model of condensation, so, when one adds the electronic degrees of freedom, a larger symmetry is introduced which maps the Hefner-Hensel line in the vapor into the metal-nonmetal transition line in the liquid across the coexistence curve. The spontaneous ionization process in the vapor becomes a spontaneous deionization process in the liquid. The density fluctuations (increases) induced by the electrons and by the holes become density decreases associated with small semiconducting regions. Finally, if the Hefner-Hensel

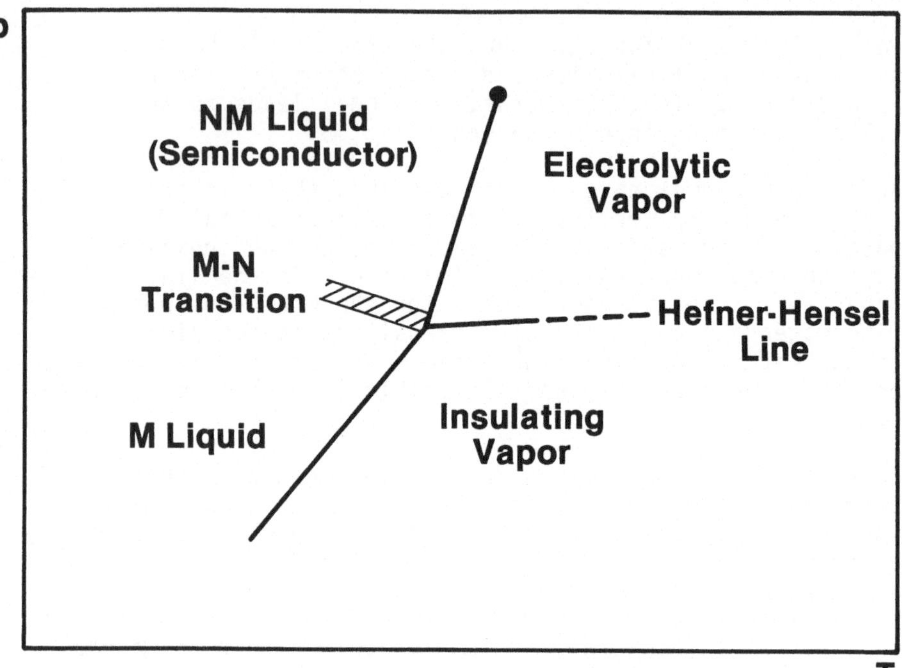

Figure 4. Phase diagram of Hg with the electronic character of the liquid and vapor shown for the different regions. There must be a transition from electrolytic vapor to semiconducting liquid somewhere around the critical point.

transition is first-order and maps into a first-order transition in the liquid one would have the microscopic inhomogeneities proposed by Cohen and Jortner if the surface tension remained small, as near a critical point. Turkevich and I are now attempting to develop our theory to the stage where testing of these speculations can be carried out.

3. CONCENTRATION FLUCTUATIONS IN METAL-AMMONIA SOLUTIONS

Recent heat capacity measurements above the consolute point in $NaNH_3$ solutions by Steinberg, et al.[13] have turned up strong evidence for a weakly first-order phase transition, as shown in Fig. 5. The character of the heat capacity traces suggest the existence of a two-phase region bounded by first-order lines, and the very long equilibration times suggest that the first-order character may be weak.

Cohen and Jortner[14] had proposed an inhomogeneous metal-nonmetal transition in $NaNH_3$ solutions along a line 10K above the consolute point starting at 2.3 MPM and finishing at 9 MPM. Within this range they proposed the existence of clusters of solvated electrons and solvated Na ions with a radius of 30 Å which are locally metallic. The volume fraction of these clusters increases linearly from zero to unity as the value of MPM increases from 2.3 to 9. The remainder of the solution remains at a concentration of 2.3 MPM, decreasing linearly in volume fraction to zero. Thus the probability distribution of concentrtion fluctuations proposed is bimodal.

The physical origin of such a bimodal distribution has remained obscure until the work of Steinberg, et al. It has already been proposed that the concentration fluctuations introduced by Cohen and Jortner arise from proximity to the consolute point and are therefore unimodal in character. Undoubtedly, proximity to the critical point has an important effect on the concentration fluctuations. However, the metal-nonmetal transition is closely similar in all of the metal-ammonia solutions while the position of the consolute point varies considerably, disappearing below the liquids in the C_s solutions. Moreover, bimodality of concentraiton fluctuations is a natural consequence of a first-order phase transition if it is sufficiently weakly first order. The macroscopic two-phase region one normally expects in a first-order phase transition is then replaced by a corresponding region of microscopic phase separation, that is, bimodal fluctuations. The reason is that the surface tension can be too small to prevent the breaking up of the second phase into microscopically small droplets in order to increase the entropy. The microscopic state proposed by Cohen and Jortner thus becomes, in view of the findings of Steinberg et al.,

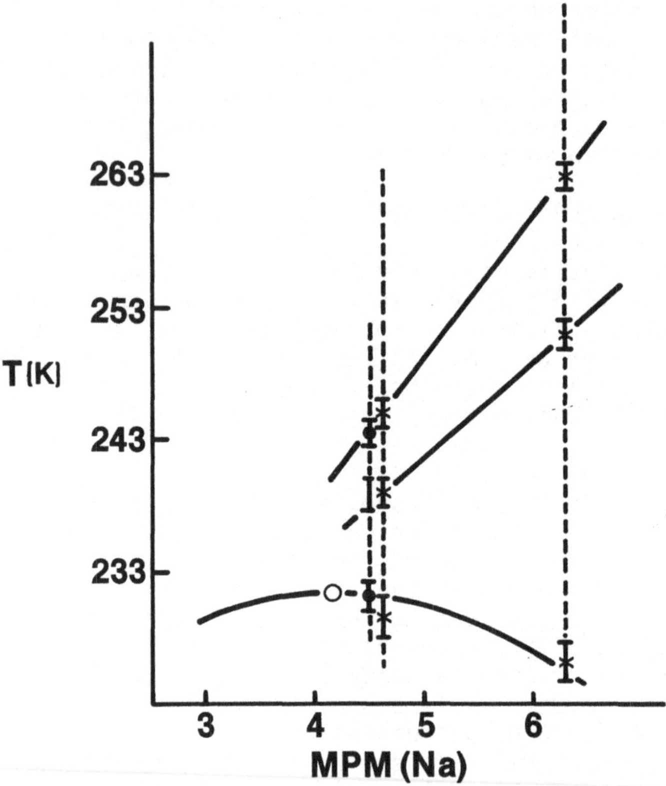

Figure 5. Phase diagram of the liquid Na NH_3 system. The lower solid curve is the coexistence curve, and the ○ indicates the consolute point. The upper solid lines bound the two-phase region proposed by Steinberg, et al. on the basis of their experimental data, shown with error bars.

analogous to the droplet state proposed to exist near but below a critical point.

Gitterman[15] has shown how such additional lines of phase transition can arise when additional degrees of freedom exist. In this case, the additional degree of freedom arises from the distribution of the sizes of complexes formed by the association of solvated electrons and solvated metal ions.

As in the Hg case, the driving force for clustering is electronic, associated with the metal-nonmetal transition. The relationship of the phase diagram of Fig. 4 to that of Fig. 5 is quite interesting. It is as though the intermediate semiconducting state proposed for liquid Hg to intervene between the metallic and the electrolytic states in Hg disappears in the sodium ammonia solution so that the transition is directly from metallic to electrolytic or to a hybrid state with both electronic and electrolytic conduction.

4. VALENCE FLUCTUATIONS IN Te, Se AND SeTe ALLOYS

At low temperatures, the chalcogens, Te, Se and their binary alloys are semiconductors, whether they are crystalline or glassy. The structure consists of chains or twofold-coordinated atoms (C_2) with, roughly speaking, Van der Waals coupling between the chains. As the glass is heated above its glass transition temperature or the crystal above its melting point, chain breaking occurs. The process that occurs may be represented as a chemical reaction

$$C_2 \rightleftarrows 2C_1 \tag{7}$$

$$C_1 + C_2 \rightleftarrows C_3 \tag{8}$$

in which one chain breaks, producing two singly coordinated chain ends, $2C_1$ Eq. (7), and one of these bonds to an atom in a neighboring chain, producing a C_3, a triply coordinated atom, Eq. (8). Kastner, Adler, and Fritzsche[16] have shown that charge transfer occurs either simultaneously or subsequently,

$$C_1 + C_3 \rightleftarrows C_1^- + C_3^+, \tag{9}$$

producing what they have termed a valence alternation pair (VAP). A vast array of experimental measurements of the semiconducting properties of the glasses was explained thereby, and there is little doubt that VAP's exist in the glasses, frozen in at the glass transition temperature, Tg.

If one studies the melt above Tg, one finds remarkable property changes. For example, the thermal expansion coefficient α decreases, becomes negative, and then increases back towards normal values[17] so that a sharp negative peak occurs in α. This phenomenon is intimately connected, in our view, both with the formation of VAP's and with the metal-nonmetal transition which occurs at still higher temperatures. Indeed, the VAP's and the dip in α provide the key to understanding the valence fluctuations proposed by Cohen and Jortner[18] to explain the metal-nonmetal transition.

The anomalous thermal expansion can be explained as follows. The charged centers in the VAP's have electrostrictive interaction with the surrounding neutral chains, causing a local density increase. The C_3's, because of the replacement of Van der Waals bonds by valence bonds, have an additional density increase associated with them. Thus each VAP has a negative activation volume, and the exponential increases in their numbers with increasing temperature ultimately overwhelms the normal thermal expansion.

Having attributed the anomalous negative thermal expansion to the acivation of VAP's, we must ask what terminates the process with increasing T. The equilibrium (9) implies that there will also be an increasing number of C_3 with temperature because of the reverse reaction. The C_1 will largely disappear because of the back reaction in (7). These C_3 will cluster as their concentration increases because an additional stabilization energy derives from electron transfer within the clusters, which may be regarded as protometallic. The C_3^+ can become incorporated into the cluster for the same reason. The extra stabilization energy for the C_3 then displaces the equilibrium from the charged species C_3^+ and C_1^- back towards C_3 and C_1. The C_1^- can give up its charge to a positively charged cluster of C_3's and join the cluster by combining with a neighboring C_2. Thus there will be a maximum in the concentration of charged species at some temperature above which they will disappear in favor of neutral C_3 clusters. The electrostrictive effect disappears, and the C_3 are expanded relative to the C_3^+. Thus the negative activation volume disappears, and normal thermal expansion resumes as temperature increases further. At this point, the structure of the system can be described as a matrix of C_2's in which clusters of C_3 occur.

The above argument then sets the stage for the Cohen-Jortner proposal of an inhomogeneous metal-nonmetal transition for these chalcogens.

5. DENSITY FLUCTUATIONS AS CLUSTERING IN GLASSES

In the systems discussed in the last three sections, clustering occurs in the density, in the composition, and in the coordination, respectively, but in all cases, the clustering is electronically driven. We now present a quite different example of clustering, clustering of density fluctuations in supercooling liquids and glasses driven by entropy.

To discuss the entropy associated with atomic movement, we must have a simple theory of atomic movement. Turnbull and Cohen[19] have developed such a theory based on the free-volume model. Cohen and Grest have extended that theory to apply to thermodynamics and relaxation processes by giving a microscopic explanation of the nature of the free volume[20].

Their key idea is that two time scales exist in dense liquids at low temperatures. One, T_v, is the period of vibration of each atom or molecule about a position of secular equilibrium within a cage or cell formed by its nearest neighbors. The other, T_D, is the time for the rearrangement of the cage structure by atomic or molecular diffusion. At low temperatures and high densities, T_v greatly exceeds T_D, the cells are well defined, and there is a clean separation between contributions to the free energy of the two kinds of motion.

One can associate a cell volume v with each cell. Cohen and Grest show how, when the cell volume exceeds some critical value v_c, the excess

$$v_f = v - v_c \tag{10}$$

can be considered as free. A cell for which $v_f \neq 0$ is called a liquid-like cell. The others are solid-like cells. Cohen and Turnbull argue diffusion can occur only when a void of volume exceeding a molecular volume forms by free-volume exchange. Cohen and Grest argue that free volume can be exchanged only between nearest neighbor liquid-like cells which themselves have a sufficient number of liquid-like cells as nearest neighbors.

This criterion for free-volume exchange is in fact a criterion for the definition of clusters which leads in turn to a percolation problem.

Let there be ν cells in a cluster. The total free volume in that cluster will be $\bar{v}_f$, where $\bar{v}_f$ is the average free volume of a liquid-like cell. For diffusion to occur within a cluster, its total free volume must exceed v_m,

$$\nu\overline{v}_f > v_m \tag{11}$$

Thus, clusters with

$$\nu > \nu_m = v_m / \overline{v}_f \tag{12}$$

support diffusion. These are liquid clusters. Within liquid clusters, diffusion gives rise to communal entropy. The communal entropy enormously enhances the fraction of liquid-like cells and suppresses clusters smaller than ν_m. The detailed theory shows that the distribution of cluster sizes is otherwise like that in percolation theory.

Direct observation of the clustering is difficult because of the small difference in density between clusters of liquid-like cells and the solid-like matrix. Accordingly, experimental confirmation of the proposed clustering comes instead indirectly from analysis of relaxation processes.

Relaxation of the thermodynamic processes occurs by diffusion across the interface between the clusters and the solid-like matrix. Accordingly, the relaxation rate of a cluster of size ν is

$$\tau_\nu = \tau_{o\nu} e^{v_m / \overline{v}_f} \tag{13}$$

$$\tau_{o\nu} = a(S/V)_\nu \tag{14}$$

In (14), a is a simple kinetic factor and $(S/V)_\nu$ is the surface to volume ratio of a cluster of size ν. $v_m / \overline{v}_f$ is a function of temperature which is obtained from fitting the viscosity data to the theory within experimental error over fourteen orders of magnitude. The theory then gives a linear relationship between the glass transition temperature and the log of the heating rate times an appropriate average of $\tau_{o\nu}$. The slope agrees with that observed to a few percent, implying that (13) accurately represents the temperature/dependence of all the τ's, and the correct magnitude for the average of $\tau_{o\nu}$ is also obtained.

More important, the dispersion of the relaxation times is also correctly given.

The existence of a probability distribution of ν, $P(\nu)$, implies through (14) that there is a probability distribution P(W) of relaxation rates $W = 1/\tau$. The relaxation function R(t) is then given by

$$R(t) = \int P(W)e^{-Wt}dw = \int P(\nu)e^{-W(\nu)t}d\nu \qquad (15)$$

W is proportional to $(S/V)_\nu$ and the latter to a negative power of ν,

$$\frac{S}{V} \alpha \nu^{-x}, \quad 0 < x < \frac{1}{3}. \qquad (16)$$

Similarly, percolation theory gives

$$P(\nu) \alpha\ e^{-A\nu^{-y}}, \quad \frac{2}{3} < y < 0.9 \qquad (17)$$

It is easy to show, from (15)-(17) that

$$R(t) \sim e^{-t/\tau_s} \quad t \to o \qquad (18)$$

$$\frac{1}{\tau_s} = \langle \frac{1}{\tau} \rangle \qquad (19)$$

and

$$R(t) \sim e^{-(t/\tau_L)^\beta} \quad t \to \infty \qquad (20)$$

$$\beta = \frac{y}{x + y}, \quad \frac{2}{3} < \beta < 0.9 \qquad (21)$$

The changeover between the short and long time behavior occurs approximately at

$$t \cong \frac{\tau_s}{(1-\beta)^2} \qquad (22)$$

Eq. (20) is the form of R(t) actually observed, and values found for β fill the predicted range, but do not fall outside[21].

Thus we have detailed agreement between theory and experiment for the magnitude, temperature dependence, and especially the dispersion of the relaxation times. The inescapable conclusion is that the density fluctuations are there, as predicted. The special character of the clustering of these density fluctuations is that it is entropically driven.

References

1. M. H. Cohen and J. Jortner, Phys. Rev. A 10, 978 (1974); Phys. Rev. B 15, 1227 (1977).

2. R. W. Schmutzler, to be published.

3. S. R. Hubbard and R. G. Ross, to be published.

4. W. Hefner and F. Hensel, Phys. Rev. Lett. 48, 1026 (1982).

5. I. M. Lifshitz and S. A Gredescul, Zh. Eksp. Teor. Fiz. 57, 2209 (1969) [Sov. Phys. JETP 30, 1197 (1970).]

6. J. P. Hernandez, Phys. Rev. Lett. 48, 1682 (1982).

7. J. Popielawsky, to be published.

8. L. Turkevich, unpublished.

10. D. Emin, T. Holstein, Phys. Rev. Lett. 36 323 (1976).
9. Y. Toyozawa, Prog. Theor. Phys., 26, 29 (1961).
11. H. H. Michels, R. H. Hobbs, J. W. D. Connolly, Chem. Phys. Lett. 68 549 (1979).

12. F. L. Arnot, M. B. Merven, Proc. Roy. Soc. A 165 133 (1938).

13. V. Steinberg et al., Phys. Rev. Lett 45, 1338 (1980).

14. M. H. Cohen and J. Jortner, J. Phys. Chem. 79, 2900 (1975); J. Jortner and M. H. Cohen, Phys. Rev. B 13, 1548 (1976); M. H. Cohen, I. Webman, and J. Jortner, J. Chem. Phys. 64, 2013 (1976).

15. M. Gitterman, V. Steinberg, Phys. Rev. A 20 1236 (1979); M. Gitterman, V. Steinberg, J. Chem. Phys. 69 2763 (1978).

16. M. Kastner, D. Ader and H. Fritzsche, Phys. Rev. Lett, 1504 (1976); M. Kastner and H. Fritzache, Phil. Mag. B 37, 199 (1978); M. Misonofu and H. Endo J. Phys. Soc. Jap.

17. H. Endo, Suppl. to Prog. Theor. Phys. (in press).

18. M. H. Cohen and J. Jortner, Phys. Rev. B 13, 5255 (1976).

19. M. H. Cohen and D. Turnbull, J. Chem. Phys. 31, 1164 (1959); D. Turnbull and M. H. Cohen, ibid 34, 120 (1961).

20. M. H. Cohen and G. S. Grest, Phys. Rev. B 20, 325 (1979); B24 4091 (1981); Phys. Rev. B26 (1982). G. S. Grest and M. H. Cohen, Phys. Rev. B 21, 4113 (1980), and in Advances in Chemical Physics, ed. by I. Prigogine and S. A. Rice (Wiley, N.Y., 1981), Vol. 48, p. 55.

21. M. A. DeBolt, et al., J. Am. Ceram. Soc. 59, 16 (1976); C. T. Moynihan, et al., Ann. N.Y. Acad. Sci. 279, 15 (1976); C. T. Moynihan and A. V. Lesikan, ibid 371, 151 (1981).

OPTICAL PROPERTIES AND SOLAR SELECTIVITY OF METAL-INSULATOR COMPOSITES

G.A. Niklasson and C.G. Granqvist

Physics Department, Chalmers University of Technology,
S-412 96 Gothenburg, Sweden

We derive a number of theoretical expressions for the average dielectric permeability of composite materials from a model which treats scattering from random unit cells chosen so as to represent the basic microstructure. The limits of validity of these expressions are given. The general bounds on the average dielectric permeability which hold irrespective of microstructure are presented. The theoretical expressions are used to explain the measured optical properties for two types of samples: Co-Al_2O_3 produced by coevaporation, and Ni-pigmented anodic Al_2O_3. Applications of these coatings for efficient photothermal conversion of solar energy are discussed in detail.

1 INTRODUCTION

In this paper we discuss the optical properties of metal-insulator composites. We present general theories for the spatially averaged dielectric permeability and treat the application of these theories to two types of samples: coevaporated Co-Al_2O_3 films and Ni-pigmented anodic Al_2O_3 coatings. The results are employed to discuss the utilization of metal-insulator composites for efficient photothermal conversion of solar energy.

It is convenient to divide the composite materials into two classes. The first one regards dilute suspensions of one (or several) component(s) in another uniform medium. Here the single-particle aspects dominate. The optical properties of this class of material is the main topic of the earlier contribution by Kreibig (1). The second class comprises mixtures of components which are concentrated, at least locally, so that interaction among neighbouring particles is essential for governing the optical properties. This class of material will be discussed in some detail below 1).

The scientific study of the optical behaviour of composite materials have a respectable history, and as early as in 1857 it was proposed by Faraday (2) that the brilliant colours observed by him in extremely thin silver and gold films could be ascribed to their aggregated nature. Much of the early work on photography also, in retrospect, belongs to the field of our present discussion. For more than a century there has been a continued scientific interest in the optical properties of composite materials. In recent years this interest has been spurred by the realization that this class of materials is of great importance for developing surfaces which efficiently convert solar energy into useful heat (3,4). Another reason for an increased interest is the connection to surface enhanced Raman scattering (5). In fact, the optical properties of composite materials nowadays constitute a very active field of research involving basic physics and chemistry as well as several important applications. Other disciplines where the optical properties of (mainly dielectric) granular matter is investigated include geology, meteorology, astronomy, biology and medicine. An incomplete and subjective listing of books, review articles and conference proceedings which cover a significant part of this vast and multidisciplinary field of research is provided by Refs. 3-24.

Figure 1 serves as a background for discussing the connection between solar energy utilization and the optical properties of coatings consisting of metal-insulator composites. In part (b) the solid curve depicts a typical solar spectrum (25), and the dashed curves refer to the emitted radiation from blackbodies at three temperatures. It is seen that the solar radiation is confined to the $0.3 < \lambda < 2$-μm wavelength range, while the thermal radiation occurs at wavelengths longer than 2 μm for the temperatures of interest. This negligible overlap between the solar and thermal spectra yields that it is possible to absorb most of the solar radiation by having <u>low</u> reflectance at $\lambda < 2$ μm and at the same time minimizing radiative heat losses by having <u>high</u> reflectance at $\lambda > 2$ μm. This desired reflectance profile characterizes the spectrally selective solar absorbing surface (26-30). Five categories of surfaces have been investigated for developing materials which approximate the ideal reflectance: these are intrinsic absorbers, metal-insulator composite coatings, multilayer antireflection coatings, semiconductor-metal tandems, and textured surfaces (27-31). The approach which has proven to be most successful so far is that of metal-insulator composites, and two of the three spectrally selective surfaces that are now produced commercially in Europe and the U.S. (30) definitely belong to this class. One of these is the Ni-pigmented anodic Al_2O_3, for which a typical spectral reflectance is given by

1) Metal-insulator composites are frequently called "cermets".

the solid curve in Fig. 1a; we return to this kind of surface in Ch. 6 below. A survey of many selectively solar absorbing surfaces is provided by the sampling of books, review articles and conference proceedings in Refs. 3, 4, 26-46.

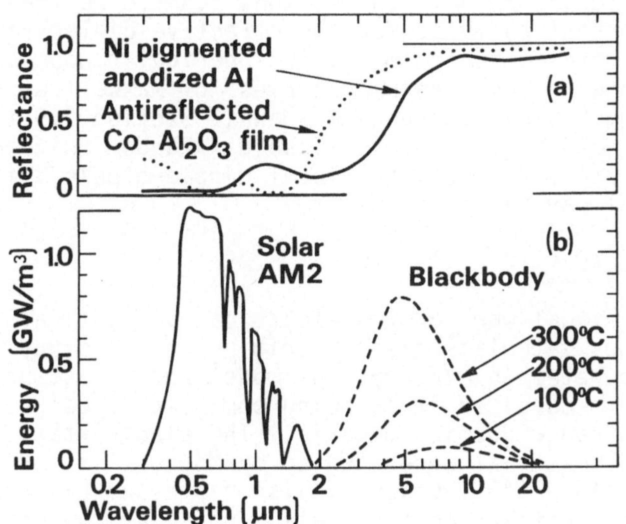

Fig. 1 Part (a) shows reflectance spectra for two selectively solar absorbing surfaces. These data will be discussed in detail in Chs. 5 and 6 below.
Part (b) reproduces the solar spectrum AM2 (for clear weather and the sun standing 30^0 above the horizon; cf. Ref. 25) together with three blackbody spectra.

This paper is organized as follows: In Chapter 2 we present a number of theories which are useful for describing the effective dielectric permeability of composite materials having different kinds of microstructure. Chapter 3 treats some recent results for the general bounds on the dielectric permeability which hold irrespective of microgeometry. Size limits of the theories are discussed in Chapter 4. The next two chapters present two case studies in which the theories are compared with experimental data for coevaporated Co-Al_2O_3 composite films (Ch. 5) and for Ni-pigmented anodic Al_2O_3 coatings (Ch. 6). Applications of these kinds of surface coatings to photothermal conversion of solar energy are treated.

2 EFFECTIVE MEDIUM THEORIES

In this chapter we present the theoretical basis for understanding the optical properties of inhomogeneous materials. Section 2.1 gives a historical overview of the subject. Section 2.2 discusses the fundamental connection between the microgeometry of the material and the theoretical model required for treating the optical properties. In Section 2.3 we briefly discuss the Lorenz-Mie theory as a necessary prerequisite for our later derivations. Section 2.4 presents a general definition of an "effective medium" in terms of classical scattering theory. In Sections 2.5-2.8 we then derive a number of different formulas for the effective dielectric permeability from a unified framework encompassing spherical random unit cells; these are the "effective medium theories" of Maxwell Garnett, Bruggeman, Ping Sheng and Bruggeman-Hanai. Section 2.9 treats the extension of these theories to ellipsoidal random unit cells. Section 2.10, finally, discusses local dipole-dipole interactions among spherical particles.

2.1 Some Early Work

The theoretical work on the electrical and optical properties of inhomogeneous materials has its origins in the study of dielectric media in the nineteenth century. One model for the structure of a dielectric was to consider it as being composed of conducting bodies separated by a nonconducting material. The electrostatic properties of such a model material were studied by Mossotti in 1850 (47) and by Clausius in 1879 (48). Clausius also expressed his results in terms of the dielectric constant (assuming spherical bodies) as

$$\bar{\varepsilon} = \frac{1 + 2f}{1 - f} , \tag{1}$$

where $\bar{\varepsilon}$ is the dielectric constant of the composite and f is the volume fraction of conducting material, the so called "filling factor". This formula is valid for perfectly conducting spheres in a medium with unity dielectric constant.

In his Treatise on Electricity and Magnetism published in 1873, Maxwell (49) derived a formula for the resistance of an inhomogeneous material composed of spheres of a medium with resistance r_A placed in a medium of resistance r_B. He compared the potential from a sphere of resistance $\bar{r}$ with the potential from a similar sphere containing the inhomogeneous material. Upon equating these he obtained

$$\bar{r} = \frac{2r_A + r_B + f_A(r_A - r_B)}{2r_A + r_B - 2f_A(r_A - r_B)}\, r_B , \tag{2}$$

and remarked that the derivation assumes that f_A (i.e., the volume fraction of spheres of material "A") is small.

A related relation for the refractive index of a medium composed of spherical molecules in "aether" was derived by Lorentz (50) and by Lorenz (51) in 1880. They obtained

$$\frac{\bar{n}^2 - 1}{\bar{n}^2 + 2} = f_A \frac{n_A^{\,2} - 1}{n_A^{\,2} + 2} , \qquad (3)$$

where $\bar{n}$ is the refractive index of the composite medium and n_A is that of the molecules. These authors employed the so called quasi-static approximation, i.e., they assumed that the wavelength of light is much larger than the dimensions of the particles and the distance between the particles. Their relation is equivalent to that of Maxwell if the latter is written in terms of conductivity.

The properties of a medium containing cylindrical or spherical inclusions arranged in rectangular order were studied by Rayleigh in 1892 (52). He extended Equation (3) by deriving a correction for higher filling factors, i.e., the next term in a series expression in f_A.

In 1904 and 1906 Maxwell Garnett (53) rederived the results of Maxwell, Lorentz and Lorenz, and used the theory to study the optical properties of glasses containing metal particles, thin metal films, and metallic sols. Maxwell Garnett explicitly treated the optical properties of particles dispersed in a matrix, and therefore his name has become attached to the theory when it is used in the study of optical properties of composite materials.

A new approach to the theory was taken in 1935 by Bruggeman (54). In contrast to previous work, he assumed that the two components of the inhomogeneous material should be treated on an equal basis, and hence he obtained equations that were symmetric in the two components. He also treated the opposite situation, when one component is completely surrounded by the other, and derived a new equation for this case.

The Maxwell Garnett theory and the Bruggeman theory have subsequently been rederived several times. A multitude of other theories have also been proposed for inhomogeneous materials with different geometries of the constituents. They are discussed in several reviews (55-59). A review which includes a discussion of the historical development of the theories was recently published by Landauer (60). It should be remarked that the same basic theory can be used for various properties of an inhomogeneous material, such as electrical conductivity, refractive index, dielectric permeability, magnetic permeability and mechanical properties.

2.2 Basis of the Effective Medium Theories

The theoretical description of the optical properties of inhomogeneous materials is given in terms of a complex dielectric permeability of an "effective medium". It is assumed that the inhomogeneous material possesses macroscopic uniformity. The effective dielectric permeability can formally be written as

$$\bar{\varepsilon} = \frac{\sum_i f_i \varepsilon_i E_i}{\sum_i f_i E_i} , \qquad (4)$$

where the sums are over the components of the inhomogeneous medium. This averaging process includes the electric fields, E_i, in the component materials and is therefore very dependent on the detailed microstructure of the composite. Thus we realize immediately that one cannot expect to find any universal effective medium theory, but different formulas will be applicable to different microstructures. This was clearly stated by Reynolds and Hough (61), who gave a unified derivation of the Maxwell Garnett and Bruggeman theories differing only in the assumptions made to calculate the fields E_i. Another valuable discussion on this issue was given recently by Aspnes (62).

The effective medium approach is not meaningful for all particle sizes and separations. The inhomogeneities must be taken as large on an atomic scale, so that each point in the material can be associated with a macroscopic dielectric permeability. On the other hand, in order to get tractable formulas, the inhomogeneities must not exceed certain size limits to be discussed in Section 4.1 below. It is also very difficult to take the detailed interactions between the inhomogeneities into consideration. In the effective medium theories the interactions are only taken account of in an average sense, and therefore it is expected that the theories will be less accurate when the filling factor f of the particles becomes large, or if there is a marked clustering of particles. However, if f becomes exceedingly small one can enter a regime where diffraction effects become important. These are not included in the effective medium theories, which consequently should be applied only if typical particle separations are much less than the wavelengths of the light.

The quasistatic approximation has been used in many calculations of the effective dielectric permeability. This means that an electrostatic calculation is extended to the optical properties by the assumption that the particle radii, a_i, are much less than the wavelengths of the light, or more specifically that

$$\sqrt{\varepsilon_i}\, \frac{2\pi a_i}{\lambda} \ll 1 , \qquad (5)$$

where λ is the wavelength of light in vacuum. If, in addition, the particles are clustered, Eq. (5) must be satisfied for a length characterizing the clusters instead of for the radius of the individual particle, as discussed by Lamb et al. (63). Thus this approach gives little more than a vague feeling of how large the particles are allowed to be, if the theory is to be valid.

It is therefore useful to consider a derivation of the effective medium theories which is directly applicable to the optical properties. It turns out that one can then also obtain better estimates of the particle size limits . It is of prime importance that proper account be taken of the relevant microstructure of the composite material. Figures 2(a) and (b) depict two possibilities: a separated grain structure in which particles of material "A" are dispersed in a continuous host of material "B", and an aggregate structure being a space-filling random mixture of the two components. Random unit cells are specified so that they account for the essential features of the microstructures. The random unit cell is taken to be embedded in an effective medium, whose properties are as yet undefined. We first assume that the inhomogeneous material on the average possesses full rotational and translational symmetry. The inhomogeneities should then on the average have a spherical shape, so the random unit cells are spherical. In Sec. 2.9 below we will relax this condition and discuss also ellipsoidal random unit cells. The concept of the random unit cell has earlier been used by Smith (64,65), Lamb et al. (63,66) and Niklasson et al. (67).

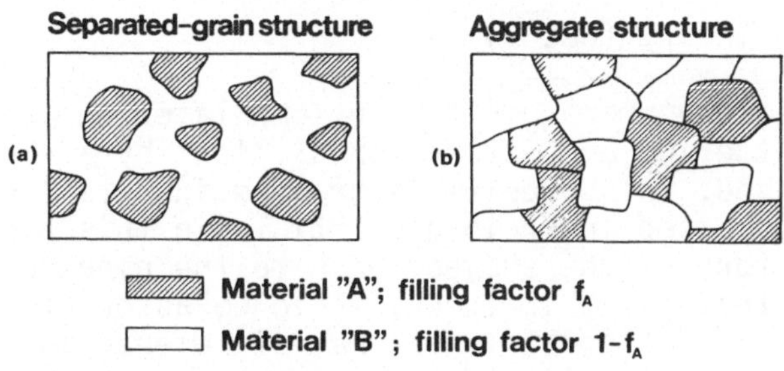

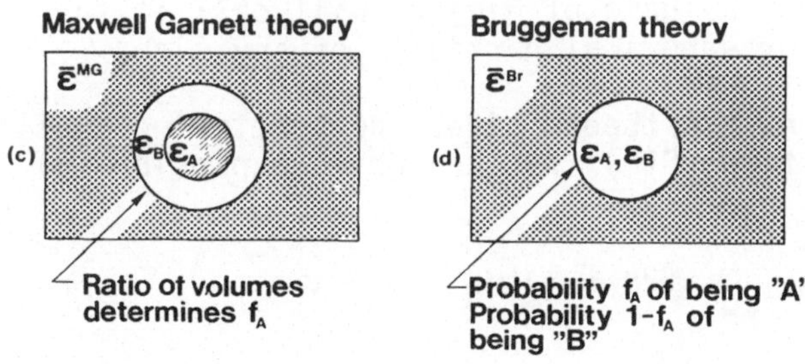

Fig. 2 Parts (a) and (b) depict two microstructures for composite two-phase materials. Parts (c) and (d) show the corresponding random unit cells used to derive the Maxwell Garnett (MG) and Bruggeman (Br) effective medium theories. The random unit cells are taken to be embedded in the effective media.

The separated-grain structure can be represented by a random unit cell being a coated sphere consisting of a core with dielectric permeability ε_A surrounded by a shell with dielectric permeability ε_B. The volume ratio of the concentric spheres is chosen so as to correspond to the overall filling factor, f_A, of the composite material. This random unit cell - shown in Fig. 2(c) - will allow us to derive an equation for the embedding effective dielectric permeability which is identical to the Maxwell Garnett expression; hence this quantity is denoted $\bar{\varepsilon}^{MG}$.

The aggregate structure, on the other hand, demands a random unit cell which guarantees the structural equivalence of the two constituents. Therefore this cell is taken to be a sphere whose dielectric permeability is ε_A with probability f_A and ε_B with probability $1-f_A$. This cell - shown in Fig. 2(d) - when embedded in an effective medium will be used to derive the Bruggeman expression for the effective dielectric permeability, $\bar{\varepsilon}^{Br}$.

Two other effective medium theories, which correspond to structures that are in a way intermediate between those in Fig. 2, can also be derived from random unit cell arguments. These are the Ping Sheng theory and the Bruggeman-Hanai theory; they will be derived in Sections 2.7 and 2.8 from the appropriate random unit cells.

2.3 Lorenz-Mie Theory (68,69)

As a necessary prerequisite for some later derivations, the problem of scattering and extinction of light by a single sphere must be addressed. This problem is treated in detail by Kerker (8) and is also discussed in Kreibig's article in this book (1). Kerker traced the history of the subject back to the nineteenth century and concluded that "if this theory is to be associated with the name or names of individuals, at least that of Lorenz, in whose paper are to be found the practical formulas so commonly used today, should not be omitted" (8). Therefore, although the theory is ordinarily connected with the name of Mie, we will call it the Lorenz-Mie theory. It is valid for spherical particles of any size.

The Lorenz-Mie theory gives the scattering amplitude in the forward direction, S(0), by the series expansion (8)

$$S(0) = \frac{1}{2} \sum_{n=1}^{\infty} (2n + 1)(\alpha_n + \beta_n) , \tag{6}$$

where α_n and β_n are complicated expressions containing Bessel functions and their derivatives. This series can be rewritten as a series in terms of (ka), where a is the sphere radius and k is the wavevector amplitude. The analogous problem for a coated sphere was first solved by Aden and Kerker (70) and Güttler (71). A series expansion like Eq. (6) holds also for this case, but the parameters α_n and β_n are defined differently. This series can also be rewritten in terms of (ka).

We consider now a dispersion of small spherical particles of dielectric permeability ε_A in a medium of dielectric permeability ε_B and assume that the particles are so far from one another that they do not interact, i.e., that f_A is very small. The effective dielectric permeability can then be obtained by the Lorenz-Mie theory. Van der Hulst (6) gives a formula for the effective complex refractive index, $\overline{n}$, according to

$$\frac{\overline{n}}{n_B} = 1 - iS(0)\,\frac{3f_A}{2(ka)^3}\ . \tag{7}$$

If the particles are very small it is sufficient to consider the first term in S(0), and the effective dielectric permeability then becomes

$$\overline{\varepsilon} = \varepsilon_B\,(1 + 3f_A\,\frac{\varepsilon_A - \varepsilon_B}{\varepsilon_A + 2\varepsilon_B})\ . \tag{8}$$

This is actually the small-f_A-limit of the various effective medium theories that will be derived shortly.

2.4 Definition of an Effective Medium

The basic definition of an effective medium is that the random unit cell, when embedded in the effective medium, should not be detectable in an experiment using electromagnetic radiation confined to a specific wavelength range (67). In other words, the extinction of the random unit cell should be the same as if it were replaced with a material with the effective dielectric permeability. This criterion makes it fruitful to use an optical theorem for absorbing media, which was recently derived by Bohren and Gilra (72). The theorem relates the extinction of the cell, compared to that of the surrounding medium, C_{ext}, with the scattering amplitude in the direction of the impinging beam, S(0), by

$$C_{ext} = 4\pi\ \mathrm{Re}[S(0)/k^2]. \tag{9}$$

Here k denotes the wave vector in the effective medium, i.e.,

$$k = 2\pi\overline{\varepsilon}^{\frac{1}{2}}/\lambda\ . \tag{10}$$

Equation (9) is clearly a generalization of the usual optical theorem for nonabsorbing media, but in the present case C_{ext} can be either positive or negative.

From the definition of an effective medium it follows that $C_{ext} = 0$, i.e.,

$$S(0) = 0 \ , \tag{11}$$

which states the fundamental property of an effective medium. This condition has been proposed earlier in somewhat different contexts (73,74). Equation (11) also indicates (6) that a plane wave entering the inhomogeneous material will propagate with an undeformed plane wave front. Fresnel's equations then apply at the boundaries of the effective medium. This is clearly an important point for any effective medium theory to be practically useful.

2.5 Maxwell Garnett Theory

In order to derive the Maxwell Garnett (MG) theory we write the scattering amplitude in the forward direction for the random unit cell in Fig. 2(c) as a power series in (kb), where b is the outer radius of the random unit cell. For the coated sphere (cs) geometry Eq. (6) then becomes (8)

$$S^{cs}(0) = i(kb)^3 \frac{(\varepsilon_B - \bar{\varepsilon})(\varepsilon_A + 2\varepsilon_B) + f_A(2\varepsilon_B + \bar{\varepsilon})(\varepsilon_A - \varepsilon_B)}{(\varepsilon_B + 2\bar{\varepsilon})(\varepsilon_A + 2\varepsilon_B) + f_A(2\varepsilon_B - 2\bar{\varepsilon})(\varepsilon_A - \varepsilon_B)} + 0[(kb)^5] \ . \tag{12}$$

The filling factor is given by

$$f_A = \frac{a^3}{b^3} \ , \tag{13}$$

where a is the radius of the inner sphere in the random unit cell.

In the small sphere limit the effective medium condition can be satisfied simply by putting the leading term in Eq. (12) equal to zero. One then obtains (with $\bar{\varepsilon} \equiv \bar{\varepsilon}^{MG}$),

$$\frac{\bar{\varepsilon}^{MG} - \varepsilon_B}{\bar{\varepsilon}^{MG} + 2\varepsilon_B} = f_A \frac{\varepsilon_A - \varepsilon_B}{\varepsilon_A + 2\varepsilon_B} \ . \tag{14}$$

Solving for $\bar{\varepsilon}^{MG}$ it is found that

$$\bar{\varepsilon}^{MG} = \varepsilon_B \frac{\varepsilon_A + 2\varepsilon_B + 2f_A(\varepsilon_A - \varepsilon_B)}{\varepsilon_A + 2\varepsilon_B - f_A(\varepsilon_A - \varepsilon_B)} \ . \tag{15}$$

By making the replacement A → B and B → A an analogous relation for the inverted structure can be obtained. Equation 15 is equivalent to the formulas obtained by Maxwell (49), Lorentz (50), Lorenz (51), Rayleigh (52) and Maxwell Garnett (53). The same expression has been given by Wagner (75), Kerner (76), Higuchi (77), Barker (78), Genzel and Martin (79), Smith (64), Lamb et al. (66) and others. Hayashi et al. (80) and Nagatani (81) have given formulas for the case of randomly arranged particles with optical anisotropy. Granqvist and Hunderi (82) have included retardation effects.

The above derivation does not require the filling factor to be small. However, it is clear that for a sufficiently large filling factor one reaches a point where particle-particle interactions, or multiple scattering, must be taken into account in a more elaborate way. Obviously such structural multipole features cannot be treated within our theoretical model, but supplementary information on inter-particle separation etc. is demanded. It is interesting to compare the Maxwell Garnett theory with the exact theories for the cubic lattices of spheres, which have recently been worked out (83-85). It has been shown that the Maxwell Garnett formula gives a good approximation to the exact formula for filling factors up to about 0.40 (86). For aperiodic systems of spheres, structural multipoles should be more important for small filling factors because close approaches between the particles are permitted. Structural multipoles in aperiodic arrangements of spheres have been studied in detail theoretically by Lamb et al. (63). The exact filling factor where multipole effects set in is dependent on the detailed microstructure of the composite medium.

2.6 Bruggeman Theory

The random unit cell in Fig. 2(d), being a sphere (s), can be used to derive the Bruggeman (Br) theory. The scattering amplitude is written as a series in (kb), analogous to Eq. (12), according to

$$S^{s}(0) = i(kb)^3 \frac{\varepsilon - \bar{\varepsilon}}{\varepsilon + 2\bar{\varepsilon}} + 0[(kb)^5] , \qquad (16)$$

where ε denotes ε_A or ε_B. Considering again the small sphere limit it is found that (with $\bar{\varepsilon} \equiv \bar{\varepsilon}^{Br}$)

$$f_A \frac{\varepsilon_A - \bar{\varepsilon}^{Br}}{\varepsilon_A + 2\bar{\varepsilon}^{Br}} + (1 - f_A) \frac{\varepsilon_B - \bar{\varepsilon}^{Br}}{\varepsilon_B + 2\bar{\varepsilon}^{Br}} = 0, \qquad (17)$$

where we have invoked the probability f_A for the random unit cell of having $\varepsilon = \varepsilon_A$ and the probability $f_B \equiv 1 - f_A$ of having $\varepsilon = \varepsilon_B$. Equation (17) was first given by Bruggeman (54) and is also found in works by Böttcher (87), Polder and van Santen (88), Landauer (89), Kerner (76), Elliott et al. (73), Wood and Ashcroft (90), Stroud (91), Berthier and Lafait (92) and others.

The difference between the Maxwell Garnett and Bruggeman theories was discussed by Stroud (93) and Webman et al. (94). Hori and Yonezawa (95-97) studied the mathematical structure of the effective medium theories and suggested an improved formula instead of Eq. (17). Stroud and Pan (74) included higher order terms in Eq. (16) into the theory.

For higher filling factors structural multipole effects are expected to play a role just as for the Maxwell Garnett theory, but a detailed study of this has not yet appeared. Ping Sheng (98) has taken account of pair-cluster interactions in a simplified way.

2.7 Ping Sheng Theory

The theory of Ping Sheng (PS) (99) is a symmetrical generalization of the Maxwell Garnett theory. The random unit cells, which are depicted in Fig. 3(a), are coated spheres and each component can be either the central core or the coating. The filling factor of material "A" is determined by the ratio of the volume of "A" to the cell volume. There are two types of the random unit cell, and the relative occurrence of them must be determined. Ping Sheng (99) did this by counting the number of equally possible configurations corresponding to different positions of the inner sphere in the random unit cell. For the case of a sphere of material "A" surrounded by a shell of material "B" this number is

$$v_1 = (1 - f_A^{1/3})^3 \ . \tag{18}$$

For the opposite situation we obtain

$$v_2 = (1 - (1 - f_A)^{1/3})^3 \ . \tag{19}$$

This argument requires that the inner sphere in Fig. 3(a) is placed eccentrically in the random unit cell. We approximate the ensemble of various eccentric stratified spheres, in the average, by random unit cells being concentric stratified spheres. Taking the small sphere limit of Eq. (12) for each of the two types of random unit cell and putting $\Sigma S^{cs}(0) = 0$ it is found that (with $\bar{\varepsilon} = \bar{\varepsilon}^{PS}$)

$$v_1 \frac{(\varepsilon_B - \bar{\varepsilon}^{PS})(\varepsilon_A + 2\varepsilon_B) + f_A(2\varepsilon_B + \bar{\varepsilon}^{PS})(\varepsilon_A - \varepsilon_B)}{(\varepsilon_B + 2\bar{\varepsilon}^{PS})(\varepsilon_A + 2\varepsilon_B) + 2f_A(\varepsilon_B - \bar{\varepsilon}^{PS})(\varepsilon_A - \varepsilon_B)} +$$

$$+ v_2 \frac{(\varepsilon_A - \bar{\varepsilon}^{PS})(\varepsilon_B + 2\varepsilon_A) + (1 - f_A)(2\varepsilon_A + \bar{\varepsilon}^{PS})(\varepsilon_B - \varepsilon_A)}{(\varepsilon_A + 2\bar{\varepsilon}^{PS})(\varepsilon_B + 2\varepsilon_A) + 2(1 - f_A)(\varepsilon_A - \bar{\varepsilon}^{PS})(\varepsilon_B - \varepsilon_A)} = 0\,. \quad (20)$$

This is clearly a symmetrization of the Maxwell Garnett theory and Eq. (20) approaches the Maxwell Garnett results when f_A is close to zero or unity. It should be remarked that Ping Sheng (99) made his derivations for the more general case of spheroidal particles, but we have chosen to give here the more tractable formula appropriate to spherical particles.

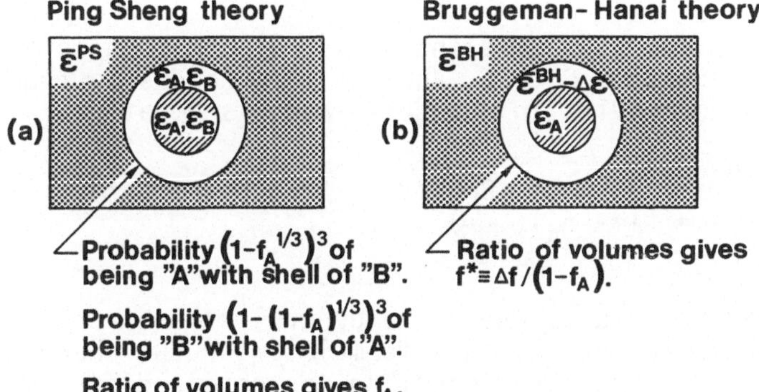

Fig. 3 Parts (a) and (b) show the random unit cells used to derive the Ping Sheng (PS) and Bruggeman-Hanai (BH) effective medium theories.

2.8 Bruggeman-Hanai Theory

The theory by Bruggeman and Hanai (BH) (54,100) can be derived from the random unit cell depicted in Fig. 3(b). It consists of a spherical core of material "A" surrounded by a shell whose effective dielectric permeability is $\bar{\varepsilon}^{BH}$ minus the contribution $\Delta\bar{\varepsilon}$ from the core itself. This is an intermediate situation between those of the Maxwell Garnett and Bruggeman theories. Rewriting Eq. (12) with the notation in Fig. 3(b) and taking the small sphere limit, it is found that (with $\bar{\varepsilon} \equiv \bar{\varepsilon}^{BH}$)

$$-\Delta\bar{\varepsilon}(\varepsilon_A + 2\bar{\varepsilon}^{BH} - 2\Delta\bar{\varepsilon}) + f^*(3\bar{\varepsilon}^{BH} - 2\Delta\bar{\varepsilon})(\varepsilon_A - \bar{\varepsilon}^{BH} + \Delta\bar{\varepsilon}) = 0 \quad (21)$$

where f^* is the ratio of the volume of the inner sphere to that of the whole random unit cell. The volume fraction of material "A" in the cell must be f_A, and the volume fraction of the shell is $f_A - \Delta f$. Thus the relation

$$f_A = (f_A - \Delta f)(1 - f^*) + f^* \tag{22}$$

is obeyed, which is equivalent to

$$f^* = \frac{\Delta f}{1 - f_A + \Delta f} \; . \tag{23}$$

Because $\Delta\bar{\varepsilon} \ll \bar{\varepsilon}^{BH}$ and $\Delta f \ll f_A$, Eq. (21) goes over into

$$\frac{\Delta\bar{\varepsilon}(\varepsilon_A + 2\bar{\varepsilon}^{BH})}{3\bar{\varepsilon}^{BH}(\varepsilon_A - \bar{\varepsilon}^{BH})} = \frac{\Delta f}{1 - f_A} \; . \tag{24}$$

We now let $\Delta\bar{\varepsilon}$ and Δf become infinitesimal and then integrate the right hand side from 0 to f_A and the left hand side from ε_B to $\bar{\varepsilon}^{BH}$. This procedure results in Bruggeman-Hanai's equation.

$$\frac{\varepsilon_A - \bar{\varepsilon}^{BH}}{\varepsilon_A - \varepsilon_B} = (1 - f_A)\left(\frac{\bar{\varepsilon}^{BH}}{\varepsilon_B}\right)^{1/3} \; . \tag{25}$$

The inverted structure, where the central sphere is of material "B", can be represented by the same formula after the changes A → B and B → A have been made. This equation was obtained by Bruggeman (54) and has subsequently been studied by Hanai (100) and Grosse (101). The theory is sometimes called the unsymmetrical Bruggeman theory.

2.9 Ellipsoidal Particles

For ellipsoidal particles there has not yet appeared a general theory analogous to that of Lorenz-Mie for spheres. Only the problem of light scattering and extinction by spheroids has recently been solved (102). We will therefore confine the discussion to the limit of small particles. Two cases will be treated. The first is when the light is incident along one of the three principal axes of an ellipsoid. This treatment will then be extended by averaging over the three principal directions to give a formula for randomly oriented ellipsoidal particles.

We first consider a generalization of the Bruggeman theory and replace the spherical random unit cell in Fig. 2(d) by an ellipsoidal one for which a, b and c denote the lengths of the three semiaxes. The scattering amplitude in the forward direction for light incident along the axis j is given by (6,103)

$$S_j(0) = ik^3 \frac{abc}{3} \frac{\varepsilon - \bar{\varepsilon}}{\bar{\varepsilon} + L_j(\varepsilon - \bar{\varepsilon})} , \tag{26}$$

where ε denotes ε_A or ε_B and L_j is the pertinent depolarization factor. The derivation of the Bruggeman theory for ellipsoids that are aligned with their j-axes parallel now proceeds as for the case of spheres. The effective medium criterion gives (with $\bar{\varepsilon} = \bar{\varepsilon}^{BR}$)

$$f_A \frac{\varepsilon_A - \bar{\varepsilon}^{Br}}{\bar{\varepsilon}^{Br} + L_j(\varepsilon_A - \bar{\varepsilon}^{Br})} + (1 - f_A) \frac{\varepsilon_B - \bar{\varepsilon}^{Br}}{\bar{\varepsilon}^{Br} + L_j(\varepsilon_B - \bar{\varepsilon}^{Br})} = 0, \tag{27}$$

where we have assumed that the same L_j is appropriate for both types of random unit cell so that the symmetry of the theory is preserved. The extension to randomly oriented ellipsoids can be obtained immediately by averaging the two terms of Eq. (27) over the three principal axes of the ellipsoid. Such a generalization was already made by Polder and van Santen (88). An alternative formula corresponding to ellipsoids of one material and spheres of another is sometimes used (104,105).

The Maxwell Garnett theory for ellipsoids is more complicated. The question of the shape of the Lorentz cavity (i.e., the random unit cell) for this case has attracted some interest (60,106). Cohen et al. (106) have shown that the formula for an ellipsoidal particle in a spherical cavity leads to inconsistencies when $L_j = 0$ or 1 and have proposed that the cavity should be ellipsoidal instead. We will follow this suggestion and replace the random unit cell of Fig. 2(c) by an ellipsoidal core with dielectric permeability ε_A surrounded by an ellipsoidal shell with dielectric permeability ε_B. The ellipsoids are taken to be confocal, since this situation is amenable to an analytic treatment (107,108). We let the light be incident along the principal axis j of the confocal ellipsoid, and take L_j^A and L_j^B to be the corresponding depolarisation factors for the inner and outer ellipsoid, respectively. We then obtain from the work of Bilboul (107)

$$S_j(0) = \\ = ik^3 \frac{abc}{3} \frac{(\varepsilon_B - \bar{\varepsilon})[(\varepsilon_A - \varepsilon_B)(f_A L_j^B - L_j^A) - \varepsilon_B] - f_A \varepsilon_B(\varepsilon_A - \varepsilon_B)}{[(\varepsilon_A - \varepsilon_B)(f_A L_j^B - L_j^A) - \varepsilon_B][\bar{\varepsilon} + L_j^B(\varepsilon_B - \bar{\varepsilon})] - f_A L_j^B \varepsilon_B(\varepsilon_A - \varepsilon_B)} , \tag{28}$$

where a, b, and c are the lengths of the semiaxes of the random unit cell. Denoting the lengths of the semiaxes of the inner ellipsoid a_1, b_1 and c_1, the filling factor becomes

$$f_A = \frac{a_1 b_1 c_1}{abc} . \tag{29}$$

The depolarisation factors L_j^A and L_j^B are complicated expressions containing the semiaxes a_1, b_1, c_1 and a, b, c, respectively (109). It is desirable to specify these values in terms of the semiaxes of the inner ellipsoid and the filling factor, since these are the quantities that are experimentally obtainable. This can be done by noting that (107)

$$a^2 - a_1^2 = b^2 - b_1^2 = c^2 - c_1^2 = \kappa , \tag{30}$$

where κ is a parameter that is obtained in terms of a_1, b_1 and c_1 from the equation

$$(\kappa + a_1^2)(\kappa + b_1^2)(\kappa + c_1^2) = \frac{a_1^2 b_1^2 c_1^2}{f_A^2} . \tag{31}$$

By applying the effective medium criterion for identical ellipsoids that are aligned with their principal axes j along the incident light, the Maxwell Garnett theory for ellipsoids is obtained (with $\bar{\varepsilon} = \bar{\varepsilon}^{MG}$) according to

$$\frac{\bar{\varepsilon}^{MG} - \varepsilon_B}{\varepsilon_B + (\bar{\varepsilon}^{MG} - \varepsilon_B)L_j^B} = f_A \frac{\varepsilon_A - \varepsilon_B}{\varepsilon_B + (\varepsilon_A - \varepsilon_B)L_j^A} , \tag{32}$$

or, upon solving for $\bar{\varepsilon}^{MG}$,

$$\bar{\varepsilon}^{MG} = \varepsilon_B \frac{\varepsilon_B(1-L_j^A) + L_j^A\varepsilon_A + f_A(1-L_j^B)(\varepsilon_A-\varepsilon_B)}{\varepsilon_B(1-L_j^A) + L_j^A\varepsilon_A - f_A L_j^B(\varepsilon_A-\varepsilon_B)} . \tag{33}$$

This equation was obtained by Bilboul (107) from a similar argument.

When f_A becomes large, the shapes of the inner and outer ellipsoids become similar, i.e., L_j^A and L_j^B will be nearly equal. Thus in the limit $f_A \to 1$, Eq. (32) reduces to an expression where $L_j^A = L_j^B$; this is the equation of Cohen et al. (106). The same relation can also be obtained from a quasistatic calculation as shown by Berthier and Lafait (110). As the filling factor becomes smaller, the shape of the outer ellipsoid will more and more approach that of a sphere. This means that L_j^B approaches one third as f_A goes to zero. The following formula is then obtained:

$$\frac{\bar{\varepsilon}^{MG} - \varepsilon_B}{\bar{\varepsilon}^{MG} + 2\varepsilon_B} = \frac{f_A}{3} \frac{\varepsilon_A - \varepsilon_B}{\varepsilon_B + (\varepsilon_A - \varepsilon_B)L_j^A} . \tag{34}$$

This equation, or its generalization to randomly oriented ellipsoids, has been used by Galeener (111), Granqvist and Hunderi (104) and Asami et al. (112). It can be rewritten as (104,113)

$$\bar{\varepsilon}^{MG} = \varepsilon_B \frac{1 + \frac{2}{3} f_A \alpha}{1 - \frac{1}{3} f_A \alpha} , \tag{35}$$

where

$$\alpha = \frac{\varepsilon_A - \varepsilon_B}{\varepsilon_B + L_j^A(\varepsilon_A - \varepsilon_B)} \tag{36}$$

is proportional to the polarizability of an ellipsoidal particle. The generalization to randomly oriented ellipsoids is simply obtained by averaging α over the three principal axes of the ellipsoid, viz.

$$\alpha = \frac{1}{3} \sum_{j=1}^{3} \frac{\varepsilon_A - \varepsilon_B}{\varepsilon_B + L_j^A(\varepsilon_A - \varepsilon_B)} . \tag{37}$$

It is also possible to treat ellipsoids which are oriented so that one of their semiaxes are parallel, but which are otherwise random (114,115), and cube-shaped particles (116-120). A compilation of formulas is given in Ref. 4.

It can be seen from the derivation above, or from the arguments of Cohen et al. (106), that Eqs. (35) and (36) are valid only in the limit of small filling factor. For randomly oriented ellipsoids it is an open question whether the ellipsoidal random unit cell is more justified than the spherical one. However, in any case Eqs. (35) and (37) are valid at least for small filling factor.

Equation (37) can be extended to randomly oriented confocally coated ellipsoids by applying again the results of Bilboul (107). We obtain

$$\alpha = \frac{1}{3} \sum_{j=1}^{3} \frac{(\varepsilon_C - \varepsilon_B)[\varepsilon_C + (\varepsilon_A - \varepsilon_C)(L_j^A - \Omega L_j^C)] + \Omega\varepsilon_C(\varepsilon_A - \varepsilon_C)}{[\varepsilon_C + (\varepsilon_A - \varepsilon_C)(L_j^A - \Omega L_j^C)][\varepsilon_B - L_j^C(\varepsilon_B - \varepsilon_C)] + \Omega L_j^C \varepsilon_C(\varepsilon_A - \varepsilon_C)} , \tag{38}$$

where ε_A, ε_C and ε_B are the dielectric permeabilities of the inner ellipsoid, coating and surrounding medium, respectively. Ω is the ratio between the volumes of the inner and outer ellipsoid, whose depolarisation factors are L_j^A and L_j^C. When $L_j^A = L_j^C = 1/3$ the equation for coated spheres is obtained as a special case of Eq. (38). Equation (38) should be valid at least for small filling factor, while the special case of a coated sphere should have a similar range of validity as that of the Maxwell Garnett theory for spheres.

Reynolds and Hough (61) have extended Bruggeman-Hanai's theory to randomly oriented spheroids. Ping Sheng's original derivation of his theory employed random unit cells consisting of confocally coated spheroids (99).

2.10 Local Dipole-Dipole Interactions Between Spherical Particles

The earlier discussion in this chapter has taken account of the interactions between neighbouring particles only in an average sense. When the particles are very close to one another such a treatment is no longer adequate since dipole-dipole interaction as well as higher multipole effects become important. Such clustering can occur even if the overall filling factor is very low, particularly in gas-evaporated specimens. The local dipole-dipole interactions can be taken into account approximately by a scheme developed by Granqvist and Hunderi (104). More general treatments have been given by Brako (121) and Persson (122).

The resonance frequencies of several kinds of geometrically well defined aggregates composed of touching identical spherical particles were calculated by Clippe et al. (123) and Ausloos et al. (124) from a model including nonretarded dipole-dipole coupling. This issue is discussed also in Ref. 1. The results by Clippe et al. were used by Granqvist and Hunderi (104) to derive triplets of effective depolarisation factors, L_j^*. These quantities obey the relation

$$\sum_{j=1}^{3} L_j^* = 1. \quad (39)$$

The effective depolarisation factors are not connected to any ellipsoidal shapes, but are only related to the resonance frequencies of the appropriate cluster of particles. By using them it is possible to approximately describe the effects of local dipole-dipole interactions by the equations that were derived for ellipsoidal particles in the previous section, the only difference being that L_j is replaced by L_j^*. We note in passing that effective depolarization factors have been used to explain the anomalous absorption in discontinuous noble metal films for many years (115, and references therein). They have also been invoked to account for microwave absorption in powders (124a).

An alternative description can be found in terms of a general relation giving the effect of dipole-dipole coupling on a polarisability (10,125-128) according to

$$\alpha^* = \frac{1}{3} \sum_{j=1}^{3} \frac{\alpha}{1 - \frac{Q_j}{4}\alpha} , \tag{40}$$

where the Q_j's are parameters which depend on the detailed state of aggregation. This formalism is identical to the formalism of effective depolarization factors if we identify

$$Q_j = 4(\tfrac{1}{3} - L_j^*). \tag{41}$$

Until now we have considered dipole-dipole coupling only. Higher multipole effects have been studied by Gérardy and Ausloos (129) and Brako (121). Lamb et al. (63) described the effects of structural multipoles by a complex and wavelength-dependent depolarisation factor.

3 BOUNDS ON THE EFFECTIVE DIELECTRIC PERMEABILITY

It was found in Chapter 2 that different prescriptions for the geometry of the random unit cell lead to different effective medium theories, and that there is no such thing as a general effective medium theory which would be valid for all microgeometries of an inhomogeneous material. A complete calculation of the effective dielectric permeability requires not only knowledge of the dielectric permeabilities and filling factors of the different phases, but also knowledge of their detailed geometrical configuration. In practice the precise microgeometry is not known, and it is then only possible to state that the effective dielectric permeability must have values between certain bounds, which can be theoretically derived. The bounds depend on the amount of geometrical information that is available and become narrower as more structural information is added.

Below we consider an inhomogeneous two-phase material for which the dielectric permeabilities and filling factors of the constituents are taken to be known. A quasistatic situation is assumed to be valid. The bounds for such a material were first derived by Wiener (130) and later, for isotropic materials, by Hashin and Shtrikman (130a). They regarded only the case of real dielectric permeabilities. During the last few years, complex dielectric permeabilities have also been thoroughly investigated by Schulgasser and Hashin (131), Bergman (132-134), Milton (135-138), McPhedran and Milton (139) and by Antonov and Pshentisyn (140). A similar formalism has been presented for mechanical properties by Milton (141). The development in this area has been very rapid.

Section 3.1 outlines the formulas for the Bergman-Milton bounds. In Section 3.2 we further discuss these bounds with regard to a specific example chosen to be an isotropic Co-Al_2O_3 composite.

3.1 Formulas for the Bergman-Milton Bounds

Bergman (142,143) has recently carried out a detailed investigation of the effective dielectric permeability of a two-phase composite material. He describes its properties in terms of a function of a complex variable,

$$F(S) = \frac{\varepsilon_B - \bar{\varepsilon}}{\varepsilon_B} , \qquad (42)$$

where

$$S = \frac{\varepsilon_B}{\varepsilon_B - \varepsilon_A} , \qquad (43)$$

and the subscripts "A" and "B" refer to the two types of materials. This function can be represented by the expression (142)

$$F(S) = \sum_n \frac{F_n}{S-S_n} , \qquad (44)$$

where S_n and F_n are the poles and the residues of the function, respectively.The positions of the poles and the residues are completely determined by the microgeometry of the composite, while ε_A and ε_B enter only through the variable S. To calculate S_n and F_n exactly, the precise microgeometry of the sample must be known. When only limited information on the geometry is available, one can vary S_n and F_n in order to maximize or minimize F(S) for any S. The available geometrical information enters as constraints on the poles and residues. In this way bounds on F(S) for different amounts of geometrical information can be derived.

We will now follow Bergman's formalism and present some examples of these bounds. Assuming that only the dielectric permeabilities and filling factors are known, the parametric representations of the bounds can be written (132,133,135,137)

$$F_c(S) = \frac{f_A}{S - S_o} , \quad 0 < S_o < f_B , \qquad (45)$$

and

$$F_d(S) = \frac{f_A S - S_o + f_B}{S(S - S_o)} \quad , \; f_B < S_o < 1 \; . \tag{46}$$

If we add the requirement that the composites should also be statistically isotropic or cubic, the corresponding bounds are given by

$$F_e(S) = \frac{f_A(S - S_o)}{S(S - S_o - \frac{1}{3}f_B)} \quad , \text{ for } 0 < S_o < \frac{2}{3} \; , \tag{47}$$

and

$$F_f(S) = \frac{f_A(S - S_o)}{(S - S_o)(S - \frac{1}{3}f_B) - \frac{2}{3}f_B(1 - S_o)} \; , \text{ for } \frac{2}{3} < S_o < 1. \tag{48}$$

These relations were obtained independently by Bergman (132) and Milton (135,138). Equations (47) and (48) are the parametrical representations of two circular arcs which cross each other in two points. The area bounded by these arcs in the complex F-plane specify the allowed values of $\bar{\varepsilon}$. Bergman (134) showed very recently that the bound $F_f(S)$ can in fact be improved. For a system where material "B" percolates $F_f(S)$ is replaced by

$$F_g(S) = \frac{f_A(S - S_o)}{(S - S_o)(S - \frac{1}{3}f_B) - f_B S_o(1 - S_o)/3(2S_o - 1)}$$

$$\text{for } \frac{2}{3} < S_o < 1 \; , \tag{49}$$

while for a system where material "A" percolates $F_f(S)$ is replaced by

$$F_h(S) = \frac{f_A(S - S_o)(S - \frac{2}{3}) - f_A S_o(1 - S_o)/3(1 - 2S_o)}{S[(S - S_o)(S - 1 + \frac{f_A}{3}) - f_A S_o(1 - S_o)/3(1 - 2S_o)]}$$

$$\text{for } 0 < S_o < \frac{1}{3} \; . \tag{50}$$

The two pairs of lines governed by F_e and F_g and by F_e and F_h determine two overlapping regions in the F-plane, whose union in fact yields the allowed region (134). The areas in the F-plane which lie between these latter bounds and those defined by $F_c(S)$ and $F_d(S)$ of course correspond to anisotropic structural arrangements. Finally we note that Milton (136,138) has analyzed the situation when more structural information than just the knowledge of statistical isotropy is available. All of the above bounds can easily be transformed from the F-plane to the ε-plane by Eq. (42).

We remind that Eqs. (45)-(50) hold only in the quasi-static case (cf. Eq. 5). Deviations are known to occur when the particles are sufficiently large that this condition is invalidated (144).

3.2 Application of the Bergman-Milton Bounds to a Specific Example: Co-Al_2O_3

We will now discuss the magnitude of the Bergman-Milton bounds in more detail. We do this by addressing two questions: Do the predictions of the earlier derived effective medium theories lie inside the bounds (as they should do); and how different can the optical properties be depending on the choice of microgeometry? The analyses are performed with regard to a specific example: a Co-Al_2O_3 composite (145). Experimental coatings consisting of this material will be discussed in Chapter 5 below.

Figure 4 presents results of a calculation of F_e, F_f, F_g and F_h from Eqs. (45)-(48) for an isotropic Co-Al_2O_3 composite. Specifically, we considered the wavelength $\lambda = 0.5$ µm and the filling factors 0.18 and 0.50 for cobalt. The bulk dielectric permeabilities for Co (146) and for electron-beam evaporated amorphous Al_2O_3 (147) were used as input data for the calculations. We note immediately that the allowed region of the complex plane - lying between F_e and F_g - is very substantial so that widely different values of $\bar{\varepsilon}$ are possible for materials with the same overall composition but different microstructures.

The different symbols inside the areas signifying the Bergman-Milton bounds in Fig. 4 were obtained by computing $\bar{\varepsilon}$ from several different effective medium theories. In addition to the Maxwell Garnett, Bruggeman, Ping Sheng and Bruggeman-Hanai theories for spheres, we also carried out computations for three other theories which are not based on random unit cell arguments: these are the theories by Grosse and Greffe (GG)(59; part VII), Lichtenecker (Li) (148) and Looyenga (Lo) (149). The constitutive equations for the latter three theories are

$$\frac{(\bar{\varepsilon}^{GG} - \varepsilon_B)(2\bar{\varepsilon}^{GG} + \varepsilon_A)}{3\bar{\varepsilon}^{GG}(\varepsilon_A - \varepsilon_B)} = f_A - \frac{2(\bar{\varepsilon}^{GG} - \varepsilon_B)(\varepsilon_A - \bar{\varepsilon}^{GG})}{3\bar{\varepsilon}^{GG}(2\varepsilon_B + \varepsilon_A)} f_A^{\,2} , \quad (51)$$

$$\ln \bar{\varepsilon}^{Li} = f_A \ln\varepsilon_A + f_B \ln\varepsilon_B , \quad (52)$$

$$\bar{\varepsilon}^{Lo} = (f_A \varepsilon_A^{\,1/3} + f_B \varepsilon_B^{\,1/3})^3 . \quad (53)$$

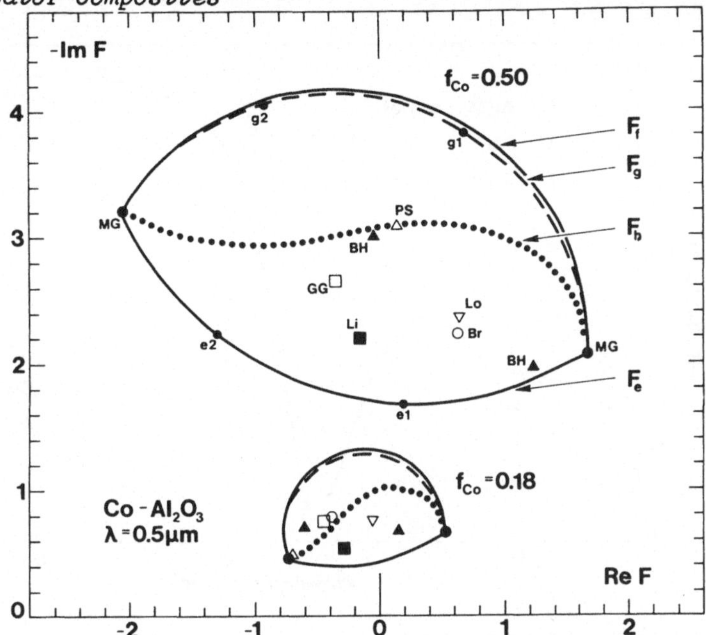

Fig. 4 Curves denote the rigorous Bergman-Milton bounds on the effective dielectric permeability for the case of Co-Al_2O_3 composites with two metal contents (f_{Co}) at 0.5 μm wavelength. Symbols represent computed results for the effective medium theories of Maxwell Garnett (MG), Ping Sheng (PS), Bruggeman-Hanai (BH), Bruggeman (Br), Grosse-Greffe (GG), Lichtenecker (Li) and Looyenga (Lo). The points e1, e2, g1, g2 will be used in the computations presented in Fig. 5.

The expression in Eq. (53) was given also by Landau and Lifshitz (109). Two data points are shown for the Maxwell Garnett and Bruggeman-Hanai theories: the points to the left hold for Co particles in an Al_2O_3 matrix and the points to the right for the inverted structure. It is found that the computed results from all effective medium theories lie within the rigorous bounds for isotropic materials. The Maxwell Garnett data are "extreme" though, being located at the intersections of the arcs defining the allowed region of $\bar{\varepsilon}$.

It is believed (132) that each allowed value of $\bar{\varepsilon}$ corresponds to a definite microstructure. Thus it appears that each of the effective medium theories would hold for some specific arrangement of conducting and non-conducting components - an arrangement which may or may not be the one underlying the derivation of the constitutive equation for the particular effective medium theory. In this sense any of the theories is equally "good". However, this statement should not be misinterpreted since (i) only certain microstructures may be common in experimental samples so that some effective medium theories may be of wider applicability than others and (ii) the relation

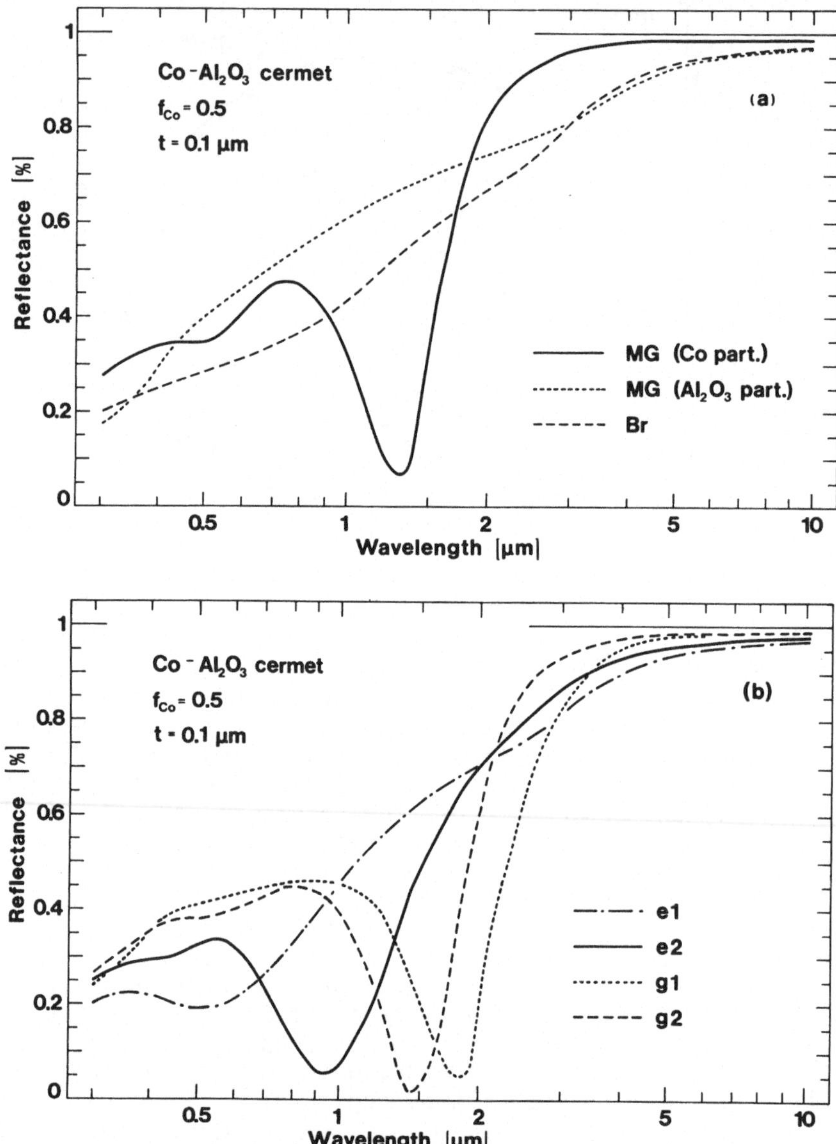

Fig. 5 Spectral reflectance of a 0.1-µm-thick Co-Al_2O_3 composite film containing 50 vol.% metal. The film is backed by an ideally reflecting substrate. Part (a) shows results from the theories of Bruggeman and Maxwell Garnett (with Co particles in a continuous Al_2O_3 matrix as well as with the inverted structure). Part (b) shows spectra corresponding to the points e1, e2, g1 and g2 in Fig. 4.

between microstructure and effective medium theory has a sound theoretical basis only in some cases and these are the only ones which allow a fundamental understanding of the optical properties.

The large allowed range for $\bar{\varepsilon}$ which was found in Fig. 4 makes it obvious that the optical properties of a composite material are very dependent on the microstructure. This dependence was further investigated by model calculations for a 0.1 µm thick Co-Al_2O_3 layer with 50 vol.% cobalt placed on an ideally reflecting substrate. Bulk data for the dielectric functions of Co (146,150) and Al_2O_3 (147) were used. Figure 5 shows spectral reflectance in the interval 0.3 µm $< \lambda <$ 10 µm. Part (a) refers to the Maxwell Garnett theory with both of the possible microstructures, and to the Bruggeman theory. The computed data are qualitatively very different, which agrees with earlier observations for other composite materials (151-156). Figure 5(b) contains supplementary spectra for a sampling of four points on the boundary curves in Fig. 4.

We stress once again that an accurate structural characterization is necessary for a detailed theoretical understanding of the optical properties of a composite material. There is no justification in picking an effective medium theory at random and using it for predicting the optical properties. We also observe that some of the curves in Fig. 5 show a spectral selectivity approaching the one required for efficient photothermal conversion of solar energy (cf. Fig. 1). Thus certain microstructures are more advantageous for this purpose than others.

4 SIZE LIMITS FOR VALIDITY OF EFFECTIVE MEDIUM THEORIES

The particles comprising the composite materials must have sizes which lie between certain limiting values if the effective medium theories are to be useful. Section 4.1 discusses the large-size-limits for the Maxwell Garnett and Bruggeman theories by drawing on the equations in Sections 2.5 and 2.6. Possible extensions of the theories are treated in Section 4.2. To provide an example of the magnitude of the large-size-limits, Section 4.3 discusses a specific example, which is again chosen to be a Co-Al_2O_3 composite. Section 4.4, finally, gives some general arguments about small-size limits.

4.1 Large-Size-Limits

As discussed in Chapter 2, the Maxwell Garnett and Bruggeman effective medium theories can be derived by considering the scattering amplitude in the small sphere limit. For particles of larger size, higher-order terms in the expansion series of $S^{cs}(0)$ and $S^{s}(0)$ begin to play a role. Within the Maxwell Garnett formalism we may then write

$$\frac{(\varepsilon_B-\bar{\varepsilon})(\varepsilon_A+2\varepsilon_B) + f_A(2\varepsilon_B+\bar{\varepsilon})(\varepsilon_A-\varepsilon_B)}{(\varepsilon_B-2\bar{\varepsilon})(\varepsilon_A+2\varepsilon_B) + f_A(2\varepsilon_B-2\bar{\varepsilon})(\varepsilon_A-\varepsilon_B)} +$$

$$+ \delta^{MG}(\varepsilon_A,\varepsilon_B,\bar{\varepsilon},f_A,b,\lambda) = 0 , \qquad (54)$$

where the magnitude of δ^{MG} can be computed from the full series indicated in Eq. (6). From Eq. (54) one obtains

$$\bar{\varepsilon} = \varepsilon_B \frac{[\varepsilon_A + 2\varepsilon_B + 2f_A(\varepsilon_A - \varepsilon_B)](1 + \delta^{MG})}{[\varepsilon_A + 2\varepsilon_B - f_A(\varepsilon_A - \varepsilon_B)]\ (1 - 2\delta^{MG})} , \qquad (55)$$

which is the ordinary Maxwell Garnett formula (cf. Eq. 15) multiplied by a factor $\sim(1 + 3\delta^{MG})$.

In the same way, there is for the Bruggeman theory an expression similar to Eq. (54), viz.

$$f_A \frac{\varepsilon_A - \bar{\varepsilon}}{\varepsilon_A + 2\bar{\varepsilon}} + (1 - f_A) \frac{\varepsilon_B - \bar{\varepsilon}}{\varepsilon_B + 2\bar{\varepsilon}} +$$

$$+ \delta^{Br}(\varepsilon_A,\varepsilon_B,\bar{\varepsilon},f_A,b,\lambda) = 0 . \qquad (56)$$

Now it is not possible to associate a certain magnitude of δ^{Br} precisely with a unique error in $\bar{\varepsilon}^{Br}$, because the relation between the two quantities depends on ε_A, ε_B and f_A. However, it can be shown that using $\bar{\varepsilon}^{Br}(1 + 3\delta^{Br})$, in analogy with the case of the Maxwell Garnett theory, one obtains a conservative estimate of the error in the average dielectric permeability.

In order to avoid misinterpretations of the above results, we will now look at how the errors enter into the real and imaginary parts of the effective dielectric permeability. If δ is small the following equation is obtained for the Maxwell Garnett case:

$$\bar{\varepsilon} = (\varepsilon_1{}^{MG} + i\varepsilon_2{}^{MG})(1 + 3\delta_1{}^{MG} + 3i\delta_2{}^{MG}) , \qquad (57)$$

where the subscripts "1" and "2" denote real and imaginary parts, respectively. This is equivalent to

$$\bar{\varepsilon}_1 = \bar{\varepsilon}_1{}^{MG}(1 + 3\delta_1{}^{MG} - 3\delta_2{}^{MG} \frac{\bar{\varepsilon}_2{}^{MG}}{\bar{\varepsilon}_1{}^{MG}}) , \qquad (58)$$

and

$$\bar{\varepsilon}_2 = \bar{\varepsilon}_2^{MG}(1 + 3\delta_1^{MG} + 3\delta_2^{MG} \frac{\bar{\varepsilon}_1^{MG}}{\bar{\varepsilon}_2^{MG}}) . \tag{59}$$

Thus the errors in the real and imaginary parts do not only depend on δ^{MG} but also on the ratio of the real and imaginary parts of $\bar{\varepsilon}^{MG}$. Expressions analogous to Eqs. (58) and (59) should give a conservative estimate of the errors also for the Bruggeman theory.

Other large-size criteria have earlier been proposed by Granqvist and Hunderi (104) and by Ruppin (157). These criteria are based on the Lorenz-Mie theory, which is compared to the Maxwell Garnett theory in the limit $f_A \to 0$, and thus they should give correct results only in this limit. Equations (58) and (59) give a more general criterion, though, since in these expressions there is no restriction on the value of the filling factor. Smith (65) has used an approach similar to the present one to derive large-size-limits. His criterion was based on the magnetic dipole contribution to δ and required that it should be less than a certain amount. This approach neglects the contribution of the ratio of the real and imaginary parts of $\bar{\varepsilon}^{MG}$ to the error in $\bar{\varepsilon}^{MG}$ as well as higher multipole effects.

4.2 Extensions of the Effective Medium Theories: Some Remarks

Extended effective medium theories can be formulated by retaining more than the first term in the series expressions for $S^{cs}(0)$ and $S^{s}(0)$ (74,158-161). Such extended theories have been invoked to account for far infrared absorption in ultrafine metal particles, which may be dominated by magnetic dipole effects. There is, however, a fundamental difficulty associated with such extended theories - a difficulty which is connected with our definition of an effective medium, which required that the scattering amplitude in the forward direction should be zero in order that the random unit cell should not be detectable by electromagnetic radiation. We could, however, equally well have made the definition that the backscattering from the random unit cell should be zero. The backscattering is proportional to $|S(\pi)|$, and thus this alternative definition becomes

$$S(\pi) = 0 . \tag{60}$$

Retaining only the electric and magnetic dipole terms in the series expression for $S(\pi)$ (8), the following result is obtained:

$$S(\pi) = \frac{3}{2}(\alpha_1 - \beta_1) . \tag{61}$$

But retaining the dipole terms in Eq. (6) we get another result, viz.

$$S(0) = \frac{3}{2}(\alpha_1 + \beta_1) . \tag{62}$$

As long as only the term α_1 plays a role $S(0)$ is equal to $S(\pi)$ and the two definitions of an effective medium are equivalent. This is the case for the derivations of the various theories in Chapter 2. However, when the higher terms begin to play a role $S(0)$ is no longer equal to $S(\pi)$, and the same effective medium formulation does not apply to both transmittance and reflectance determinations on a composite medium. The definition $S(0) = 0$ then applies only to transmittance measurements.

4.3 Computation of Large-Size-Limits for a Specific Example: Co-Al_2O_3

This section presents calculations for a model material taken to be a Co-Al_2O_3 composite. The purpose is to show quantitatively how the limits of validity depend on f_A and λ. Co-Al_2O_3 by no means has unique properties, but very similar results could have been reached with other transition metal composites. In the calculations we used the published bulk dielectric permeabilities for Co (146, 162) and for electron-beam evaporated Al_2O_3 (147).

The full series expansions of $S^{cs}(0)$ were employed to compute the particle radii which would yield a 1,2 and 5% correction to the average dielectric permeability within the Maxwell Garnett theory for Co spheres in Al_2O_3. Results obtained for a fixed wavelength equal to 1 μm are shown in Fig. 6. They are confined to $f_A < 0.7$, since the separated-grain structure cannot be maintained to too large filling factors. For the real part of $\bar{\varepsilon}$, the limiting radius is seen to decrease monotonically when the filling factor is increased, and, expectedly, a smaller radius is required the higher the accuracy we impose. For the imaginary part of $\bar{\varepsilon}$, the required limiting radii are in general smaller than those found for the real part. For example, the radius is only 5 to 6 nm over the main part of the f_A-range when 1% accuracy is demanded. An unexpected peak in the limiting radius of $\bar{\varepsilon}_2$ is found at small f_A's. The reason for this behaviour is possibly that the description of the material in terms of a random unit cell, which underlies the Maxwell Garnett theory, is not appropriate as $f_A \to 0$. In this latter case the Lorenz-Mie derivation may be more applicable.

Figure 7 shows results of a calculation of the limiting radii for a 1% correction to the Maxwell Garnett expression for $\lambda = 0.4$, 10 and 40 μm. Corresponding results for $\lambda = 1$ μm are shown by the dashed curves in Fig. 6. We note that the limiting radii show quite a different behaviour depending on whether the oxide is non-absorbing ($\lambda = 0.4$ and 1 μm) or absorbing ($\lambda = 10$ and 40 μm). The limits of validity are more generous in the latter case, which is understandable since the multipole interaction between the particles then becomes less important. The peculiar peak of the dashed curve in the left hand part of Fig. 7 is conceivably due to an accidental cancellation of high-order terms in $S^{cs}(0)$.

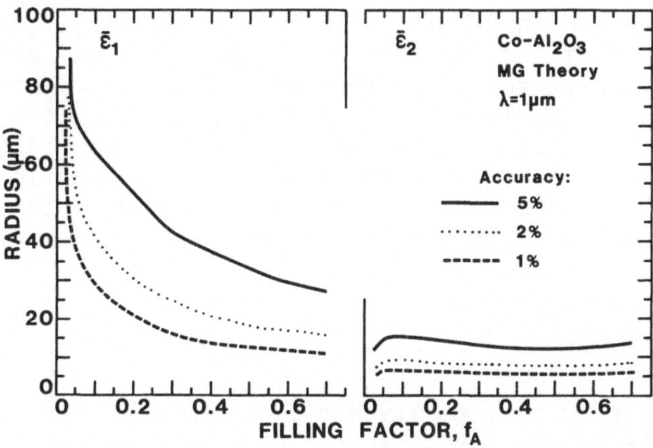

Fig.6 Limiting radii that are not to be exceeded if the real and imaginary parts of the Maxwell Garnett expression for the effective dielectric permeability are to be accurate to within the shown values. The data pertain to $Co-Al_2O_3$ composites with different compositions at $\lambda = 1$ μm.

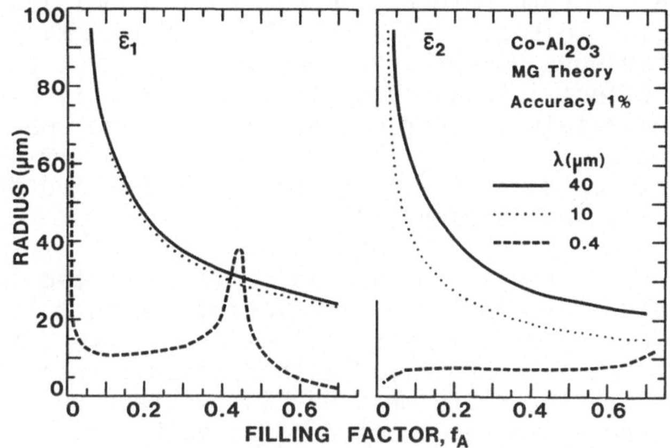

Fig.7 Limiting radii that are not to be exceeded if the real and imaginary parts of the Maxwell Garnett expression for the effective dielectric permeability are to be accurate to within 1% at the shown wavelengths. The data pertain to $Co-Al_2O_3$ composites with different compositions.

4.4 Small-Size-Limits

The smallest particle size for which the effective medium theories are practically useful is set by the limit in which the components of the heterogeneous material can no longer be described by macroscopic dielectric permeabilities. This limit does not presume the use of the ordinary bulk dielectric permeability, because a size dependence of the electron relaxation time can easily be incorporated at least for free-electronlike metals. When metal particles become smaller than 10 to 20 nm, a correction to the bulk dielectric permeability has to be applied because scattering of the conduction electrons against the surface of the particle becomes important. This scattering gives a contribution to the electron relaxation time, τ, which can be written (104, 163, 164).

$$\frac{1}{\tau_{part}} = \frac{1}{\tau_{bulk}} + \frac{v_F}{l} , \tag{63}$$

where v_F is the Fermi velocity and l is the mean free path of the conduction electrons. For spherical particles and diffuse boundary scattering, the mean free path is equal to the particle radius, a, (165). If the scattering is isotropic the mean free path is instead $\frac{4}{3}a$ (164).

The quantum mechanical analogue to this effect is the so called Quantum Size Effect. Because of the limited particle size, the energy bands will be discrete rather than quasicontinuous and for small enough particles this discreteness will introduce marked deviations from bulk behaviour. Several theories have been put forward in order to take this effect into account (161, 166-171). It has been described approximately by introducing an effective mean free path different from the classical one. However, there seems to be no great difference between this effective mean free path and the mean free path of the classical theory (172).

The effects surveyed so far can be incorporated into the usual effective medium formulation by correcting the bulk dielectric permeability. For very small particle sizes, however, a breakdown of the bulk bandstructure connected with a gradual transition toward the molecular cluster state must set in. A structural transition, probably between a cluster state and the bulk structure, for gold occurs at a particle diameter of $\sim$ 3 nm (173,174). A similar effect has also been observed in small silver particles (175). Photoemission experiments on various metal particles have shown that the valence bandwidth becomes narrower than in the bulk for particles below a certain size (176-179). The particle sizes necessary to obtain bulk-like behaviour in these cases have been measured to be in the range 1.5 to 4 nm. Such structural transitions and alterations of the bandstructure puts a practical small-size-limit, and we conclude that the effective medium theories should be used with much caution for samples with particle sizes below a few nanometers.

5 CASE STUDY ONE: OPTICAL PROPERTIES AND SOLAR SELECTIVITY OF COEVAPORATED Co-Al_2O_3 FILMS

We now turn to experimental data for the optical properties of composite materials and discuss how these can be analyzed in terms of effective medium theories. This Chapter is devoted to Co-Al_2O_3 composite films prepared by coevaporation under vacuum. Section 5.1 treats film production and characterization. Section 5.2 presents the dielectric function of the Co-Al_2O_3 films and its relation to the Maxwell Garnett effective medium theory and the Bergman-Milton bounds. In Section 5.3 we then employ the experimental data of the dielectric function in a computer optimization of solar absorptance and thermal emittance of Co-Al_2O_3 films on metal substrates. Experimental coatings designed so that they approximate the optimized design are then discussed in Section 5.4. Some preliminary reports on the material covered in this Chapter have appeared elsewhere (180-182).

5.1 Production and Characterization of the Films

The Co-Al_2O_3 composite films were produced by simultaneous evaporation of the two species onto a substrate. This technique is very versatile and is useful for preparing composites of various kinds. We used the system outlined in Fig. 8. It consists of an oil diffusion pumped stainless steel container with two electron-beam guns powered from a common supply. The gun used for deposition of Al_2O_3 was equipped with an automatic beam sweep in order to permit uniform evaporation for extended periods of time. The depositions of Co and Al_2O_3 were monitored independently by two vibrating quartz microbalances. As indicated in the figure, these are connected to computerized deposition controllers which maintained the evaporations at preset rates (between 0.5 and 2.5 nm/s for the films to be discussed) via feedback control of the electron-beam gun power supply. The substrates were clamped to a holder placed 30 cm above the guns. The substrate temperature could be stabilized at temperatures up to 300^oC but was always kept below 100^oC for the present experiments. A mechanical shutter was used to prevent deposition during warming-up and outgasing of the evaporants.

A reliable comparison between experiments and effective medium theories is contingent on a sufficiently accurate characterization of the specimens. Generally speaking, this issue has only seldom received proper attention (4), and as a result several of the published optical spectra for composite materials are susceptible to interpretation within different effective medium theories (153). To avoid such ambiguities as far as possible, we characterized the present Co-Al_2O_3 specimens by a variety of experimental techniques.

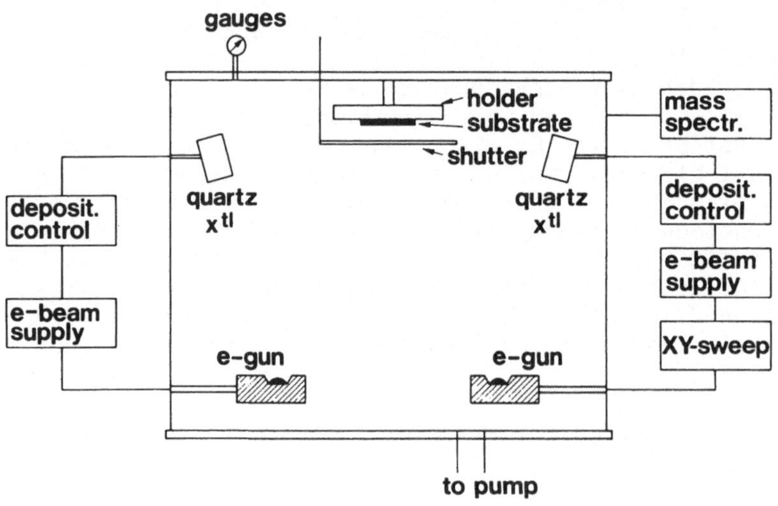

Fig. 8 Schematic view of the evaporation system and process controls. The system is described in detail in Ref. 183.

Transmission electron microscopy was used to determine the microstructure. Figure 9 shows typical electron micrographs taken on a composite film with 30 vol.% metal in bright-field and dark-field modes. We see that the composite comprises well defined single crystal particles in a continuous matrix. Electron diffraction analysis, performed within the electron microscope, proved that the particles are hexagonal close-packed Co and that the matrix is amorphous Al_2O_3. Size distributions were obtained from measurements on the micrographs. These were consistent with the log-normal distribution function (184-186). The median diameter was found to vary from ~ 1.5 nm in films with low Co content to ~ 2.5 nm in films containing 50 vol.% Co, and the geometric standard deviation was typically 1.6. Scanning electron microscopy showed no surface roughness on the micron scale.

The composition of the film as a function of depth below its surface is an important property. It can be determined by Auger Electron Spectroscopy combined with sputtering, Secondary Ion Mass

Spectroscopy, and by Rutherford Back-Scattering. A useful comparison of these methods is given by Evans (187). We used all of these depth profiling techniques on our Co-Al_2O_3 samples and verified that the composition remained constant to within about ±10%. We also attempted to use Field Ion Microscopy and Atom Probe techniques in order to obtain data with atomic resolution; this work has given encouraging results and is continuing.

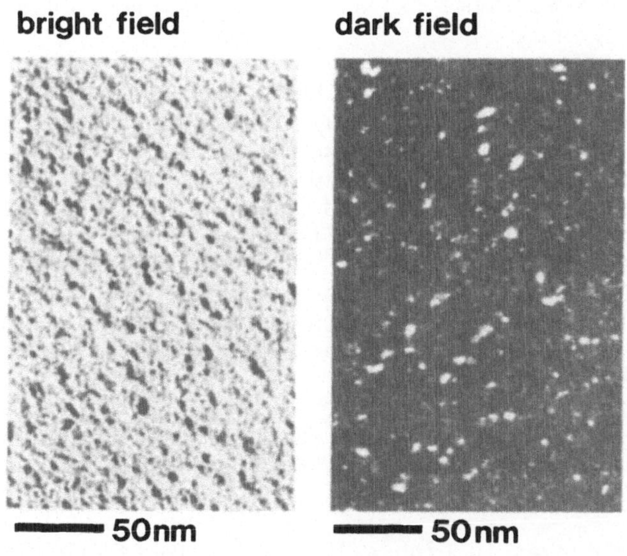

Fig. 9 Bright-field and dark-field electron micrographs for a co-evaporated Co-Al_2O_3 composite film with $f_A = 0.3$. The dark spots in the left-hand part and the bright spots in the right-hand part result from Co. The film was condensed onto a carbon film covering a Cu grid simultaneously with the production of a film for optical studies.

Film thickness and overall filling factor were determined by use of a simple surface profiler technique (188). Thicknesses were found also by optical methods to be discussed below.

The electrical properties measured as a function of filling factor give important information which is supplementary to the optical data (189). The electrical conductivity (σ_e) was determined for Co-Al_2O_3 films with $f_A \lesssim 0.3$ by capacitance measurements at 1 kHz and for $f_A \gtrsim 0.3$ by direct dc measurements between the ends of the films. It was found that, typically, $\sigma_e = 10^{-6}$ $(\Omega m)^{-1}$ at $f_A = 0.2$; the conductivity then increased sharply at $f_{Co} \approx 0.3$ and reached ~ 10^2 $(\Omega m)^{-1}$ at $f_A = 0.4$ and ~ 10^4 $(\Omega m)^{-1}$ at $f_A = 0.6$. The temperature coefficient of resistivity, determined at near-ambient temperature, went from ~ - 10^{-2} K^{-1} for small f_A to zero at $f_A \approx 0.7$.

5.2 Dielectric Permeability in the 0.3-40-μm Range

The spectral optical properties were determined for the Co-Al_2O_3 composites in the 0.3-40-μm range using double-beam spectrophotometers with reflectance attachments. Normal transmittance (T) was measured for films on substrates of glass (Corning 7059), silicon, and thallium-bromide-iodide (KRS-5). Near-normal reflectance was measured on the same films (R) as well as on films on silver-coated glass (R_m).

There is a large number of schemes for determining the complex dielectric permeability (or optical constants) of thin films (190-203). Below we will follow an approach which was recently put forward and analyzed by Hjortsberg (203). The dielectric permeability and the film thickness were first computed by applying Fresnel's equations to the data for T, R and R_m. This technique, which was normally applicable over a narrow spectral range only, yielded thickness values which agreed well with those recorded with the afore-mentioned surface profiler technique. Once the thickness was known, $\bar{\varepsilon}_1$ and $\bar{\varepsilon}_2$ were obtained by combining (T,R) and (T,R_m) and using the most accurate of these two evaluations (203). In this way we could determine the dielectric permeability over the whole solar- and thermal ranges, which, as far as we know, was not done in any earlier work on composite materials produced by coevaporation or cosputtering (154-156,204-214).

The solid curves in Figure 10 show evaluated results of $\bar{\varepsilon}_1$ and $\bar{\varepsilon}_2$ for Co-Al_2O_3 films with three magnitudes of the experimentally determined overall filling factor of Co (denoted f_{exp}; as we will see shortly this quantity can be larger than the "theoretical" filling factor $f_{th} \equiv f_A$ used in the effective medium theories). Two or three films with the same composition but different thicknesses had to be evaluated to cover the whole 0.3-40-μm interval. The curves in Fig. 10 are drawn as "best fits" to the extracted data with one exception: $\bar{\varepsilon}_2$ in Fig. 10(c) is given as two separate branches pertaining to $f_{exp} = 0.58$ and $f_{exp} = 0.61$. The uncertainties in $\bar{\varepsilon}_1$ and $\bar{\varepsilon}_2$ from a conservatively taken one-percent-error in the spectrophotometric data are, typically, ±0.2 to ±0.6 for the two lowest filling factors, and ±0.4 (at $\lambda < 2.5$ μm) to ±16 (at the largest wavelengths) for $f_{exp} \approx 0.60$. Additional uncertainties stem from errors in the determination of film thickness. The internal consistency in the data for $\bar{\varepsilon}_1$ and $\bar{\varepsilon}_2$ was verified by Kramers-Kronig analysis. We note from Fig. 10 that the $\bar{\varepsilon}_2$-curves increase rapidly at $\lambda \gtrsim 10$ μm; this is caused by absorption in Al_2O_3 (147).

The experimental dielectric permeability was compared with computations based on the Maxwell Garnett theory. This particular theory is expected to be adequate on account of the sufficiently small particle size (cf. Sec. 4.3) and the fact that a separated-grain structure was found by electron microscopy (cf. Fig. 9). As input data we used the bulk dielectric functions by Johnson and Christy (146) and by Bolotin et al. (162) for Co and by Eriksson et al. (147) for electron-beam evaporated Al_2O_3. In fact, the determination for Al_2O_3 was carried out as part of the present work. The bulk dielectric function for Co has been measured several times, and somewhat differing results have been stated (146,150,162,215-220). However, the differences in the Maxwell Garnett calculations induced by chosing different sets of input data were found to be insignificant for our comparison with the experiments. We earlier noted that the Co particles were mostly single crystals and therefore they are optically anisotropic; the differences in the dielectric function for various crystal orientations are small, though (217, 220). Finally, we investigated the change in the dielectric function introduced by boundary scattering of the conduction electrons in the Co particles (cf. Sec. 4.4) by using the extracted Drude parameters for this metal (162). Again the effects on the Maxwell Garnett results were small. Summarizing the last paragraph, we find that our choice of input parameters should make a valid comparison of theory and experiments possible.

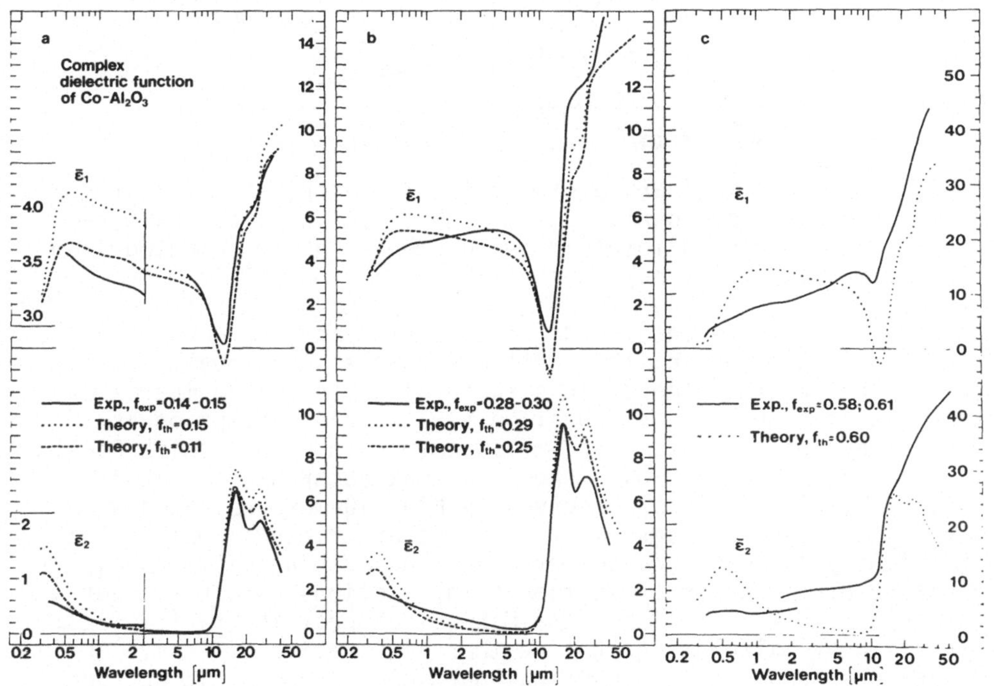

Fig. 10 Real and imaginary parts of the dielectric function for co-evaporated $Co\text{-}Al_2O_3$ films with three different compositions as evaluated from spectrophotometric measurements and as computed from the Maxwell Garnett theory. Note the different vertical scales in (a), (b), and (c).

The dashed and dotted curves in Fig. 10 show theoretical results for $\bar{\varepsilon}^{MG} \equiv \bar{\varepsilon}_1^{MG} + i\,\bar{\varepsilon}_2^{MG}$. The dotted curves refer to $f_A = f_{th} = f_{exp}$, i.e., assuming that all of the evaporated Co is included in the particles. These curves are seen to deviate systematically from the measurements, the discrepancies being most striking for low filling factor. We found that, for example, a specimen with $f_{exp} = 0.07$ could be reconciled with the Maxwell Garnett theory only by taking $f_{th} \approx 0.02$. Similarly, a composite film with $f_{exp} = 0.11$ required $f_{th} = 0.06$. This latter example is elaborated in Figure 11, in which the good agreement obtained by taking $f_{th} = f_{exp} - 0.05$ is evident.

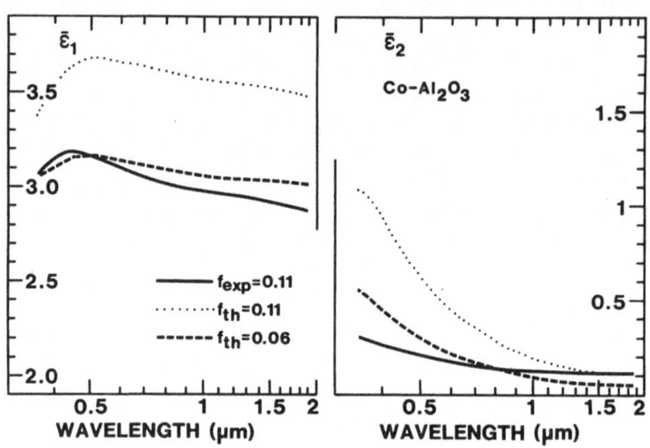

Fig. 11 Real and imaginary parts of the dielectric function for a coevaporated $Co\text{-}Al_2O_3$ film with $f_{exp} = 0.11$ as evaluated from spectrophotometric measurements and as computed from the Maxwell Garnett theory.

This consistent difference leads us to postulate that about 5 vol.% of the matrix surrounding the particles consists of cobalt 1). In fact, it appears that the basic structural units of the matrix are approaching the spinel configuration (221). This assumption is supported by spectrophotometric measurements on micron-thick $Co\text{-}Al_2O_3$ films with $f_{exp} = 0.05$ which showed a weak absorptance peak at $\lambda \approx 0.6$ μm (221). The dashed curves in Fig. 10 were obtained with $f_{th} = f_{exp} - 0.04$. The agreement with the experimental data is now significantly improved. We believe that the ensuing differences are indicative of higher order multipole interactions, which are not properly accounted for within the Maxwell Garnett theory (63,122). We verified by calculation that neither alternative effective medium theories nor presumed non-spherical particle shapes could account for the measured dielectric function.

---

1) We can compare this result with measurements by Devenyi et al. on electron-beam coevaporated $Nb\text{-}Al_2O_3$ films (220a). They found that only about 10% of the total amount of Nb was concentrated in particles, the rest being dispersed in the Al_2O_3 matrix.

The measured dielectric function for Co-Al_2O_3 was compared also with the Bergman-Milton bounds. Three kinds of behaviour were encountered: (i) for $\lambda \leq 10$ µm and $f_{exp} \leq 0.4$, the $\bar{\varepsilon}$'s lay inside the bounds for statistically isotropic materials; (ii) for $\lambda \lesssim 10$ µm and $0.4 < f_{exp} < 0.7$, the $\bar{\varepsilon}$'s were consistent only with the wider bounds for anisotropic materials; and (iii) for $\lambda \gtrsim 10$ µm, the experimental $\bar{\varepsilon}$'s sometimes fell outside even the generous bounds appropriate when only the dielectric permeabilities for the constituents and the filling factor are taken to be known. The reason for this latter discrepancy is not yet known.

5.3 Computer Optimization of Solar Absorptance and Thermal Emittance

The efficiency by which a surface can convert solar energy into useful heat is governed by two parameters: the normal solar absorptance, a_s, and the hemispherical thermal emittance, e_T (26-30). As discussed in the Introduction, the solar absorptance should be as close to unity as possible, whereas the thermal emittance should be as close to zero as possible. A practical solar converting surface of course also requires sufficient durability, etc. The evaluated dielectric permeability of Co-Al_2O_3 composite films makes it possible to perform a valid optimization of the magnitudes of a_s and e_T that can be achieved with coatings of this material onto metal substrates.

The normal solar absorptance is defined by

$$a_s = \int_0^\infty d\lambda\ \phi(\lambda)[1 - R(0,\lambda)] / \int_0^\infty d\lambda\ \phi(\lambda)\ , \tag{64}$$

where $\phi(\lambda)$ is the spectral solar irradiance and $R(0,\lambda)$ is the normal reflectance. The hemispherical thermal emittance is given by

$$e_T = \frac{1}{\sigma T^4} \int_0^\infty d\lambda\ W(\lambda,T)e(\lambda), \tag{65}$$

$$e(\lambda) = 1 - \tfrac{1}{2} \int_0^{\pi/2} d(\sin^2\Theta)[R_{TE}(\Theta,\lambda) + R_{TM}(\Theta,\lambda)]\ , \tag{66}$$

where $W(\lambda,T)$ is the blackbody exitance for temperature T, TE(TM) denotes transverse electric (transverse magnetic) polarisation, and σ, the Stefan-Boltzmann constant, is equal to 5.67×10^{-8} $Wm^{-2}K^{-4}$. $R_{TE}(\Theta,\lambda)$ and $R_{TM}(\Theta,\lambda)$ are conveniently computed from $\bar{\varepsilon}$ by use of a matrix technique (222-224). Useful treatments of the radiative properties of surfaces are given in Refs. 3 and 225-227. In our computations, to be presented below, we set $\phi(\lambda)$ equal to the AM2 spectrum (25; cf. Fig. 1b) and T = 100°C.

Figure 12 shows computed results of a_s and e_T for Co-Al_2O_3 films

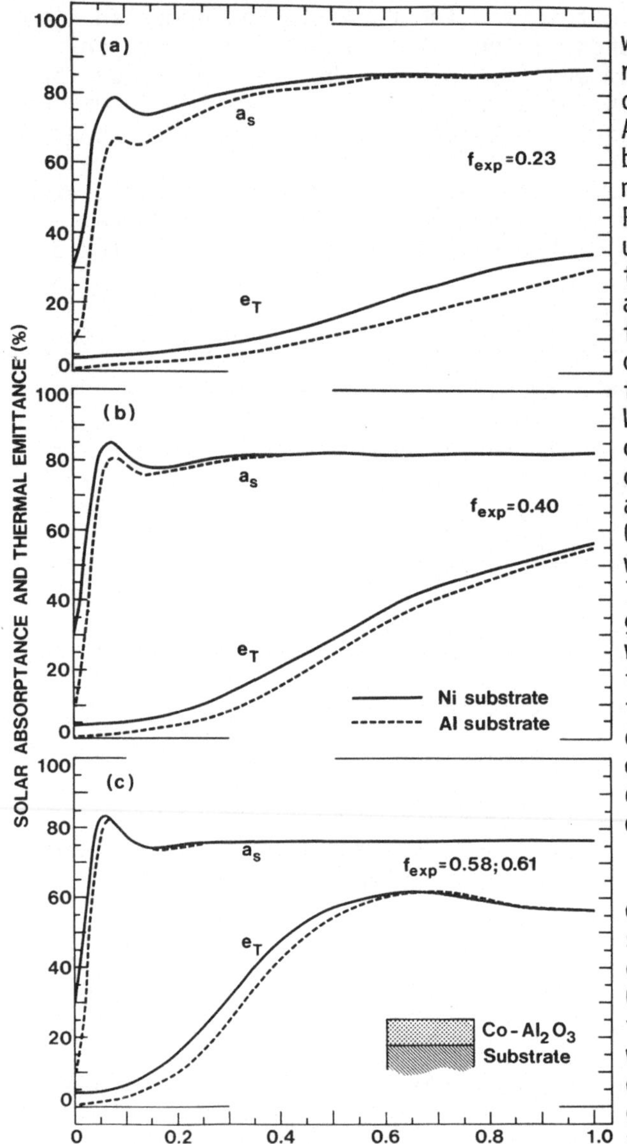

with different thicknesses and filling factors on substrates of Ni and Al. The dielectric permeabilities of the substrate metals were taken from Ref. 228, which is a very useful compilation of optical data. In Fig. 12(c) a_S was obtained from $\bar{\varepsilon}$ for f_{exp}=0.58 whereas e_T was obtained from $\bar{\varepsilon}$ for f_{exp}=0.61 (cf. Fig. 10c). We observe that a_S increases rapidly with increasing film thickness and reaches a maximum of 0.85 at 70 nm in the films with 40 and ∿ 60 vol.% Co laid on Ni. Al substrates give somewhat lower a_S's which is expected from the smaller absorptance in this metal. The thermal emittance is 0.06 for 70 nm of Co-Al_2O_3 on Ni and is even lower for such films on Al.

The maximum values of solar absorptance, shown in Fig. 12, are undesirably low for the bare Co-Al_2O_3 films, and with the goal of enhancing a_S without significantly increasing e_T we investigated the role of an antireflecting overcoat of

Fig. 12 Calculated results for solar absorptance and thermal emittance of Co-Al_2O_3 with different thicknesses and compositions laid on Ni and Al.

Al_2O_3. Figure 13 shows the effect of putting an Al_2O_3 film on top of a 70-nm-thick Co-Al_2O_3 layer with f_{exp} = 0.58. A maximum solar absorptance as high as 0.95 is found with 70 to 80 nm of Al_2O_3 and with a Ni substrate.

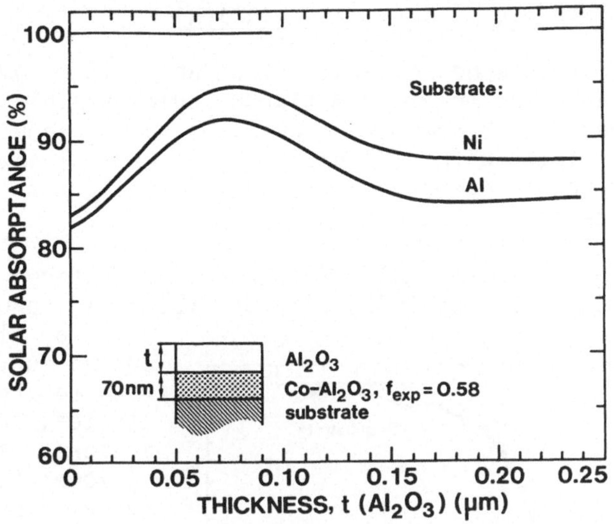

Fig. 13 Calculated solar absorptance for substrates of Ni and Al coated with a $Co-Al_2O_3$ film antireflected by Al_2O_3.

Figure 14 shows results of some further optimization studies of coatings onto Ni. Now the Al_2O_3-thickness was kept at 80 nm while the $Co-Al_2O_3$-thickness was varied for films with three compositions. The most favourable case occurred for 70 nm of $Co-Al_2O_3$ with $f_{exp} = 0.58$ for which $a_s = 0.95$ was found. The e_T-value was 0.07 for this configuration.

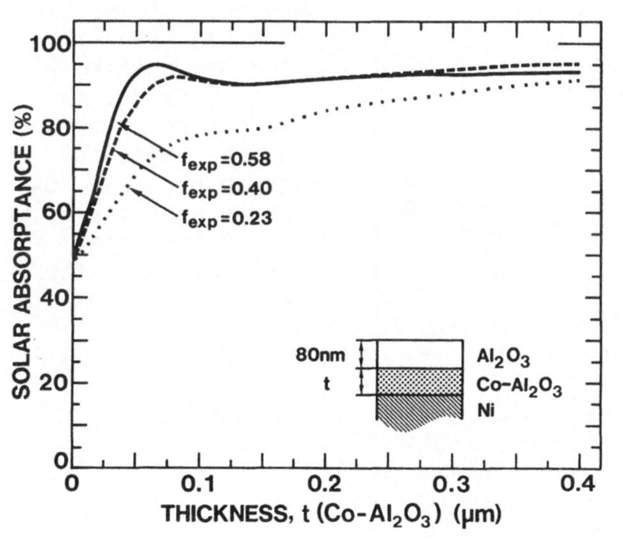

Fig. 14 Calculated solar absorptance for Ni coated with $Co-Al_2O_3$ films having different thicknesses and compositions and antireflected by 80 nm of Al_2O_3.

5.4 Experimental Studies of Optimized Coatings

We found in the preceding section that an optimum combination of solar absorptance and thermal emittance could be achieved with Ni substrates coated with ~ 70 nm of Co-Al_2O_3 having about 60 vol.% metal and antireflected with ~ 70 nm of Al_2O_3. We will now discuss experimental samples produced so as to approximately comply with this most favourable design.

Figure 15 shows near-normal reflectance for coatings on substrates being glass (Corning 7059) with an evaporated opaque layer of Ni or Al. The dotted curves, referring to the bare substrates, show that the measured reflectance is lower for Ni than for Al. The parallel-band absorption in Al is evident at $\lambda \approx 0.8$ µm (229-231). After deposition of 65 to 70 nm of Co-Al_2O_3 with $f_{exp} = 0.6$, the dashed curves show a rather high normal solar absorptance and a thermal normal emittance which is only slightly higher than that of the bare substrate. Finally, after overcoating with 70 nm of Al_2O_3, the solid curves prove that the region of high absorptance now extends over the main part of the solar spectrum, while the thermal normal emittance is practically unchanged. The small

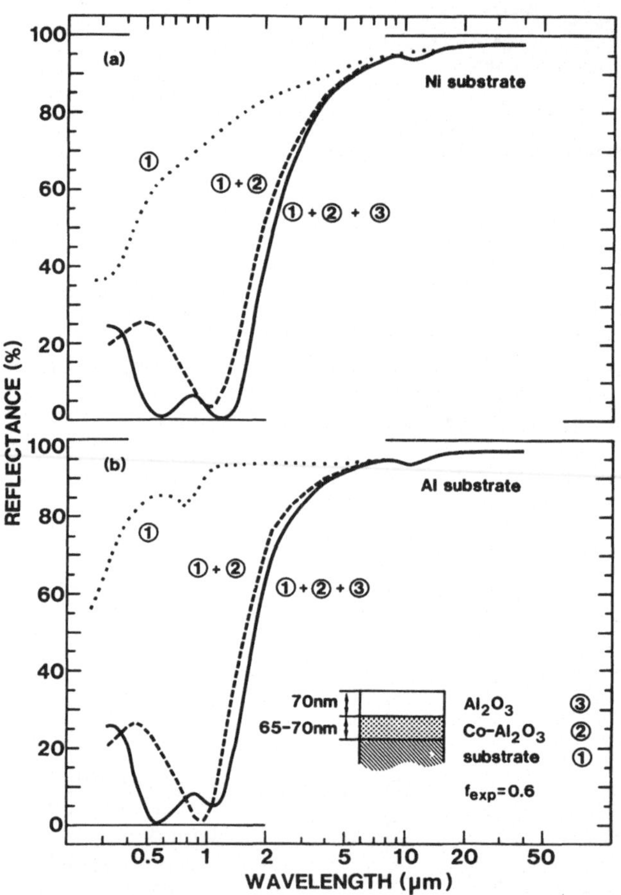

Fig. 15 Measured spectral reflectance of Ni and Al surfaces (dotted curves), after deposition of Co-Al_2O_3 (dashed curves), and after overcoating with Al_2O_3 (solid curves). The different layers, and the corresponding curves, are marked by the encircles numbers.

reflectance minima at $\lambda \approx 11$ µm are indicative of an emittance peak at this wavelength which shows up distinctly in the hemispherical emittance (232). This feature is caused by the emission of TM-polarised light, which is strong particularly at large off-normal angles (233). The measured spectral data were found to agree excellently with computations for near-normally incident light as well as for TE- and TM-polarised light with 25^0, 45^0 and 60^0 incidence angle.

The normal solar absorptance was evaluated from the spectral reflectance by use of the selected ordinate technique (234,235). The solid curve in Fig. 15(a) corresponds to $a_s = 0.94$ and the solid curve in Fig. 15(b) corresponds to $a_s = 0.91$. We found that a_s remained larger than 0.90 at incidence angles up to 60^0 for films on Ni; this is an important property for a practical solar absorbing surface. The hemispherical thermal emittance was measured by transient calorimetry at 60 to 100^0C. This type of measuring device is described in Refs. 236-239. We found $e_T = 0.095 \pm 0.01$ for coatings on Ni and $e_T = 0.085 \pm 0.01$ for coatings on Al. We therefore conclude that the favourable spectral selectivity found in the computer optimization can be realized with experimental samples.

Summarizing, we have found that surfaces which absorb ~ 95% of the incoming solar radiation and emit only ~ 10% of the blackbody radiation can be constructed by use of Co-Al_2O_3 composite films. The most favourable composition is one where the metal content is 50 to 60%. This filling factor is much larger than in most earlier work on metal-dielectric films for selective absorption of solar energy (206-210,214,240,241), even if some reports also on concentrated composites have appeared recently (242,243). In several earlier papers the needed high solar absorptance was achieved by varying the composition during the deposition ("graded-index films") or by application of surface layers with controlled roughness (155,212,244-248). One important result of the present work is therefore to demonstrate that the extra complexity introduced for obtaining composition-grading and/or roughening may not be necessary for most applications related to photothermal conversion of solar energy.

6 CASE STUDY TWO: ELECTROLYTICALLY COLOURED ANODIC Al_2O_3 COATINGS

In this Chapter we go from laboratory-scale experiments on metal-insulator composite films to coatings of this kind produced commercially for solar energy utilization on a scale of tens of thousands of square meters per year. We will now treat anodic Al_2O_3 coatings coloured by electrolytic means so that they contain a fine metallic pigment. The analysis of these coatings is not yet as elaborate as for the coevaporated Co-Al_2O_3 composites, but anyway their structure is sufficiently well known that effective medium theories are of much value for understanding the optical properties.

In Section 6.1 we treat the preparation, characterization and reflectance spectra of Ni-pigmented anodic Al_2O_3. Section 6.2 then gives an analysis of the reflectance in terms of the Maxwell Garnett and Bruggeman effective medium theories. Much of the material covered below has been presented in detail in earlier reports (4,249-251) and for this reason the present discussion is rather brief.

6.1 Preparation, Characterization and Spectral Reflectance

Oxide films are known to be formed on Al by anodizing in acid electrolytes (252,253). The main part of those films contain fine pores with a net orientation perpendicular to the surface of the sheet and extending from the outer surface as far as to a compact scalloped "barrier" layer which separates the porous part from the metal (254-257). The porous Al_2O_3 can be coloured by several different methods; we used electrolytic colouration since this technique is known to be able to yield extremely lightfast coatings (258-266). The coatings to be analyzed below were formed by dc anodization of 99.5% pure Al sheet in dilute phosphoric acid followed by ac electrolysis in a bath containing $NiSO_4$. By this technique a metallic Ni pigment was occluded in the oxide matrix. The desired metal content could be obtained by various combinations of process parameters such as anodizing voltage, time, bath temperature, and $NiSO_4$ concentration.

The coatings were analyzed by several experimental techniques: Scanning electron microscopy was used to study specimens which had been fractured by excessive bending. The typical film thickness was found to be 0.7 μm. Some small-scale surface roughness was seen on the micrographs. A columnar structure of the coatings was evident with a typical separation of 50 nm between the pores. Auger electron spectroscopy combined with depth profiling by sputtering was used to study the cross-sectional composition. It was found that the Ni concentration was significant only in a band close to the Al interface. Following earlier investigations (258-263,266), we interpret this as an effect of a metallic Ni pigment accumulated at the bottoms of the pores during the electrolysis. Atomic absorption analysis on dissolved coatings was employed to quantitatively determine the amount of Ni per unit area. This number is important since it gives the filling factor.

Figure 16 shows our structural model for the coatings. The Al base metal is covered with an extremely thin "barrier" layer of compact Al_2O_3 which plays little role for the optical performance. The barrier layer is coated with 0.7 μm of porous Al_2O_3, whose lowest portion is impregnated with metallic Ni. The filling factor is typically 0.23. The topmost part is taken to consist solely of porous Al_2O_3; this part is divided into two, which is motivated by an expected widening of the pores towards their openings.

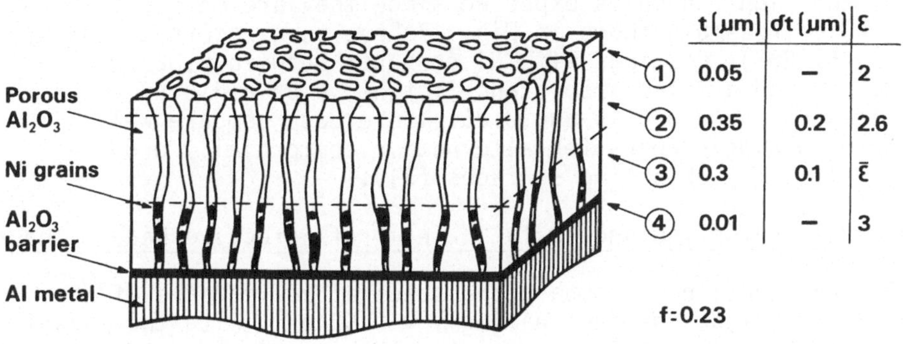

	t [μm]	σt [μm]	ε
①	0.05	–	2
②	0.35	0.2	2.6
③	0.3	0.1	$\bar{\varepsilon}$
④	0.01	–	3

Fig. 16 Structural model for Ni-pigmented anodic Al_2O_3 produced for selective absorption of solar energy. The inset table shows typical numbers for film thickness, roughness, and dielectric permeability of the four constituent layers.

The spectral reflectance was measured by spectrophotometry. Figure 17 shows results for four coatings whose Ni contents, γ, are different. The reflectance is low in the solar range and high in the thermal-infrared range, so that the surfaces have the spectral

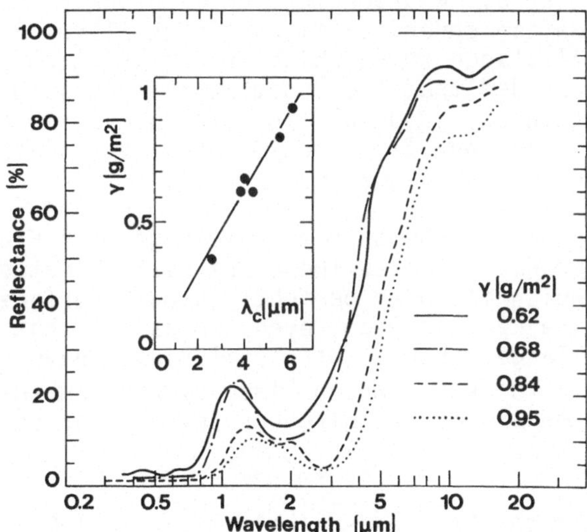

Fig. 17 Measured spectral reflectance for films with different Ni contents (γ). The inset figure shows the relation between γ and the wavelength corresponding to 50% reflectance (λ_c).

selectivity desired for photothermal conversion of solar energy. Films with higher Ni contents have their reflectance curves shifted towards the right which is expected since they are more absorbing. Curves which resemble those in Fig. 17 have been recorded also by others (267,268). Coatings with Ni contents between 0.62 and 0.95 gm^{-2} had a solar absorptance between 0.92 and 0.97 and a thermal emittance between 0.10 and 0.26. These data compare well with the corresponding numbers for alternative commercially produced selectively solar absorbing surfaces (31).

6.2 Analysis of Reflectance Data in Terms of Effective Medium Theories

In this section we compare the spectral reflectance of Ni-pigmented anodic Al_2O_3 coatings having $\gamma = 0.62\ gm^{-2}$ with the Maxwell Garnett and Bruggeman effective medium theories. Both formulations will be applied since the pertinent microstructure of the metal-insulator composites has not been resolved 1). As input data we used for Ni the published bulk dielectric permeability (146,218,269) and for Al_2O_3 the numbers shown in the inset in Fig. 16. These data for Al_2O_3 are sufficiently accurate for $\lambda \lesssim 10$ µm (147). No size dependence was invoked for the Ni particles. The spectral reflectance of the four-layer stack shown in Fig. 16 was then computed by use of a computer-oriented matrix technique (222-224). The layer thicknesses and the filling factor are those shown in that figure.

Figure 18 serves to demonstrate the significance of the individual layers in the structural model. The reflectance of the bare Al surface exceeds 85% at all wavelengths. Application of the pigmented layer assumed to contain spherical Ni particles (layer 3) is seen to give a spectral reflectance curve which is beginning to resemble the measured one provided that the Bruggeman theory is used, whereas the Maxwell Garnett theory is unable to give such agreement. The role of the other layers, as seen from the solid curve, is essentially to decrease the reflectance at $\lambda \lesssim 3$ µm.

The multiple peaks at the shorter wavelengths in Fig. 18 indicate strong interference effects. These are overestimated by the structural model, which includes perfectly smooth surfaces. Real surfaces, though, are certainly not even. We assume that different parts (taken to be larger than λ) of the ith layer have equal probability of having a thickness which lies anywhere in the interval $t_i \pm \delta t_i/2$. It is then simple to introduce the role of roughness into

1) From Fig. 16 it might be guessed that the Maxwell Garnett approach would be superior. However, one should not jump to conclusions from this simple model, and we have not so far been able to conclude whether the separated-grain or the aggregate structure (cf. Fig. 2) is most adequate.

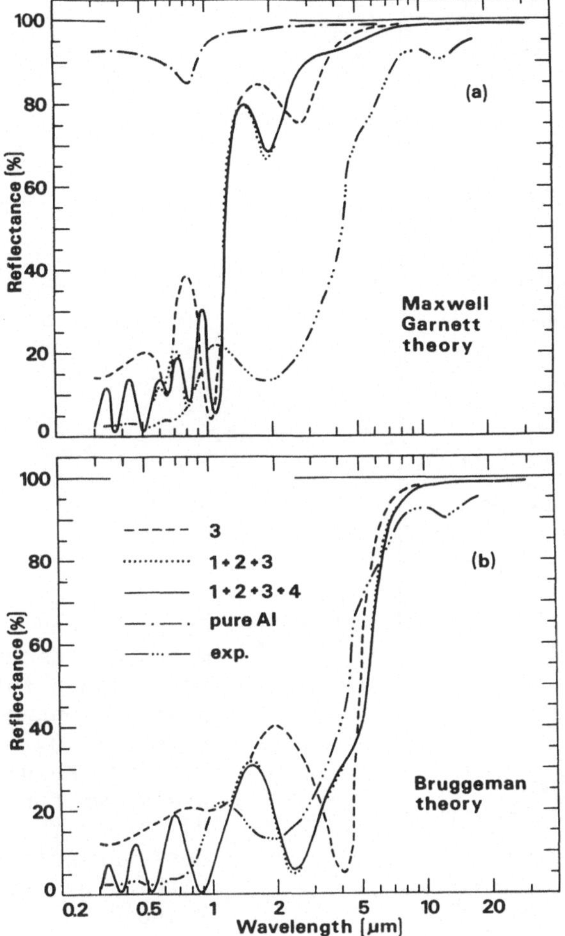

Fig. 18 Computed and measured spectral reflectance for a Ni-pigmented anodic Al_2O_3 coating. The numbers refer to the layers in the structural model in Fig. 16.

the computations. Figure 19 proves that with reasonable magnitudes of δt_2 and δt_3 only the main features of the interference peaks remain. The agreement with the experiments is thus improved.

The role of non-spherical particles was investigated in computations for which the Ni particles were taken to be spheroidal with their symmetry axes perpendicular to the surface of the coating (i.e., along the pore axes). This geometry makes it possible to characterize the shapes by one unique depolarisation factor, L, for light with normal incidence. By choosing L slightly larger than 1/3 we could improve the agreement between the experimental and theoretical curves somewhat provided that the Bruggeman theory was applied. This choice amounts to having prolate spheroidal particles elongated in the pore direction - a geometry which is plausible considering the nature of the anodic Al_2O_3 layers. In no case could a reasonable particle shape produce agreement between experiments and computations based on the Maxwell Garnett theory. We are then led to the conclusion that the basic microgeometry of the Ni-pigmented anodic Al_2O_3 coatings is the aggregate structure underlying the Bruggeman theory rather than the separated-grain structure of the Maxwell Garnett approach.

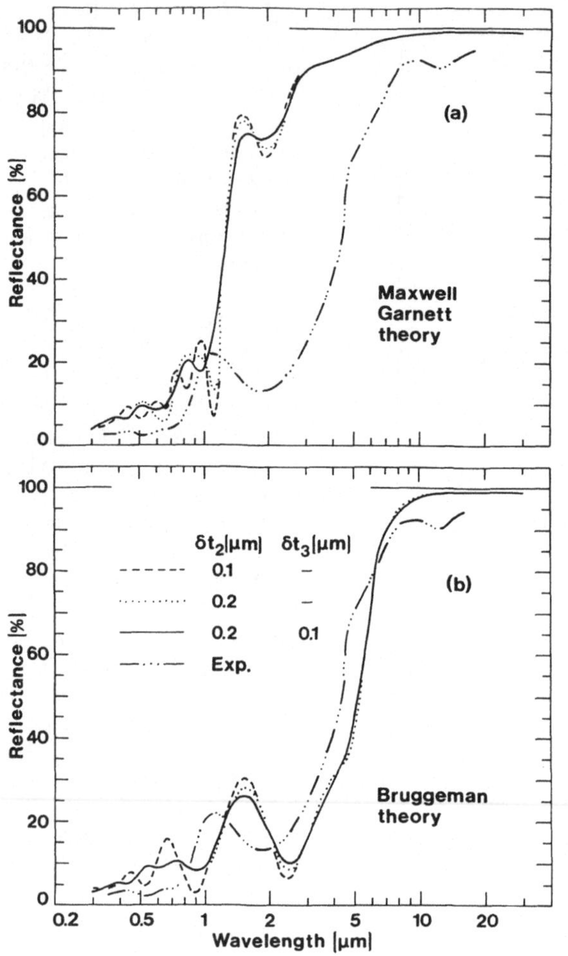

Fig. 19 Computed and measured spectral reflectance for a Ni-pigmented anodic Al_2O_3 coating. Results are shown for different thickness variations within the layers 2 and 3 (cf. Fig. 16).

Figure 20 shows our final comparison of the measured reflectance of Ni-pigmented anodic Al_2O_3 coatings and the Bruggeman theory. The calculations assume prolate spheroidal particles with major axis a, perpendicular to the surface of the coating, and minor axes c. The inset shows a computation of the wavelength corresponding to 50% reflectance, λ_c, versus a/c. The experimental value λ_c = 4 μm was obtained for a/c = 1.24. The main part of Fig. 20 shows the satisfactory agreement between experimental and theoretical data that can be achieved with this particular choice of a/c-ratio. Thus an effective medium theory is capable of providing at least a semi-quantitative description of the optical properties of a practically useful surface for photothermal conversion of solar energy.

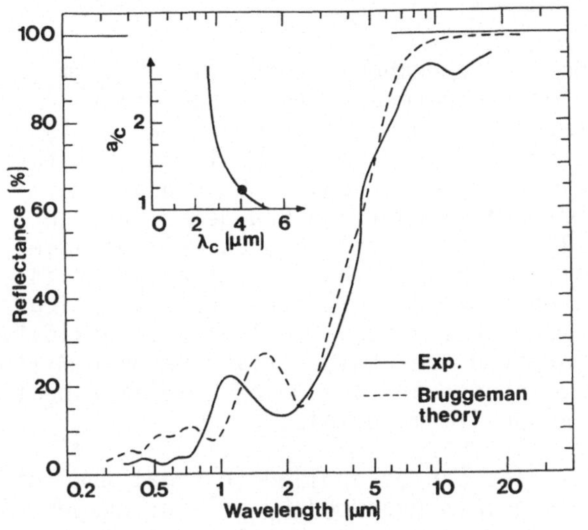

Fig. 20 Computed and measured spectral reflectance for a Ni-pigmented anodic Al_2O_3 coating. The theoretical curve refers to prolate spheroidal particles with a ratio between major and minor axes as indicated in the inset.

Our last point is to prove that optical properties similar to those for Ni-pigmented anodic Al_2O_3 can be obtained also by use of other transition-metal pigments. The solid curve in Figure 21 is identical to the solid curve in Fig. 19(b) and serves as a reference. The dashed and dotted curves in Fig. 21 were obtained by inserting the dielectric functions for Cr and Co (228), respectively, into the computations and retaining all other parameters. The three curves are indeed similar to one another, and hence the selection of a suitable pigment can be performed with regard to long-time durability of the coating rather than with regard to its specific optical properties.

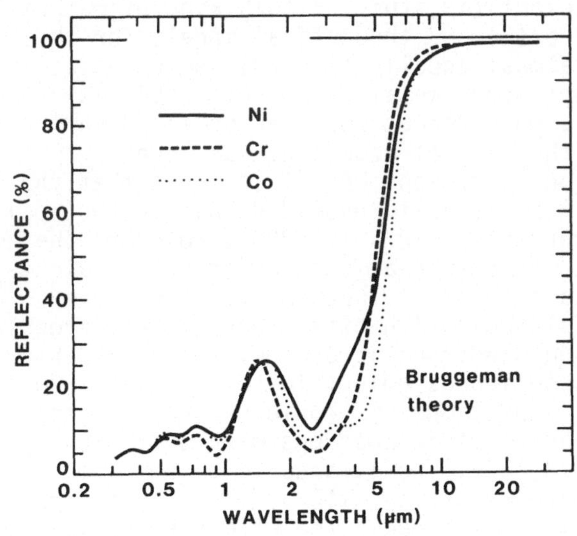

Fig. 21 Computed spectral reflectance for anodic Al_2O_3 coatings with pigments of Ni, Cr and Co.

7 SUMMARY AND REMARKS

In this paper we have derived a number of effective medium theories for the average dielectric permeability of composite materials from a unified framework in which scattering theory is applied to random unit cells defined so as to properly account for the basic microstructure. We were able to regain the old and well known expressions from the Maxwell Garnett and Bruggeman theories, and extensions of these to ellipsoidal particles, as well as some more recent theories. The main virtues of the present approach is that the fundamental connection between microstructure and optical properties is made explicit, and that large-size-limits of validity are obtained. We have also surveyed the newly derived Bergman-Milton bounds on the average dielectric permeability which hold irrespective of the microgeometry of the composite materials.

The effective medium theories can be applied to treat the optical properties of inhomogeneous materials produced by a large number of techniques for various purposes. However, we want to warn against an uncritical use of the theories: they can be applied with confidence only if the microstructure of the composite is known. This shows the importance of accurate characterizations of the microstructure - a need that is most acute for metal-insulator composites. This issue has received proper attention only in recent years. We will finish this paper by a survey over various composite material systems and point at the most severe characterization problems. The discussion is limited to concentrated composites (according to the distinction made in the Introduction).

Discontinuous metal films are our first example. These are two-dimensional arrays of islands distributed over a dielectric substrate. As earlier mentioned, the scientific study of this kind of materials began more than a century ago (2) and theoretical models for their optical properties go back almost equally far (53). Work on the optical properties of discontinuous metal films until 1977 is covered by Refs. 10 and 115; more recent work can be found in Refs. 175 and 270-287. The theoretical modelling has recently reached a high degree of sophistication, and it is known that the basic Maxwell Garnett formalism is valid provided that an "optical" film thickness is invoked correctly (127, 288-290). However, the application of this model to experimental samples is not straight-forward. The main problem is to properly account for the three-dimensional shapes of the islands (which often look rather irregular when viewed in a transmission electron microscope) and their inter-island separations. For discontinuous gold films evaporated onto glass under well controlled conditions, we could reach good agreement between theory and experiments in coatings with sufficiently small mass thicknesses (4,113,127). The islands in these films could be well represented by prolate spheroids with symmetry axes along the substrate plane (127,291), which makes the effective medium computations tractable. It is not known whether this spheroidal approximation has a more general applicability.

Gas evaporated coatings are produced by vaporization of a material in the presence of a reduced atmosphere of an inert gas 1). The evaporated species undergoes collisions with the gas atoms and is thereby cooled so that a supersaturated state is reached from which tiny nuclei are formed by homogeneous nucleation (186). These primary nuclei then grow in the gas via multiple coalescence events. The particles are transported by gas convection until they collide with some surface (substrate) where they stick and form a loosely packed composite structure. Gas evaporated coatings have been studied since the early 1930's. Recently, the interest in this kind of materials has been revived since it is realized that they can provide simple "model systems" which simulate the optical performance ofselectively solar absorbing surfaces. Work in this field until 1976 is cited in Ref. 104; more recent research is reported in Refs. 120 and 292-307. The overall filling factor of the coatings is of the order of 1%, and therefore the different effective medium theories are not expected to give results which significantly differ from one another. However the gas evaporated particles exhibit a strong local clustering into various kinds of aggregates, so that local-field effects are very significant for governing the optical properties. The pertinent theoretical analysis for local dipole-dipole interaction was outlined in Sec. 2.10. We showed earlier (104) that by taking gas evaporated gold coatings to consist of only three different types of cluster (i.e., introducing two adjustable parameters), we could bring computed transmittance spectra into excellent agreement with measured data. The main problem with regard to sample characterization is to account for the pronounced clumping in a reliable fashion. As a result of this difficulty, theoretical predictions of the optical properties of gas evaporated specimens are at best semi-quantitative.

Codeposited coatings can be produced by simultaneous evaporation or sputtering of two or more components. Composite coatings of almost any combination of materials can be made. Mechanical and thermal stability can be achieved by chosing the proper constituents. Coatings of this kind therefore have a great potential for practical applications; with the advent of large magnetron sputtering plants, one can expect that even large-scale uses of codeposited composite films will be possible. In Chapter 5 we discussed our work on co-evaporated Co-Al_2O_3 films, and we also gave references to earlier work on this type of material. We found that the most important uncertainty in the theoretical analysis came from the fact that the isolating "Al_2O_3" matrix which surrounds the Co particles contained

---

1) Particles from combustion sources can also be regarded as "gas evaporated".

several percent of unprecipitated cobalt. Possibly, this amount could have been diminished by annealing. It is difficult to assess the detailed influence of the unprecipitated Co on the optical properties. Nevertheless, it is our opinion that the theoretical analysis can be performed with more confidence on codeposited solid coatings than on any other type of coating discussed in this paper 1).

Anodized coatings on aluminium can be produced cheaply and efficiently by industrial techniques. Chapter 6 treated our work on Ni-pigmented anodic Al_2O_3 and gave references to earlier work. A detailed sample characterization is difficult for this class of materials, and we were not able to judge from direct experimental evidence (such as high resolution transmission electron micrographs) whether the separated-grain or the aggregate microstructure prevailed. The optical data could be reconciled with effective medium calculations only by assuming the latter situation, though. We also note that integral colouring techniques (252) can produce dilute metal-insulator composites (309-312) which may be of interest for selective absorption of solar energy (313).

Electroplating is a well known technique for producing surface coatings. "Black chrome", which is today's most widely used selectively solar absorbing surface, is prepared by this method. These coatings have a composite character and consist mainly of Cr and Cr_2O_3 (314-321). The detailed microstructure is not known and the problems associated with a wholly reliable modelling appear to be formidable. Nevertheless it is interesting to observe that this very important coating belongs to the class of metal-insulator composites.

Summing up, we have demonstrated in this paper that effective medium theories can be used to compute the optical properties and the solar selectivity for several different kinds of composite materials. Good agreement with experimental data is generally achieved provided that the sample characterization is sufficiently reliable that an adequate effective medium formulation can be chosen.

---

1) Recent work by Kreibig et al. (308) indicates that aqueous colloidal Ag coatings can be characterized in great detail.

REFERENCES

1. Kreibig, U. Earlier chapter in this volume.
2. Faraday, M. Philos. Trans. R. Soc. Lond. 147 (1857) 145.
3. A.J. Sievers. Spectral Selectivity of Composite Materials, in B.O. Seraphin, ed., Solar Energy Conversion: Solid State Physics Aspects, Topics in Applied Physics, vol. 31 (Berlin, Springer, 1979), p. 57.
4. Granqvist, C.G. Optical Properties of Cermet Materials. J. Phys. (Paris) 42 Colloque C-1 (1981) 247.
5. Chang, R.K. and T.E. Furtak, eds. Surface Enhanced Raman Scattering (New York, Plenum, 1982).
6. van de Hulst, H.C. Light Scattering by Small Particles (New York, Wiley, 1957).
7. Goody, R.M. Atmospheric Radiation (London, Oxford University Press, 1964).
8. Kerker, M. The Scattering of Light and Other Electromagnetic Radiation (New York, Academic, 1969).
9. Proceedings of the International Meeting on the Small Particles and Inorganic Clusters. J. Phys. (Paris) 38 Colloque C2 (1977).
10. Rouard, P. and A. Meessen. Optical Properties of Thin Metal Films. Prog. Opt. 15 (1977) 79.
11. Huffman, D.R. Interstellar Grains: The Interaction of Light with a Small-Particle System. Adv. Phys. 26 (1977) 129.
12. Ishimaru, A. Wave Propagation and Scattering in Random Media, vols. 1 and 2 (New York, Academic, 1978).
13. Garland, J.C. and D.B. Tanner, eds. Electrical Transport and Optical Properties of Inhomogeneous Media. AIP Conf. Proc. 40 (1978).
14. Martin, P.G. Cosmic Dust: Its Impact on Astronomy (Oxford University Press, 1978).
15. Egan, W.G. and T.W. Hilgeman. Optical Properties of Inhomogeneous Materials: Applications to Geology, Astronomy, Chemistry, and Engineering (New York, Academic, 1979).
16. Savage, B.D. and J.S. Mathis. Observed Properties of Interstellar Dust. Ann. Rev. Astron. Astrophys. 17 (1979) 73.
17. Proceedings of the Workshop on Thermodynamics and Kinetics of Dust Formation in the Space Medium. Astrophys Space Sci. 65 (1979).
18. McNulty, P.J., H.W. Chew and M. Kerker. Inelastic Light Scattering, in W.H. Marlow, ed., Aerosol Microphysics I: Particle Interaction, Topics in Current Physics, vol. 16 (Berlin, Springer, 1980).
19. Digest of Technical Papers Presented at the Topical Meeting on Optical Phenomena Peculiar to Matter of Small Dimensions (Optical Society of America, 1980).
20. Schuerman, D.W. ed. Light Scattering by Irregularly Shaped Particles (New York, Plenum, 1980).

21. Bourdon, J., ed. Growth and Properties of Metal Clusters: Applications to Catalysis and the Photographic Process, Studies in Surface Science and Catalysis, vol. 4 (Amsterdam, Elsevier, 1980).
22. Borel, J.-P. and J. Buttet, eds. Proceedings of the Second International Meeting on the Small Particles and Inorganic Clusters, Surface Sci. 106 (1981).
23. Perenboom, J.A.A.J., P. Wyder and F. Meier. Electronic Properties of Small Metallic Particles. Phys. Rep. 78 (1981) 173.
24. Gerber, H.E. and E.E. Hindman. Light Absorption by Aerosol Particles (Hampton, Spectrum Press, 1982); Appl. Opt. 21 (3) (1982).
25. Moon, P. J. Franklin Inst. 230 (1940) 583.
26. Granqvist, C.G. Radiative Heating and Cooling with Spectrally Selective Surfaces. Appl. Opt. 20 (1981) 2606.
27. Seraphin, B.O. and A.B. Meinel. Photothermal Solar Energy Conversion and the Optical Properties of Solids, in B.O. Seraphin, ed., Optical Properties of Solids - New Developments (Amsterdam, North-Holland, 1976), p. 927.
28. Hahn, R.E. and B.O. Seraphin. Spectrally Selective Surfaces for Photothermal Solar Energy Conversion. Phys. Thin Films 10 (1978) 1.
29. Seraphin, B.O. Spectrally Selective Surfaces and Their Impact on Photothermal Solar Energy Conversion, in Solar Energy Conversion: Solid State Physics Aspects, Topics in Applied Physics, vol. 31 (Berlin, Springer, 1979), p. 5.
30. Seraphin, B.O. Spectrally Selective Surfaces in Photothermal Solar Energy Conversion, in A.E. Dixon and J.D. Leslie, eds., Solar Energy Conversion: An Introductory Course (New York, Pergamon, 1979), p. 287.
31. Herzenberg, S.A. and R. Silberglitt. Low Temperature Selective Absorber Research. Proc. SPIE 324 (1982) 92.
32. Meinel, A.B. and M.P. Meinel. Applied Solar Energy: An Introduction (Reading, Addison-Wesley, 1976).
33. Zerlaut, G.A. Fundamental Materials Considerations for Solar Collectors, in C. Stein, ed., Critical Materials Problems in Energy Production (New York, Academic, 1976), p. 389.
34. Melamed, L. and G.M. Kaplan. Survey of Selective Absorber Coatings for Solar Energy Technology. J. Energy 1 (1977) 100.
35. de Winter, F, and M. Cox, eds. Proceedings of the International Solar Energy Society Congress (New York, Pergamon, 1978).
36. Böer, K.W. and B.H. Glenn, eds. Sun II: Proceedings of the International Solar Energy Society Silver Jubilee Congress (New York, Pergamon, 1979).
37. Tabor, H. Selective Surfaces, in A.E. Dixon and J.D. Leslie, eds., Solar Energy Conversion: An Introductory Course (New York, Pergamon, 1979) p. 253.

38. Koltun, M.M. Selektivnye Opticheskie Poverkhnosti Preobrazovatelei Solnechnoi Energi (Moscow, Nauka Press, 1979). English translation: Selective Optical Surfaces for Solar Energy Converters (New York, Allerton, 1981).
39. Lampert, C.M. Coatings for Enhanced Photothermal Energy Collection I. Selective Absorbers. Solar Energy Mater. 2 (1979) 1.
40. Koltun, M.M. Present State of Research on Selective Coatings for Solar Energy Converters. Geliotekh. 16 (6) (1980) 34. English translation: Appl. Solar Energy 16 (6) (1980) 30.
41. Pillai, P.K.C. and R.C. Agarwal. Spectrally Selective Surfaces for Photothermal Conversion of Solar Energy. Phys. Stat. Sol. A 60 (1980) 11.
42. Murr, L.E., ed. Solar Materials Science (New York, Academic, 1980).
43. Agnihotri, O.P. and B.K. Gupta. Solar Selective Surfaces (New York, Wiley, 1981).
44. Proceedings, Conférence Internationale sur les Matériaux pour la Conversion Photothermique de l'Energie Solaire. J. Phys. (Paris) 40 Colloque C-1 (1981).
45. Lampert, C.A., ed. Optical Energy Coatings for Energy Efficiency and Solar Applications. Proc. SPIE 324 (1982).
46. Niklasson G.A. and C.G. Granqvist. Surfaces for Selective Absorption of Solar Energy: An Annotated Bibliography. To be published.
47. Mosotti, O.F. Mem. di Matem. e Fisica della Soc. Ital. della Sci. Residente in Modena 24 (1850) 49.
48. Clausius, R. Die Mekanische Wärmetheorie, 2nd edition, vol. 2, section 3 (Braunschweig, Vieweg, 1879), p. 62.
49. Maxwell, J.C. A Treatise on Electricity and Magnetism, vol. 1 (Oxford, Clarendon Press, 1873), p. 360.
50. Lorentz, H.A. Ann. Phys. Chem. 9 (1880) 641.
51. Lorenz, L. Ann. Phys. Chem. 11 (1880) 70.
52. Rayleigh, J.W.S. Phil. Mag. 34 (1892) 481.
53. Garnett, J.C.M. Phil. Trans. R. Soc. Lond. A 203 (1904) 385; 205 (1906) 237.
54. Bruggeman, D.A.G. Ann. Phys. (Leipzig) 24 (1935) 636.
55. Lowry, H.H. J. Franklin Inst. 203 (1927) 413.
56. van Beek, L.K.H. Progr. Dielectrics 7 (1967) 69.
57. Tinga, W.R., W.A.G. Voss and D.R. Blossey. J. Appl. Phys. 44 (1973) 3897.
58. Böttcher, C.J.F. and Bordewijk, P. Theory of Electric Polarization, 2nd edition, vol. 2 (Amsterdam, Elsevier, 1978), p. 476.
59. Grosse, C. and J.-L. Greffe. J. Chim. Phys. 76 (1979) 305.
60. Landauer, R. AIP Conf. Proc. 40 (1978) 2.
61. Reynolds, J.A. and J.M. Hough. Proc. Phys. Soc. Lond. 70 (1957) 769.
62. Aspnes, D.E. Thin Solid Films 89 (1982) 249.
63. Lamb, W., D.M. Wood and N.W. Ashcroft, Phys. Rev. B21 (1980) 2248.

64. Smith, G.B. J. Phys. D: Appl. Phys. 10 (1977) L39.
65. Smith, G.B. Appl. Phys. Lett. 35 (1979) 668.
66. Lamb, W., D.M. Wood and N.W. Ashcroft, AIP Conf. Proc. 40 (1978) 240.
67. Niklasson, G.A., C.G. Granqvist and O. Hunderi. Appl. Opt. 20 (1981) 26.
68. Mie, G. Ann. Phys. 25 (1908) 377.
69. Lorenz, L. Videnskap. Selskab Skrifter 6 (1890).
70. Aden, A.L. and M. Kerker. J. Appl. Phys. 22 (1951) 1242.
71. Güttler, A. Ann. Phys. 11 (1952) 65.
72. Bohren, C.F. and D.P. Gilra. J. Colloid Interface Sci. 72 (1979) 215.
73. Elliott, R.J., J.A. Krumhansl and P.L. Leath. Rev. Mod. Phys. 46 (1974) 465.
74. Stroud, D. and F.P. Pan. Phys. Rev. B17 (1978) 1602.
75. Wagner, K.W. Arch. Electrotechn. 2 (1914) 371.
76. Kerner, E.H. Proc. Phys. Soc. B69 (1956) 802.
77. Higuchi, W.I. J. Phys. Chem. 62 (1958) 649.
78. Barker Jr., A.S. Phys. Rev. B7 (1973) 2507.
79. Genzel, L. and T.P. Martin. Surface Sci. 34 (1973) 33.
80. Hayashi, S., N. Nakamori and H. Kanamori. J. Phys. Soc. Japan 46 (1979) 176.
81. Nagatani, T. J. Appl. Phys. 51 (1980) 4944.
82. Granqvist, C.G. and O. Hunderi. Phys. Rev. B16 (1977) 1353.
83. Doyle, W.T. J. Appl. Phys. 49 (1978) 795.
84. McPhedran, R.C. and D.R. McKenzie. Proc. R. Soc. Lond. 359 (1978) 45.
85. McKenzie, D.R., R.C. McPhedran and G.H. Derrick. Proc. R. Soc. Lond. 362 (1978) 211.
86. Perrins, W.T., R.C. McPhedran and D.R. McKenzie. Thin Solid Films 57 (1979) 321.
87. Böttcher. C.J.F. Rec. Trav. Chim. 64 (1945) 47.
88. Polder, D. and J.H. van Santen. Physica (Utrecht) 12 (1946) 257.
89. Landauer,R. J.Appl. Phys. 23 (1952) 779.
90. Wood, D.M. and N.W. Ashcroft. Phil. Mag. 35 (1977) 269.
91. Stroud, D. Phys. Rev. B19 (1979) 1783.
92. Berthier, S. and J. Lafait. Opt. Commun. 33 (1980) 33.
93. Stroud, D. Phys. Rev. B12 (1975) 3368.
94. Webman, I., J. Jortner and M.H. Cohen. Phys. Rev. B15 (1977) 5712.
95. Hori, M. J. Math. Phys. 14 (1973) 514; 14 (1973) 1942; 16 (1975) 1772; 18 (1977) 487.
96. Hori, M. and F. Yonezawa. J. Math. Phys. 15 (1974) 2177; 16 (1975) 352; 16 (1975) 365.
97. Hori, M. and F. Yonezawa. J. Phys. C: Solid State Phys. 10 (1977) 229.
98. Ping Sheng. Phys. Rev. B22 (1980) 6364.
99. Ping Sheng. Phys. Rev. Lett. 45 (1980) 60.

100. Hanai, T. Kolloid Z. 171 (1960) 23.
101. Grosse, C. J. Chim. Phys. 76 (1979) 153.
102. Asano, S. and G. Yamamoto. Appl. Opt. 14 (1975) 29.
103. Gans, R. Ann. Phys. 37 (1912) 881.
104. Granqvist, C.G. and O. Hunderi. Phys. Rev. B16 (1977) 3513.
105. Granqvist, C.G. and O. Hunderi. Phys. Rev. B18 (1978) 1554.
106. Cohen, R.W., G.D. Cody, M.D. Coutts and B. Abeles. Phys. Rev. B8 (1973) 3689.
107. Bilboul, R.R. Brit. J. Appl. Phys. (J. Phys. D) 2 (1969) 921.
108. Donnadieu, A. Thin Solid Films 6 (1970) 249.
109. Landau, L.D. and E.M. Lifshitz. Electrodynamics of Continuous Media (Oxford, Pergamon, 1960).
110. Berthier, S. and J. Lafait. J. Phys. (Paris) 40 (1979) 1093.
111. Galeener, F.L. Phys. Rev. Lett. 27 (1971) 421.
112. Asami. K., T. Hanai and N. Koizumi. Japan. J. Appl. Phys. 19 (1980) 359.
113. Granqvist, C.G. AIP Conf. Proc. 40 (1978) 196.
114. Hunderi, O. Phys. Rev. B7 (1973) 3419.
115. Norrman, S., T. Andersson, C.G. Granqvist and O. Hunderi. Phys. Rev. B18 (1978) 674.
116. van Gelder, A.P., J. Holvast, J.H.M. Stoelinga and P. Wyder. J. Phys. C (Solid State Phys.) 5 (1972) 2757.
117. Fuchs, R. Phys. Rev. B11 (1975) 1732.
118. Langbein, D. J. Phys. A (Gen. Phys.) 9 (1976) 627.
119. Brako, R. and M. Sunjić, in R. Dobrozemsky, F. Rüdenauer, F.P. Viehböck and A. Breth, eds., Proceedings of the Seventh International Conference on Solid Surfaces (Vienna, Dobrozemsky et al., 1977), p. 1281.
120. Granqvist, C.G. and G.A. Niklasson. J. Appl. Phys. 49 (1978) 3512.
121. Brako, R. J. Phys. C (Solid State Phys.) 11 (1978) 3345; 12 (1979) 1139.
122. Persson, B. Unpublished results.
123. Clippe, P., R. Evrard and A.A. Lucas. Phys. Rev. B14 (1976) 1715.
124. Ausloos,M., P.Clippe and A.A. Lucas, Phys.Rev. B18(1978) 7176
124a. Sansonetti, J,E. and J,K. Furdyna. Phys. Rev. B22 (1980) 2866.
125. Yoshida, S., T. Yamaguchi and A. Kinbara. J. Opt. Soc. Am. 61 (1971) 62.
126. Meessen,A. J.Phys. (Paris) 33 (1972) 371.
127. Bedaux, D. and J. Vlieger. Physica (Utrecht) 73 (1974) 287.
128. Dignam, M.J. and J. Fedyk. J. Phys. (Paris) 38 (1977) C5-57.
129. Gérardy, J.M. and M. Ausloos, Phys. Rev. B22 (1980) 4950; B25 (1982) 4204.
130. Wiener, O. Abh. Sächs. Akad. Wiss. Leipzig Math.-Naturwiss. Kl. 32 (1912) 509.
130a. Hashin, Z. and S. Shtrikman. J. Appl. Phys. 33 (1962) 3125.
131. Schulgasser, K. and Hashin, Z. J. Appl. Phys. 47 (1976) 424.
132. Bergman, D.J. Phys. Rev. Lett. 44 (1980) 1285.

133. Bergman, D.J. Phys. Rev. B 23 (1981) 3058.
134. Bergman, D.J. Ann. Phys. (N.Y.) 138 (1982) 78.
135. Milton, G.W. Appl. Phys. Lett. 37 (1980) 300.
136. Milton, G.W. Phys. Rev. Lett. 46 (1981) 542.
137. Milton, G.W. J. Appl. Phys. 52 (1981) 5286.
138. Milton, G.W. J. Appl. Phys. 52 (1981) 5294.
139. McPhedran, R.C. and G.W. Milton. Appl. Phys. A26 (1981) 207.
140. Antonov, V.A. and V.I. Pshenitsyn. Opt. Spektrosk. 50 (1981) 362 (English translation Opt. Spectrosc. 50 (1981) 195).
141. Milton, G.W. Appl. Phys. A26 (1981) 125.
142. Bergman, D.J. Phys. Rep. 43 (1978) 377.
143. Bergman, D.J. Phys. Rev. B19 (1979) 2359.
144. Aspnes, D.E. Phys. Rev. B25 (1982) 1358.
145. Niklasson, G.A. and C.G. Granqvist. Solar Energy Mater. 5 (1981) 173.
146. Johnson, P.B. and R.W. Christy. Phys. Rev. B9 (1974) 5056.
147. Eriksson, T.S., A. Hjortsberg, G.A. Niklasson and C.G. Granqvist. Appl. Opt. 20 (1981) 2742.
148. Lichtenecker, K. and K. Rother. Phys. Z. 32 (1931) 255.
149. Looyenga, H. Physica (Utrecht) 31 (1965) 401.
150. Kirillova, M.M. and B.M. Charikov. Opt. Spektrosk. 17 (1964) 254 (English translation Opt. Spectrosc. 17 (1964) 134).
151. Abeles, B. and J.I. Gittleman. Appl. Opt. 15 (1976) 2328.
152. Gittleman J.I. and B.Abeles. Phys. Rev. B15 (1977) 3273.
153. Granqvist, C.G. and O. Hunderi. Phys. Rev. B18 (1978) 2897.
154. Granqvist, C.G. J. Appl. Phys. 50 (1979) 2916.
155. Buhrman, R.A. and H.G. Craighead, in L.E. Murr, ed., Solar Materials Science (New York, Academic, 1980), p. 277.
156. Gibson, U.J., H.G. Craighead and R.A. Buhrman. Phys. Rev. B25 (1982) 1449.
157. Ruppin, R. Phys. Stat. Sol. B 87 (1978) 619.
158. Tanner, D.B., A.J. Sievers and R.A. Buhrman. Phys. Rev. B 11 (1975) 1330.
159. Granqvist, C.G., R.A. Buhrman, J. Wyns and A.J. Sievers. Phys. Rev. Lett. 37 (1976) 625.
160. Granqvist, C.G. Z. Phys. B 30 (1978) 39.
161. Genzel, L. and U. Kreibig. Z. Phys. B 37 (1980) 93.
162. Bolotin, G.A., M.M. Noskov and I.I. Sasovskaya. Fiz. Metal. Metalloved. 35 (1973) 699 (English translation Phys. Metals Metallogr. 35 (4) (1973) 24).
163. Doyle, W.T. Phys. Rev. 111 (1958) 1067.
164. Kreibig, U. J. Phys. F (Metal Phys) 4 (1974) 999.
165. Euler, J. Z. Phys. 137 (1954) 318.
166. Gorkov, L.P. and G.M. Eliashberg. Zh. Eksp. Teor. Fiz. 48 (1965) 1407 (English translation Soviet Phys. - JETP 21 (1965) 940).
167. Kawabata, A. and R. Kubo. J. Phys. Soc. Japan 21 (1966) 1765.
168. Cini, M. and P. Ascarelli. J. Phys. F (Metal Phys.) 4 (1974) 1998.

169. Genzel, L., T.P. Martin and U. Kreibig. Z. Phys. B 21 (1975) 339.
170. Ruppin, R. and H. Yatom. Phys. Stat. Sol. B 74 (1976) 647.
171. Cini,M. J. Opt. Soc. Am. 71 (1981) 386.
172. Kreibig, U. Z. Phys. B 31 (1978) 39.
173. Kreibig, U. J. Phys. (Paris) 38 (1977) C2-97.
174. Kreibig, U., in J. Bourdon, ed., Growth and Properties of Metal Clusters (Amsterdam, Elsevier, 1980), p. 371.
175. Shinozaki, K. and T. Yamaguchi. Surface Sci. 103 (1981) L 97.
176. Baetzold, R.C., M.G. Mason and J.F. Hamilton. J. Chem. Phys. 72 (1980) 366.
177. Roulet, H., J.-M. Mariot, G. Dufour and C.F. Hague. J. Phys. F (Metal Phys.)10 (1980) 1025.
178. Apai, G., S.-T. Lee and M.G. Mason. Solid State Commun. 37 (1981) 213.
179. Lee, S.-T., G. Apai, M.G. Mason, R.Benbow and Z. Hurych. Phys. Rev. B 23 (1981) 505.
180. G.A. Niklasson and C.G. Granqvist, in D.O. Hall and J. Morton, eds., Solar World Forum: Proceedings of the International Solar Energy Society Congress, vol. 1 (Oxford, Pergamon, 1982), p. 221.
181. Niklasson, G.A. and C.G. Granqvist, in Proceedings of the Seventh International Conference on Vacuum Metallurgy (Tokyo, 1982), to be published.
182. Niklasson, G.A. and C. G. Granqvist, to be published.
183. Niklasson, G.A. Ph. D. Thesis (Gothenburg, Chalmers University of Technology, 1982), unpublished.
184. Aitchison, J. and J.A.C. Brown. The Lognormal Distribution (Cambridge, Cambridge University Press, 1957).
185. Smith, J.E. and M.L. Jordan. J. Colloid Sci. 19 (1964) 549.
186. Granqvist, C.G. and R.A. Buhrman. J. Appl. Phys. 47 (1976) 2200.
187. Evans, Jr., C.A. J. Vac. Sci. Technol. 12 (1975) 144.
188. Niklasson, G.A. and C.G. Granqvist. Thin Solid Films 74 (1980) L5.
189. Abeles, B. Appl. Solid State Sci. 6 (1976) 1.
190. Hunter, W.R. J. Opt. Soc. Am. 55 (1965) 1197.
191. Bennett, J.M. and M.J. Booty. Appl. Opt. 5 (1966) 41.
192. Abeles, F. and M.L. Theye. Surface Sci. 5 (1966) 32.
193. Ward, L. and A. Nag. Brit. J. Appl. Phys. 18 (1967) 277; 18 (1967) 1629.
194. Nilsson, P.O. Appl. Opt. 7 (1968) 435.
195. Johnson, K.W. and E.E. Bell. Phys. Rev. 187 (1969) 1044.
196. Nestell, J.E. and R.W. Christy. Appl. Opt. 11 (1972) 643.
197. Miller, R.F. and A.J, Taylor. J. Phys. D (Appl. Phys.) 4 (1971) 1419.
198. Denton, R.E., R.D. Campbell and S.G. Tomlin. J. Phys. D (Appl. Phys.) 5 (1972) 852.

199. Hunter, W.R. and G. Hass. J. Opt. Soc. Am. 64 (1974) 429.
200. Bringans, R.D. J. Phys. D (Appl. Phys.) 10 (1977) 1855.
201. Parker, T.J., W.G. Chambers, J.E. Ford and C.L. Mok. Infrared Phys. 18 (1978) 571.
202. Carey, R., B.W.J. Thomas and D.M. Newman. Thin Solid Films 66 (1980) 139.
203. Hjortsberg, A. Appl. Opt. 20 (1981) 1254.
204. Lissberger, P.H. and R.G. Nelson. Thin Solid Films 21 (1974) 159.
205. Lissberger, P.H. and P.W. Saunders. Thin Solid Films 34 (1976) 323.
206. Fan, J.C.C. and P.M. Zavracky. Appl. Phys. Lett. 29 (1976) 478.
207. Fan, J.C.C. and S.A. Spura. Appl. Phys. Lett. 30 (1977) 511.
208. Gittleman, J.I., B. Abeles, P. Zanzucchi and Y. Arie. Thin Solid Films 45 (1977) 9.
209. McKenzie, D.R. and R.C. McPhedran. AIP Conf. Proc. 40 (1978) 283.
210. Okuyama, M., K. Furusawa and Y. Hamakawa. Solar Energy 22 (1979) 479.
211. Gittleman, J.I., E.K. Sichel and Y. Arie. Solar Energy Mater. 1 (1979) 93.
212. Craighead, H.G., R. Bartynski, R.A. Buhrman, L. Wojcik and A.J. Sievers. Solar Energy Mater. 1 (1979) 105.
213. Craighead, H.G. Ph. D. Thesis (Ithaca, Cornell University, 1980), unpublished.
214. Berthier, S. and J. Lafait. Thin Solid Films 89 (1982) 213.
215. Afanas'yeva, L.A. and M.M. Kirillova. Fiz. Metal. Metalloved. 23 (1967) 472 (English translation Phys. Metals Metallogr. 23 (8) (1967) 80).
216. Yu, A.Y.-C., T.M. Donovan and W.E. Spicer. Phys. Rev. 167 (1968) 670.
217. Lenham, A.P. and D.M. Treherne, in F. Abelès, ed., Optical Properties and Electronic Structure of Metals and Alloys (Amsterdam, North Holland, 1966), p. 196.
218. Siddiqui, A.S. and D.M. Treherne. Infrared Phys. 17 (1977) 33.
219. Gushchin, V.S., K.M. Shvarev, B.A. Baum and P.V. Geld. Dokl. Akad. Nauk. SSSR 240 (1978) 320 (English translation Soviet Phys. Doklady 23 (1978) 344).
220. Weaver, J.H., E. Colavita, D.W. Lynch and R. Rosei. Phys. Rev. B 19 (1979) 3850.
220a Devenyi, A., W. Theiner, S.K. Sharma and R. Manaila-Devenyi. Thin Solid Films 15 (1973) 39.
221. Balmer, P., H. Blum, M. Forster, A, Schweiger and H.H. Günthard, J. Phys. C (Solid State Phys.) 13 (1980) 517.
222. Heavens, O.S. Optical Properties of Thin Solid Films (New York, Dover, 1965).

223. Born, M. and E. Wolf. Principles of Optics (Oxford, Pergamon, 1975).
224. Fowles, G.R. Introduction to Modern Optics, 2nd edition (New York, Holt, Rinehart and Winston, 1975).
225. Touloukian, Y.S., D.P. DeWitt and R.S. Hernicz. Thermal Radiative Properties (New York, IFI/Plenum, 1972).
226. Sievers, A.J. J. Opt. Soc. Am. 68 (1978) 1505.
227. Sievers, A.J., in L.E. Murr, ed., Solar Materials Science (New York, Academic, 1980), p. 229.
228. Weaver, J.H., C. Krafka, D.W. Lynch and E.E. Koch. Optical Properties of Metals, Pts. I and II, Physics Data, vols. 18-1 and 18-2 (Karlsruhe, Fachinformationszentrum Energie Physik Mathematik GmbH, 1981).
229. Ehrenreich, H., H.R. Philipp and B. Segall. Phys. Rev. 132 (1963) 1918.
230. Harrison, W.A. Phys. Rev. 147 (1966) 467.
231. Ashcroft, N.W. and K. Sturm. Phys. Rev. B 3 (1971) 1869.
232. Eriksson, T.S., A. Hjortsberg and C.G. Granqvist. Solar Energy Mater. 6 (1982) 191.
233. Berreman, D.W. Phys. Rev. 130 (1963) 2193.
234. Olson, O.H. Appl. Opt. 2 (1963) 109.
235. Sheklein, A.V. Geliotekh. 3 (2) (1967) 28 (English translation Appl. Solar Energy 3 (1967) 68).
236. Ramanathan, K.G., S.H. Yen and E.A. Estalote. Appl. Opt. 16 (1977) 2810.
237. Smalley, R. and A.J. Sievers. J. Opt. Soc. Am. 68 (1978) 1516.
238. Willrath, H. and R.B. Gammon. Solar Energy 21 (1978) 193.
239. Smith, G.B. and Willrath, H. J. Phys. E (Sci. Instrum.) 12 (1979) 813.
240. Zeller, H.R. and D. Kuse. J. Appl. Phys. 44 (1973) 2763.
241. Sella, C., T.K. Vien, J. Lafait and S. Berthier, Thin Solid Films 90 (1982) 425.
242. Gaziev, U. Kh., Sh. A. Faiziev, V.V. Li and V.S. Trukhov. Geliotekh. 16 (1980) 30 (English translation Appl. Solar Energy 16 (1980) 30).
243. Thornton, J.A. and J.L. Lamb, Thin Solid Films 83 (1981) 377.
244. Bastien, R.C., R.R. Austin and T.P. Pottenger. Proc. SPIE 140 (1978) 140.
245. McKenzie, D.R. Appl. Phys. Lett. 34 (1979) 25.
246. Craighead, H.G., R.E. Howard, J.E. Sweeney and R.A. Buhrman. Appl. Phys. Lett. 39 (1981) 29.
247. Nyberg, G.A., H.G. Craighead and R.A. Buhrman. Proc. SPIE 324 (1982) 117.
248. Nyberg, G.A. and R.A. Buhrman. Appl. Phys. Lett. 40 (1982) 129.
249. Granqvist, C.G., Å. Andersson and O. Hunderi. Appl. Phys. Lett. 35 (1979) 268.

250. Granqvist, C.G., Å. Andersson and O. Hunderi, in K.W. Böer and B.H. Glenn, eds., Sun II: Proc. Int. Solar Energy Soc. Silver Jubilee Congress, vol. 3 (New York, Pergamon, 1979), p. 1955.
251. Andersson, Å., O. Hunderi and C.G. Granqvist. J. Appl. Phys. 51 (1980) 754.
252. Wernick, S. and R. Pinner, The Surface Treatment and Finishing of Aluminium and Its Alloys, 4th edition, vols. 1 and 2 (Teddington, Robert Draper Ltd., 1972).
253. Brace, A.W. and P.G. Sheasby. The Technology of Anodizing Aluminium (Stonehouse, Technicopy Ltd., 1979).
254. Keller, F., M.S. Hunter and D.L. Robinson. J. Electrochem. Soc. 100 (1953) 411.
255. O'Sullivan, J.P. and G.C. Wood. Proc. R. Soc. Lond. A 137 (1970) 511.
256. Wood, G.C., in J.W. Diggle, ed., Oxides and Oxide Films, vol. 2 (New York, Marcel Dekker, 1973), p. 167.
257. Thompson, G.E., R.C. Furneaux, G.C. Wood, J.A. Richardson and J.S. Goode. Nature 272 (1978) 433.
258. Endtinger, F. Schweizer Aluminiumrundschau 62 (1970) 268.
259. Sautter, W. Galvanotechnik 62 (1971) 454.
260. Läser, L. Aluminium (Düsseldorf) 48 (1972) 169.
261. Sandera, L. Aluminium (Düsseldorf) 49 (1973) 533.
262. Sautter, W., G. Ibe and J. Meier. Aluminium (Düsseldorf) 50 (1974) 143.
263. Sheasby, P.G. and W.E. Cooke. Trans. Inst. Metal Finish. 52 (1974) 103.
264. Sato, T. Metalloberfläche 31 (1977) 290.
265. Goad, D.G.W. and M. Moskovits. J. Appl. Phys. 49 (1978) 2929.
266. Csanády, A., I. Imre-Baán, E. Lichtenberger-Bajza, E. Szontágh and F. Dömölki. J. Mat. Sci. 15 (1980) 2761.
267. Tsuda, S. and Y. Asano. Belg.-Ned. Tijdschr. Oppervlaktetechn. Met. 22 (1978) 33.
268. Uchino, H., S. Aso, S. Hozumi, H. Tokumasu and Y. Yoshioka. Matsushita Electr. Ind. Nat. Techn. Rep. 25 (1979) 994.
269. Kirillova, M.M. Zh. Eksperim. Teor. Fiz. 61 (1971) 336 (English translation Soviet Phys. JETP 34 (1972) 178).
270. Rasigni, M. and G. Rasigni. J. Opt. Soc. Am. 67 (1977) 510.
271. Gasparini, J.P., R. Fraisse and R. Philip. Thin Solid Films 41 (1977) 179.
272. Marton, J.P. and B.D. Jordan. Phys. Rev. B 15 (1977) 1719.
273. Shklyarevskii, I.N., E. Anachkova and G.S. Blyashenko. Opt. Spektrosk. 43 (1977) 723 (English translation Opt. Spectrosc. 43 (1977) 427).
274. Shklyarevskii, I.N. and G.S. Blyashenko. Opt. Spektrosk. 44 (1978) 545 (English translation Opt. Spectrosc. 44 (1978) 315).
275. Shklyarevskii, I.N., G.S. Blyashenko and V.P. Kostyuk. Opt. Spektrosk. 44 (1978) 962 (English translation Opt. Spectrosc. 44 (1978) 585).

276. Yamaguchi, T., H. Takahashi and A. Sudoh. J. Opt. Soc. Am. 68 (1978) 1039.
277. Truong, V.V. and G.D. Scott. J. Opt. Soc. Am. 68 (1978) 189.
278. Carlan, A. and G. Desrousseaux. J. Opt. Soc. Am. 68 (1978) 1019.
279. Breuer, H.D. and W. Bach. Z. Phys. Chem. 110 (1978) 267.
280. Chavaux, R. and A. Meessen. Thin Solid Films 62 (1979) 125.
281. Gadenne, P. Thin Solid Films 57 (1979) 77.
282. Al-Abdella, R.B., V.P. Kostyuk and I.N. Shklyarevskii. Opt. Spektrosk. 48 (1980) 1143 (English translation Opt. Spectrosc. 48 (1980) 625).
283. Hayashi, S., T. Yamada and H. Kanamori. Opt. Commun. 36 (1981) 195.
284. Garron, R. Appl. Opt. 19 (1980) 554.
285. Anno, E. and R. Hoshino. J. Phys. Soc. Japan 50 (1981) 1209.
286. Chee, K.T., F.E. Girouard and V.V. Truong. Appl. Opt. 20 (1981) 404.
287. Al-Abdella, R.B., V.P. Kostyuk and I.N. Shklyarevskii. Opt. Spektrosk. 51 (1981) 316 (English translation Opt. Spectrosc. 51 (1981) 173).
288. Vlieger, J. Physica (Utrecht) 64 (1973) 63.
289. Bedaux, D. and J. Vlieger. Physica (Utrecht) 67 (1973) 55.
290. Vlieger, J, and D. Bedaux. Thin Solid Films 69 (1980) 107.
291. Andersson, T. and C.G. Granqvist. J. Appl. Phys. 48 (1977) 1673.
292. McKenzie, D.R. J. Opt. Soc. Am. 66 (1976) 249.
293. Harding, G.L. Thin Solid Films 38 (1976) 109.
294. O'Neill, P., C. Doland and A. Ignatiev. Appl. Opt. 16 (1977) 2822.
295. Doland, C., P. O'Neill and A. Ignatiev. J. Vac. Sci. Technol. 14 (1977) 259.
296. O'Neill, P. and A. Ignatiev. Phys. Rev. B18 (1978) 6540.
297. O'Neill, P., A. Ignatiev and C. Doland. Solar Energy 21 (1978) 465.
298. Granqvist, C.G., N. Calander and O. Hunderi. Solid State Commun. 31 (1979) 249.
299. Niklasson, G.A. and C.G. Granqvist. J. Appl. Phys. 50 (1979) 5500.
300. Cocks, F.H. and M.J. Peterson. J. Vac. Sci. Technol. 16 (1979) 1560.
301. Peterson, M.J. and F.H. Cocks. Mater. Sci. Engr. 41 (1979) 143.
302. Granqvist, C.G. and Hunderi, O. J. Appl. Phys. 51 (1980) 1751.
303. Ignatiev. A., in L.E. Murr, ed., Solar Materials Science (New York, Academic, 1980), p. 151.
304. Fantini, M.C.A., J.R. Moro and M. Abramovich. J. Phys. (Paris) 42 (1981) C1-317.
305. Donn, B., J. Hecht , R. Khanna, J. Nuth, D. Stranz and A.B. Anderson. Surface Sci. 106 (1981) 576.

306. Santucci, S., P. Picozzi, L. Paoletti and F. Tangucci. Thin Solid Films 79 (1981) 133.
307. Picozzi, P., S. Santucci, M. Diociaiuti and L. Paoletti. Opt. Acta 29 (1982) 511.
308. Kreibig, U., A. Althoff and H. Pressman. Surface Sci. 106 (1981) 308.
309. Wefers, K. and W.T. Evans. Plating and Surface Finishing 62 (1975) 951.
310. Wefers, K. and P.F. Wallace. Aluminium (Düsseldorf) 52 (1976) 485.
311. Chang, R. and W.F. Hall. Thin Solid Films 46 (1977) L5.
312. Granqvist, C.G. J. Appl. Phys. 51 (1980) 3359.
313. Cochran, W.C. and J.H. Powers. Aluminium (Düsseldorf) 54 (1978) 147.
314. Chang, R. and W.F. Hall. AIP Conf. Proc. 40 (1977) 305.
315. Hogg, S.W. and G.B. Smith. J. Phys. D (Appl. Phys.) 10 (1977) 1863.
316. Ignatiev, A., P. O'Neill and G. Zajac. Solar Energy Mater. 1 (1979) 69.
317. Lampert, C.M. and J. Washburn. Solar Energy Mater. 1 (1979) 81.
318. Driver, P.M., in K.W. Böer and B.H. Glenn, eds., Sun II: Proc. Int. Solar Energy Soc. Silver Jubilee Congress (New York, Pergamon, 1979), p. 1887.
319. Zajac, G., G.B. Smith and A. Ignatiev. J. Appl. Phys. 51 (1980) 5544.
320. Berthier, S. and J. Lafait. J. Phys. (Paris) 40 (1979) 1093; 42 (1981) C1-285.
321. Sweet, J.N. and R.B. Pettit. Sandia Report SAND 82-0964 (1982).

ADHESION AND SINTERING OF SMALL PARTICLES

A.R. Thölén

Laboratory of Applied Physics I
Technical University of Denmark
2800 Lyngby, Denmark

1. INTRODUCTION

Coalescence is the term used when single atoms or groups of atoms react and form a larger entity. Sintering on the other hand describes the process when macroscopic particles start to amalgamate and can perhaps better be defined as "heat treatment of powders and porous compacts" (1). Although the underlying physical processes have much in common or are identical the two phenomena are generally described in different vocabulary. Another concept describing a related subject in macroscopic terms is adhesion although the interaction between surfaces is due to forces between individual atoms.

Using the electron microscope as a tool there exists a possibility to shed light on adhesion, coalescence and sintering down to atomic dimensions. However, there are some restrictions on the thickness of samples suitable to transmission electron microscopy and it will here be reported about adhesion between small metal particles.

When two surfaces adhere the amount of energy released per unit area is $w_{12} = \gamma_1+\gamma_2-\gamma_{12}$, where γ_1 and γ_2 are the surface energy values of the two original surfaces and γ_{12} is the interface energy of the newly formed boundary. Consequences of this adhesion can be studied at contacts in a particle system and such observations will now be described.

2. PRODUCTION OF PARTICLES

The metal particles are conveniently made by evaporation in a gas (2,3). The gas used here is argon and the particles can be made bigger simply by increasing the pressure. The pressure typically used for particle of radius 30 nm is 10 torr.

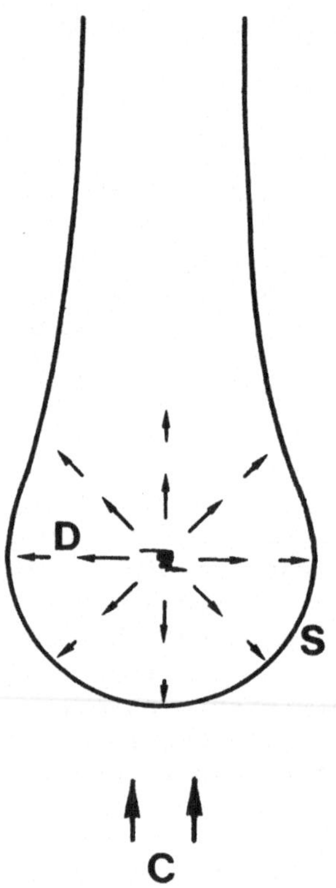

Fig. 1. Evaporation of a metal in a gas atmosphere. The metal atoms from the central source diffuse out along the paths D, coalesce and form a smoke at S. The smoke is mainly confined in a tubelike shell, where the particles ascend in the convection current C. (4)

The metal atoms diffuse radially out from the source and coalesce when they reach a lower temperature (Fig. 1). In case of gold particles the coalescence occurs at a temperature of about 0.35-0.4 T_m (< 300 ^oC), where T_m is the absolute melting temperature. A thermodynamic calculation of the nucleation gives an extremely limited nucleation rate and the fact is that thermodynamics is pushed beyond its limitations with too minute aggregates as critical nuclei. The question of particle growth is not completely clear either, being it a coalescence process of encountering particles or a further growth from the vapour phase. Much of the evidence, however, points in the direction of coalescence.

The small particles produced adhere to form a smoke which is collected on carbon covered electron microscope grids and then observed in the electron microscope (JEM 200A).

Some investigations have also been done on extracted particles, but the observations are similar in the two cases.

3. GENERAL OBSERVATIONS

The first observation in the electron microscope relates to the size and arrangement of particles. The size distribution is lognormal which has been taken as an indication that the particles coalesce (5) and the particles are further arranged three-dimensionally like the branches on a tree except for the magnetic material (iron, nickel and cobalt) where the particles form long chains (4). Observations can be made of the geometrical shape of the particles and the precision of this determination can be improved by using the weak beam technique (6,7). No dislocations are observed in small particles with diffraction contrast in spite a large number of metals have been investigated including gold, silver, copper, magnesium, aluminium, iron, cobalt, nickel and zink. Planar faults (twin boundaries) were, however, readily seen in many of the particle systems. Facetting is observed on many particles but the true nature of this process remains unknown. It seems however, to be tied both to the history of the particles, to their size and to foreign atoms at the surface.

The most remarkable factor observed was the strain contrast at some particle-particle contacts (Fig. 2) and this strain contrast was found to be due to adhesion and have the following background (8). When two particles come together it is energetically favorable for them to make contact over an area with radius a, rather than at one point (Fig. 3) as this diminishes the outer surface area. On the other hand, work is required to bring the originally curved surfaces together and a balance is reached with a minimum in energy and a finite contact radius.

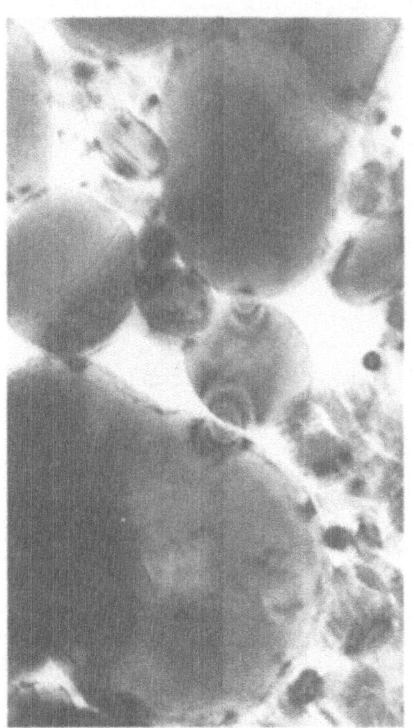
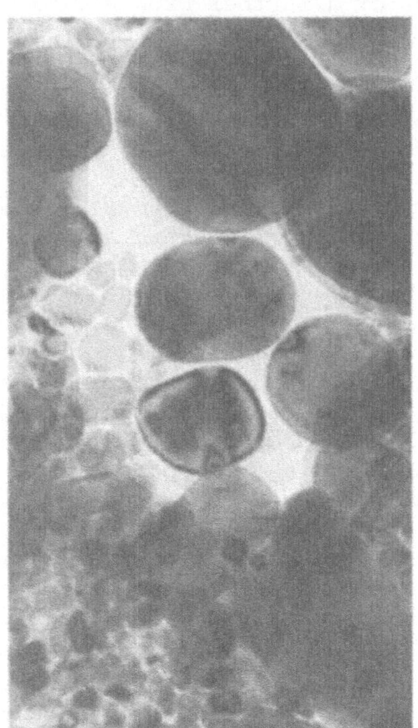

Fig. 2. Extracted cobalt particles (from a Cu-Co alloy) with strain contrast at contacting points.

This radius a is given by the following expression

$$a = \left(\frac{R^2\ \gamma_{eff}}{0{,}16\ E}\right)^{1/3} \quad \text{where}$$

R = particle radius

E = Young's modulus

$\gamma_{eff} = 2\gamma_s - \gamma_{gb}$

γ_s = surface energy of the particles (assumed to be the same for the two particles)

γ_{gb} = grain boundary energy of particles.

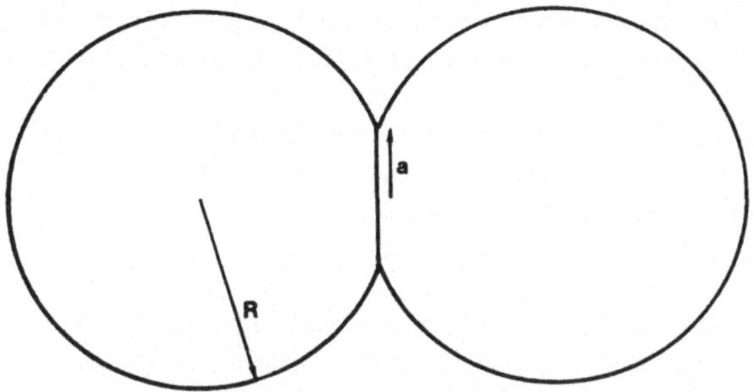

Fig. 3. Contact geometry of two particles. Due to surface energy the particles attract over a circular area with radius a.

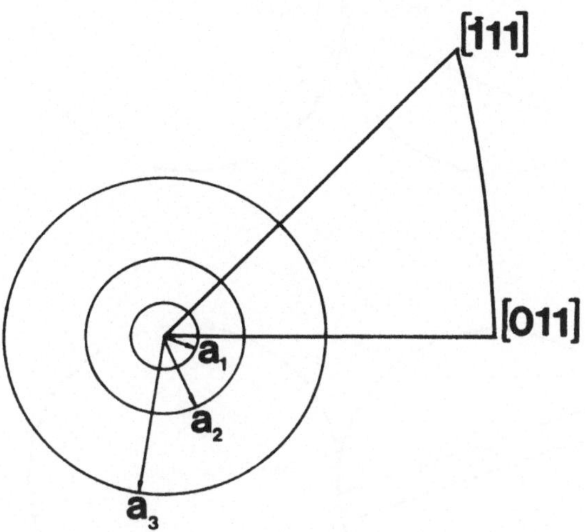

Fig. 4. The relative size of the contact circle for gold. The contact radii a_1, a_2 and a_3 refers to particles with radii 300, 30 and 3 nm respectively. It is assumed that [001] is perpendicular to the centre of contact and thus [011] is a radius R 45° off [001].

This expression was originally used by Johnson, Kendall and Roberts (9) to explain the contact between rubber balls and is an extension of the Hertz theory of contact. In its final form the theory also bears a great resemblance to fracture mechanics.

The observed stress fringes are due to bending of the atomic planes near the contact area and this influences the Bragg-reflection of electrons. The contact area increases with particle size, but the relation a/R decreases (Fig. 4).

It was mentioned above that twinning was frequent in many particle systems, though with the notable exceptions of aluminium and iron. It was also observed that twinning was heavily associated with the contacts (4,10,11,12). Twins were very often found starting at the rim of the contact circle (Figs. 5, 6).

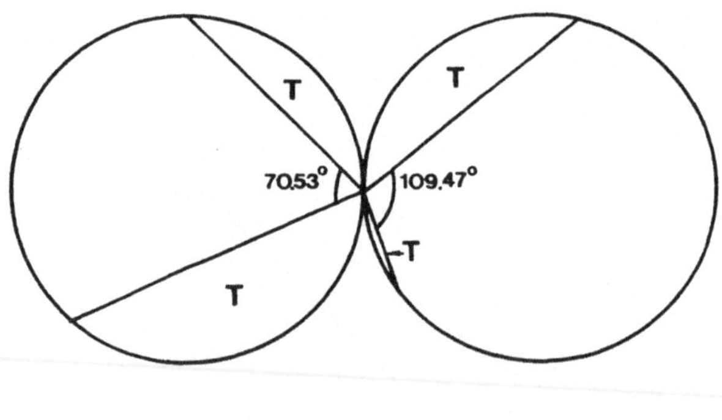

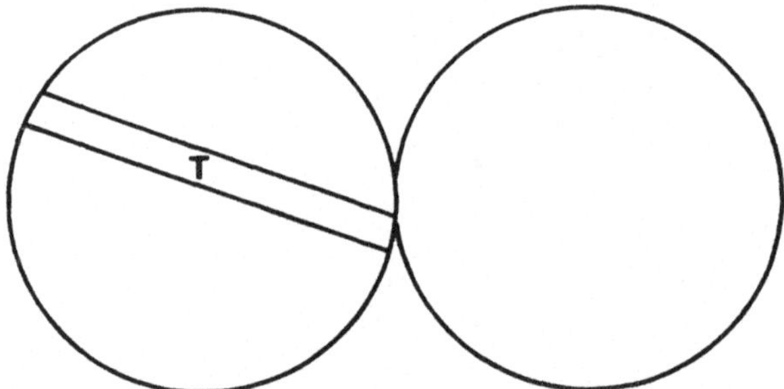

Fig. 5. Twins (T) are often found associated with contacts. In the spheres above twins on different twin systems are indicated. The figure below shows a common situation with a thin twin lamella starting in the contact zone.

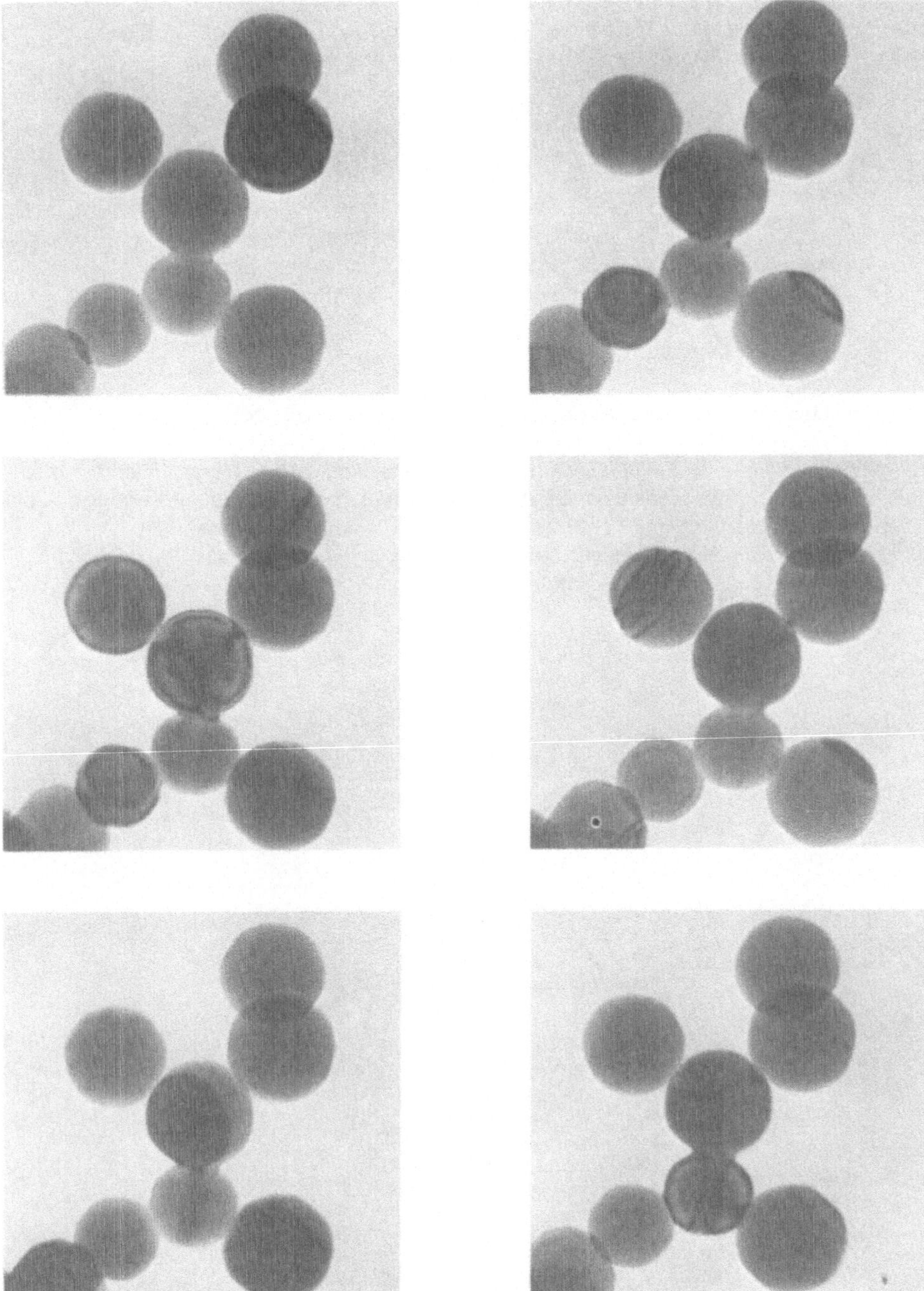

Fig. 6. Gas evaporated copper particles taken with different diffraction conditions. Twins associated with contacts can clearly be seen, but also other twin families. Contact strain contrast is observed in the central particle.

These twins are most likely a consequence of the high contact stress, which is given by the following expression

$$\sigma = \frac{2\,Ea}{3\pi(1-\upsilon^2)R} \cdot \frac{2-3\,\frac{r^2}{a^2}}{\sqrt{1-\frac{r^2}{a^2}}}$$

υ = Poisson's ratio

σ = normal stress across the boundary

r = the radial distance from the centre of contact.

This stress field of dipole character is drawn in Fig. 7 and it reaches beyond the elastic limit for small particles. Although, it in a mathematical sense reaches infinity, the integral $\int_A \sigma dA$ taken over the contact area is zero.

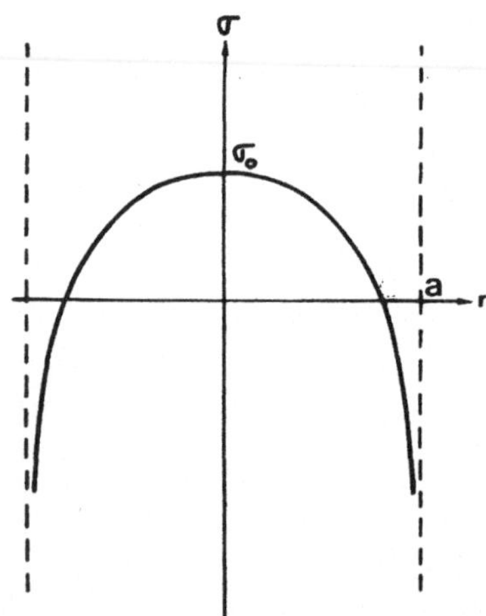

Fig. 7. Elastic stress σ across contact. The contact radius is a and the stress at the centre of the contact is $\sigma_o = \frac{4\,Ea}{3\pi(1-\upsilon^2)R}$

4. DYNAMICS ASPECTS OF CONTACT

The description above is based on a static picture of contact. However, when two atoms start to interact they will be subjected to vibrations which is also true at large scale impact. There will of course be a dynamic situation when our small metal particles hit each other and we will now consider this problem.

The JKR-theory (9) gives the static contact radius. If one reconsiders the problem in the light of the two encountering particles it is found (12) that the total potential energy curve (surface energy + elastic energy) is close to a parabola (Fig. 8). This of course leads to harmonic vibrations of the particles before they come to rest. The frequency of the dumb-bell vibrations is

$$f = \frac{1}{2\pi}\sqrt{\frac{9}{8\pi}\cdot\frac{Ea}{\rho R^3}}$$

a = equilibrium contact radius

ρ = density of the particles.

For gold particles of radius 30 nm the frequency is $2.9 \cdot 10^9$ Hz.

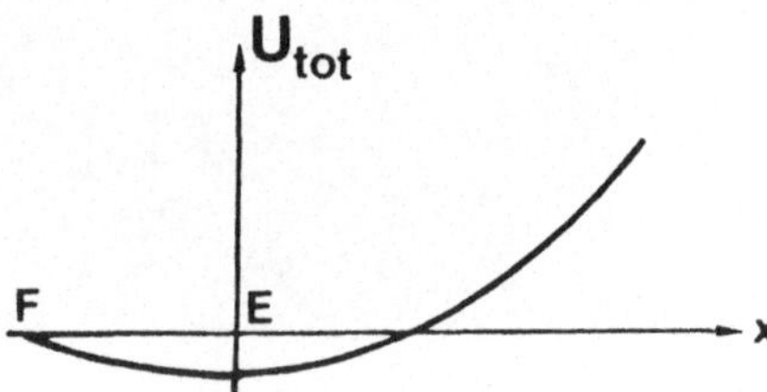

Fig. 8. The potential energy (surface and elastic energy) for two impinging particles. x is a coordinate along the centre to centre line. F is the point of first contact, and E is the equilibrium position.

These vibrations are then most likely coupled to internal vibrations in the particles before they die out. The frequencies for internal vibrations in the gold particles lie in the 10^{10} Hz range and are thus not far from the dumb-bell vibration frequency.

Internal vibrations with an associated stress field was proposed to explain the very regular twin structure observed in many materials which bears a resemblance to a standing wave pattern (13). Although vibrations are bound to exist there has recently been proposed a different mechanism for parallel twins. Here the twins are a part of the growth history (14,15), but this theory can however not explain the regular spacing often observed.

In discussing vibrations it should also be mentioned that a very regular twin structure is often formed in cobalt particles and this is found associated with vibrations and the martensitic transformation (16). The bandwidth of the course regular, twin structure is on the average found to be 15 nm and these coarser bands do often contain extremely fine twins (Fig. 9). The same bandwidth is, curiously enough, also found in small iron particles in a copper foil, which have undergone martensitic transformation (17).

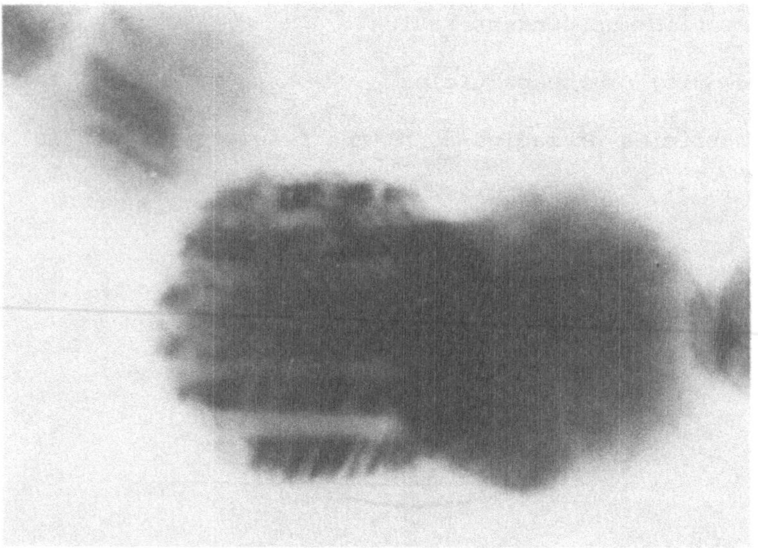

Fig. 9. Cobalt particle which has undergone a martensitic transformation. Note the regular twin banding and also the finer twinning within the bands.

5. OTHER ASPECTS OF ADHESION

The adhesion problem has been attacked theoretically using the JKR-approach for a long series of different geometries (18). The force of separation using a fine needle-flat surface geometry in an UHV-system has been measured (19,20) and indications have been found that the originally elastic contact is accompanied by a plastic deformation in the same way as seen in small particles.

6. SINTERING OF SMALL PARTICLES

Sintering is a concept borrowed from the macroscopic world to describe plastic deformation and densification of powder materials. The processes which are most important in sintering are volume diffusion, surface diffusion and grain boundary diffusion. Other processes which occur are evaporation-condensation and also perhaps some dislocation motion near the contact area (Fig. 10). The sintering rate depends on particle size, temperature, material etc.

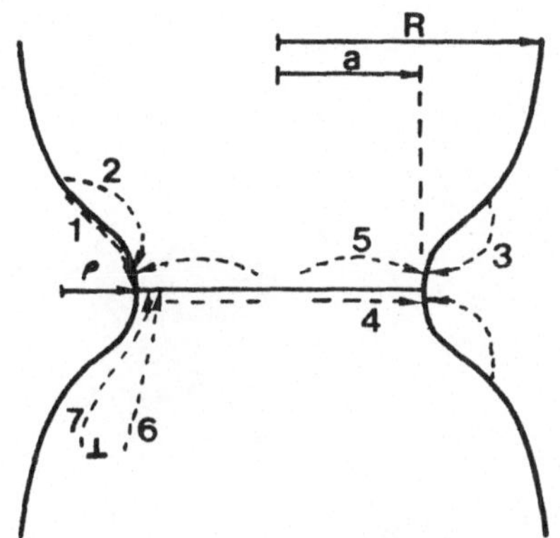

Fig. 10. Possible sintering mechanisms in a two particle contact. 1: Surface diffusion. 2: Volume diffusion. 3: Evaporation-condensation. 4: Grain boundary diffusion. 5: Volume diffusion. 6: Volume diffusion. 7: Dislocation motion. (21,22)

Large scale sintering is more easily grasped since the introduction of sintering diagrams (21,22), although they do not contain any new information about the fundamental processes. These sintering diagrams show the sintering rate as a function of temperature and relative neck size.

The excess energy which governs the sintering processes comes from the surfaces and this forms the basic of all existing sintering models. They mostly cover simple geometries but one starts now

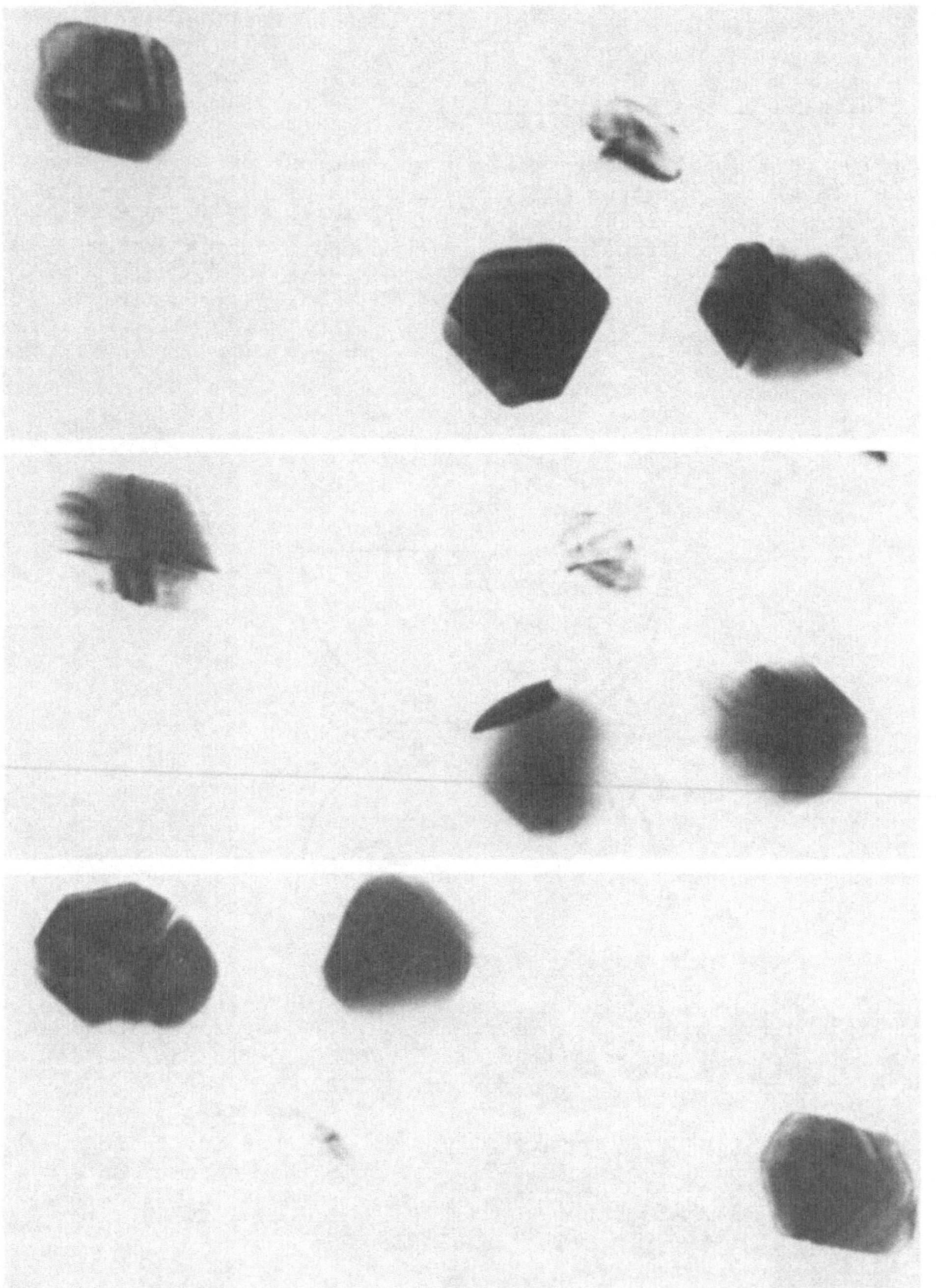

Fig. 11. Small gold particles sintered in the electron microscope. Note facetting and different twin systems in the end result.

more and more to realise the importance of rearrangement of particles and Exner in a recent article has pushed this view as well as the importance of a chemical driving force (1).

Sintering seen on a finer scale is of course a question of transport of single atoms, but when two small bodies of atomic size come together one usually call that coalescence. Movement of small particles on a substrate has been observed in the electron microscope but the particles must be subjected to some outer force in order to cover long distances. Short covered distances can be thought of as a local rolling to minimize energy for flat or anisotropic particles (11).

There has also been some observations of sintering in the electron microscope but there are some associated difficulties. Firstly the temperature is quite undefined in the case of loosely hanging particles in the microscope and it is also difficult to measure. There are indications that the temperature even locally can vary quite a lot. The second difficulty is perhaps more severe. While observing particles in the electron microscope a thin layer of carbon (contamination) is often found on the particles due to decomposition of oil vapour from the diffusion pumps. This layer will greatly change the pattern of sintering as surface processes are totally dominant at this particle size and it can therefore be extremely difficult to observe sintering as it proceeds. One possibility is of course to sinter with no beam on and then to look at the specimen. This technique has been used in connection with sintering of gold particles and the result of such a sintering experiment is shown in Fig. 11. The well separated sintered particles are typically very faceted and contain a lot of internal twins on different twin systems (12), and this result should be compared with a recent article (14,15), which treats coalesced gold particles in a SiO_2 surrounding. Although the original particles as well as the final particles are faceted, the intermediate product shows a rounded shape in the electron microscope.

7. CONCLUSION

Sintering on a larger scale seems to be grasped with a greater confidence today but in the case of small particle sintering we are only on our way to a better understanding. No doubt that twinning plays an important role in contact and early sintering but as the processes are extremely fast it is difficult to envisage the true mechanism. What is needed in all sintering is more experiments and also experiments which are made under defined conditions.

REFERENCES

1. Exner, H.E. Role of interfaces in sintering, Metal Sc. J. 16 (1982) 451-455.
2. Kimoto, K., Kamiya, Y., Nonoyama, M. and Uyeda, R. Jap. J. appl. Phys. 2 (1963) 702.
3. Kimoto, K. and Nishida, I. Jap. J. appl. Phys. 6 (1967) 1047.
4. Thölén, A.R. On the formation and interaction of small metal particles. Acta Metall. 27 (1979) 1765-1778.
5. Granqvist, C.G. and Buhrman, R.A. Ultrafine metal particles, J. Appl. Phys. 47 (1976) No 5, 2200-2219.
6. Yacamań, M.J. and Ocańa, Z.T. High-Resolution Dark-Field Electron Microscopy of Small Metal Particles, phys. stat. sol. (a) 42 (1977) 571-577.
7. Yacamań, M.J., Romeu, L.D., Gómez, A. and Munir, Z.A. Topographical interpretation of vacuum-deposited fine metallic particles. Phil. Mag. A 40 (1979) 645.
8. Easterling, K.E. and Thölén, A.R. Surface energy and adhesion at metal contacts. Acta Metall. 20 (1972) 1001-1008.
9. Johnson, K.L., Kendall, K. and Roberts, A.D. Surface energy and the contact of elastic solids. Proc. R. Soc. A 324 (1971) 301-313.
10. Hansson, I. and Thölén, A. Adhesion Between Gold Particles. Scand. J. Met. 6 (1977) 27-28.
11. Hansson, I. and Thölén, A. Adhesion and Twin Formation in Fine Particle Systems. Scand. J. Met. 7 (1978) 33-35.
12. Hansson, I. and Thölén. A. Adhesion between fine gold particles. Phil. Mag. A 37 (1978) No 4, 535-559.
13. Thölén, A.R. On Possible Vibrations in Small Metal Particles. phys. stat. sol. (a) 60 (1980) 153-166.
14. McGinn, J.T., Greenhut, V.A., Tsakalakos, T. and Blanc, J. Formation of fault structures during coalescence and growth of gold particles in a fused silica matrix - I. Acta Metall. 30 (1982) 2093-2102.
15. McGinn, J.T., Greenhut, V.A. and Tsakalakos, T. A mechanism for fault formation in fine particles and implications for theories of annealing twins in f.c.c. metals - II. Acta Metall. 30 (1982) 2103-2110.
16. Thölén, A.R. Vibrations and martensitic transformation in small cobalt particles. Surface Sci. 106 (1981) 70-78.
17. Kato, M., Monzen, R. and Mori, T. A stress-induced martensitic transformation of spherical iron particles in a Cu-Fe alloy. Acta Metall. 26 (1978) 605-613.
18. Maugis, D. Adherence of solids. Microscopic Aspects of Adhesion and Lubrication (ed. J.M. Georges) Elsevier (1982) 221-252.
19. Pollock, H.M. and Chowdhury, S.K.R. A study of metallic deformation, adhesion and friction at the thousand-angstrom level. Microscopic Aspects of Adhesion and Lubrication. (ed. J.M. Georges) Elsevier (1982) 253-262.

20. Pethica, J.B. and Tabor, D. Surface Science 89 (1979) 182.
21. Ashby, M.F. A first report on sintering diagrams. Acta Metall. 22 (1974) 275-289.
22. Swinkels, F.B. and Ashby, M.F. A second report on sintering diagrams. Acta Metall. 29 (1981) 259-281.

PHYSICS ON THE BEACH or THE THEORY OF WINDSURFING

A.R. Thölén, Laboratory of Applied Physics I
Technical University of Denmark, 2800 Lyngby, Denmark

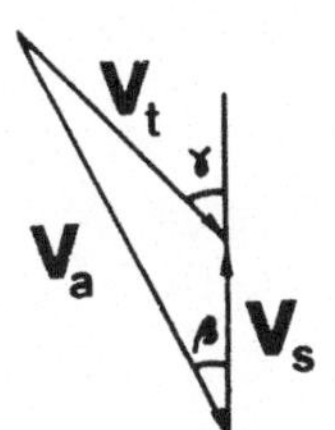

List of symbols

v_t = true wind velocity

v_s = boat speed

v_a = apparent wind velocity

β = apparent course angle

γ = true course between v_t and v_s

α = angle of incidence of the sail

δ = sheet angle

λ = leeway angle

S = sail area

A = area of fin and rudder

According to the figures

$$\alpha = \beta-\lambda-\delta$$

$$v_a^2 = v_t^2+v_s^2+2v_tv_s \cos\gamma$$

$$v_a \sin\beta = v_t \sin\gamma$$

Forces acting on the sail

$$F_L = \tfrac{1}{2}\, \rho_{air}\, v_a^2\, c_L\, S$$

$$F_D = \tfrac{1}{2}\, \rho_{air}\, v_a^2\, c_D\, S$$

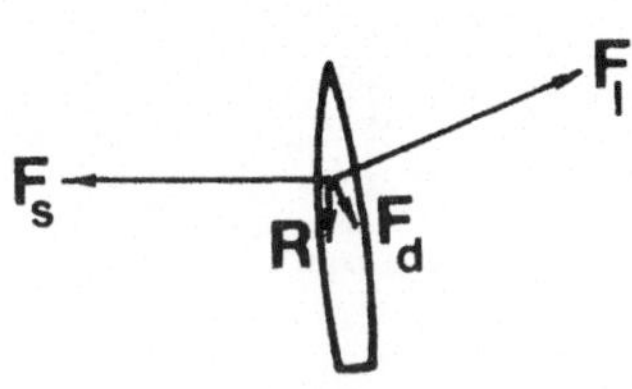

Forces acting on the underwater body

$$F_s = F_s(\lambda,\ v_s,\ A)$$

$$R = R\ (\lambda,\ v_s,\ A)$$

The important factor is now only to keep the forces in two orthogonal directions in balance

$$\Sigma\ F_h = 0$$

$$\Sigma\ F_v = 0$$

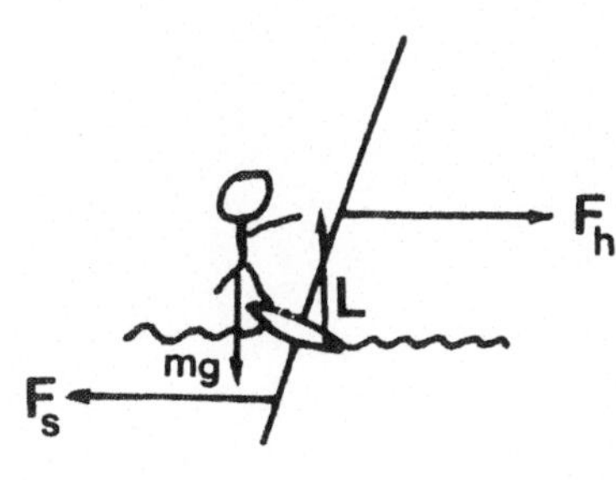

But, please, do not forget the equation for the rolling (righting) moment

$$\Sigma\ M = 0$$

Ref. 1. Marchaj, C.A., Aero-Hydrodynamics of sailing, Granada Publ. 1979.

INDEX

A

B

C

D

E

F

G

H

I

J

K

L

M

N

O

P

Q

R

S

T

U

V

W

X

Z